THE CREATIONISTS

UNIVERSITY OF CALIFORNIA PRESS

BERKELEY · LOS ANGELES · LONDON

Recipient of the
1991 Albert C. Outler Prize
in Ecumenical Church History of
the American Society of Church History

THE CREATIONISTS

Ronald L. Numbers

University of California Press
Berkeley and Los Angeles, California

University of California Press, Ltd.
London, England

First paperback printing 1993

Library of Congress Cataloging-in-Publication Data

Numbers, Ronald L.
    The creationists : the evolution of scientific creationism /
Ronald L. Numbers.
        p.   cm.
    Originally published: New York : Knopf : Distributed by Random
House, 1992
    Includes bibliographical references and index.
    ISBN 0-520-08393-8
    1. Creationism.   2. Evolution—Religious aspects—Christianity.
I. Title.
[BS651.N85   1993]   93-15804
231.7'65—dc20   CIP

Manufactured in the United States of America

    2   3   4   5   6   7   8   9

The paper used in this publication meets the minimum requirements
of American National Standard for Information Sciences—Permanence
of Paper for Printed Library Materials, ANSI Z39.48–1984. ∞

CONTENTS

INTRODUCTION

Within a couple of decades of the publication of Charles Darwin's landmark book *Origin of Species* (1859), the idea of organic evolution had captivated most British and American scientists and was beginning to draw favorable comment from religious leaders on both sides of the Atlantic. By the late nineteenth century, evolutionary notions were infiltrating even the ranks of evangelical Christians, and, in the opinion of many observers, belief in special creation seemed destined to go the way of the dinosaur. But contrary to the hopes of liberals and the fears of conservatives, creationism did not become extinct. Many English-speaking Christians, particularly in North America, remained true to a traditional reading of Genesis and from time to time, most notably in the 1920s and since the 1960s, mounted campaigns to contain the spread of evolutionary theory. An overwhelming majority of Americans saw no reason to oppose the teaching of creationism in public schools, and according to a 1991 Gallup poll 47 percent, including a fourth of the college graduates, continued to believe that "God created man pretty much in his present form at one time within the last 10,000 years."[1] Two states, Arkansas and Louisiana, passed laws mandating equal treatment for "creation science" and "evolution science." Although the courts subsequently ruled these particular statutes to be unconstitutional, the creationist movement showed few signs of slackening as it entered the 1990s.

Besides the unexpected revival in recent years, which caught even enthusiasts by surprise, the most striking development in the history of twentieth-century creationism is the ascendancy since the early 1960s of a distinctive brand of creationism known as "scientific creationism" or "creation science." As defined in the 1981 Arkansas law,

> Creation-science includes the scientific evidences and related inferences that indicate: (1) Sudden creation of the universe, energy, and life from nothing; (2) The insufficiency of mutation and natural selection in bringing about development of all living kinds from a single organism; (3) Changes only within fixed limits of originally created kinds of plants and animals; (4) Separate ancestry for man and apes; (5) Explanation of the earth's geology by catastrophism, including the occurrence of a worldwide flood; and (6) A relatively recent inception of the earth and living kinds.

Advocates of this view—essentially biblical creationism stripped of explicit references to God, Adam, and Noah—read the first chapters of Genesis in a way that allows for no life on earth before Eden and no death before the Fall.[2]

Until the last few decades most creationists would have regarded such notions as unnecessarily extreme. By the late nineteenth century even the most conservative Christian apologists readily conceded that the Bible allowed for an ancient earth and pre-Edenic life. With few exceptions, they accommodated the findings of historical geology either by interpreting the days of Genesis 1 to represent vast ages in the history of the earth (the so-called day-age theory) or by separating a creation "in the beginning" from a much later Edenic creation in six literal days (the gap theory). Either way, they could defend the accuracy of the Bible while simultaneously embracing the latest geological and paleontological discoveries. William Jennings Bryan, the much misunderstood leader of the post–World War I antievolution crusade, not only read the Mosaic "days" as geological "ages" but allowed for the possibility of organic evolution—so long as it did not impinge on the supernatural origin of Adam and Eve. Harry Rimmer, the flamboyant evangelist who occupied center stage on the creationist platform between the great wars, squeezed millions of years into the presumed gap in the Genesis narrative and drained the deluge story of all but local significance.

The creation scientists, by contrast, compress the history of life on earth into less than ten thousand years. To accomplish this, they attribute most of the fossil record to the brief period of the flood and its aftermath. They believe that the majority of plants and animals buried sequentially in the stratified rocks once lived together in the antediluvian world; thus these relics do not represent successive populations of flora and fauna spanning millions of years, as evolutionists and most other creationists would assert. In such classics of creation science as George McCready Price's *New Geology* (1923) and John C. Whitcomb, Jr., and Henry M. Morris's *Genesis Flood* (1961), geological issues push biological ones to the periphery, and the Noachian deluge tends to eclipse the Adamic creation in importance. "The Genesis Flood is the real crux of the conflict between the evolutionist and creationist cosmologies," declares the text *Scientific Creationism* (1974).

> If the system of flood geology can be established on a sound scientific basis, and be effectively promoted and publicized, then the entire evolutionary cosmology, at least in its present neo-Darwinian form, will collapse. This, in turn, would mean that every anti-Christian system and movement (communism, racism, humanism, libertinism, behaviorism, and all the rest) would be deprived of their pseudo-intellectual foundation.[3]

The chief architect of flood geology, a term virtually synonymous with creation science and scientific creationism, was the self-described geologist George McCready Price, who during the early decades of the twentieth century stood virtually alone in insisting on the recent appearance of life and on a flood that rearranged the features of the earth. Although his "new catastrophism" received nearly universal acclaim from fellow creationists, he won few true converts to flood geology outside his own small Seventh-day Adventist sect. It was not until the creationist renaissance of the 1960s, marked by the publication of Whitcomb and Morris's *Genesis Flood* and the subsequent birth of the Creation Research Society, that fundamentalists in large numbers began to read Genesis in the Pricean manner and to equate his views with the intended message of Moses. By the 1980s the flood geologists had virtually co-opted the name creationism to describe the once marginal views of Price.[4] This remarkable shift in the prevailing meaning of creationism—from the theologically orthodox day-age and gap theories that allowed the history of life on earth to span millions of years to a doctrine of suspect provenance (because of its Adventist origins) that compressed earth history into no more than ten thousand years— serves as the focus of my study.

CREATIONIST INTERPRETATIONS OF GENESIS

Interpretations

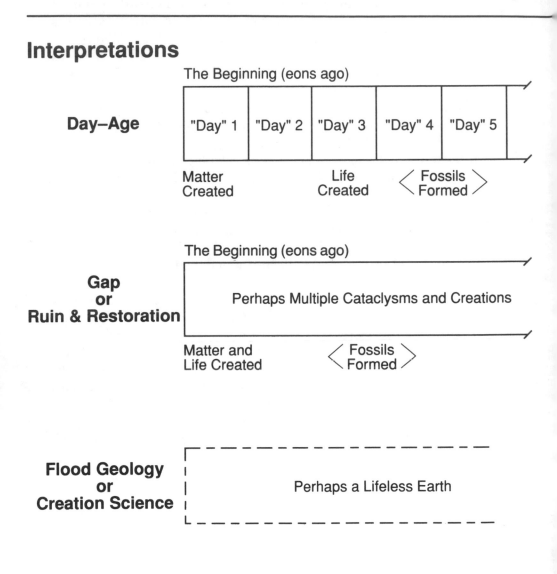

Advocates

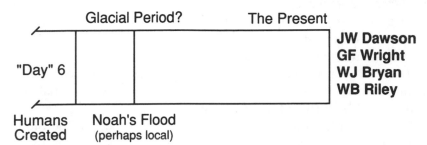

Glacial Period? The Present

"Day" 6

Humans Created Noah's Flood (perhaps local)

JW Dawson
GF Wright
WJ Bryan
WB Riley

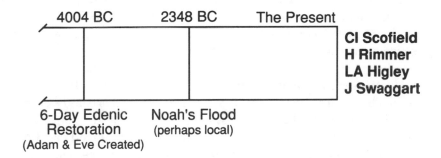

4004 BC 2348 BC The Present

6-Day Edenic Restoration
(Adam & Eve Created) Noah's Flood (perhaps local)

CI Scofield
H Rimmer
LA Higley
J Swaggart

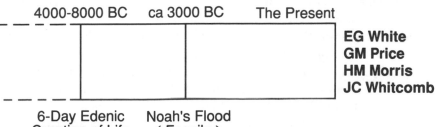

4000-8000 BC ca 3000 BC The Present

6-Day Edenic Creation of Life and Humans
(and perhaps even matter) Noah's Flood ⟨Fossils Formed⟩

EG White
GM Price
HM Morris
JC Whitcomb

In writing this history, I have chosen to concentrate on those creationists who possessed, or claimed to possess, scientific credentials. This might strike some readers as an odd choice of topic for a historian of science, but I would submit that one of the best ways to learn about the history of "science" is to explore how interested parties have contested its boundaries. Many books in recent years have sought to discredit creationism scientifically or theologically, but only a few have examined the movement historically, and then primarily from a legal or pedagogical perspective.[5] None has looked carefully at the intellectual origins of scientific creationism. Consequently, even relatively informed persons tend to overlook the substantial changes in creationist thought during the twentieth century and the intense controversies precipitated by those changes. The common assumption seems to be that one creationist is pretty much like another.[6] As we shall see, nothing could be further from the truth.

Although scientifically trained creationists, especially in the biological and earth sciences, had become an endangered species by the early years of the century, they gradually reappeared as more and more fundamentalist youth sought higher education. During the 1920s, creationists relied for scientific expertise on a few teachers in Christian colleges (none with even a master's degree in biology or geology), on a physician or two, and on a medical-school dropout. In 1963, when the Creation Research Society was organized, five of its ten founders held earned doctorates in biology from major universities, and two others possessed Ph.D.s in science or engineering. Not surprisingly, these scientifically credentialed creationists frequently enlisted scientific arguments to support their views. But to a man they embraced creationism primarily from religious conviction. To illuminate these beliefs and to trace the tangled religious roots of creationism—from Baptist and Presbyterian to Lutheran and Adventist—I have included biographical sketches of a number of the leading creationists.

Readers acquainted with such works as Andrew Dickson White's influential *History of the Warfare of Science with Theology in Christendom* (1896) will not be surprised to learn that the history of modern creationism includes some of the fiercest skirmishes in the annals of science and religion. But only those familiar with the recent critiques of the warfare thesis will be prepared to accept my claim that the creationist conflicts rarely conformed to the battle lines drawn by White.[7] Rather than finding clerics arrayed in simple opposition to scientists, we discover conflicts of a different sort: psychological, as creationists struggled to reconcile the apparently conflicting claims of science and Scripture; and social, as they quarreled with one another over competing scientific and biblical interpretations or contested the boundaries of science and religion with evolutionists in courthouses, legislative halls, and school-board rooms. In virtually every public battle, even when creationists squared off against evolutionists, scientists and preachers could be found on both sides, and sometimes in unexpected numbers. For example, in the Arkansas creation-evolution trial in 1981, the plaintiffs, who opposed creation science, came overwhelmingly from the ranks of religious organizations, while virtually all of the experts testifying in support of creationism possessed graduate degrees in science. The irony prompted the Protestant theologian Langdon Gilkey, who served as a witness for the plaintiffs, to observe that the only "warfare" in Little Rock found "liberal religion and liberal science on the one side, and absolutist religion and its appropriate 'science' on the other."[8]

I should also warn readers that my aim is not to expose the scientific defenders of creationism as "pseudoscientists." Although such efforts no doubt have their place—one of my favorite journals is the *Skeptical Inquirer*—as a historian I am much more interested in how persons and parties *used* "science" and "pseudoscience" to further their ends than in judging whether they employed these labels appropriately by the standards of the 1990s. Lately many scholars, including the philosopher of science Larry Laudan and the sociologist of science Thomas F. Gieryn, have shown the sterility of efforts to demarcate between science and pseudoscience on analytical grounds. Laudan has gone so far as to dismiss the demarcation problem as "a pseudo-problem."[9] I agree, but hasten to add that this says nothing about the *practical* and *historical* significance of attempts at demarcation. On the basis of criteria (including falsifiability, testability, tentativeness, and naturalness) suggested by the philosopher-historian Michael Ruse, the federal judge in the Arkansas case declared creation science to lie outside the domain of science and within the realm of religion. This allowed him to rule that the mandatory teaching of creationism was unconstitutional because it violated the First Amendment requirement that church and state be separate.[10] Merely showing creation science to be "bad science" would have been insufficient in this case, because the constitution does not ban the teaching of bad science in public schools.

Over the past decade or so I have lectured on the history of creationism to many audiences, both academic and general. On almost every occasion someone has asked me to reveal my own position on the scientific merits of creationism or to disclose my religious beliefs. I would like to think that readers of this book will accept or reject my rendering of the creationist story on the basis of the evidence and arguments I present, not because of my personal convictions. But as a concession to the universal curiosity about such matters, I offer a brief autobiography.

Born and reared in a fundamentalist Seventh-day Adventist family of ministers, I learned Price's version of earth history at my parents' knees. I subsequently attended Adventist church schools from first grade through college, and though I majored in science, I saw no reason to question the claims of strict creationism. In fact, I do not recall ever doubting the recent appearance of life on the earth until the late 1960s, while studying the history of science at the University of California at Berkeley. I vividly remember the evening I attended an illustrated lecture on the famous sequence of fossil forests in Yellowstone National Park and then stayed up much of the night with a biologist friend of like mind, Joe Willey, first agonizing over, then finally accepting, the disturbing likelihood that the earth was at least thirty thousand years old. Having thus decided to follow science rather than Scripture on the subject of origins, I quickly, though not painlessly, slid down the proverbial slippery slope toward unbelief. In 1982, when attorneys for both sides in the Louisiana creation-evolution trial requested my services as a possible expert witness, I elected to join the ACLU team in defending the constitutional wall separating church and state. In taking my pretrial deposition, Wendell R. Bird, the creationist lawyer who had tried to recruit me for his side, devoted two lengthy sessions to probing the limits of my historical knowledge and the thinness of my religious beliefs. On the basis of this inquisition Bird publicly labeled me an "Agnostic."[11] The tag still feels foreign and uncomfortable, but it accurately reflects my theological uncertainty.

Although I no longer believe in creationism of any kind, I am strongly committed to treating its advocates with the same respect I might accord evolutionists. (As a constant reminder to do so, I have kept above my desk a framed handbill from the early 1940s announcing a public lecture on "God's Answer to Evolution: Are Men and Monkeys Relatives?" The featured speaker was my father, Raymond W. Numbers, then holding a series of evangelistic meetings in the Kansas City Canvas Tabernacle.) For too long now students of science and religion have tended to grant the former a privileged position, often writing more as partisans than historians and grading religious "beliefs" by how much they encouraged or retarded the growth of scientific "knowledge." Recently we have heard persuasive calls for a more even-handed treatment.[12] But even academics who would have no trouble empathetically studying fifteenth-century astrology, seventeenth-century alchemy, or nineteenth-century phrenology seem to lose their nerve when they approach twentieth-century creationism and its fundamentalist proponents. The prevailing attitude, colorfully expressed at one professional meeting I attended, is that "we've got to stop the bastards." In other words, although many scholars seem to have no trouble respecting the unconventional beliefs and behaviors of peoples chronologically or geographically removed from us, they substitute condemnation for comprehension when scrutinizing their own neighbors. I think it is profitable to get acquainted with the neighbors, especially so if we find them threatening.

R. L. N.
Madison, Wisconsin
July 1, 1991

THE CREATIONISTS

ONE

Creationism in the Age of Darwin

W ithin twenty years after the publication of Charles Darwin's *On the Origin of Species by Means of Natural Selection* in 1859, nearly every naturalist of repute in North America had embraced some theory of organic evolution. The situation in Great Britain looked equally bleak for creationists, and on both sides of the Atlantic liberal churchmen were beginning to follow their scientific colleagues into the evolutionist camp. Although the majority of Bible-believing Christians undoubtedly remained true to the idea of a specially created world, evolution was infiltrating even the ranks of evangelicals by the closing years of the nineteenth century. As early as 1880 the editor of one American religious weekly estimated that "perhaps a quarter, perhaps a half of the educated ministers in our leading Evangelical denominations" believed "that the story of the creation and fall of man, told in Genesis, is no more the record of actual occurrences than is the parable of the Prodigal Son."[1] When in 1910 the early fundamentalists sought a scientific champion to meet the threat of evolution, the best they could find was George Frederick Wright, a cleric-geologist who had risen to prominence a few decades earlier as an apologist for Christian Darwinism.

This chapter surveys the scientific and religious responses to organic evolution in the half-century following the appearance of Darwin's famous book, focusing in particular on scientific and theological resistance to evolution among conservative American Protestants. Contemporary readers who associate creationism with the teachings of the so-called scientific creationists will no doubt be surprised by the small number of nineteenth-century creationist writers who subscribed to a recent creation in six literal days and the even greater rarity of those who attributed the fossil record to the Noachian flood. Creationists of the Victorian era generally assimilated the findings of historical geology to such an extent that today they seem intellectually closer to the theistic evolutionists of their time than to the scientific creationists of the late twentieth century.

EVOLUTION COMES TO AMERICA

Confusion about the distinction between creation and evolution dates back at least to the publication of Darwin's *Origin of Species* (1859), which explained speciation in terms of the struggle to survive and propagate among organisms that differ in their abilities because of randomly occurring variations. Although one of Darwin's principal goals was "to overthrow the dogma of separate creations," he invoked at least one creative act for the purpose of getting life going and allowed for the possibility of several more interventions. "I believe that animals have descended from at most only four or five progenitors, and plants from an equal or lesser number," he wrote in a widely quoted passage, adding that analogy would lead him to believe "that probably all the organic beings which have ever lived on this earth have descended from some one primordial form, into which life was first breathed." This concession to conventional views prompted the religiously orthodox American botanist Asa Gray (1810–1888), Darwin's foremost American disciple, to suggest that because his British friend had accepted "a supernatural beginning of life on earth," he should be willing to allow another "special origination" in connection with the appearance of humans. Darwin, who soon came to regret his use of "Pentateuchal" language in the *Origin of Species*, rejected not only this advice but also Gray's proposal to attribute the inexplicable variations in Darwin's scheme to divine providence. In *Variation of Plants and Animals under Domestication* (1868), Darwin announced that "however much we may wish it, we can hardly follow Professor Asa Gray in his belief" in divinely guided evolution.[2]

By the time Darwin penned his treatise *The Descent of Man* (1871), he no longer felt the need to truckle to public opinion by including veiled references to the Creator. In uncompromisingly naturalistic language that contrasted sharply with the biblical story of Adam and Eve, Darwin offered his readers a new genealogy:

> Man is descended from a hairy quadruped, furnished with a tail and pointed ears, probably arboreal in its habits, and an inhabitant of the Old World. This creature, if its whole structure had been examined by a naturalist, would have been classed among the Quadrumana, as surely as would the common and still more ancient progenitor of the Old and New World monkeys. The Quadrumana and all the higher mammals are probably derived from an ancient marsupial animal, and this through a long line of diversified forms, either from some reptile-like, or some amphibian-like creature, and this again from some fish-like animal.

Darwin tried to soften this blow to human pride by pointing out that if humans no longer laid claim to a pedigree "of noble quality," at least they could take comfort from having one "of prodigious length." But some of his readers reacted less cheerfully to the news of tailed ancestors. As one critic caustically complained, in contrast to the biblical record, which "places a crown of honor and dominion on the brow of our common humanity . . . Darwinism casts us all down from this elevated platform, and herds us all with four-footed beasts and creeping things. It tears the crown from our heads; it treats us as bastards and not sons, and reveals the degrading fact that man in his best estate—even Mr. Darwin—is but a civilized, dressed up, educated monkey, who has lost his tail."[3]

In evaluating the response of Americans to the *Origin of Species*, we must distinguish between Darwin's twin goals of showing that species had not been supernaturally created and of demonstrating that natural selection had been the chief agent in effecting evolutionary change. By the mid-1870s the majority of professional naturalists in America had embraced the evolutionary origin of species, but many—perhaps most—of them remained skeptical about the primacy of natural selection in the evolutionary process, emphasizing instead such factors as the inheritance of environmentally induced characteristics.[4]

The rapid conversion from special creation to evolution occurred in part because of the empirical evidence Darwin and others marshaled in favor of the transmutation of species, but equally important was the growing self-consciousness among biological scientists over their continued reliance on miraculous explanations at a time when natural laws had supplanted divine activity in virtually every other area of science.

In reviewing the *Origin of Species* for the *Atlantic Monthly* in 1860, Asa Gray addressed the question of why most naturalists no longer felt comfortable relying on the Genesis account of creation to explain speciation:

> Sufficient answer may be found in the activity of the human intellect, "the delirious yet divine desire to know," stimulated as it has been by its own success in unveiling the laws and processes of inorganic Nature. . . . Surely the scientific mind of an age which contemplates the solar system as evolved from a common revolving fluid mass—which, through experimental research, has come to regard light, heat, electricity, magnetism, chemical affinity, and mechanical power as varieties or derivative and convertible forms of one force, instead of independent species—which has brought the so-called elementary kinds of matter, such as the metals, into kindred groups, and pertinently raised the question, whether the members of each group may not be mere varieties of one species—and which speculates steadily in the direction of the ultimate unity of matter . . .—the mind of such an age cannot be expected to let the old belief about species pass unquestioned.[5]

In similar manner, the American astronomer Simon Newcomb (1835–1909) reduced the reasons for choosing evolution over creation to one basic principle: "We are not to call in a supernatural cause to account for a result which could have been produced by the action of the known laws of nature."[6] Thus, despite continuing debates about the exact mechanism responsible for organic development, the negative sentiment against special creation, combined with the positive evidence for the transmutation of species, created an intellectual climate favorable to the acceptance of evolution.

After a slow start in the early 1860s, the idea of organic evolution, effectively promoted by Gray, quickly won the allegiance of American biologists and geologists. In 1872, less than thirteen years after the appearance of the *Origin of Species*, the paleontologist Edward Drinker Cope (1840–1897) observed that "the modern theory of evolution has been spread everywhere with unexampled rapidity, thanks to our means of printing and transportation. It has met with remarkably rapid acceptance by those best qualified to judge of its merits, viz., the zoologists and botanists." Just before his death in 1873, even Louis Agassiz (1807–1873), the archcritic of developmental theories, conceded that the idea of organic development had won "universal acceptance." The next year James Dwight Dana (1813–1895), editor of the *American Journal of Science* and the country's best-known geologist, publicly announced his lukewarm conversion to evolution, although he clung to the conviction that a special creative act had introduced the first humans. The nearly unanimous acceptance of evolution within the American scientific community was spotlighted in 1880, when the editor of the *Independent*, after surveying the teaching of evolution in the colleges of the North, challenged a rival religious weekly, the *Observer*, to name just "three working naturalists of repute in the United States—or two (it can find one in Canada)—that is not an evolutionist." But besides the Canadian John William Dawson (1820–1899) of McGill, only one could be found: Princeton's Arnold Guyot (1807–1884).[7]

SCIENTIFIC CREATIONISTS

During the lifetime of Harvard's Louis Agassiz, antievolutionists could cloak themselves in the respectability of America's most famous scientist. Already an internationally acclaimed authority on fossil fishes and glaciers when he emigrated from Switzerland to the United States in 1846, Agassiz used his scientific standing to defend the idea of special creation. His creationism, however, bore little resemblance to the narrative found in Genesis. Agassiz adamantly refused to let religion determine the course of his science, whether it be with respect to the age of the earth, the appearance of humans, or the reality of the Noachian flood. The scion of a long line of Huguenot ministers, Agassiz by 1859 had drifted into nominal Unitarianism and sporadic churchgoing. His ice-age theory had helped, in the 1840s, to drain the last drops of geological significance from the Noachian deluge, and his espousal of the plural origin of the human races in opposition to the biblical account of Adam and Eve had, in the 1850s, aroused the enmity of many devout Christians. Instead of a single Edenic creation in six days, he taught that the geological evidence indicated a series of catastrophes and creations by which the earth had been repeatedly depopulated and repopulated. In his *Essay on Classification* (1857) he explained that "species did not originate in single pairs, but were created in large numbers," in the habitats they were intended to populate, and he dismissed the antiquated notion that fossils were "the wrecks of the Mosaic deluge." Living species thus had no genetic connection with previous inhabitants of the earth—and might not even be related, except ideally, to members of the same species now living. His creationism owed more to philosophy than to revelation. Nevertheless, religiously orthodox antievolutionists repeatedly claimed him as one of their own during the early years of the Darwinian debates.[8]

Within his scientific domain at Harvard, Agassiz showed little tolerance for the mixing of science and dogmatic theology. In the spring of 1873 he invited a former student, John McCrady (1831–1881), a marine biologist then teaching at the College of Charleston, to join him on the staff of the Museum for Comparative Zoology. McCrady, an outspoken man possessed of strong political, religious, and scientific convictions, including a burning antipathy to Darwinism, gratefully accepted the position, noting that it offered him the opportunity to advance "genuine Science, and at the same time strike a sturdy blow for Religion." The prospect of a religious zealot in his own laboratory alarmed the secular-minded zoologist. "My convictions are recorded in the Essay on Classification & there need be no question about them," wrote Agassiz in a letter asking McCrady to clarify his intentions, "but I hardly think that these topics belong to the lecture room beyond a general discussion of principles, which do not yet constitute Religion; at least as practiced in this country, where every body is expected to subscribe to some particular tenets, & tenet teaching is no part of science." McCrady's assurance that he intended only to combat "the current erroneous views of (so-called) 'Evolution' " associated with Darwin and other heterodox scientists and his promise to refrain from "tenet teaching" apparently assuaged Agassiz's concern; at any rate the southerner soon joined his mentor in Cambridge, where each in his own way sought to stamp out the Darwinian heresy.[9]

Agassiz's death soon after McCrady's arrival left creationists without a world-famous leader. By 1873 even most of Agassiz's own students had forsaken creationism, and, as mentioned earlier, Dana was shifting sides. McCrady, who succeeded Agassiz as professor of zoology at Harvard, might have continued the fight against Darwinism in Cambridge, but in 1877 President Charles Eliot (1834–1926) forced him to resign, ostensibly because McCrady's teaching was too specialized for Harvard students and because he had failed to publish any new research in recent years. McCrady chose instead to blame his dismissal on such factors as local favoritism, his religious orthodoxy, his refusal to teach "Darwinism, Huxleyism and Spencerism," and the machinations of Darwin's friend Asa Gray. Whether the retention of his Harvard platform would have helped the creationist cause is, of course, impossible to determine, but it seems unlikely that it would have. Although an implacable foe of Darwinism and skeptical of "the common descent of species," McCrady refused to treat creationism and evolutionism as mutually exclusive explanations. He devoted the last decades of his life to formulating an esoteric Universal Law of Development that would harmonize "the apparently antagonistic views of Agassiz & Darwin," which he regarded as being "only the opposite poles of one great truth." Despite presenting his views at such institutions as the Harvard Theological School and the Johns Hopkins University, McCrady seems to have exerted little influence—either for or against creationism—outside of a small intellectual circle. Several factors contributed to this fate: the difficulty in understanding his views, the loss of his book manuscript in a fire, and his isolation at the University of the South, an Episcopal school carved into the hills of southeastern Tennessee, where he spent the last years of his life as professor of biology and the relation of science and religion.[10]

With McCrady unavailable to fill Agassiz's shoes, the scientific defense of creationism in the late nineteenth century fell largely to Agassiz's friend and countryman Arnold Guyot and to the Canadian geologist and educator John William Dawson. Guyot, who had abandoned theology for science while a student at the University of Berlin, followed Agassiz to the United States in 1848 and six years later landed a permanent job at the College of New Jersey (now Princeton University) in Princeton as professor of physical geography and geology, a position he occupied until his death. Unlike Agassiz, Guyot, an active member of the Presbyterian church, sought to harmonize science and Scripture. By interpreting the "days" of Genesis 1 as epochs in cosmic history, he was able to correlate the earth's physical, geological, and biological development with the sequence of creative events sketched by Moses, a scheme popularized by Dana and Dawson and widely embraced by orthodox Christians. Also unlike Agassiz, Guyot tended to minimize the number of special creations, delegating most of the work of creation to the divinely ordained laws of nature. In the fullest statement of his views, *Creation; or, The Biblical Cosmogony in the Light of Modern Science* (1884), completed shortly before his death, he expressed reservations about the common ancestry of the invertebrates and about the descent of vertebrates from invertebrates, but he insisted only on the special creation of matter, life, and humans. "Evolution from one of these orders into the other—from matter into life, from animal life into the spiritual life of man—is impossible," he declared, adding that "the question of evolution within each of these great systems—of matter into various forms of matter, of life into the various forms of life, and of mankind into all its varieties—remains still open." He was sure, however, that "any length of time that Darwin might desire for his transformations, would never suffice to make of the monkey a civilizable man." Despite such statements, one of

Guyot's closest scientific friends at the time of his death, James Dwight Dana, believed that the Princeton geographer had come "to accept, though with some reservation, the doctrine of evolution through natural causes." In today's parlance he probably would be called a progressive creationist.[11]

In the antievolutionist writings of the late nineteenth century, no creationist was appealed to more frequently than the distinguished Canadian geologist John William Dawson, who during the 1880s served as president of both the American Association for the Advancement of Science and the British Association for the Advancement of Science, a unique achievement. A native of Nova Scotia, he studied geology in Edinburgh in the 1840s and, through a chance meeting in Canada, became a protégé of the British geologist Charles Lyell (1797–1875). He regularly taught Sunday school at his Presbyterian church in Montreal, and he was much in demand in orthodox circles as a lecturer and writer on science and religion. In 1878, when declining health prompted Guyot to ask that his duties in geology be given to another professor, President James McCosh (1811–1894) of Princeton, himself a theistic evolutionist, urged Dawson to join the faculty because he did not know where else "to look for a geologist of repute who is not a Darwinian."[12]

Dawson remained a special creationist until his death, but his opposition to evolution noticeably softened over the years as he came to acknowledge that evolution was not inconsistent with design. Never a biblical literalist, he readily granted that the days of creation represented long periods of time, that the Noachian deluge was universal only in the sense that it encompassed the narrator's experience, and that the earth—but not man—was of great antiquity. In *The Story of Earth and Man* (1873), one of Dawson's many books, he spelled out what "the theory of creation" meant to him:

> Simply this; that all things have been produced by the Supreme Creative Will, acting either directly or through the agency of the forces and materials of His own production. This theory does not necessarily affirm that creation is miraculous. . . . It does not contradict the idea of successive creations. . . . It does not even exclude evolution or derivation to a certain extent: anything once created may, if sufficiently flexible and elastic, be evolved or involved in various ways.

Although Dawson deplored invoking the Creator as frequently as Agassiz, he seems never to have gone as far as Guyot in reducing the required number of special creative acts.[13]

Beyond Dawson and to a lesser extent Guyot, one searches in vain for a prominent late nineteenth-century North American scientist who actively opposed organic evolution. Certainly there were some like J. Lawrence Smith (1818–1883), the Baptist chemist from Kentucky, who in his 1873 address as retiring president of the American Association for the Advancement of Science could not resist a parting shot at Darwinism, but Smith seems to have said little else on the topic, and he died in the early 1880s. Joseph Henry (1797–1878), the churchgoing Presbyterian physicist who headed the Smithsonian Institution and served as a trustee of the Princeton Theological Seminary, sometimes appeared with his friend Agassiz in lists of prominent creationists, but his true sympathies seem to have been on the other side of the issue. "I have given the subject of evolution much thought," he wrote Asa Gray, "and have come to the conclusion that it is the best working hypothesis which you naturalists have ever had." When Agassiz heard of his friend's defection, he apparently made a special trip from Cambridge to Washington to reconvert him. But Agassiz succeeded only in talking Henry into maintaining a discreet silence on the subject, a silence that apparently led some to believe that he opposed evolution.[14]

As the preceding discussion has shown, the intellectual differences between creationists and evolutionists were not always as great as one might assume. Late nineteenth-century creationists such as Guyot and Dawson granted the progressive nature of the fossil record, the antiquity of the earth, the figurative language of Genesis, and the possibility of at least limited organic development, while evolutionists such as Gray and Dana tended to insist on at least some divine involvement, especially with respect to the appearance of the first humans. At times Gray and Dana sounded as much like creationists as Guyot, who with Dawson was so moderate in opposing evolution that one historian has recently concluded that "on the most antievolutionary reading of their position, they can at most be described as doubtful detractors."[15]

To find a creationist who insisted on the recent appearance of all living things in six literal days, who doubted the evidence of progression in the fossil record, and who attributed geological significance to the biblical deluge, one has to look far beyond the mainstream of scientific thought. Even at the margins of the scientific enterprise occupied by clerical professors of science who rejected evolution, there were few advocates of a young earth. For example, the Congregational clergyman Edward Hitchcock (1793–1864), a respected geologist at Amherst College and one of the country's oracles on Genesis and geology, saw no theological barriers to admitting the antiquity of the earth, a local flood, or even an Adamic creation limited to particular species. He rejected organic evolution, but for scientific as much as religious reasons. Writing in *Bibliotheca Sacra* four years after the publication of the *Origin of Species*, he listed a number of religious objections to developmental hypotheses, including the allegations that they made God unnecessary and promoted materialism. "But after all," he added, "the real question is, not whether these hypotheses accord with our religious views, but whether they are true." And it seemed preposterous to him that anyone would claim that humans were "merely the product of transformation of the radiate monad through the mollusk, the lobster, the bird, the quadruped, and the monkey."[16]

Enoch Fitch Burr (1818–1907), another scientifically trained Congregational clergyman, has been described as "one of the most influential . . . writers against evolution." For a number of years after Hitchcock's death, Burr drew on his extensive training in mathematics and astronomy to teach Amherst students about the scientific evidences of religion, thus partially filling the void created by the departure of Hitchcock. Although he vigorously opposed evolution on scientific and theological grounds—one religious journal hailed his book *Pater Mundi; or, Doctrine of Evolution* (1873) as a work that "not only knocks the breath of life out of the theory, but chops it in pieces and throws it to the dogs"—he neither tried to fit the history of the earth into six literal days nor doubted the evidence of past geological ages. Convinced that "evolutionism means *materialism*," he directed his antipathy more toward grand schemes of evolution by natural law than toward Darwin's particular theory of the origin of species.[17]

George D. Armstrong (1813–1899), the leading spokesman on science and religion among Southern Presbyterians in the late nineteenth century, assumed an even less strident stance in his criticism of evolution. A Princeton alumnus who taught chemistry and geology at Washington College in Virginia for thirteen years before taking the pulpit of the First Presbyterian Church in Norfolk, he rejected "the hypothesis of evolution in all its forms," but for what he regarded as purely scientific reasons. In *The Two Books of Nature and Revelation Collated* (1886) he developed a line of argument similar to Guyot's, granting the flexibility of the language of Genesis to accommodate the findings of geologists while insisting on the special creation of life and humans. If evolution excluded the transition from inorganic to organic, on the one hand, and the transition from animals to humans, on the other, and if it were taken simply as "a mode of creation," he thought it was neither atheistic nor irreconcilable "with the Bible account of the origin of plants and animals in the world." Its unfavorable reception among many Christians he attributed to the suspicion that it fostered "the conception of our world as 'an automatic machine,' running itself."[18]

The Reverend Herbert W. Morris (1818–1897), sometime professor of mathematics in the Newington Collegiate Institution and author of the best-selling *Science and the Bible; or, The Mosaic Creation and Modern Discoveries* (1871), came as close as any of the clerical professors of science to defending a literal reading of Genesis. But although he insisted on a miraculous creation in six *"literal and natural days"* and argued that human evolution was "irreconcilable alike with the testimony of Scripture and the facts of nature," he allowed geologists to fill in the sequence of events that Moses allegedly skipped between the first two verses of Genesis, and he suspected that the deluge had been universal only in the sense that it drowned all humans not saved in the ark.[19]

CLERICAL CREATIONISTS

In general, Protestant ministers lagged behind their scientific brethren in debating the issues raised by Darwin. In a recent study of the response of Protestant intellectuals to the theory of organic evolution, based on a comprehensive analysis of late-century religious quarterlies and books, Jon H. Roberts detects a watershed in the mid-1870s. Before that time Protestant intellectuals critical of evolution tended to dismiss the Darwinian hypothesis as an unscientific theory exposed as such by Agassiz and other scientific luminaries. After about 1875, by which time it was becoming clear that the majority of American naturalists had embraced organic evolution, Protestant leaders "sought to prove the theory of organic evolution invalid by showing its inconsistency with central elements of the Christian message" rather than by stressing its scientific liabilities. During the last quarter of the century, says Roberts, "a sizable minority" of the Protestant contributors to religious quarterlies rejected organic evolution because of a conviction that it was "simply impossible to endorse the theory while remaining committed to the veracity of the Scriptures," particularly to the biblical account of human origins. As one of America's most popular preachers, the Presbyterian T. De Witt Talmage (1832–1902), put it, "The moment you begin to explain away the miraculous and supernatural, you surrender the Bible."[20]

But in spite of the overriding biblical concerns of the leading clerical critics of evolution, virtually none of them insisted on compressing the history of life on earth into a mere six thousand years or invoked the Noachian deluge to explain the fossil record. Even Princeton Seminary's rock-ribbed Charles Hodge (1797–1878), who concluded that Darwinism was atheism because it banished God from the world and enabled one "to account for design without referring it to the purpose or agency of God," conceded the great antiquity of the earth and gave his imprimatur to Guyot's and Dana's interpretation of the days of Genesis as geological epochs. And his disciple Robert Lewis Dabney (1820–1898), the scourge of evolutionists in the Presbyterian church of the South, withheld judgment on the question of geological ages and a pre-Adamite earth.[21]

Some of the most strident critics of organic evolution were premillennialists, whose predictions of Christ's imminent return depended on a quasi-literal reading of the Scriptures. Because of their conviction that one error in the Bible invalidated the entire book—and their fear that "if they allowed, in deference to modern science, that the Mosaic account of creation was mythic," they would find it difficult to defend a literal Second Coming—they had little patience with scientists who, as described by the evangelist Dwight L. Moody (1837–1899), "dug up old carcasses . . . to make them testify against God."[22]

Although Moody himself often consorted with such prominent theistic evolutionists as Henry Drummond (1851–1897) and Joseph Cook (1838–1901), his circle of associates included several of the most outspoken critics of evolution in the late nineteenth and early twentieth centuries: H. L. Hastings, Luther T. Townsend, and Alexander Patterson. As a youth in the 1840s, Hastings (1833?–1899) had joined an apocalyptic band of millenarians known as the Millerites, followers of the Baptist farmer-preacher William Miller, whose reading of biblical prophecy had led him to predict the Second Coming of Christ in 1843 or 1844. After the so-called Great Disappointment associated with the failure of Christ to appear as predicted, Hastings sided with a faction of the Millerite movement that evolved into the Advent Christian Church, and in time he became one of that sect's leading preachers and editors. Eventually ambition drove him to establish his own millenarian ministry, which included editing the *Christian*, a popular nondenominational paper, publishing "probably 100 tons" of religious literature, and occasionally assisting Moody in his crusades.[23]

In an effort to warn the public about the skeptical tendencies of evolution, Hastings included in his "Anti-Infidel Library" a series of four creationist pamphlets—subsequently published in book form as *The Errors of Evolution* (1885)—by Robert Patterson (1821–1885), the

Irish-born pastor of the East Oakland Presbyterian Church in California, who shared Hastings's hope in the soon return of Christ. Later Hastings contributed a tract of his own to the library, *Was Moses Mistaken? or, Creation and Evolution* (1896), in which he addressed the "delicate" issue of human ancestry, by far the topic of greatest concern to turn-of-the-century creationists:

> I do not wish to meddle with any man's family matters, or quarrel with any one about his relatives. If a man prefers to look for his kindred in the zoological gardens, it is no concern of mine; if he wants to believe that the founder of his family was an ape, a gorilla, a mud-turtle, or a monar [moner], he may do so; but when he insists that *I* shall trace *my* lineage in that direction, I say No sir! . . . I prefer that my genealogical table shall end as it now does, with "Cainan, which was the son of Seth, which was the son of Adam, which was *the son of* God," rather than invent one which reads, "Which was the son of skeptic, which was the son of monkey, which was the son of oyster, which was the son of monar, which was the son of *mud!*"—a genealogical table which begins in the mud and ends in the gravel, which has a monar at the head, a monkey in the middle, and an infidel at the tail.

Despite his obvious doubts about the wisdom of many scientists, Hastings, like Robert Patterson, freely granted them all the time they needed before the Edenic creation to accommodate geological "convulsions, changes, and catastrophes": "There may have been time enough after [the] 'Beginning' for all the geologists of the ages to have floundered in primeval mud until they had worked out all the problems concerning which they now can but theorize and guess."[24]

The Methodist minister and author Luther T. Townsend (1838–1922), though not a millenarian, emerged in the late nineteenth century as one of the most visible of the clerical opponents of evolution. A graduate of Dartmouth College and the Andover Theological Seminary, Townsend served as pastor of several Methodist Episcopal churches before becoming, in 1868, professor of Hebrew and New Testament Greek at the Boston Theological Seminary, which later became the Boston University School of Theology. He remained at the seminary in various capacities until 1893, when he resigned his professorship in order to devote full time to writing and lecturing. Active in the Chautauqua movement, he also served on the board of directors of the Bible League of North America, founded in 1903, which claimed to be "the only Organized Movement of any kind . . . that stands for the Defense and Confirmation of the Faith of the People in the Bible as the inspired and authoritative Word of God." In such works as *Evolution or Creation* (1896), *Adam and Eve* (1904), and *Collapse of Evolution* (1905), a pamphlet that circulated widely for a couple of decades, he defended the first chapters of Genesis as "a simple, straightforward narrative of the facts as they actually occurred." His interpretation, however, was anything but straightforward. In order to accommodate the findings of geologists with a literal reading of the word "day," he argued that "the six vast geological epochs which science describes are types and prophecies of the six ordinary days of which Moses writes, a thousand years being as one day and one day as a thousand years." In other words, "in six literal days, and in the order given in the Bible, the Creator brought the world out of the chaos of the glacial wreck, made it habitable, created modern flora and fauna and gave them life and power to propagate themselves until the end of time." Unlike some creationists who could not imagine being wrong, Townsend modestly acknowledged that time might necessitate the revision of some of his conclusions, that paleontological research, for example, might require the modification of his views on early humans and missing links.[25]

Another clerical creationist connected with the Moody empire was Alexander Patterson, a Presbyterian evangelist and longtime friend of Moody's who taught and lectured at the Moody Bible Institute. His book *The Other Side of Evolution* (1903) so impressed the pastor of the Moody Church in Chicago, A. C. Dixon (1854–1925), that Dixon arranged for the Bible Institute Colportage Association, forerunner of the Moody Press, to purchase the plates for a revised edition, which students from Moody's Bible Institute would then sell from their horse-drawn "Gospel Wagons." Dixon considered Patterson's treatise to be "about the best thing" he had read on the subject, and he thought that the book would "do an immense amount of good if sent to all the preachers, theological professors, theological students, Y.M.C.A. secretaries and Sunday School superintendents of the English-speaking world."[26]

Like Hastings and Townsend, Patterson believed that a historical hiatus occurred between the first and second verses of Genesis 1: "In that interval lies all geology tells us of. The history of the formation and after progress of the earth is there. This includes all up to the beginning of the Six Days' Creation. Here is where the fossil creatures lived and died. All this is passed over in silence in the Bible account." Although inclined toward a literal reading of the Bible, he did not think it necessary to interpret the days of creation as "our short days of twenty-four hours" or to insist on the universality of the flood. And whether or not the Edenic creation made use of preexisting species was of little concern to him—so long as there was no connection between humans and brutes, which he regarded as "the central point" of the doctrine of evolution. Adopting what Jon H. Roberts has called the "populist conception of validation," he demanded the right to participate in the evaluation of evolution, because "the questions involved are too important to be left to the scientist alone." Evolution, he insisted, must appear "before the judgment seat of Christian Common Sense," where "the best juryman will be the intelligent non-scientific mind." In this respect he found it heartening to remember "that there are thousands of quiet thinkers who have never given in their adhesion to this startling theory, and more, that the great masses of the church at least, have no confidence in it." Nevertheless, he feared that evolution was sweeping the country, and he predicted ominously that "if this theory is accepted, we must look for widespread lapse from all Christian faith and, as conduct follows belief in all intelligent creatures, we shall see also great moral declension."[27]

CONSERVATIVE CREATIONISTS

Thus far our survey of post-Darwinian creationist opinion, based in large part on the views of the antievolutionists most frequently cited and quoted by their contemporaries, has failed to turn up a single scientist or cleric who rejected the antiquity of the earth, denied the progressive nature of the fossil record, or attached geological significance to the Noachian flood. The most famous creationist of all, Agassiz, simply ignored the biblical record. Guyot, Dawson, Burr, Armstrong, and Hodge—as well as Dana before his conversion—cherished the Bible as God's inspired word but were willing nevertheless, in the interest of accommodating the findings of science, to adopt a figurative reading of the first chapter of Genesis. Even Hitchcock, Morris, Hastings, Townsend, and Alexander Patterson, who adhered to a more traditional interpretation of the creative week, managed to squeeze the ages of geology into an assumed gap in the revealed record; and Hitchcock and Patterson, at least, were willing to entertain the possibility that the days of Genesis 1 were not literal twenty-four-hour periods. Dabney refused to choose between the day-age and gap theories but condemned neither. No doubt many Christians, perhaps most, remained unpersuaded by the geological evidence of the earth's great age and continued to believe in a recent creation in six literal days, but these people rarely expressed their views in books and journals. Of those who did, only a tiny minority invoked the deluge to explain the fossil record, the most compelling evidence of an ancient earth.

One of the last nineteenth-century advocates of flood geology was Eleazar Lord (1788–1871), who with his brother David Nevins Lord (1792–1880) rose to prominence in conservative religious circles for his articulate and passionate advocacy of a premillennial eschatology, a view somewhat tarnished by the recent Millerite fiasco. Both brothers as youths aspired to careers in the church. Eye trouble, however, compelled Eleazar to terminate his studies at Princeton and take up a business career, while poor health forced David, after his graduation from Yale, into the dry-goods business in New York City, where Eleazar was already president of the Manhattan Fire Insurance Company and would later preside over the New York and Erie Railroad. Throughout their lives both brothers remained active Christian laymen, Eleazar as a Presbyterian, David as a Congregationalist.[28]

Eleazar Lord's last major contribution to the literature of creationism, *The Epoch of Creation* (1851), appeared eight years before Darwin's *Origin of Species*. Believing that "the inspired account literally and unequivocally asserts that in six days the Lord made heaven and earth, the sea, and all that in them is," Lord rejected the "gratuitous assumptions of the geologists" and their efforts to make the language of Genesis accommodate the findings of so-called science. Instead he resurrected the antiquated notion, seldom encountered during the last two-thirds of the nineteenth century, that the fossil-bearing strata, which gave the earth the appearance of great age, had been deposited during the Noachian deluge. The apparent progression of organisms in these fossiliferous beds he explained by the rising floodwaters: "What can be more obvious than to suppose that as the waters of the deluge rose, and progressively became charged with sedimentary matter, shell-fish should have been stifled and buried, before the placoids, ganoids, or other species of fish, fishes before reptiles, reptiles before birds, and birds before quadrupeds?"[29]

David Lord concurred with his elder brother on the need to limit earth history to about six thousand years and on the gravity of the threat to Christianity posed by modern geology. "It is clear," he wrote in a manner that would later characterize much creationist thinking, "that if the geological theory is true, the Mosaic history is not, and thence that the sacred volume at large, which everywhere recognizes that history as divine, cannot be from God." In defense of scriptural geology he wrote a book titled *Geognosy* (1855), which students in at least one millenarian college were using as a geology text as late as 1875. And from 1848 to 1861 he edited the *Theological and Literary Journal,* from which he launched repeated attacks on historical geology and developmental biology. In attempting to answer "the great question in theoretical geology," the origin of the fossil-bearing strata, he discarded his brother's diluvial solution in favor of a scheme that emphasized stratification before and after the time of Noah. Many strata, he thought, had been "thrown up from the depths of the earth" during the interval between the creation and the flood; others had been formed after the deluge, when God had miraculously restocked the remote and inaccessible regions of the globe. Lord, like Agassiz, adopted the extrabiblical supposition that animals had been created "not in single pairs, like the progenitors of the human race, but in crowds." Given his antipathy toward historical geology, it is not surprising that in 1860 he greeted the publication of Darwin's *Origin of Species* with disapproval, calling it an affront to both science and Scripture.[30] Many conservative Christians no doubt concurred with this judgment, but for years to come the Lords stood virtually alone among creationist writers in limiting the history of the earth to a mere six thousand years.[31]

TWO

George Frederick Wright: From Christian Darwinist to Fundamentalist

A s the Reverend A. C. Dixon laid plans for *The Fundamentals* (1910–1915), the series of twelve mass-produced booklets that launched the fundamentalist movement, he asked the cleric-geologist George Frederick Wright (1838–1921) for a chapter called "Evolution from the Christian Point of View" that would discredit both materialistic and theistic varieties of the theory. In making his pitch as editor, Dixon noted the strengths and weaknesses of other candidates for the job. He greatly admired Alexander Patterson's *Other Side of Evolution* for the factual evidence it presented, but he recognized that no one would consider the Chicago evangelist authoritative on the subject. Luther T. Townsend had proved himself to be "strong along these lines," but he, too, had no scientific qualifications. The same was true of James Orr (1844–1913), the conservative Scottish theologian who taught at the United Free Church College in Glasgow and who occasionally addressed scientific issues. Orr saw little wrong with divinely guided evolution by saltation, a fact Dixon did not mention. In view of these impediments, Dixon thought Wright, who possessed both scientific and theological credentials, might be just the person to contribute an article that would "make the matter so clear that the vagaries of Evolution shall be driven from the minds of thousands."[1] Ironically, Dixon had picked to be the first fundamentalist spokesman on science a man who just a few decades earlier had been one of Darwin's most enthusiastic American advocates.

CHRISTIAN DARWINIST

G. Frederick Wright, as he styled himself in later years, was born into a pious family of farmers in upstate New York in 1838. At the age of seventeen he left home for Oberlin College, where President Charles G. Finney (1792–1875) was preaching an evangelical Christianity that stressed the authority of the Bible and the factual, evidentiary nature of religious belief. On graduating from college in 1859, Wright enrolled in the theological course at Oberlin, which he completed, after a brief hitch in the Union army, three years later. For the next decade he pastored a Congregational church in the small Vermont town of Bakersfield, at the foot of the Green Mountains.[2]

During his tenure at Bakersfield, Wright's attention turned increasingly to scientific matters. Because his meager income as a minister did not permit expensive vacations, he spent many of his leisure hours exploring the surrounding countryside and acquiring a modest reputation as an expert on local geology, particularly glacial deposits. He also took time to read such provocative works as Darwin's *Origin of Species* and Charles Lyell's *Geological Evidences of the Antiquity of Man* (1863), which clashed with the views he had been taught as a youth. Wright's autobiographical writings do not reveal the extent to which these books may have precipitated a crisis of faith; they do indicate, however, that he found in Asa Gray's theistic interpretation of Darwinism a compromise that allowed him simultaneously to embrace organic evolution and to retain his belief in a divinely designed and controlled universe.[3] As Gray realized, such views departed considerably from Darwin's own. Although Gray described himself as "one who is scientifically, and in his own fashion, a Darwinian," he confessed to a friend that his ideas on divinely directed variations were "very anti-Darwin."[4]

Wright especially appreciated a passage in which Gray described "the popular conception" of efficient cause: "Events and operations in general go on in virtue simply of forces communicated at the first, but that now and then, and only now and then, the Deity puts his hand directly to the work." This view of God's relationship to the natural world appealed to Wright as an ideal solution to the problem of reconciling the respective demands of science and Scripture. As he later wrote, it "allows us to retain our conceptions of reality in the forces of

nature, makes room for miracles, and leaves us free whenever necessary, as in the case of the special endowments of man's moral nature,
to supplement natural selection with the direct interference of the
Creator."[5]

In 1872 Wright accepted a call to serve as pastor of the Free Church
in Andover, Massachusetts, home of the orthodox Andover Theological
Seminary and the equally sound *Bibliotheca Sacra*. By this time American theologians were beginning to take notice of Darwin's theory, and
Wright was, as he put it, "soon drawn into the vortex of discussion."
The year after moving to Andover he contributed a review of "Recent
Works on Prehistoric Archaeology" to *Bibliotheca Sacra*, in which he
attempted, in a manner that would become his trademark, to stake out
a middle position on the question of human antiquity—between the
vast ages assigned by "reckless" uniformitarian geologists and the brief
six-thousand-year history commonly believed by the intellectual heirs
of Archbishop James Ussher (1581–1656). To Wright, who regarded
the Bible as the inspired Word of God and who preferred literal readings where possible, even such moderate tampering with received
opinion demanded an explanation. Thus he reminded his readers that
"the integrity of the Bible depends only upon the truth of those doctrines and interpretations which are woven into the very woof and
warp of the book." Because the genealogies of Genesis had no such
status, Wright happily predicated that there would be "little difficulty
in adjusting the Bible history to any facts of chronology that archaeol-
ogists may fairly establish."[6]

While residing in Andover, Wright continued his geological investigations, often devoting Mondays to tracing the line of gravel ridges
left by the receding ice sheet that had once covered the area. His
careful work brought him recognition as a glacial geologist and won
him entrée to the scientific circles of nearby Boston, where he was
befriended by Gray. The eminent botanist, then in his sixties, became
"like a father" to the young minister, tutoring him in the ways of
science and sharing his vision of a Christianized Darwinism. Together
they sought to fashion "a right *evolutionary teleology*" that would provide an alternative interpretation to that offered by the "infidel class of
Darwinian expositors" on the one hand and such antievolutionists as
the theologian Charles Hodge and the geologist John William Dawson
on the other.[7]

Their collaboration produced two of the most important theological
contributions to the Darwinian debates: Gray's *Darwiniana* (1876),
which, at Wright's urging, brought together the botanist's previously
anonymous and scattered essays; and Wright's five-part series of arti-

cles in *Bibliotheca Sacra,* "Recent Works Bearing on the Relation of Science to Religion," which later formed the heart of his book *Studies in Science and Religion* (1882).

In his series Wright argued that scientists and theologians shared a common methodology, inductive reasoning, which placed a premium on interpreting rather than merely collecting facts. In contrast to the "scientific book-keeping" practiced by the followers of Francis Bacon (1561–1626), modern scientists such as Darwin were aiming "to do far more than observe and classify. They are seeking the deeper meaning of the facts which they observe. They are endeavoring to trace out the bond of order which all believe to reign supreme in nature." Like students of the Scriptures, they offered probability, not certainty.[8]

In making the case for the natural origin of species, Wright blunted the possible psychological shock of Darwin's theory by retaining such familiar concepts as God, miracles, and the special creation of humans. He also repeatedly used language that seemed to restrict natural selection to the lower end of the taxonomic scale, while attributing kingdoms and the broader taxonomic groupings to special creation. According to Wright's version of Darwin's theory, "The Creator first breathed life into one, or, more probably, four or five, distinct forms," after which a process combining miraculous variations and natural selection split each "order" into families, genera, and species. He thought the appearance of humans might legitimately remain outside the evolutionary process, writing that "the miraculous creation of man might no more disprove the general theory of natural selection than an ordinary miracle of Christ would disprove the general reign of natural law." Darwin himself described Wright's account of his views as "powerfully written and most clear" but apparently said nothing about its accuracy.[9]

Wright, like Gray, derived great comfort from Darwin's inability to explain the origin of the variations preserved by natural selection, because this limitation seemed to open the door for divine intervention. It "rob[bed] Darwinism of its sting," "left God's hands as free as could be desired for contrivances of whatever sort he pleased," and preserved a "reverent interpretation of the Bible." Although Wright judiciously refused to identify himself as "a disciple of Mr. Darwin or as a champion of his theory," only the myopic reader could have failed to see that he was a disciple of Gray and a champion of theistic evolution.[10]

Wright rarely alluded to the Scriptures in his essays; instead, he emphasized that Darwinism, as interpreted by men like Gray, did not invalidate the design argument for God's existence. However, when he collected his essays for publication in book form, he added a chapter devoted specifically to the subject, "The Relation of the Bible to Science." Wright expressed no doubts about the inspiration of the Scriptures, but he allowed that "the utterances of the Bible are not infallible except as pertaining to things 'necessary to be known, believed, and observed for salvation.' " As authority for his position he cited the conservative Princeton theologian Charles Hodge, who, Wright claimed, maintained that the sacred writers "stood on the same level with their contemporaries" when they wrote about science, history, and philosophy.[11]

Freed from the burden of defending biblical science, Wright saw no reason to look for correspondences between Genesis and geology. In fact, he viewed such efforts with scarcely veiled scorn, writing, for example, that "in many of these attempts it is difficult to tell which has been most distorted, the rocks or the sacred record." The ancient authors of Genesis had included the creation story as "a protest against polytheism," not as a natural history; thus it seemed to Wright "an impertinence to endeavor to find all modern science in the document, however easy it may be for science to find shelter under the drapery of its rhetoric."[12]

Because he believed that the inspired writers intended only to state the *"fact of creation by Divine agency,"* Wright professed to see "no difficulty at all in adjusting the language of the first chapter of Genesis to that expressing the derivative origin of species." But he remained too much of a biblical literalist simply to dismiss the story of Eve's creation from one of Adam's ribs. And, though he readily accepted the natural evolution of the human body, he insisted on a supernatural infusion of the soul. "No! man is not merely a developed animal; but the inventive genius displayed in the rudest flint implement stamps him as a new creation," he declared. "The new creation, however, is spiritual rather than material or physical."[13]

To harmonize the evidence of human antiquity with the apparently short chronology given by the Old Testament genealogies, Wright again appealed to Hodge, who had argued that the genealogical tables "were intended to prove that Christ was the son of David, and of the seed of Abraham, and not how many years had elapsed between the creation and the advent." Wright also quoted the Princeton biblical scholar William Henry Green (1825–1900), who had written that "nothing can be plainer than that in the usage of the Bible, 'to bear' and 'to beget' are used in a wide sense to indicate descent, without restricting this to immediate offspring." The elasticity of these records,

Wright concluded, allowed "the geologist and the archaeologist and the Egyptologist and the linguist to work peaceably at their several problems. When these investigators have created a real demand for more time than is allowed by the discordant schemes of chronology which men have upon their own responsibility drawn out of the Bible, we shall remove the misleading dates of Archbishop Usher [sic] from the margins of our 'authorized' translation: and it is already high time this was done."[14]

CHRISTIAN APOLOGIST

In 1881 Wright abandoned the stimulating intellectual environment of New England for the relative isolation of north-central Ohio, where he assumed the professorship of New Testament language and literature in the Oberlin Theological Seminary. While living in the environs of Boston, home of one of the most active scientific communities in America, he had become a recognized expert on the ice age in New England, had served as a director of the Boston Society of Natural History, and had enjoyed the confidence of the country's foremost botanist. His growing reputation in science had led to offers of a chair in biology at Western University in Pittsburgh and a professorship of geology at Colorado College. In recent years, however, he had become alarmed by the growing popularity of biblical criticism in America, and he saw in the invitation from his alma mater a chance to defend orthodoxy.[15]

Wright's return to Oberlin signaled an intellectual drift to the right that continued until his death. During the 1870s, when the leaders of American Protestantism had first faced the issue of evolution, Wright had fought for a modified version of Darwinism, laboring mightily to show that an evolutionary account of the origin of species left the tenets of natural and revealed religion unscathed.[16] However, beginning in the 1880s with his removal from New England and the Boston scientific scene, the thrust of his efforts shifted from defending evolution and the scientific enterprise against biblical literalists to defending the historical accuracy of the Bible against critics who applied evolution to the making of the Bible itself. At the same time, his attitude toward science changed from enthusiasm to skepticism. By the early 1900s he had allied himself with the conservative religious movement that came to be known as fundamentalism.

Wright's refusal to compromise the authority of the Bible stemmed from his conviction that the very existence of Protestant Christianity, to which he and countless other ministers had devoted their lives, was at stake:

> Of all the questions facing a Protestant clergyman none presses more persistently and imperatively for an answer than that pertaining to the relation of the Bible to the true religious faith of himself and of his parish and of the world. The ground upon which the Protestant pulpit is permitted to speak with its present authority is the general belief that the Bible is the word of God, and that it is not a sealed book even to ordinary readers, much less to men with the experience and the advantages for study of the ordinary parish clergyman. Any other position than this drives us logically to rationalism or to the Papacy.

Wright was fighting to save not only orthodoxy but professional identities and careers as well.[17]

Evidence of his growing conservatism can be seen in his changing attitude toward the first chapter of Genesis. As we have seen, in 1882 he had dismissed the efforts of apologists to correlate the Mosaic days of creation with the epochs of earth history as an "impertinence." Only two years later, however, in *The Divine Authority of the Bible* (1884) he cited the harmonizing scheme of Arnold Guyot as a plausible interpretation of the Mosaic account. Shortly thereafter James Dwight Dana, whom Wright regarded as "the foremost scientific man of America," began urging Wright to accept the Mosaic account of creation as an inspired cosmogony; to do otherwise, Dana argued, would endanger "the credibility of all that follows in the Bible." Dana's persuasion—and the unrelenting aggression of biblical critics—apparently convinced Wright that this was indeed an accurate assessment. In an 1886 address defending the inspiration of the Bible against the attacks of the higher critics, he expressed his unwillingness to "surrender any portion" of the Scriptures or to "label it as 'unessential.' "[18]

Wright's final break with the notion that Genesis was merely a protest against polytheism did not come until the 1890s, when he publicly admitted his earlier error and embraced Guyot's widely held interpretation. While claiming to have no objection "to being held in the main to the brief statements" he had made in the early 1880s about the meaning of Genesis, he noted that

> prolonged attention to the subject has increased my respect for those who have sought a positive harmony between the geological history and the system unfolded in the first chapter of Genesis. It has seemed to me that the opponents of all attempts at positive

harmonization have too generally failed to appreciate the peculiarities of popular literature as distinguished from scientific, and have assumed that the freer handling of language, appropriate to popular presentation, is incapable of giving any expression to general truths.

A year later he went even further, confessing that "in writing upon this subject at previous times I have dwelt, I now believe, somewhat too exclusively upon the adaptation of the document to the immediate purpose of counteracting the polytheistic tendencies of the Israelites and, through them, of the world." The story of a six-day creation might not be literally true, but at least it was scientifically accurate. [19]

Wright's views on science also underwent a noticeable transformation following his return to Oberlin. In the 1870s he had striven to convince his ministerial colleagues of the value of science and the harmlessness of theistic evolution—and as late as 1883 he had held literal-minded theologians primarily responsible for what John William Draper (1811–1882) had called the "conflict between science and religion." But as his alarm grew about the materialistic direction of evolutionary thought and the uses made of it by his archenemies, the higher critics and liberal theologians, his writings took on an increasingly negative tone. As his friend Charles H. Hitchcock (1836–1919), Edward's son, observed, many of the country's intellectual leaders seemed to have lost their balance: "They have been well instructed in their youth, but seem now to have thrown off all religious restraint, and to enjoy a sense of freedom like boys in vacation. Evolution has given rise to this license: and it is evidently proper to inquire what grains of truth may be sifted out from its principles." [20]

Even before leaving Andover, Wright had expressed his disgust with "*a priori* evolutionary universe-maker[s]," such as Herbert Spencer (1820–1903) and John Fiske (1842–1901), who rashly pushed beyond Darwin's "limited conclusions" to construct a system of cosmic evolution. In the ensuing years Wright frequently contrasted the modest, cautious Darwin, who had sought to explain only the origin of species and who had limited his theory of descent to no more than "all the members of the same great class or kingdom," with the impetuous— and often impious—souls who tried to explain the evolution of the entire world and who described development from "the first jelly speck of protoplasm to the brain of a Newton or a Gladstone" without any direct reference to the Creator. This, he declared, was "Darwinism gone to seed in barren soil." [21]

The common conflation of such views with "Darwinism," together with the strident materialism of such Darwinists as the zoologist Thomas H. Huxley (1825–1895), made Wright's defense of Darwin increasingly difficult. Darwin's own public rejection in the 1870s of Gray's theistic version of evolution also increased Wright's burden, forcing him to distinguish between Darwin's science and his theology. At a time when the leading Darwinists were expanding their theory to embrace all living things, Wright insisted on limiting Darwinism, as its founder had done in 1859, to the origin of species from perhaps four or five primordial forms, presumably created by God. Wright also began to emphasize the unproven status of natural selection and to repeat Darwin's description of himself—in the context of his continuous efforts to explain away the difficulties of his theory—as the prince of "wrigglers," a confession Wright put to apologetical use by noting that the "wrigglings of the biblical interpreters are by no means so conspicuous as those of the defenders of the continuity of species."[22]

Wright's long-festering fears about the implications of higher criticism for an orthodox view of the Bible reached crisis proportions in the early 1890s, when the controversial Charles A. Briggs (1841–1913), a Presbyterian theologian whose heresy trial rocked the religious world, caused Wright temporarily to question his own belief in the Mosaic authorship of the Pentateuch. For years Briggs's rejection of the inerrancy of the original scriptural autographs and his iconoclastic attacks on the Mosaic authorship of the first five books of the Bible had nettled Wright. Then early in 1891, while listening to Briggs in person, Wright fell "under the spell of his eloquent and positive affirmations" and almost abandoned his long-standing convictions. "So violent had been the shock," he reported, "that out of self-respect I have found it necessary to turn a little aside from my main studies to examine anew the foundations of my faith, and see to what extent the sappers and miners, who have been so diligently at work the last quarter of a century, have really disturbed the foundations." Wright emerged from this period of soul-searching convinced more firmly than ever in the Mosaic authorship of the Pentateuch and in a supernatural view of history. He noted, however, that only his belief in miracles and "external confirmatory evidence" had saved him from succumbing to Briggs's seductive arguments.[23]

CHARLATAN?

In 1880, while still living in the East, Wright visited Trenton, New Jersey, to inspect some undisturbed gravel deposits along the banks of the Delaware River, reputed to contain rude stone implements of human origin. Because the deposits had been formed by the drainage of the melting ice-sheet that had once covered the region, the presence of these implements suggested that humans had lived there during the closing stages of the glacial period, approximately eight thousand to ten thousand years ago. Although Wright himself failed to find any implements in place, his visit to Trenton, he later observed, "gave new zest to my investigation, since now glacial studies touched on the theological and Biblical questions in which I was primarily interested." In the following years he traveled widely throughout North America, tracing the southern boundary of the glaciated area and searching for evidence of human antiquity. In 1889 he brought out his magnum opus, *The Ice Age in North America: And Its Bearing upon the Antiquity of Man*, which passed through five editions before his death.[24]

By this time Wright was sufficiently committed to the historical accuracy of the Bible to be disturbed by the fact that his estimates of human life on earth exceeded the six thousand years commonly attributed to the Old Testament genealogies. Seeking a way out of his dilemma, Wright paid a visit to Professor B. B. Warfield (1851–1921) of the Princeton Theological Seminary, who shared his belief in the trustworthiness of the Bible. Warfield took his troubled visitor to see the impeccably orthodox Green, who had long been teaching that the genealogies of Genesis allowed the insertion of as much time as science required in the interval between Adam and Abraham. Furthermore, he had concluded that actual violence was done to the Scriptures by imposing "upon the genealogical tables of the Old Testament an inelastic chronology." Wright conversed with the Princeton scholar until "the small hours of the night" and left reassured that the biblical genealogies did indeed provide the wriggling space he needed to harmonize his findings about human antiquity with his belief in the inspiration and inerrancy of the Genesis record.[25]

In 1892 Wright published the most controversial work of his life, *Man and the Glacial Period*, a condensed version of the much longer *Ice Age in North America* that incorporated data he had subsequently gathered on field trips to Europe and to the American West. As in his earlier book, he argued for a single, relatively brief ice age that coincided with the appearance of humans in North America. Though this view was falling out of favor with some geologists, he had no reason to expect anything but a civil reception. He had long enjoyed a cordial relationship with professional scientists, and just three years earlier the *American Geologist* had greeted his book on the ice age with the comment that "geologists in America will welcome this excursion of the theologian into actual science." But as the reviews of his latest book began to appear, it seemed as though Wright had worn out his welcome.[26]

The attack on Wright came largely from men associated with the United States Geological Survey, particularly Thomas C. Chamberlin (1843–1928) and W J McGee (1853–1912), who disagreed with Wright's contention that there had been only one ice age. The former, chief of the Pleistocene division and an authority on glaciers, also resented Wright's growing fame as a glacialist. In reviewing *Man and the Glacial Period* for the *Dial*, Chamberlin accused Wright of stupidly misreading the evidence regarding humans and the ice age and of deliberately misleading readers by identifying himself on the title page as "Assistant on the United States Geological Survey," when he no longer was. McGee, writing for *Science*, called Wright a "knave" and a "dupe." Addressing him as "the Reverend Professor," the ethnologist advised Wright to stick to theology, adding that "the world would be wiser if the book were not written." In a subsequent piece for the *American Anthropologist*, McGee described Wright as "a betinseled charlatan whose potions are poison." "Would that science might be well rid of such harpies," he concluded.[27]

Although Wright's defense of a single ice age flew in the face of mounting scientific evidence to the contrary—and his theological convictions undoubtedly colored his scientific conclusions—at least one geologist, Dana, regarded the public abuse of Wright as "a disgrace to American Science." Another, Edward W. Claypole (1835–1901), thought the attack on Wright savored "too strongly of the old-time, intolerant, theological method of crushing a formidable rival by dint of concerted action or force in default of reason." Notwithstanding these expressions of support, Wright's public humiliation at the hands of some of the nation's leading geologists could only have strengthened his growing suspicion about the arrogance of scientists and fostered a sense of alienation from the scientific community.[28]

Wright continued after this unpleasant episode to contribute to scientific journals, but increasingly his energy went into vindicating the Bible against claims that it contained scientific errors. His efforts along these lines were facilitated by his appointment in 1892 to a specially created professorship of the harmony of science and revelation, which freed him from his teaching duties at Oberlin for half of each year and allowed him over the next fifteen years, until his retirement, to travel the world in search of scientific evidence that would "reestablish confidence in the historical statements of the Old Testament." His research culminated in the publication of *Scientific Confirmations of Old*

Testament History (1906), which one colleague described as "the standard work on the important subject of Pentateuchal physics." In it Wright sought to corroborate such biblical occurrences as an actual but nonuniversal deluge, the crossing of the Red Sea, the parting of the River Jordan, and the destruction of Sodom and Gomorrah—"events which are peculiarly open to scientific cross-examination."[29]

As Wright's espousal of the Noachian catastrophe suggests, his allegiance to biblical history was fast driving him from the mainstream of science. Even as a Christian Darwinist he had excluded the origin of matter, life, and the human soul from the rule of natural law; by the late 1890s he was sounding more and more like a special creationist. Like many late nineteenth-century scientists, he had come to reject not only Darwin's estimates of the antiquity of the earth but also his theory that evolution resulted from the natural selection of minute variations over long periods of time. Instead, Wright embraced the theory of "paroxysmal evolution" advocated by the American scientists Clarence King (1842–1901) and Joseph Le Conte (1823–1901), who argued in neo-Lamarckian fashion that natural catastrophes precipitated rapid evolutionary change. This theory, admitted Wright in a particularly revealing statement, did "not differ much in its phenomena from the old-style theory of special creation." Indeed, it was becoming difficult, if not impossible, to distinguish Wright's evolutionism from the creationism of someone like Guyot, who had always tried to keep the number of supernatural interventions to a minimum—and whose views on matters pertaining to science and religion Wright now frequently adopted.[30]

In discussing the origin of humans, Wright emphasized the great gap between "the highest animal and the lowest man," though he allowed that a divine miracle might have bridged the gap, thereby joining humans and animals. In fact, he proposed a way by which such a development could be harmonized with the statement in Genesis that God had formed man "of the dust of the ground." Since the word *dust* could signify more than mere dirt, he reasoned that

> the advocates of derivative origin may properly say that the ape-like progenitor of the human race spoken of by Darwin is still but dust of the earth, and may feel that [they satisfy] all the conditions of the problem when [they propound] the hypothesis that when through the slow processes of evolution a bodily form with sensational and intellectual endowments approaching those of man had been reached, the Creator miraculously added to an embryonic germ the spiritual potencies of the human race.

But regardless of how much physical preparation had preceded the appearance of humans, Wright insisted that "there came a time when both the body and the soul were transformed into a new creature bearing the image of God."[31]

The opening years of the twentieth century found Wright not only publicly damning "the antiquated Uniformitarian geology of Lyell and Darwin" and arguing for "the traditional view that man originated, through supernatural interference, at a comparatively recent time, somewhere in Central Asia," but denigrating the very scientific method that had previously grounded his confidence in Darwin. In language reminiscent of the Baconians he had once scorned, he now called attention to "the infirmity of all inductive reasoning which makes any advance upon the barest facts of observation." Indeed, the "uncertainties of science" emerged as a common theme running through his later writings.[32]

By this time Wright's status as an evolutionist had become so problematic that he felt compelled in 1902 to address the question of whether or not he was an "evolutionist."

> If I should answer, "Yes," I should instantly add, as the witness in a celebrated trial a few years ago did, "but I am not a fool." But in view of the vagueness of the term "evolution," and the extensive misuse to which it is put, I should object to answering this question at all. A man may believe in the origin of species through natural selection, and still not be an evolutionist.[33]

This evasive answer hardly clarified the situation.

FUNDAMENTALIST

If Wright's identity as an evolutionist was uncertain at the turn of the century, it practically disappeared during the next decade, when he joined forces with the leaders of the emerging fundamentalist movement. Symbolic of his shifting allegiance was a complimentary, though cautious, introduction he wrote in 1903 for Alexander Patterson's *Other Side of Evolution*. "While not saying that all the points in this little volume are well taken, I can say that I disagree with fewer things in it than with those in almost any other on the subject, and that it is fitted to serve as a very needful tonic in these days of the confusion of bad Philosophy and fragmentary Science." Describing the popular theory of evolution as "one-tenth bad Science and nine-tenths bad Philosophy," Wright sought to lay most of the blame on Spencer rather than on Darwin. The latter, he argued, "was not strictly an Evolutionist, and rarely used the word. He endeavored simply to show that Species were enlarged varieties. . . . On the larger questions of the origin of genera and the more comprehensive orders of plants and animals, he spoke with great caution and only referred to such theories as things 'dimly seen in the distance.' "[34]

By 1903 Wright had achieved such prominence as a spokesman for conservative Christianity—as editor of *Bibliotheca Sacra* and as a contributor to such staunchly antimodernist journals as the *Sunday School Times* and the *Homiletic Review*—that he was invited to serve as a director of the newly formed American Bible League, which engaged "in an active warfare against unbelief in many of its most dangerous forms," but especially against higher criticism. The very first volume of the league's magazine, the *Bible Student and Teacher*, carried an address by Wright in which he dismissed Darwinism as "a theory which is well described as a series of 'loopholes' and 'may-bes.' " Later, in 1909, he contributed an essay, "The Mistakes of Darwin and His Would-be Followers," in which he criticized Darwin for making "two great mistakes": taking too much time because of his unwarranted assumption of uniformity in nature and thinking that minute variations would ever produce something beneficial.[35]

In assessing Wright's attitude, we must bear in mind that he was writing during a period when many biologists believed that Darwin's theory of natural selection could not adequately account for organic evolution. The criticism of Darwinism by scientists misled many laypersons into concluding that the scientific community had lost faith in organic evolution generally and prompted a spate of publications with such titles as "The Collapse of Evolution" (by Luther T. Townsend, another of the league's directors). Given this state of affairs, it is understandable that some evangelical Christians who had formerly condoned biological evolution withdrew their endorsements.[36]

In view of Wright's exposé of "The Mistakes of Darwin" and his introduction to Patterson's antievolutionary tract, Dixon had ample reason for believing that the former Christian Darwinist shared his feelings toward evolution. And Wright did not disappoint him. In his essay for *The Fundamentals*, titled "The Passing of Evolution," Wright lashed out at contemporary evolutionists while emphasizing the creationist elements in Darwin's own early writings. Unlike his modern disciples, who were irresponsibly teaching that all forms of life had arisen by strictly natural processes from one primordial speck, Darwin, said Wright, had postulated a Creator who breathed the force of life into several forms of plants and animals "and at the same time endowed them with the marvelous capacity for variation which we know they possess." And instead of attributing "all the differences between animals or between plants" to natural selection, the great English naturalist had taught "simply that species may reasonably be supposed to be nothing more than enlarged or accentuated varieties, which all admit are descendants from a common ancestry." Man, continued Wright, differed so greatly from the higher animals it was "necessary to suppose that he came into existence as the Bible represents, *by the special creation of a single pair*, from whom all the varieties of the race have sprung."[37]

Wright's article, which seems to limit evolution to species and to argue for the special creation not only of humans but of the earliest plants and animals, reveals the extent to which the former Christian Darwinist had abandoned his earlier beliefs. Or does it? Other evidence suggests that the discontinuity in Wright's thinking may have been more apparent than real. In 1912, about the same time his essay in *The Fundamentals* appeared, Wright published a book, *Origin and Antiquity of Man*, in which he argued for a theistic, paroxysmal history of the earth. On the question of human origins, he sounded like the antithesis of the person who wrote "The Passing of Evolution." "Upon comparing the bodily structure of man with that of the higher animals associated with him," he wrote, "the argument in favor of a common origin, so far as physical structure is concerned, becomes almost overwhelming." And again: "In view of all these facts, it is difficult to resist the conclusion that, so far as his physical organism is concerned, man is genetically connected with the highest order of the Mammalia, but it is equally evident that he is not descended from any existing species of that order." Wright granted that God had put the finishing touches on man's physical organization and had imparted to him "his higher mental qualities," but this still seemed to leave him far short of "the special creation of a single pair" as generally understood.[38]

The apparent discrepancy between Wright's two statements was not lost on his contemporaries. Late in 1914 Dixon, now pastoring the Metropolitan Tabernacle in London, received a letter from a confused reader of *Origin and Antiquity of Man*, inquiring about the harmony of the views presented there with those appearing in *The Fundamentals*. "I must confess," wrote Dixon to Wright, "that they are quite puzzling to me. Have you changed your view as to Darwinian Evolution, or will the context in your book so modify these quotations as to make them harmonize with your article in the 'Fundamentals'?" Unfortunately, Wright's reply has not survived; thus it is impossible to tell how he explained the apparent contradiction. (A reading of Wright's book eliminates the possibility that Dixon's correspondent took Wright's comments out of context.) Wright's final statement on the origin of humans, in his autobiography, published in 1916, provides little clarification: "I believe that, whatever may be true about some organic connection between man and some unknown species of anthropoid apes, man with his present physical and spiritual characteristics appeared suddenly on the earth, at no very distant period, as geologists count time."[39]

Wright's true views on evolution, especially on the origin of humans, remain as puzzling today as they did to Dixon in 1915. Did he believe, as he wrote in *The Fundamentals*, that humans "came into existence as the Bible represents, by the special creation of a single pair," or, as he wrote in *Origin and Antiquity of Man*, that "man is genetically connected with the highest order of the Mammalia"? As a veteran of the Darwinian debates, Wright certainly knew that the term *special creation*, as commonly used, ruled out a genetic connection. How could such a clear-thinking, knowledgeable person make such apparently contradictory statements about the most emotionally laden issue connected with evolution? Was he merely tailoring his language to meet the expectations of different audiences or to camouflage his true views?

In attempting to reconcile his apparently contradictory remarks, it may be helpful to recall that Wright, especially in his later years, refused to regard creation and evolution as mutually exclusive explanations. And even during his days as a Christian Darwinist, he espoused a view of organic evolution that, in Michael McGiffert's words, "opened the backdoor to the doctrine of special creation." Later, he abandoned Darwin's hypothesis of minute variations for a paroxysmal theory of development, which, in his own words, did "not differ much in its phenomena from the old-style theory of special creation." Thus, although creation and evolution came to represent polar opposites in the minds of many people, for Wright they were virtual synonyms, equally appropriate for describing a process of emergent evolution in which God "created" humans from previously existing apelike creatures.[40]

Wright was not alone in describing the divinely assisted evolution of humans as an act of creation; indeed, such ambiguous phraseology seems to have been in vogue among evangelical intellectuals who accepted evolution. For example, after abandoning belief in the fixity of species, Dana explained to an inquiring clergyman that "while admitting the derivation of man from an inferior species, I believe that there was a Divine creative act at the origin of man; that the event was as truly a creation as if it had been from earth or inorganic matter to man." Similarly, Orr insisted that " 'Evolution,' in short, is coming to be recognized as but a new name for 'creation,' only that the creative power now works from *within*, instead of, as in the old conception, in an *external*, plastic fashion. It is, however, creation none the less."[41]

The reasons for Wright's transformation from Christian Darwinist to fundamentalist can only be surmised. Undoubtedly his near conversion to higher criticism, his rough treatment at the hands of some scientists, and his professorial obligation to demonstrate "the harmony of science and revelation" all played a role. In some ways his changing views on evolution simply reflected the disillusionment that many turn-of-the-century scientists felt toward Darwinism. But this factor does not explain his growing tendency to judge science by its correspondence with the Bible. The likely source of this practice was his growing obsession with higher criticism, which he feared would leave ministers of the gospel unemployed, turn the Bible into a collection of fables, undermine modern civilization, and, certainly not least, make a mockery of his lifework. Ironically, textual criticism had provided him in the 1860s with the intellectual freedom he needed to accept a baptized version of Darwinism; when he turned his back on the findings of modern biblical scholarship, he considerably reduced his intellectual options.

Like many fundamentalists, Wright believed that evolution served as the theoretical basis for the "so-called science" of the higher critics. It required "but a cursory examination of their work," he claimed, "to see that *a priori* objections to miracles, and to the whole supernatural system of the Bible, is a controlling element in a large part of their reasoning, and that a rigid theory of evolution, such as is unsupported even in the natural world, is brought into the realm of historical investigation, and made the determining factor in solving its delicate and complicated problems."[42] Given his convictions, it is not surprising that Wright's enthusiasm for biological evolution waned as he became persuaded of its support for the abhorrent notion that the Bible, like living organisms, had evolved naturally over time. And when we recall that most early fundamentalists accepted a long earth history and that many even embraced a nonliteral reading of Genesis, it is also not surprising that they looked to Wright as an authority on matters pertaining to evolution and Christian faith.

THREE

Creationism in the Fundamentalist Controversy

S o long as discussions of evolution remained confined mostly to scholarly circles, Christians who objected to evolution on biblical
grounds saw little reason to speak up. However, as the debate spilled
over into the public arena during the 1880s and 1890s, creationists
grew increasingly alarmed. "When these vague speculations, scattered
to the four winds by the million-tongued press, are caught up by
ignorant and untrained men," warned one alarmed premillennialist in
1889, "it is time for earnest Christian men to call a halt." The aggressive declarations of a few biologists, who announced their determination to drive the last vestiges of supernaturalism from science, also
aroused fear and anger among the orthodox. "Every preacher of the
present hour is compelled to deal with the theory of evolution," complained an irritated Baptist cleric in 1909. "Its advocates have invaded
his realm." As evidence of the invasion he cited a text written by
Maynard M. Metcalf (1868–1940), professor of biology in the Woman's
College of Baltimore, who, though a Christian himself, applauded efforts to cleanse religion from the life sciences, "the last stronghold of
the supernaturalist."[1]

Turn-of-the-century debates within the scientific community over
the validity of Darwinism undoubtedly encouraged such critics to speak
out. Although the overwhelming majority of scientists after 1880 ac-

cepted some form of organic evolution, by the late nineteenth century many were expressing skepticism about the ability of Darwin's theory of natural selection to account for the origin of species. By the early twentieth century, criticism of Darwinism had swelled to such proportions, especially in Germany, that the Stanford biologist Vernon L. Kellogg opened his book *Darwinism To-Day* (1907) with a chapter called "The 'Death-Bed of Darwinism,' " the title of a recent German publication. However, in contrast to Germany, where even reputable biologists had joined the chorus of criticism, in America Kellogg saw little evidence of intellectual unrest: "Our bookshop windows offer no display, as in Germany, of volumes and pamphlets on the newer evolutionary study; our serious-minded quarterlies, if we have any, and our critical monthlies and weeklies contain no debates of discussions over '*das Sterbelager des Darwinismus.*' " But, he predicted, "just as certainly as the many material things 'made in Germany' have found their way to us so will come soon the echoes and phrases of the present intellectual activity in evolutionary affairs."[2]

Some Americans had already begun to note developments abroad. Two years before the appearance of Kellogg's book, the American Bible League had published a pamphlet by Luther T. Townsend, *Collapse of Evolution* (1905), in which he told of the mounting anti-Darwinian sentiment in Europe and challenged American evolutionists to be "as honest and manly" as their German colleagues in admitting the failure of Darwinism. As word spread of the alleged demise of Darwinism, some American antievolutionists came to suspect that perhaps liberal Christians had capitulated to evolution too quickly. In view of the recent turn of events it seemed likely to one observer that those who had "abandoned the stronghold of faith out of sheer fright will soon be found scurrying back to the old and impregnable citadel, when they learn that 'the enemy is in full retreat.' "[3]

EARLY FUNDAMENTALISM

For the time being, however, those conservative Christians who soon marched under the banner of fundamentalism perceived a greater threat to orthodox faith than evolution: higher criticism, which treated the Bible more as a historical document than as God's inspired Word. Their relative apathy toward evolution is evident in *The Fundamentals*, the manifesto of militant evangelicals. A. C. Dixon, founder and first editor of the series, confessed to feeling "a repugnance to the idea that an ape or an orang-outang was my ancestor" but expressed a willingness "to accept the humiliating fact, if proved." Reuben A. Torrey (1856–1928), who brought out the last two volumes, took an even more conciliatory position, perhaps as a result of having studied at Yale under James Dwight Dana. Although Torrey himself gave up believing in evolution "for purely scientific reasons," he acknowledged—to the consternation of some of his less tolerant fundamentalist friends—that a man could "believe thoroughly in the absolute infallibility of the Bible and still be an evolutionist of a certain type."[4] The essays in *The Fundamentals*, roughly one-fifth of which touched on the issue of evolution, covered the entire spectrum of evangelical opinion: from George Frederick Wright's progressive creationism and James Orr's admission of "some genetic connection of higher with lower forms" of life to one author's identification of evolution as "the inspiration of the Higher Criticism" and the enemy of the Christian faith. But despite frequent disapproval of the theory of evolution and premature celebration of its imminent death, the collection as a whole lacked the strident attacks that would soon characterize the fundamentalist movement. Fundamentalists may not have liked evolution, but at this time few, if any, saw the necessity or desirability of launching a crusade to eradicate it from the schools and churches of America.[5]

The notion of organic evolution was, meanwhile, rapidly infiltrating those institutions. The early twentieth century witnessed the unprecedented growth of public high schools, most of which used biology texts that presented evolution favorably. More than anything else, argues historian Edward J. Larson, it was this phenomenon "that carried evolution to an increasing number of America's youth for the first time, including the children of countless fundamentalist parents." In some instances the purveyors of evolution may have provoked retaliation by their intolerance and insensitivity. Shailer Mathews (1863–1941), one of the leading modernist theologians of the day, believed that "the teachers of science themselves were partially to blame because of sometimes a 'smart Alec attitude' toward religion." Academic freedom, he argued, did not grant teachers "license to insult other people's convictions." The biologist Charles B. Davenport (1866–1944) agreed. "A little humanizing of the biologist," he thought, "would go a long way toward calming the fears of those who believe that society is being undermined by the biologist who uses his knowledge to attack a social arrangement which, on the whole, plays a useful role." Another contemporary observer blamed the creation-evolution controversy in part on the "intellectual flapperism" of irresponsible and poorly informed teachers who delighted in shocking naïve students with unsupportable statements about evolution. It was understandable, wrote an Englishman, that American parents would resent sending their sons and daughters to public institutions that exposed them to "a multiple assault upon traditional faiths."[6]

The extent to which biological evolution had penetrated the academic institutions of America by the end of the First World War is evident in a contemporary survey conducted by Charles A. Blanchard (1848–1925), president of Wheaton College. In 1919 Blanchard circulated a questionnaire among fellow presidents of midwestern colleges, many of them religiously affiliated, asking, among other things, how they handled the issue of evolution. Of the fifty-three who responded, nearly three-fourths (thirty-nine) indicated that their science faculty taught evolution, often identified as "theistic"; ten (18.9 percent) gave ambiguous responses or none at all; and only four (7.5 percent) claimed to be strictly creationist. As more than one president pointed out, the schools often had little choice but to teach evolution because virtually all science teachers and textbooks assumed "a process of development or evolution in nature."[7]

Even in the theologically conservative South, dominated culturally by Bible-believing Baptists, a number of church-related colleges had been teaching the theory of evolution for decades. William Louis Poteat (1856–1938), a Berlin-trained biologist, reputedly inaugurated the teaching of scientific evolution in the South in the 1880s while serving on the faculty of Wake Forest College, a Southern Baptist school in North Carolina. Shortly after the turn of the century, Baylor University in Texas called one of Poteat's students, John Louis Kesler (1861–1956), to organize that school's biology department. He in turn recruited Lulu Pace (1868–1925), a theistic evolutionist who a few years later received a doctorate in biology from the University of Chicago, to join him in the department. But despite such appointments academic biology grew slowly in the church colleges of the South. On the eve of World War I Kesler could think of only two Southern Baptist biologists, Poteat and Pace, who "would be thought of out of their own neighborhood when biology is mentioned."[8]

Before 1920 neither Poteat nor Kesler and Pace heard more than murmurings about their teaching Baptist students the theory of evolution. Within a short time, however, they were fighting for their careers and reputations, besieged by angry fundamentalists who increasingly in the years after World War I identified organic evolution as the cause of the social ills plaguing modern civilization.[9]

THE ANTIEVOLUTION CRUSADE

Early in 1922 William Jennings Bryan (1860–1925), Presbyterian layman and thrice-defeated Democratic candidate for the presidency of
the United States, heard of an effort in Kentucky to ban the teaching
of evolution in public schools. "The movement will sweep the country," he predicted hopefully, "and we will drive Darwinism from our
schools." His prophecy proved overly optimistic, but before the end of
the decade more than twenty state legislatures debated antievolution
laws, and three—Tennessee, Mississippi, and Arkansas—banned the
teaching of evolution in public schools. A fourth, Oklahoma, prohibited the adoption of evolutionary textbooks; a fifth, Florida, condemned
the teaching of Darwinism as "improper and subversive." Even the
United States Senate debated—but eventually rejected—an amendment that would have banned radio broadcasts favorable to evolution.
At times the controversy became so tumultuous that it looked to some
as though "America might go mad." Many persons shared responsibility for these events, but none more than Bryan. His entry into the fray
had a catalytic effect and gave antievolutionists what they needed most:
"a spokesman with a national reputation, immense prestige, and a loyal
following."[10]

The development of Bryan's own attitudes toward evolution closely
paralleled that of the fundamentalist movement's attitudes. Since early
in the century he had occasionally alluded to the silliness of believing
in monkey ancestors and to the ethical dangers of thinking that might
makes right, but until the outbreak of World War I he saw little reason
to quarrel with those who disagreed with him. The war, however,
exposed the darkest side of human nature and shattered his illusions
about the future of Christian society. Obviously something had gone
awry, and Bryan soon traced the source of the trouble to the paralyzing
influence of Darwinism on the human conscience. As he explained to
one young correspondent, "The same science that manufactured poi-
sonous gases to suffocate soldiers is preaching that man has a brute
ancestry and eliminating the miraculous and the supernatural from the
Bible." By substituting the law of the jungle for the teaching of Christ,
it threatened the principles he valued most: democracy and Christianity. Two books in particular confirmed his suspicion. The first,
Kellogg's *Headquarters Nights* (1917), recounted firsthand conversations with German officers that revealed the role Darwin's biology had

supposedly played in persuading the Germans to declare war. The second, Benjamin Kidd's *Science of Power* (1918), purported to demonstrate the historical and philosophical links between Darwinism and German militarism. Both fundamentalists and scientists took these revelations seriously. But whereas Bryan responded by attempting to quash the offending doctrine, a number of American biologists sought to restore the tarnished reputation of science by stressing the importance of cooperation rather than conflict in the evolutionary process.[11]

About the time that Bryan discovered the alleged Darwinian origins of the war, he also became aware, to his great distress, of unsettling effects the theory of evolution was having on America's own young people. From frequent visits to college campuses and from talks with parents, pastors, and Sunday-school teachers, he heard about an epidemic of unbelief that was sweeping the country. Upon investigating the cause, his wife reported, "he became convinced that the teaching of Evolution as a fact instead of a theory caused the students to lose faith in the Bible, first, in the story of creation, and later in other doctrines, which underlie the Christian religion." Again Bryan found confirming evidence in a recently published book, *Belief in God and Immortality* (1916) by the Bryn Mawr psychologist James H. Leuba (1868–1946), who demonstrated statistically that college attendance endangered traditional religious beliefs. After the much-publicized 1924 trial of Nathan Leopold and Richard Loeb for the kidnapping and murder of Robert Franks, Bryan and other antievolutionists commonly added this heinous crime to their list of charges against the teaching of Darwinism, prompting one frustrated critic to point out that Jesse James was a fundamentalist and that "Memphis, Tennessee, where Evolution is outlawed, . . . has more murders in proportion to population than any other city in the United States."[12]

Armed with information about the cause of the world's and the nation's moral decay, in 1921 Bryan launched a nationwide crusade against the offending doctrine. In one of his most popular and influential lectures, "The Menace of Darwinism," he summed up his case against evolution, arguing that it was both un-Christian and unscientific. Darwinism, he declared, was nothing but "guesses strung together," and poor guesses at that. Borrowing an illustration from Alexander Patterson's *Other Side of Evolution* (1903), Bryan explained how the evolutionist accounted for the origin of the eye:

The evolutionist guesses that there was a time when eyes were unknown—that is a necessary part of the hypothesis. . . . A piece of pigment, or, as some say, a freckle appeared upon the skin of an animal that had no eyes. This piece of pigment or freckle converged the rays of the sun upon that spot and when the little animal felt the heat on that spot it turned the spot to the sun to get more heat. The increased heat irritated the skin—so the evolutionists guess, and a nerve came there and out of the nerve came the eye!

"Can you beat it?" he asked incredulously—and that it happened not once but twice? As for himself, he would take one verse in Genesis over all that Darwin wrote.[13]

Bryan was far from alone in balking at the evolutionary origin of the eye. Christian apologists had long regarded the intricate design of the eye as "a cure for atheism," and Darwin himself had readily conceded his vulnerability on this point. "To suppose that the eye, with all its inimitable contrivances for adjusting the focus to different distances . . . could have been formed by natural selection, seems, I freely confess, absurd in the highest possible degree," he wrote in the *Origin of Species*. But logical consistency impelled him to extend "the principle of natural selection to such startling lengths." When his American friend Asa Gray wrote a letter candidly describing the section dealing with the making of the eye as "the weakest point in the book," Darwin confided in reply that "the eye to this day gives me a cold shudder." That Bryan, too, shuddered says nothing about his scientific sophistication.[14]

Like many other nonscientific critics of evolution, Bryan rankled at allegations that he lacked the competence to judge the merits of the case. Thus despite his denial that formal schooling was essential for understanding evolution and his frequent disparaging remarks about the value of higher education, he could not on occasion resist the temptation to parade his own academic qualifications, which included a B.A., an M.A., an LL.B., and at least seven honorary doctorates. If people would not quit calling him an ignoramus, he threatened to print his degrees all over his business cards and then "challenge any son of an ape to match cards" with him.[15]

Though one would scarcely have guessed it from some of his public pronouncements, Bryan was not a strict biblical literalist. In fact, his beliefs regarding both evolution and the Bible diverged considerably from those of his more hermeneutically conservative supporters. Shortly before his death he confided to one friend that he had no objection to "evolution before man but for the fact that a concession as to the truth of evolution up to man furnishes our opponents with an argument which they are quick to use, namely, if evolution accounts for all the species up to man, does it not raise a presumption in behalf of evolution to include man?" Until biologists could actually demonstrate the evolution of one species into another, he thought it best to keep them on the defensive. As for the meaning of Genesis 1, Bryan had long held to the day-age theory. "The only persons who talk about a twenty-four-hour day in this connection do so for the purpose of objecting to it," he wrote in 1923; "they build up a straw man to make the attack easier, as they do when they accuse orthodox Christians of denying the roundness of the earth, and the law of gravitation." During the Scopes trial Bryan readily conceded that the world was far older than six thousand years and that the six days of creation had probably been longer than twenty-four hours each. Yet Bryan did not hesitate to appeal to miracles to explain puzzling scientific phenomena. For example, in discussing the geographical distribution of animals, he noted that the creationist who believes that "God can create anything that He likes and put it where He pleases" could always invoke "a creative act when there is no reasonable evidence of distribution by physical causes."[16]

Throughout his political career Bryan had placed his faith in the common people, and he resented the attempt of a few thousand elitist scientists "to establish an oligarchy over the forty million American Christians," to dictate what should be taught in the schools. To a democrat like Bryan it seemed preposterous that this "scientific soviet" would not only demand to teach its insidious philosophy but impudently insist that society pay its salaries. Confident that nine-tenths of the Christian citizens agreed with him, he decided to appeal directly to them, as he had done so successfully in fighting the liquor interests. "Commit your case to the people," he advised creationists. "Forget, if need be, the highbrows both in the political and college world, and carry this cause to the people. They are the final and efficiently corrective power."[17]

THE ANTIEVOLUTIONISTS

Who were the people who joined Bryan's crusade? As recent studies have shown, they came from all walks of life and from every region of the country. They lived in New York, Chicago, and Los Angeles as well as in small towns and in the country. Few possessed advanced degrees, but many were not without education. Nevertheless, Bryan undeniably found his staunchest supporters and won his greatest victories in the conservative and still largely rural South, described hyperbolically by one fundamentalist journal as "the last stronghold of orthodoxy on the North American continent," a region where the "masses of the people in all denominations 'believe the Bible from lid to lid.' "[18]

The relative strength of Bryan's following within the churches is difficult to determine, because not all fundamentalists were creationists and many creationists refused to participate in the crusade against evolution. However, a 1929 survey of the theological beliefs of seven hundred Protestant ministers provides some valuable clues. The question "Do you believe that the creation of the world occurred in the manner and time recorded in Genesis?" elicited the following positive responses:

Lutheran	89%
Baptist	63%
Evangelical	62%
Presbyterian	35%
Methodist	24%
Congregational	12%
Episcopalian	11%
Other	60%

Unfortunately, these statistics tell us nothing about the various ways respondents may have interpreted the phrase "in the manner and time recorded in Genesis." Some perhaps believed that Genesis taught a recent special creation in six twenty-four-hour days, although William Bell Riley (1861–1947), the influential fundamentalist pastor of the First Baptist Church in Minneapolis, insisted that there was not "an intelligent fundamentalist who claims that the earth was made six thousand years ago; and the Bible never taught any such thing." Like Bryan, Riley followed Dana and John William Dawson in subscribing to the day-age view.[19]

Many antievolutionists, including Dixon, adopted the view found in The Scofield Reference Bible (1909), a fundamentalist favorite prepared by C. I. Scofield (1843–1921), a biblical literalist and sometime minister of the First Congregational Church in Dallas, Texas, who inserted a gap between the first and second verses of Genesis. This version of the King James text assigned the creation story to "B.C. 4004" in the margins but explained in an annotation that the original creation of "the heaven and the earth," mentioned in Genesis 1:1, "refers to the dateless past, and gives scope for all the geological ages." The condition of the earth portrayed in the second verse—"without form, and void"—Scofield attributed to "a cataclysmic change" produced by a divine judgment on fallen angels. In a note accompanying the eleventh verse, describing the appearance of vegetation, Scofield wrote that "it is by no means necessary to suppose that the life-germ of seeds perished in the catastrophic judgment which overthrew the primitive order. With the restoration of dry land and light the earth would 'bring forth' as described. It was *animal* life which perished, the traces of which remain as fossils. Relegate fossils to the primitive creation, and no conflict of science with the Genesis cosmogony remains." Scofield's Bible, which may have sold in excess of ten million copies, enjoyed immense respect among Christians who shared the editor's commitment to the inerrancy of the Scriptures. One historian of the fundamentalist movement has called it "perhaps the most influential single publication in millenarian and Fundamentalist historiography."[20]

The poll of preachers also fails to indicate how the level of political involvement in the campaign against evolution varied from denomination to denomination. Lutherans, for example, despite their overwhelming rejection of evolution, generally preferred education to legislation and tended to view legal action against evolution as "a dangerous mingling of church and state." Similarly, premillennialists, who saw the spread of evolution as one more sign of the world's impending end, sometimes lacked incentive to correct the evils around them.[21]

Baptists and Presbyterians, who dominated the fundamentalist movement, most visibly and vocally opposed evolution. Baptist preachers such as Riley, John Roach Straton (1875–1929) of New York City, J. Frank Norris (1877–1952) of Fort Worth, and T. T. Martin (1862–1939) of Blue Mountain, Mississippi, were among the earliest and most outspoken critics of evolution. In the minds of many contemporaries, the whole imbroglio over evolution began when the fire-breathing Martin attacked Poteat for teaching the subject at Wake Forest and Norris jumped on the bandwagon with similar charges against Baylor's biologists. Martin, an itinerant evangelist with a reputation for combining doctrinal fanaticism with compassionate Christianity, had begun his career as a teacher of the natural sciences at Baylor Female College in Belton, Texas. Although he claimed to be a longtime student of the sciences, his knowledge of organic evolution was such that he identified the inheritance of acquired characteristics and natural selection as the "two great pillars" of the evolutionary edifice, apparently ignorant that biologists had come to view these two theories as describing competing rather than complementary mechanisms. Convinced that evolution was poisoning the minds of the nation's youth, he campaigned tirelessly in the 1920s to eradicate "the scourge" from the classrooms of America. In a diatribe titled *Hell and the High Schools* (1923), written while alleged wartime atrocities were still fresh in people's minds, he argued that "the Germans who poisoned the wells and springs of northern France and Belgium" and fed little children poisoned candy were angels "compared to the text-book writers and publishers who are poisoning the books used in our schools." Martin could hardly find words strong enough to express the enormity of evolution. "Next to the fall of Adam and Eve," he insisted, "Evolution and the teaching of Evolution in tax-supported schools is the greatest curse that ever fell upon this earth."[22]

Shortly after Martin began hounding Poteat for his modernist ideas, Grove Samuel Dow (1888–?), a professor of sociology at Baylor University, published an introductory college text that briefly, but explicitly, described the appearance of the first humans "somewhere between one hundred thousand and a million years" ago. Primitive man, he wrote, "was a squat, ugly, somewhat stooped, powerful being, half human and half animal, who sought refuge from the wild beasts first in the trees and later in caves . . . he was about halfway between the anthropoid ape and the modern man." Such un-Mosaic language quickly caught the attention of Martin and Norris. The latter, whom the *Christian Century* in 1924 called "probably the most belligerent fundamentalist now abroad in the land," wasted little time in mounting an attack, using his church paper, the *Searchlight,* to expose Dow and other evolutionists at Baylor, where the teaching of evolution for the past fifteen years had "been kept under cover." Although Dow publicly confessed to having "made some blunders in expressions" and professed not to believe or teach "that man came from another species," he was forced to resign his position.[23]

Unappeased by Dow's departure, Norris and his fundamentalist friends in the Baptist General Convention of Texas next attacked Lulu Pace and O. C. Bradbury (1890–1969), a zoologist who had joined the Baylor faculty in 1917, replacing Kesler, who was driven from Waco because of his pro-German sympathies. In conversations with university president Samuel Palmer Brooks (1863–1931), both Pace and Bradbury confessed that they viewed evolution as "God's method" of creation and that they read the first two chapters of Genesis as "illustrative or allegorical"; nevertheless, they accepted an apparently contradictory report drafted by a special investigative committee that included the following statement: "We do not believe in Darwinian Evolution or any form of Evolution that leaves God out as Creator, or any theory by whatever name called, that teaches or proposes to teach that there is, or has been, such a thing in nature as the transmutation of species, or the Evolution of Life from one species to another, or that

Man came from the anthropoid ape, or any lower form of animal life."
Although he survived this inquisition, Bradbury soon resigned, and
Pace died in 1925 after a prolonged illness. The search for replacements indicated the administration's newfound sensitivity toward questions of origins. When a young botanist at the University of Iowa
inquired about the opening, Brooks informed him that Baylor would
not employ any teacher who "doubts God as the creator of the world or
discounts the Bible as God's revelation to man." The president's "arch-
enemy," Norris, whom many acquaintances rightly regarded as being
disturbed and dangerous, soon found himself under attack, facing a
murder charge against which he successfully pleaded self-defense.[24]

During the 1920s, Baptist biologists throughout the South became
objects of orthodox suspicion. In 1922 the Tennessee Baptist Convention investigated rumors that C. W. Davis of Union University in
Jackson was teaching the "Protoplasmic Evolution of man from lower
animal life." The investigators found that although Davis sometimes
used the word *evolution* to describe "variations in species and the
development of life in various spheres"—an indiscretion they attributed to his secular training—he was theologically orthodox and thus
employable. Two years later the Georgia Baptist Convention forced
Henry Fox (1875–1951), who held a doctorate from the University of
Pennsylvania, to resign as professor of biology from Mercer University.
Although his views on evolution inevitably clouded the picture, it was
his refusal unequivocally to affirm his belief in the deity of Christ, the
divine inspiration of the Bible, and the Virgin Birth that led to his
termination. In 1926 the Baptists of South Carolina drove Andrew Lee
Pickens (1890–1969), who held degrees in both theology and biology,
from Furman University after he was quoted in a newspaper as subscribing to the theory of evolution. When questioned by the trustees,
Pickens refused to back down. On submitting his resignation, he declared defiantly that "it is utterly foolish for a man to talk about teaching biology and not teaching evolution." That same year the Southern
Baptist Convention voted unanimously that "this Convention accepts
Genesis as teaching that man was the special creation of God, and
rejects every theory, evolution or other, which teaches that man originated in, or came by way of, a lower animal ancestry."[25]

Elsewhere across the religious landscape the Presbyterian church
contributed Bryan and other leaders to the creationist cause but, as the
1929 survey suggests, also harbored many evolutionists. In 1923 the
General Assembly turned back an attempt by Bryan and his fundamentalist cohorts to cut off funds to any church school found teaching
human evolution, approving instead a compromise measure that con-

demned only materialistic evolution. The other major Protestant bodies paid relatively little attention to the debate over evolution, though pockets of militant anti-Darwinism could be found in such places as the Holiness movement within the Methodist tradition. Catholics, though divided on the question, seldom favored restrictive legislation.[26]

Leadership of the antievolution movement came not from the organized churches of America but from individuals like Bryan and from interdenominational organizations such as the World's Christian Fundamentals Association (WCFA), founded by Riley in 1919. Riley grandly described the WCFA's first conference as "an event of more historical moment than the nailing up, at Wittenberg, of Martin Luther's ninety-five theses." Though originally dedicated to launching "a new Protestantism" based on premillennial interpretations of biblical prophecy, the association during the early 1920s increasingly turned its attention to combating evolution. Throughout the middle years of the decade, a host of other antievolution organizations bearing such names as the Anti-Evolution League of America, the Bryan Bible League, and the Anti-False Science League of America made brief appearances. Most were one-man bands that paraded long lists of officers drawn from the same pool of fundamentalist luminaries. Typical was the ephemeral American Science Foundation, created by the publicity-seeking evangelist Fred E. Bennett, who named himself president. Assisted by a slate of twenty-one vice presidents, including virtually every notable creationist from the late 1920s, Bennett ambitiously sought to unite scientists, clergymen, and businessmen in preventing the "DRAGONS" of atheistic communism and evolutionism from "TEARING AT THE VITALS OF THIS NATION." But aside from an occasional copy of his manifesto surviving in some obscure manuscript collection, little evidence remains that the foundation ever existed.[27]

The WCFA aimed not only at stopping the teaching of evolution but at identifying textbooks appropriate for use in Christian schools. Since the earliest days of the antievolution movement, Bryan had complained about the impossibility of finding "any text book on biology which does not begin with monkeys." "One of the greatest needs of the day," reported the Textbook Committee in 1928, "is suitable textbooks on the physical and biological sciences, written by authors who are not obsessed with theories that are harmful to the young people of our schools." The only mainstream biology text the committee felt comfortable approving was George A. Baitsell's *Manual of Biological Forms* (1923), a "purely empirical" work that indulged in "no speculation on evolution." Perhaps because so few possessed the requisite credentials, creationists themselves did little to fill the void. However, John

S. Park (1894–?), sometime Baptist pastor and high-school biology teacher who in the mid-1920s taught in the biology department at Findlay College in Ohio, operated by the Churches of God in North America, became convinced of the need for a series of textbooks written "from the fundamentalist standpoint." So far as he could tell, it had been forty years since the appearance of the last creationist textbook. His books, he explained to one publisher, would "present only actually known and observed scientific facts, and [would] not speculate on philosophical subjects such as organic evolution." Instead, he would adopt a "seasonal" approach, which would give students "the opportunity of studying the animals and plants that are most abundant in the fall and spring as they occur in nature and [would leave] the winter months for the study of microscopic organisms and the theoretical side of Biology." When a number of publishing houses—including, he claimed, Macmillan, Dorrance, Nelson, Heath, Saunders, Scribner's, and Crowells—expressed interest in the project if the demand were great enough, Park circulated a questionnaire to the presidents of hundreds of denominational colleges inquiring about the likelihood of adoption. Park departed from Findlay shortly thereafter and disappeared from the creationist movement, having apparently abandoned his textbook project. By this time, however, the need for creationist texts was abating as publishers in the wake of the Scopes trial increasingly diluted their coverage of biological evolution.[28]

DEFINING SCIENCE AND COUNTING SCIENTISTS

In 1922 Riley outlined the reasons why fundamentalists opposed the teaching of evolution. "The first and most important reason for its elimination," he explained, "is the unquestioned fact that evolution is not a science; it is a hypothesis only, a speculation." Bryan often made the same point, defining true science as "classified knowledge, . . . the explanation of facts." This view came straight from the dictionary. "What is science?" asked one Holiness preacher who opposed evolution. "To the dictionaries!" he answered. There one learned that science was "certified and classified knowledge." Although twentieth-century creationists rarely appealed to philosophical sources, their insistence on the factual, nontheoretical nature of science harmonized with the once-venerated teachings of the English philosopher Francis Bacon, whose name in nineteenth-century America had symbolized correct scientific method. By narrowly drawing the boundaries of science and emphasizing its empirical nature, creationists could at the same time label evolution as false science, claim equality with scientific authorities in comprehending facts, and deny the charge of being antiscience. "It is not 'science' that orthodox Christians oppose," a fundamentalist editor insisted defensively. "No! no! a thousand times, No! They are opposed only to the theory of evolution, which has not yet been proved, and therefore is not to be called by the sacred name of *science*." The creationists' narrow definition of science possessed the additional advantage of allowing them to deny scientific legitimacy to the various metaphysical movements that flew the banner of science. As Riley once noted, "We have so far converted this word into a mere mental commodity that a designing woman [Mary Baker Eddy] employs it for purely commercial purposes by calling her mental vaporings 'Christian Science,' and multitudes are deceived thereby."[29]

To support their claim that evolution was unscientific, creationists compiled impressive rosters of scientists who allegedly shared their point of view. Shortly after the turn of the century, Townsend assembled one of the earliest—and most frequently cribbed—lists in order to prove that "the most thorough scholars, the world's ablest philosophers and scientists, with few exceptions, are not supporters, but assailants of evolution." In addition to such late worthies as Louis Agassiz, Arnold Guyot, and John William Dawson, Townsend cited the English physiologist Lionel S. Beale (1828–1906) and the German pathologist Rudolf Virchow (1821–1902), both of whom had expressed doubts about the evidence for human evolution, as well as a dozen or so lesser lights, including a "Dr. Etheridge, of the British Museum, one of England's most famous experts in fossilology," and "Professor Fleischmann, of Erlangen, one of the several recent converts to anti-Darwinism." In 1920 Martin, responding to claims that all responsible scientists were evolutionists, borrowed generously from Townsend's booklet to draw up his own list of "twenty-one really great scientists in the world" who rejected evolution. The target of Martin's scholarship, Poteat, replied with a withering analysis of the credentials of the twenty-one:

Two do not appear in the biographical dictionaries, five are misrepresented, seven won reputation in other than biological fields, and six have been in their graves more than forty years, two of these having died long before Darwin's great book was published. One lone biologist is left to support the thesis that the doctrine of evolution is discarded by the science of today. And this man's position is so peculiar that he is usually mentioned as the single exception to the universal acceptance of evolution by biologists of responsible position.[30]

The one lone biologist identified by Poteat was Albert Fleischmann (1862–1942), a reputable but relatively obscure German zoologist who taught for decades at the University of Erlangen in Bavaria. In 1901 he published a scientific critique of organic evolution, *Die Descendenztheorie*, in which he rejected not only Darwinism but all theories of common organic descent. This placed him in a unique position among biologists. As Kellogg noted in 1907, Fleischmann seemed to be "the only biologist of recognised position . . . who publicly declared a disbelief in the theory of descent." The German creationist apparently remained of the same mind for the rest of his life. In 1933, the year of his retirement from Erlangen, he presented a paper to the Victoria Institute in London in which he dismissed the notion of a "genealogical tree" as a "fascinating dream." "No one can demonstrate that the limits of a species have ever been passed," he asserted. "These are the Rubicons which evolutionists cannot cross." In his declining years Fleischmann informed English acquaintances that he was writing a book "that will wipe evolution off the slate," but the work never appeared.[31]

The widely touted "Dr. Etheridge, of the British Museum," who always appeared in creationist literature without a given name, was quoted by Townsend as saying, "In all this great museum there is not a particle of evidence of transmutation of species. Nine-tenths of the talk of evolutionists is sheer nonsense, not founded on observation and wholly unsupported by fact. This museum is full of proofs of the utter falsity of their views." The content of Etheridge's statement varied from work to work, and its source remained unidentified, except for Alexander Patterson's comment that Etheridge was answering a question put to him by a Dr. George E. Post. When curious parties in the 1920s inquired about the identity of Etheridge, the director of the British Museum surmised that the man in question was "Robert Etheridge, Junr., who was Assistant Keeper of Geology in this Museum from 1881 to 1891," at which time he left for Australia, where he died in 1920. The director hastened to add that "Mr. Etheridge's opinion on this subject should not be considered as in any way representing scientific opinion in this Museum."[32]

The antievolutionists received an important psychological boost at the beginning of their campaign from an address in 1921 to the American Association for the Advancement of Science (AAAS). At that meeting the distinguished British biologist William Bateson (1861–1926) declared that scientists had *not* discovered "the actual mode and process of evolution." As knowledge of living things had multiplied, he said, biologists had grown increasingly agnostic about questions of origin. Although he warned against misinterpreting his comments as a rejection of evolution, creationists applauded his speech as "the swan song of Darwinism" while evolutionists rushed to control the damage. The council of the AAAS issued a statement affirming that "the evidences in favor of the evolution of man are sufficient to convince every scientist of note in the world," and Henry Fairfield Osborn (1857–1935), head of the American Museum of Natural History, contemptuously dismissed Bateson's remarks as having come from someone "living the life of a scientific specialist, out of the main current of biological discovery." To say that "we have failed to discover the origin of species," declared Osborn, "is valueless and directly contrary to the truth." Creationists, however, paid no more attention to the warnings of the AAAS and Osborn than they did to Bateson's own caveat.[33]

Creationists celebrated again in the late 1920s when the distinguished Smithsonian zoologist Austin H. Clark (1880–1954), struck by the absence of intermediate forms between the major groups of animals, challenged the conventional view of evolution that represented species as branches of a single tree. The *Christian Fundamentalist* quoted him as saying that "so far as concerns the major group of animals, the creationists seem to have the better of the argument. There is not the slightest evidence that any one of the major groups arose from any other. Each is a special animal-complex, related more or less closely to all the rest, and appearing, therefore, as a special and distinct creation." But both in correspondence with creationists and in his controversial book *The New Evolution: Zoogenesis* (1930), Clark pointedly refused to support a supernatural view of origins. As one confused and disappointed creationist explained, "Dr. Clark . . . denies the most pointed evidence of evolution, but he sticks to evolution just the same."[34] Thus during the heyday of the antievolution movement, the German zoologist Fleischmann remained the only well-credentialed biologist to lend his authority to the creationist attack on evolution.

FOUR

Scientific Creationists in the Age of Bryan

In spite of the much-heralded collapse of evolution, American creationists could name few scientists of their own: a couple of self-made men of science, one or two physicians, and a handful of professors who, as one critic described them, were "trying to hold down, not a chair, but a whole settee, of 'Natural Science' in some little institution." Among the creationists who claimed scientific expertise, several also earned their living as preachers or priests. Besides the already-mentioned John S. Park and T. T. Martin, there was, for example, Alfred Fairhurst (1843–1921), the author of three antievolution books who pastored various Disciples of Christ churches in central Kentucky while simultaneously serving as professor of natural science at Transylvania University in Lexington. After graduating from Butler College, a Disciples school in Indianapolis, he had taken a year of advanced work in science at Harvard University before returning to Butler as a teacher. In 1881, after experimenting with a legal career, he moved to Kentucky. Fairhurst followed John William Dawson in interpreting the days of Genesis as indefinite periods of time and in allowing the history of life on earth to span millions of years. He rejected theistic as well as materialistic evolution, arguing that the former "destroys the Bible as the inspired book of authority as effectually as does atheistic evolution." Although references to his works are sprinkled throughout

54

the creationist literature of the 1920s, Fairhurst's influence seems not to have been great—even within his own church and school. A sometimes difficult man, he was forced to retire from the classroom in 1914. Three years later, as a Transylvania trustee, he helped mount a campaign to stamp out liberalism and evolutionism among the faculty of religion, only to suffer additional ignominy at the hands of one of the accused professors, who brought the proceedings to a halt by tricking Fairhurst into condemning some of his own previously published statements on evolution. He died in 1921, on the eve of the crusade to ban the teaching of evolution in public schools.[1]

Probably the most influential of the clerical science professors was the Benedictine priest George Barry O'Toole (1886–1944), whose book *The Case against Evolution* (1925) won widespread praise from creationists. In this work, one of the few creationist books brought out by a major publishing house, Macmillan, O'Toole focused on the problems of greatest concern to Catholics—the evolution of life, the human body, and the soul—concluding that evolution was merely a hypothesis, not an established fact. Because the title page identified O'Toole as professor of animal biology at Seton Hill College, a small Catholic school for women in Pennsylvania, many readers understandably concluded that the author was a scientist. Indeed, he had taken courses in biology, geology, and chemistry at Columbia University during one winter and a couple of summers, but his graduate degrees, including a Ph.D. and an S.T.D., were all from the Urban University (of the Propaganda) in Rome, where he studied from 1906 to 1912. Except for a couple of years teaching biology at Seton Hill, he devoted his life primarily to theology, philosophy, and academic administration. When attorneys representing the state of Tennessee during the appeal of the Scopes case called O'Toole a "learned educator and scientist," indignant members of the Tennessee Academy of Science responded by dismissing his book as "a religious and not a scientific work" and by pointing out that the priest's name could not be found "among the 9,500 American scientists listed in American Men of Science nor in the membership list of 14,253 names of the American Association for the Advancement of Science." By the time his book actually appeared, O'Toole was in China serving as rector of the Catholic University of Peking, an institution he had helped to found. Upon O'Toole's return to America from the Orient in 1934, Pope Pius XI honored him with a domestic prelacy; he spent his last years teaching philosophy at Duquesne University and the Catholic University of America.[2]

During the 1920s S. James Bole (1875–1956), professor of biology at fundamentalist Wheaton College, Illinois, stood virtually alone as a creationist with advanced training in biology. After receiving a bachelor's degree from the University of Michigan, Bole worked in the public schools of Michigan for several years before enrolling in the University of Wisconsin to work on a master's degree in education. He left degreeless after one year, however, and it was not until 1912 that he finally obtained his A.M. in education, from the University of Illinois, having submitted a thesis on penmanship in the elementary schools of Urbana. For the next six years he pursued a Ph.D. in pomology (that is, fruit culture) in the school of agriculture at Illinois while supporting himself as a graduate instructor in such courses as "Elements of Horticulture," "Commercial Fruit Culture," and "Evolution of Horticultural Plants." In 1918, having apparently completed all the requirements for a Ph.D. but a dissertation, he accepted a position teaching science and agriculture at Missouri Wesleyan College, where as a Methodist of decidedly conservative leanings he soon became embroiled in "a little unpleasantness" with a liberal Bible teacher, a dispute that Bole took public. After his dismissal from the college "for believing and teaching that the Word of God is true," as he put it, he took a position teaching biology at Wheaton College, where he attended the Congregational College Church.[3]

During his student years Bole had embraced biological evolution, not so much by coming under the influence of some atheistic university professor as by reading a book on theistic evolution, *The Ascent of Man* (1894), by the evangelical Henry Drummond, Dwight L. Moody's friend and associate. While at Illinois, Bole, a self-described "worldly Christian" with rationalist leanings, experienced a "Damascus Road" conversion that led to his rejection of evolution and higher criticism in favor of the Mosaic story of creation and biblical inerrancy. Shortly after Bole moved to Wheaton, an acquaintance at the nearby Moody Bible Institute in Chicago urged him to use his experience and expertise to write a book against evolution. The resulting work, originally serialized in the *King's Business* under the title "Satan's Triangle: Evolution, Philosophy, Criticism," appeared in book form as *The Modern Triangle* (1926). In an unusual concession for a creationist, Bole granted that "from the point of view of specialized knowledge it would seem that the greater authority rested with the evolutionists." But because organic evolution dealt with theory, not facts, choosing between creation and evolution became a matter of faith: "The evolutionist has faith in his hypothesis; the Christian has faith in his God."[4]

In interpreting Genesis, Bole adopted the ruin-and-restoration model so popular among American fundamentalists. This allowed him simultaneously to concede the antiquity of the earth and most fossils

while insisting on a recent Edenic creation in six literal days. Archaeological evidence confirmed his belief in Noah's flood, but "since its
thick deposit of silt has not been found outside the Tigris-Euphrates
Valley," he concluded that it must have been a local event. Bole's
position in an increasingly prominent fundamentalist college and his
various publications gave him considerable standing among creationists, symbolized by his appointment to a vice presidency in the Science
Foundation and to a position on the membership committee of the
World's Christian Fundamentals Association. For a time he aspired to
write a weekly column answering questions about evolution and edu-
cating Christians on the significance of such subjects as gill-slits, cave
fish, and geographical distribution. However, his position as a fundamentalist spokesman on evolution was eroded in 1932, when as a man
in his late fifties he suffered a humiliating dismissal from Wheaton for
allegedly "showing an improper attitude toward young ladies in his
classes and in his home." Given the puritanical mores on campus, his
transgressions may have been trivial by today's standards; we simply
do not know. Two years later he finally received a Ph.D. in horticulture, from Iowa State College in Ames. From 1936 to 1943 he taught
biology at John Fletcher College, a Holiness school in Oskaloosa, Iowa;
he spent his last year in the classroom at Taylor University, a nondenominational Holiness institution in Upland, Indiana. Although he
went on to publish a second attack on evolution, *The Battlefield of
Faith* (1940), and to serve as a lay preacher in Iowa, he never recov-
ered the influence he once exerted in the world of creationism.[5]

Creationists compensated for the absence of any distinguished scientists in their ranks by making the most of the modest credentials
they possessed—emphasizing, for example, Bole's faculty status while
he was a graduate student at the University of Illinois. By their own
reckoning the greatest "scientist" in their midst—indeed, one of the
greatest scientists in the Western world—was Arthur I. Brown, M.D.,
C.M., F.R.C.S.E. (1875–1947), an American-born surgeon who had
settled in Vancouver, British Columbia. Brown received his M.D.
from Trinity Medical College, Toronto, in 1897 and became a fellow of
the Royal College of Surgeons of Edinburgh in 1913, having completed
"a very stiff post graduate course taken in Scotland." (The C.M. stands
for a British master of surgery degree.) Apparently later that same year
he began practicing surgery in Vancouver, eventually becoming, in his
own estimation, "one of the leading surgeons of the Pacific Coast" and
earning a reported (and perhaps exaggerated) $50,000 a year at a time
when most physicians in North America still made less than one-tenth
that amount.[6]

Little is known about Brown's spiritual and intellectual development, but by the early 1920s he was publishing antievolution pamphlets on such subjects as *Evolution and the Bible* (1922) and *Men,
Monkeys and Missing Links* (1923). Despite his superior knowledge of
science, at least for a creationist, he contributed few novel ideas to the
cause; instead, he tended to popularize the works of others and to use
illustrations of apparent design in the natural world to argue for the
necessity of special creation. In *Men, Monkeys and Missing Links*, for
example, he provided a synopsis of *God—or Gorilla* (1922), a sneer-
ing, indignant attack on human evolution by a muckraking Catholic
journalist, Alfred Watterson McCann (1879–1931), who reveled in exposing Piltdown man as a hoax years before the scientific community
conceded that it was. Although McCann had no sympathy for fundamentalism, Brown lauded his book as a "great work" that offered "the
most scathing and unanswerable indictment ever published against
this untenable hypothesis." Perhaps Brown's most original contribution to the arsenal of antievolution polemics was a critique of arguments for evolution based on laboratory analyses of blood serum from
animals and humans, which provided scientists with a means of identifying lines of descent that turned out to be identical to those suggested by morphological studies. In *Evolution and the Blood-
Precipitation Test* (1925) Brown argued that biologists had no
justification for assuming that animals having chemically similar blood
shared a common ancestor, because similarities in blood types could
also be explained by appealing to "an Omniscient and Omnipotent
Creator and Designer."[7]

Brown claimed to reject evolution because it seemed "contrary to
the clear facts of Science as well as to the plain statements of Scripture," which he believed to be "infallible and inerrant from Gen. 1:1 to
Rev. 22:21." In his opinion evolution was not only "the greatest hoax
ever foisted on a credulous world" but also "the most potent and
effective weapon which is being used by Satan in his present furious
attack on the Bible." However, like most fundamentalists during the
first half of the century, he had no quarrel with evidence for an ancient
earth, believing that the language of Genesis 1 allowed for the passage
of "unknown ages" between the original creation "in the beginning"
and the subsequent Edenic restoration. On the basis of an obscure
passage in Ezekiel 28:17, he surmised that the first creation had been
wiped out by a stupendous catastrophe, possibly a flood, caused by the
"Devil's hatred of God when he was thrown or 'cast' to the ground."
But although he assiduously sought to decipher biblical prophecies
that seemed to foretell the imminent return of Christ, and had no

doubt that the "days" of creation were twenty-four hours in length, for the most part he avoided discussions of the exact meaning of the first chapters of Genesis.[8]

In November 1925, "at the urgent solicitation of many prominent Fundamentalist leaders," Brown took a one-year leave of absence from his lucrative surgical practice in order to devote his full time to lecturing on science and the Bible. The demand for his services proved so great he gave up medicine permanently and lived off the offerings received during his lectures, often delivered in large Baptist churches. His life-style seems to have suffered little as a result of this decision; he did well enough financially to travel with his wife, stay in the finest hotels, and hire an advance man. For years he maintained a grueling pace, crisscrossing the country to meet engagements. Early in 1929, for example, he wrote from upstate New York about the "wonderful series of campaigns here in the East, Philadelphia, New York, Boston, Paterson, Passaic, Camden, Brooklyn, Atlantic City, Baltimore, Washington, and now am at First Baptist Church here in Buffalo." From Buffalo and Rochester he planned to head west to Cleveland, Pontiac, Racine, Chicago, Springfield, St. Louis, and Kansas City. After more than three years on the road, he showed no signs of burning out. "Crowds everywhere, intense interest, and invariable requests for return engagements," he proudly reported to a preacher in Fort Worth whom he would soon be visiting. "I can promise you audiences will be clamoring for more when I have finished."[9]

Brown's appeal, which even he and his wife professed to find surprising, stemmed more from his reputation than from his platform manner. His own handbills touted him as "one of the best informed scientists on the American continent," and hosts routinely introduced him as a famous scientist, perhaps the "greatest scientist in all the world." With such a buildup, explained one newspaper reporter, audiences expected that "at the blast of his ram's horn the walled city of modern knowledge was to tumble." Even fundamentalists who generally had little use for the trappings of higher education took pride in his professional accomplishments, mistaking them for academic achievement. "Notice the degrees attached to the writer's name," wrote one Kansas editor in introducing an antievolution piece by Brown, "realize that he was trained in some of the best Universities in Europe, and you will know why he is recognized as an authority." The president of Wheaton College, eager to improve the academic standing and visibility of his institution, coveted the services of the creationist savant, and virtually every antievolution organization sought the use of his name. The Bible Crusaders of America, founded by the Boston capi-

talist George F. Washburn (1859–1931), honored him with the title
Scientist General.[10]

Unlike some of the fire-and-brimstone preachers who damned evolution, Brown resorted to no pyrotechnics. A far-from-partisan reporter covering his 1926 campaign in Raleigh, North Carolina, observed that the visiting celebrity, who had "quit medical doctoring for divinity doseing [sic]," possessed "a great deal more sense, scholarship, personality and platform ability than most of the agitators" who sought to outlaw the teaching of Darwinism. Although the reporter thought Brown's "scientific speech" fell dead, he applauded the lecturer's gentleness and civility. Years later a graduate student at the University of Minnesota who met Brown during a crusade in Minneapolis found the itinerant surgeon to be "perhaps the most godly, gracious Christian gentleman I ever met, as well as one of the finest Bible teachers and creationist scientists." Brown was not, however, above directing some occasional humor at evolutionists. In " 'Bunk' and the 'Monk,' " a poem lampooning the efforts of scientists to explain evolution, he concluded with the following verse:

> So, we're cousins to moles, to fish and tadpoles,
>   Don't smile friends, beware,—that's called "science" today,
> We've a "common ancestor"—You've heard of the quest, sir,—
>   His old bones they do hunt night and day.
> But though hot on the trail of this mythical tail
>   There's no trace, of poor lost chimpanzee.
> And this "brain-stormy" theory can't answer my query,—
>   Not one ape roosts in my family tree!

Brown's ministry ended prematurely when he died in an automobile accident late in 1947.[11]

HARRY RIMMER

No antievolutionist reached a wider audience among American evangelicals during the second quarter of the century than Harry Rimmer (1890–1952), Presbyterian minister and self-styled "research scientist." A native Californian, young Rimmer grew up in poverty in the mining and lumber camps of the northern part of the state. While still a teenager, he worked as a blacksmith, a lumberjack, a sawyer, and a longshoreman. Although forced to quit school before completing the third grade, he continued his education informally under the tutelage of a solicitous mining engineer who prescribed a reading course in geology, biology, physics, and chemistry for his curious young friend. At age nineteen Rimmer joined the U.S. Army, where he served in the Coast Artillery and won local acclaim as the artillery's best welterweight boxer, an experience that left him with "a cauliflower ear, a badly broken nose, and bulging muscles."[12]

Following his hitch in the military, Rimmer enrolled in San Francisco's Hahnemann Medical College, a small homeopathic institution that required only a high-school diploma or its "equivalent" for admission. Although its facilities, as described by Abraham Flexner (1866–1959) in his famous exposé of American medical schools, were far from adequate for the proper training of physicians, the school did occupy "a small, well kept building containing the usual dissecting room, a laboratory for elementary chemistry, one fairly equipped laboratory in common for histology, bacteriology, and pathology, and a small orderly library." Here, supported by money earned as a prizefighter, Rimmer acquired much of his limited knowledge of science and picked up a vocabulary of "double-jointed, twelve cylinder, knee-action words" that later served to impress the uninitiated. An excellent student with an exceptionally retentive memory, Rimmer found it financially impossible, despite his boxing, to continue in school past two terms; thus one term short of receiving his M.D. degree he reluctantly dropped out.[13]

Up until this time Rimmer had shown little interest in religion. But returning home from a prizefight one night, he stopped to hear a sidewalk preacher, who convinced him to give his life to Christ. Shortly thereafter he retreated to the woods of Lake County, where alone with his Scofield Reference Bible he mastered the tenets of his new-found faith. In 1915 Rimmer and his new wife, whom he had met at the Bible College of San Francisco, moved to southern California, where during the next few years Harry studied briefly at Whittier College and at the Bible Institute of Los Angeles (present-day Biola University) and pastored a Quaker church. During the early 1920s, having abandoned the Quakers for the Presbyterians, Rimmer served as an itinerant speaker for the YMCA while maintaining a home in Los Angeles. Between trips he personally built a little workshop in the backyard next to the garage, which he liked to call his laboratory. As described by his wife, this "little laboratory . . . contained a darkroom, a sink and running water, a microscope, centrifuge and test tubes." Although he rarely conducted actual experiments, he spent considerable time in the lab taking pictures of microscopic animals and other objects, which he used to illustrate his lectures and books. On at least one occasion he invited a group of college students to his lab, where he demonstrated the "methods and patterns of mitosis," or cell division, and showed them "cross sections of the human embryo, especially of the head region." In describing this visit, Rimmer neither identified the source of his embryos nor justified their destruction for pedagogical purposes.[14]

The construction of his laboratory coincided with his awakening interest in the subject of evolution. Although skeptical of evolutionary theories since his days as a medical student, he apparently did not become actively involved in the growing agitation against evolution until the early 1920s. According to his wife, his interest in the subject was aroused when some students at Occidental College informed her husband about a science professor who was attacking the biblical story of creation. Rimmer began coaching the students, whereupon the professor turned his attacks on the preacher, who did not hesitate to fight back. Later that year, while on a speaking tour of the Rocky Mountain region, Rimmer enrolled in a correspondence course in geology offered by the University of Colorado and began collecting fossils. Before long he had added to his repertoire of lectures such titles as "The Seamy Side of Evolution" and "The Collapse of Evolution" and was acquiring a reputation among fundamentalists as a uniquely qualified critic of evolution. "Dr. Harry Rimmer," declared one fundamentalist journal in 1926, "is one of the most remarkable young men in the country. He is the rare combination of a scientist, who is a successful soul-winning evangelist." The same publication described him as "one of the most widely known of all the young men among fundamentalists."[15]

Rimmer had already by this time organized the Research Science Bureau. This institution evolved on paper from "a clearing house" for the work of Christian scientists into an organization that aspired

> to promote and conduct research and expeditions in the fields of Biology, Paleontology, Archeology, and Anthropology in the various countries of the earth where the search may lead; TO EN-COURAGE AND PROMOTE RESEARCH IN SUCH SCIENCES AS HAVE DIRECT BEARING ON THE QUESTION OF THE INSPIRATION AND INFALLIBLE NATURE OF THE HOLY BIBLE; to desseminate [sic] by means of public lectures, printed literature, and other methods, facts and information concerning the harmony of true Science and the Word of God.

In fact, the bureau existed primarily, if not exclusively, to underwrite Rimmer's ministry and occasional field trips. Incorporated in 1921 with a board of directors consisting of a handful of well-to-do supporters, the bureau offered two classes of memberships: "Associate" members paid $5.00 a year to support the work of "active" members, the most active of which by far was Rimmer. In return for their annual dues, associates—of whom there were over three thousand by the early 1930s—received the bureau's publications, twenty-five-cent pamphlets written by President Rimmer himself or occasionally by fellow creationists like Brown. While on tour Rimmer signed up as many new members as he could; on returning home to Los Angeles he and his wife and children—the bureau's staff—would sit around the family table addressing and stuffing envelopes.[16]

For a brief period in 1927 Rimmer produced monthly newsletters for the bureau, which appeared in Gerald B. Winrod's right-wing, fundamentalist magazine, the *Defender,* of which Rimmer served as associate editor. In these newsletters, as elsewhere, Rimmer unabashedly presented himself as a person "professionally engaged in scientific research"—and therefore incapable of being antagonistic to "true science." In typical fashion he dismissed the argument that "all authorities agree" about the animal origins of humans with a personal anecdote: "I looked up seven of these quoted 'authorities' one at a time, and they all quoted [the German biologist Ernst] Haekel [Haeckel]! They agreed, all right, but what did they agree upon, except that Haekel had said a certain thing!"[17]

In his first newsletter Rimmer promised that the bureau would soon be sending an expedition to Africa, "where the members will study the Gorilla in its native haunts, and greatly advance the present growing conviction that Man is not related in any way to this or any other animal. A careful and painstaking parallel dissention [dissection] of the Gorilla and Human will complete this study, and the results will be given to each member." Illness, however, forced Rimmer to cancel the trip and dashed his dream of dramatically disproving human evolution. Although he later visited Africa twice, in the late 1940s, for the time being he had to settle for a gorilla skull sent by a missionary doctor in Siam and some American Indian skulls unearthed during a series of archaeological digs in the western states. Using photographs "all taken by the Author in his own Laboratory, from specimens in his own private Museum," Rimmer prepared his first antievolution pamphlet, *Monkeyshines: Fakes, Fables, Facts concerning Evolution,* designed to illustrate to high-school students the dissimilarity between gorilla and human skulls. Evolutionists, he charged, deceptively retouched their photographs to make the two appear similar, but even a child could tell the difference. "I have three children, all under ten," he wrote, "and if I say 'One of you run out to the laboratory and get me a Gorilla skull,' even the baby knows the difference, and out of fifty skulls on the shelf she can pick out the Gorilla without hesitation." For a time Rimmer planned to write a syndicated weekly column, to edit a monthly magazine called *Science and the Sacred Scriptures,* and to build a large teaching laboratory for training young ministers how to repel the attacks of evolutionists. But the coming of the Great Depression made such schemes impossible. "Money crawled in the hole and pulled the hole in after it," he noted sadly. From the late 1920s to the early 1940s the bureau lay dormant as Rimmer channeled his creationist fervor into William Bell Riley's World's Christian Fundamentals Association, which he served as field secretary.[18]

In 1934 Rimmer accepted the pastorate of the First Presbyterian Church of Duluth, Minnesota, under an arrangement that allowed him to spend six months each year writing and lecturing on the Bible and science, particularly evolution. Although Rimmer on occasion preferred the company of humble Baptists to that of the more socially prominent Presbyterians and scored his greatest evangelistic successes among fundamentalist Baptists in the Bible Belt, he remained a Presbyterian partially in deference to his wife's desire for social standing. "If he had married differently," said his son, "he would have been an honest Baptist." He might also have spent less time on the road. Instead, by the mid-1930s he had spoken to some 3,876 student bodies throughout North America and had acquired a reputation as one of the fundamentalist movement's star attractions, known for his exuberance and wit. On the platform Rimmer exuded the self-confidence of a gifted athlete and orator, but, like many performers, his bravado on stage masked long-standing insecurities that no doubt stemmed from his impoverished, rootless childhood.[19]

Not surprisingly, judgments of his performances varied according to the scientific and theological leanings of his listeners. An enthralled antievolutionist from Tennessee wrote that "it was like the breath of a new day to hear this enthusiastic young scientist exposing from the facts of science the collapse of the evolutionary theory." During one of Rimmer's several engagements in Atlanta, the sports editor of the *Journal* wrote: "His is not only one of the most brilliant minds in the ministry today, but he is so intensely human, so virile and possesses such a delightful sense of humor. . . . He is a he-man in every sense of the word and a mighty good soldier of the Lord Jesus Christ."[20]

Critics tended to dismiss him as a charlatan and buffoon. A writer for the *Debunker* who attended one of Rimmer's lectures in Philadelphia noted the speaker's "abundant 'wit' " but added: "I am assuming it to be wit from the uproarious greeting with which it was received by the audience." Rimmer's "scientific" method "was to accept the Bible as absolute truth and discard, ignore, or hopelessly distort all facts that had been gathered, until the remaining facts in their distorted nature 'proved' the religious conclusions which had been established even before the facts had been consulted." After a two-week series of lectures in the large Central Presbyterian Church in Denver, one citizen devoted an entire pamphlet to exposing "how false, stupid and absurd" Rimmer's claims and teachings were. "As a scientist, Rimmer is a joke. . . . It is obvious . . . that he assumes the role of a scientist to enhance his prestige. As a preacher his anti-evolution twaddle would fall flat (so low is the public estimation of the clergy), but as a 'scientific research worker' he is able to impress the uninformed and collect the shekels." There was, of course, an element of truth to this charge, but Rimmer was neither the first nor the last participant in the creation-evolution debates to cloak himself in the mantle of scientific respectability. According to the biographer of the philosopher-historian John Fiske, one of the most effective popularizers of evolution in late nineteenth-century America, "Fiske's first great advantage in convincing listeners was his claim to be a scientist."[21]

Like most creationists with scientific pretensions, Rimmer displayed ambivalence toward the scientific enterprise. On the one hand, he decried the current "fetish of science," which manifested itself in "an amazing willingness to surrender the eternal verity of God's revelation for the unfounded theories propounded by men who are utterly without ability to prove their wild imaginings." On the other hand, he reveled in the prestige that he accrued by passing as a scientist. He resolved the conflict in typical fundamentalist fashion by defining science so narrowly—as "a correlated body of absolute knowledge" obtained by direct observation—that theories of evolution clearly fell outside the scientific domain. "If scientific knowledge comes only through trained observation," he asked, "how could there be knowledge of creation, or origin, which antedated the presence of man, the observer?" His quarrel was with the interpretation of facts, not the facts themselves: "*With facts we have no dispute; we accept every fact of every science known to man.*" Believing that "every fact and proof that militates against the truth of the theory of organic evolution is a tremendous argument in favor of the fact of special creation," and that the theory of evolution "admits of no active intelligence in control, and design is foreign to the entire process," he built his case for creation largely on alleged flaws in evolutionary theory and on the evidence of apparent design in the world of living things, which he regarded as being incompatible with a developmental origin.[22]

Wherever possible, especially on college campuses, Rimmer sought to engage evolutionists in public debate, confident that few, if any, could match his forensic skills and his genius for manipulating large crowds. His confidence proved to be well founded—but not because of his superior logic and evidence. As his son once observed, "Dad never won the argument; he always won the audience." And most of the time the majority was on his side from the beginning. Rimmer relished these contests, which were frequently broadcast over the radio. "His words flowed rapidly," recalled his wife. "Humor and sarcasm mixed in with the factual material, and the laughs soon were spontaneous and frequent." Rimmer often lured evolutionists into debate by offering them $100 or more for the experience. The eminent American paleontologist Henry Fairfield Osborn refused an offer of $500 plus expenses, but Samuel Christian Schmucker (1860–1940), a retired biologist from the State Normal College in West Chester, Pennsylvania, and author of two popular books on evolution, whom Rimmer grandiosely described as "just about the greatest champion of evolution in the East," agreed to match wits with Rimmer in Philadelphia in 1930. The debate attracted about twenty-five hundred paying spectators, who anticipated "a battle royal" between two scientific giants. Instead they witnessed what Rimmer described to his wife as a verbal homicide: "The debate was a walk-over, a massacre—murder pure and simple. The eminent professor was simply scared stiff to advance any of the common arguments of the evolutionists, and he fizzled like a wet fire-cracker!" A poll at the close of the debate revealed that only one-fifth of the audience professed to be evolutionists, and only seventy voted for Schmucker as the winner. "It was a mighty profitable evening for our side," declared the jubilant victor, who saw himself as a modern-day David. "The good old pebbles and sling once more!"[23]

Rimmer's most famous debate took place not with an evolutionist but with a fellow creationist, the Baptist preacher Riley, who took issue with Rimmer on the proper reading of the first chapter of Genesis. Riley, like George Frederick Wright and William Jennings Bryan, believed that the "days" represented geological ages; Rimmer, following his trusted Scofield Bible, argued that they were only twenty-four hours long. The debate first took place in the late 1920s at a summer Bible conference near Minneapolis, organized by Riley, who had invited his friend Rimmer to be the featured speaker. "We had a buster of a crowd," reported the irrepressible Rimmer.

The tabernacle wouldn't hold them, and they sat on the platform, stood at the back and down the sides, and outside looking through doors and windows. . . . We sure skinned each other without mercy whenever there was a good opening, and had a swell time. When the vote was taken I won by a majority of five to one. Riley conceded defeat and congratulated me, saying it showed what brains could do; make a man able to come into a fellow's own crowd and beat him on the wrong side of the question! I replied: "Nothing of the kind. It showed that truth is mighty and will prevail."

In an apparent response to popular demand on the West Coast, Rimmer and Riley later reenacted their debate, at the Bible Institute of Los Angeles.[24]

Although both preachers defended their respective views with conviction, neither man, as Rimmer pointed out, was "dogmatic or immovable on this issue." Rimmer confessed to having once believed in the day-age theory himself—until a thorough study of the question convinced him "that the Genesis account of creation is to be received as a literal week of time, seven days in extent, each day of twenty-four hours duration." Apparently practical as much as exegetical considerations decided the issue for him. According to Rimmer's reading of Moses, God had created plants the day before he drove away the mist shrouding the earth and allowed the rays of the previously created sun to shine upon the earth. How, asked an incredulous Rimmer, "could the plants have survived an entire geological age, of some half million years, with no sunshine at all?" Every farmer and gardener in America knew the answer. Riley, too, appealed to scientific evidence in defense of his day-age reading of Genesis, arguing that "the testimony of geology makes this interpretation certain." Appealing to such nineteenth-century authorities as Arnold Guyot, James Dwight Dana, and John William Dawson, "a fundamentalist [who] believed in God and in His Word," Riley insisted that the doctrine of geological ages was "the uniform position of the Christian geologists of the world!"[25]

As a devotee of the Scofield Reference Bible, Rimmer readily granted the antiquity of life on earth. Although he did not pretend to know exactly when life first appeared, and did not accept the notion of pre-Adamic *humans*, he thought that a "positively inconceivable" span of time had elapsed between the creation of life and the creation of Adam and Eve, which, after some uncertainty, he placed at around 5862 B.C.[26] During the enormous gap between the original creation of heaven and earth and God's subsequent "reconstruction" of the earth in six literal days, many of the fossils now found in ancient rocks may have been buried, but others probably dated from the period between the Edenic creation and the Noachian flood, which Rimmer suspected had been a local affair "limited to the world of man." Thus Noah's ark needed only to be large enough to carry animals that inhabited the flooded area. Before the deluge, wrote Rimmer in 1925,

> It never had rained. Around the globe was a protecting lense [sic] of ice that made the whole world an Eden of tropical splendor. It was a "firmament" indeed; and rain could not come thru it. The rays of the sun that cause decay were strained out, and long life was the rule. Men lived for hundreds of years. These rays also cause fermentation; and Noah, who was accustomed to drink the juice of the grape in the days before the flood, was amazed when the fermented juice made him drunk [after the flood]!

Thus did Rimmer not only explain the longevity of the antediluvians but the embarrassing episode of Noah's postdiluvian drunkenness.[27]

As the above discussion suggests, Rimmer's defense of special creation was tied to a much broader effort aimed at protecting the Bible from all charges of scientific error. Although he readily acknowledged that the Bible was not written as a scientific text, he believed that its claim to divine origin ruled out the possibility of any mistake, scientific or otherwise: "For regardless of the purpose of the Book, God knew the facts; and a revelation from Him must be perfect in all points." For Rimmer, the assertion that "all scripture is given by inspiration of God" (2 Timothy 3:16) literally meant "that the Spirit of God carefully and rigidly supervised the construction of the Holy Book word by word, and line by line, over-seeing the labors of all the writers whom He used as emanuenses [sic] of His will and purpose."[28]

So convinced was Rimmer of the scientific inerrancy of the Scriptures that beginning in the mid-1920s he offered $100 to anyone who could prove otherwise. But despite an enormous number of responses, challenging everything from the adequacy of accommodations on Noah's ark to the likelihood of apples growing in the Garden of Eden, no one ever demonstrated a biblical error to Rimmer's satisfaction. To critics who questioned the veracity of the story of Jonah and the whale—the argument "most frequently advanced by the superficial critic"—Rimmer replied, first, that the Bible specified "a great fish," not a whale, and second, that the *Rhinodon typicus* or whale-shark could have accommodated Jonah. He claimed actually to have met an English sailor who survived two days and nights inside the belly of a gigantic *Rhinodon*. To skeptics who doubted the account of Joshua's commanding the sun to stand still, Rimmer pointed out that a correct reading of the passage in Hebrew indicated that Joshua instructed the sun to be silent, not still. In response to the Israelite leader's plea for relief from the heat of the sun, God sent a hailstorm, which not only cooled down the army but destroyed the enemy.[29]

On two occasions hopeful claimants actually took Rimmer to court in attempts to win the promised reward. Both lost. In the first case, in 1929, a retired army colonel from California challenged the story of God's miraculously feeding the children of Israel in the wilderness by sending so many quail that they were piled two cubits high for a day's journey around the camp (Numbers 11:31). The skeptical colonel calculated that there had been 29,613,991,260,171 dead quail in the pile, or 12,266,171 for each Israelite to eat—a scientific impossibility! However, the judge who heard the case ruled against the plaintiff, apparently reasoning that Moses, who had actually seen the miracle, was a more reliable witness than the calculating colonel. Buoyed by their impeccable record after nearly a decade and a half, Rimmer and his fellow directors on the board of the Research Science Bureau upped the reward in 1939 to $1,000.[30]

Shortly thereafter, William Floyd (1871–1943), a New York humanist who edited the *Arbitrator*, a magazine devoted to "outlawing war, abolishing poverty and unveiling superstition," sued Rimmer on the grounds that he had discovered five scientific errors in the Bible: the six-day creation story, the contradictions between the two creation accounts found in Genesis 1 and 2, the record of Noah's ark, the alleged number of quail sent to feed the Israelites in the wilderness, and the descriptions of the camel, the coney, and the hare found in Leviticus 11:4–6. Floyd disclaimed any interest in the money; he simply wanted "to convince fundamentalists, through a court judgment[,] that there are such errors in the Bible."[31]

Rimmer won the case on a technicality: The particular newspaper offer to which Floyd had responded had not been placed by Rimmer himself. However, Rimmer, to the great irritation of his critics, widely claimed that the trial "ended in legally establishing the position of all who hold that the Word of God is inerrant" and that the court had decided "that it is *all right for Christians to go right on reading and believing the Bible!*" His widely read version of events, a frequently reprinted book first serialized in the *Sunday School Times*, dramatically described the "utter defeat" of "a great group, composed of Atheists, Freethinkers, Modernists, and Humanists" by a band of "simple, believing Christians [who] stood steadfast for the infallibility of the Book."[32]

By the time of the trial, the degreeless evangelist who claimed to have no desire for any academic honors besides the M.D. he had come so close to earning as a young man was billing himself as Harry Rimmer, Sc.D., D.D. (honorifics bestowed respectively by Wheaton College and Colquith College), and claiming membership in the American Association for the Advancement of Science (AAAS). After hearing Rimmer brag about his scientific credentials, including service on important AAAS committees, the skeptical president of Furman University in South Carolina wrote to the association inquiring about Rimmer's standing. An assistant to the secretary replied that although Rimmer had indeed been a member since 1925, he had never served on any AAAS committee. Noting that "any person interested in the advancement of science and education" could join the association, the assistant added that "so far Doctor Rimmer has avoided such drastic statements that would cause his removal from the list of members."[33]

Despite his lack of an earned academic degree, Rimmer from time to time received offers of faculty appointments. In 1924 John Brown University, a fundamentalist school in Siloam Springs, Arkansas, offered him a "chair of Research Science," which he declined, and later William Jennings Bryan College in Dayton, Tennessee, twice tried to recruit him as its president, each time without success. For years his closest academic affiliation was with the Kansas State Teachers College of Pittsburg (now Pittsburg State University), whose president, W. A. Brandenburg (1869–1940), openly favored the fundamentalist cause. When, at the president's invitation, Rimmer in 1924 first visited the small college in southeastern Kansas, he attacked the theory of evolution in a lecture titled "Modern Science and Research." Posing as a scientist—but exposing his ignorance of current scientific thinking about the mechanism of evolution—he focused his critique on the discredited Lamarckian theory of the inheritance of acquired characteristics. As reported in the college paper:

Dr. Rimmer offered as the greatest argument against evolution the fact that acquired characteristics are not inherited. For instance, he related an experiment performed on the rabbit family.

[A] rubber vest was strapped to the breast of the female rabbit to prevent the use of fur for the lining of the nest. In its place carded wool was used. After ten generations it was found that the rabbit did not inherit the acquired habit that was thrust upon its ancestors.

Dr. Rimmer cited as another instance that acquired characteristics are not inherited, the age-old Chinese custom of binding of the feet. After three thousand years of this custom we find the Chinese children are born with perfectly shaped feet. . . .

Regardless of his arguments, local antievolutionists hailed him as "a scientific Moses or Lincoln," and Brandenburg urged him to join the faculty.[34]

Although Rimmer declined repeated invitations to take a permanent position at Pittsburg, he did return from time to time to find "the door is not only wide open for me—but off the hinges!" For years he also served, apparently unofficially, as the curator of the college museum, a connection that "gave him the privilege of collecting specimens in many spots which otherwise would have been denied him." On occasion Rimmer took students from Pittsburg with him on archaeological digs, the artifacts from which he often donated to the museum. The college catalog for 1928–1929 thanked Rimmer for graciously contributing "a burial mound, numerous skulls, and stone implements," as well as a collection of "many beautiful butterflies."[35]

Rimmer's career as a creationist peaked in the 1930s. Among the scientifically marginal fundamentalists, who craved confirmation of their views, he acquired a reputation not only as a fearless and entertaining critic of evolution but also as a genuine scientist who had "made many important discoveries in physics, chemistry and biology." (As late as 1976 an antievolution tract described Rimmer as "one of the foremost scientists of this century."[36]) Although Rimmer contributed nothing novel to the creation-evolution debate, his theological and scientific authority helped to establish the gap theory as the dominant view of origins among pre–World War II fundamentalists. His books alone, to say nothing of his pamphlets, reached hundreds of thousands of readers. However, by the time of his death in 1952 both his theological and scientific luster had faded badly. The growing number of trained evangelical scientists tended to dismiss his scientific claims. His rejection of cherished fundamentalist interpretations of prophetic images in the Book of Daniel led to a flurry of criticism and reduced him to insisting pathetically "I am still a Fundamentalist." Apologetic fashions changed. In terms of influence, observed his son, Rimmer "disappeared in the Second World War. . . . The world left Dad behind."[37]

FIVE

George McCready Price and
the New Catastrophism

vents surrounding the Scopes trial in 1925 confirmed the paucity of
reputable scientists willing to defend special creationism. In the
spring of that year John Thomas Scopes (1901–1970), a high-school
teacher in Dayton, Tennessee, confessed to having violated the state's
recently passed law banning the teaching of human evolution in public
schools. His subsequent trial focused international attention on the
antievolution crusade and brought William Jennings Bryan to Dayton
to assist the prosecution. In anticipation of arguing the scientific merits
of evolution, Bryan sought out sympathetic experts, such as S. James
Bole, who might be willing to testify for the state. The responses to his
letters of inquiry surely disappointed the aging crusader. Howard A.
Kelly (1858–1943), a highly respected Johns Hopkins gynecologist who
had contributed an essay to *The Fundamentals*, reminded Bryan that
though he believed in the special creation of Adam and Eve, he remained open to the possibility of a "continuous sequence in the life
history of the lower creation."[1] Louis T. More (1870–1944), a physicist
and dean at the University of Cincinnati who had just written a book,
The Dogma of Evolution (1925), protesting the extension of evolution
from biology to philosophy, replied that he accepted evolution as a
working hypothesis.[2] The ever-pugnacious Alfred Watterson McCann,
who had ridiculed evolution in his book *God—or Gorilla* (1922), re-

sponded by chiding Bryan for supporting Prohibition in the past and for now trying "to bottle-up the tendencies of men to think for themselves."[3]

At the trial itself Bryan fared no better. Under examination by defense attorney Clarence Darrow (1857–1938), Bryan could name only two scientists whose views on earth history he respected: George Frederick Wright, who had been dead for four years, and George McCready Price (1870–1963), whose employment in England had prevented him from accepting Bryan's invitation to participate in the trial. Although Darrow dismissed Price as "a mountebank and a pretender and not a geologist at all," Price had by the mid-1920s established a reputation, among friends and foes alike, as the most innovative and influential defender of special creation. Even the unsympathetic editor of *Science* recognized him as "the principal scientific authority of the Fundamentalists." Price's voluminous writings against evolution, directed particularly at its geological foundations, spanned six decades but attracted little attention among orthodox scientists. If he was noticed at all outside of creationist circles, it was for being one of the twentieth century's most persistent scientific cranks, worthy of inclusion only in histories of pseudoscience, such as Martin Gardner's *Fads and Fallacies in the Name of Science* (1957). Gardner, one of the first critics to recognize Price's historical significance, described him in the 1950s as being "the last and greatest of the anti-evolutionists." The last he was not, but a strong case can be made that he was the greatest.[4]

THE MAKING OF A CREATIONIST

George Edward Price (he substituted his mother's maiden name, McCready, for Edward upon launching his literary career) was born on August 26, 1870, in rural New Brunswick, Canada. His mother came from a well-bred family that included at least two journalists; his father, a farmer and mill operator, died when George was about twelve, leaving him and his younger brother to run the farm. A couple of years later the widow Price, along with her two sons, joined the Seventh-day Adventist church, a small apocalyptic sect that traced its origins back to the Millerite enthusiasm of the 1840s, when followers of the farmer-preacher William Miller (1782–1849) set various dates for the return of Christ to earth. The Seventh-day Adventist offshoot of the Millerite movement was led by the charismatic Ellen G. White (1827–1915), who claimed to receive divine messages in trancelike visions and whose

pronouncements Adventists placed on a par with the Bible. Like their
Millerite forebears, Seventh-day Adventists expected the imminent
end of the world and, in accordance with God's instructions given in
the fourth commandment of the Decalogue (Exodus 20:8–11), worshiped on Saturday as a memorial to a literal six-day Creation. Because
of their distinctive Sabbath doctrine, Adventists adamantly opposed
any scientific theory that proposed interpreting the days of creation
symbolically. To follow "infidel geologists" in supposing that the events
described in Genesis 1 "required seven vast, indefinite periods for
their accomplishment, strikes directly at the foundation of the Sabbath
of the fourth commandment," argued White. "It makes indefinite and
obscure that which God has made very plain." If she harbored any
doubts about the correct reading of the first chapter of Genesis, they
were erased during one of her visions, in which she was "carried back
to the creation and was shown that the first week, in which God performed the work of creation in six days and rested on the seventh day,
was just like every other week."[5]

In a published gloss on the Mosaic cosmogony, White endorsed the
largely discarded view of Noah's flood as a worldwide catastrophe that
had buried the fossils and reshaped the earth's surface. After the floodwaters had subsided, exposing the rotting carcasses of antediluvian life,
God had buried the organic debris, she explained, by causing "a powerful wind to pass over the earth . . . in some instances carrying away
the tops of mountains like mighty avalanches, forming huge hills and
high mountains where there were none to be seen before, and burying
the dead bodies with trees, stones, and earth." The buried forests
subsequently turned into coal and oil, which God occasionally ignited
to produce "earthquakes, volcanoes and fiery issues." White's authoritative descriptions of times past made most Adventists, including
Price, unwilling to entertain interpretations of Genesis, such as the
day-age and gap theories, that allowed other fundamentalists to accommodate the findings of historical geology.[6]

At age seventeen, two years out of high school, Price married an
Adventist woman twelve years his senior, and for several years the
couple tramped across the Maritime Provinces of eastern Canada selling White's books door to door. In the fall of 1891 he matriculated in
Battle Creek College, an Adventist school in Michigan. Although the
college offered a "scientific course," Price, who since childhood had
aspired to a literary career like some of his mother's relatives, elected
instead to take the "classical course." He remained in Battle Creek for
two years before his money ran out, forcing him to return home and
resume selling books, a job he had come to hate but did well. In 1896,

frustrated by the book business, he enrolled in a one-year teacher-training course at the Provincial Normal School of New Brunswick (now the University of New Brunswick), where he took "some elementary courses in some of the natural sciences, including some mineralogy," apparently taught by one of John William Dawson's former students. This was the extent of Price's formal training in science. In the fall of 1897 he accepted the first of a series of small-town teaching positions.[7]

Two years later Price found himself teaching high school at Tracadie, a dreary, remote village on the Gulf of St. Lawrence populated mostly by French-Canadian farmers and fishermen. Alone and hungry for intellectual companionship—his wife was teaching elsewhere—he struck up a friendship with Dr. Alfred Corbett Smith (1841–1909), head of the medical department of the government leprosarium near the village and one of the few English-speaking residents in the community. During one of their many conversations, Smith, a Harvard Medical School graduate, inquired about his young friend's views on evolution. Upon learning that Price was a strict creationist, he offered to lend the unworldly teacher various works on evolution from his own library. To Price, the theory of evolution described in these books seemingly "*all turned on its view of geology, and that if its geology were true, the rest would seem more or less reasonable.*" Several times Price tottered on the brink of accepting this line of reasoning. In one of the most detailed and revealing accounts of a person struggling to reconcile the claims of science and religion, he later recalled that "on three distinct occasions, while I was making my first investigations, determined to get to the very bottom of the entire problem, I said to myself, 'Well, there must be something to this claim that the fossils do occur in a definite sequence, and thus there must be something to the geological ages.'" But how could he possibly harmonize this conclusion with the Mosaic account of creation as interpreted by White? The gap theory involved too much intellectual "dodging and twisting" to suit his tastes. The day-age theory seemed to be nothing less than "a libel on Moses" that struck "at the very basis of the Sabbath," since, according to that view, the seventh day of creation was not a literal day.[8]

After much anguish and prayer he found the solution to his dilemma in White's "revealing word pictures of the Edenic beginning of the world, of the fall and the world apostasy, and of the flood." Her suggestion that the deluge and related events had buried the fossils, a notion no longer held by any geologist, particularly intrigued him. Still, he puzzled over ways to interpret the evidence that apparently indicated the earth's antiquity, which at first glance seemed "so strong and plausible." Only after poring over the standard geology texts and "almost tons of geological documents, government reports, memoirs, and monographs on special geological topics" did he discover "how the actual facts of the rocks and fossils, *stripped of mere theories*, splendidly refute this evolutionary theory of the invariable order of the fossils, *which is the very backbone of the evolution doctrine.*"[9]

This discovery not only resolved his immediate crisis but determined his future course. Believing that he had found a fatal flaw in the logic of evolutionary geology, he grew increasingly convinced that God wanted him "to enter this unworked field; accordingly I threw myself into it with all the energy I possessed, constantly asking and receiving special help from the guiding and enlightening Spirit of God." Responding to this call not only satisfied spiritual needs but also allowed him to fulfill his dream of becoming a writer. Before leaving Tracadie, he completed a book-length manuscript, *Outlines of Modern Christianity and Modern Science* (1902), the first attempt since the beginning of the Darwinian debates, so far as he knew, to take "the real Biblical side of the pending Science-versus-Religion controversies." In contrast to theistic evolutionists and old-earth creationists, he vowed not to compromise a strictly literal reading of Genesis, which seemed unambiguously to teach that "life has been on our globe only some six or seven thousand years; and that the earth as we know it . . . was brought into existence in six literal days." His initial authorial effort may have lacked the fluid literary style of his later writings, but it nevertheless revealed the spirit of a man willing, if need be, to challenge both the scientific and religious establishments.[10]

In his 271-page work, which he later liked to call "the first Fundamentalist book," Price attempted to show that "the whole scheme of evolution"—from the nebular hypothesis and developmental geology to Darwinism—"is in the highest degree improbable and absurd." Convinced that geology furnished "nine-tenths" of the argument for organic evolution, he focused his attack on that science, arguing that Charles Lyell with his uniformitarianism and Louis Agassiz with his theory of the geological succession of life "were but the advance agents of Darwin, Spencer, Haeckel & Co., and their modern troupe playing at dethroning the Creator." Using criticisms long current among Adventists, he argued that the principle of uniformity was not only an unwarranted assumption but "a point blank denial of the record of the Deluge" and that the succession-of-life idea indicated circular reasoning "of the most glaring kind." Geologists, charged Price, dated rocks by their fossil content while simultaneously determining the age of the fossils by their location in the geological column.[11]

As an alternative to these "geological guesses," he proposed reconstructing the science of geology on the basis of events described in the Bible: a recent creation and a worldwide catastrophe. Noah's flood, especially, cleared up a host of geological problems for Price. It explained, for instance, why the fossils appear in a predictable sequence (a conclusion he later repudiated). Assuming that a sudden shift of the earth's axis had released massive subterranean reservoirs of water, he speculated on how the resulting disaster would first kill the "smaller and more helpless animals," with their burial place being determined by their specific gravities. "The vertebrate fishes, if killed in numbers, would tend to rise to the surface. The larger animals and man would flee to the hill-tops from the rising waters, and, when finally engulfed, would be simply drowned, and not immediately covered with earthy deposits." To keep these decaying bodies from polluting the air, Price, following White, invoked a miraculous "cosmic storm" to bury them. Thus to the discerning eye, the fossil record revealed not a temporal succession of life but simply a sorting out of contemporaneous antediluvian life-forms. Or, in other words, the universally accepted geological column was nothing but a cunningly devised fable.[12]

Price went on to explain that before the sedimentary strata had completely hardened, the elements had carved out such natural wonders as the Niagara River Gorge and the Grand Canyon. He admitted the existence of post-flood glaciers, but the phenomena geologists generally associated with glacial action he tended to attribute to the deluge, pointing out that "when water really gets in a hurry, it can pile things up in quite a wonderful manner." He quoted the observation of Sir Henry H. Howorth (1842–1923) that the notion of a great ice age was "the wildest dream which a fertile imagination ever imported into science." Price speculated that the great mountain ranges, such as the Alps and the Himalayas, were formed from deposits laid down during Noah's flood, then "folded and elevated to their present height by the great lateral pressure that accompanied its subsidence." He agreed with White that coal originated from forests buried during the flood and that the ignition of deep coal deposits produced volcanoes. Because a universal catastrophe accounted for so many phenomena, he concluded that its existence was "not only possible, but scientifically certain."[13]

Whatever elation Price may have felt by completing his first book soon gave way to desperation, as a sense of failure engulfed him. In the spring of 1902 he left Tracadie to serve as an Adventist evangelist on Prince Edward Island. Unfortunately, because of his schoolmarmish manner and squeaky voice he possessed neither the temperament nor the talent for success in the pulpit; thus as soon as possible church leaders reassigned him to head a new boarding academy in Nova Scotia. Within a short time they realized that he was no better as an administrator than as a preacher, but efforts to find him a full-time teaching job proved unsuccessful. Thoroughly discouraged and driven by guilt to earn a living for his wife and three children, he returned in the summer of 1904 to the one job that had brought him a measure of success: selling religious books. But as he pedaled his bicycle over the rough roads of Prince Edward Island, he continued to dream of a literary career, "the thing for which I am best fitted and which I thoroughly enjoy above everything else." He had tried various lines of church work only to find "black, dismal Failure" mocking him at every turn. By late summer he had grown so depressed by his situation that he was contemplating suicide. However, out of consideration for his family he decided instead to leave church employment and head for New York City to try his hand at writing "hack stuff for the Metropolitan newspapers and magazines." If life did not improve in the city, he planned to sell his watch, buy a revolver, and rid the world "of another useless, good-for-nothing man."[14]

In the city his circumstances only worsened. Unable to find steady work, he suffered unspeakable privations—and the torment of knowing his family was "destitute and almost starving" back in Canada. Since youth he had derived strength from his religious faith, but now in his neediest hour he quit even attending church. The only positive development during this period came from teaching himself to type in his spare time, using his shabby hotel bed "for a desk and an ordinary chair for a stool." His wife, fearing the worst, wrote to the Seventh-day Adventist headquarters in Takoma Park, Maryland, on the outskirts of Washington, DC, begging for help for her husband. Moved by the Price family's plight, the president of the church personally offered the estranged worker a temporary construction job helping build a new physical plant in Takoma Park. After a brief delay, Price gratefully accepted the offer, noting that he was willing to go anywhere and do anything, "even if it means hard manual labor."[15]

His pride tarnished but his ambition undiminished, Price spent the summer of 1905 driving a team of horses that hauled gravel and cement to a construction site and occasionally stealing a few hours to explore Washington-area libraries. In the fall he left the East Coast with a trunk full of research notes to become principal of a small Adventist school in Oakland, California. At the close of the school year he moved on to southern California, eventually settling in Loma Linda, where as a "heartbroken" thirty-six-year-old writer he found employment as a construction worker and handyman at a recently purchased Adventist sanitarium.[16]

Despite his shattered plans, Price mustered sufficient energy and borrowed enough money to self-publish his second book, a small paperback, *Illogical Geology: The Weakest Point in the Evolution Theory* (1906), the title of which he borrowed from the English evolutionist Herbert Spencer. In the years since writing his first book, Price had made two "Providential discoveries" that led him to question the alleged order of the fossil-bearing rocks, which he had earlier explained by the specific gravity of the dead animals. In a copy of an annual report of the Canadian Geological Survey, which Smith, as a government employee at the leprosarium, had received free, Price ran across

an example or two of what the geologists now call "deceptive conformity," where strata alleged to be very "young" occur in perfect conformity over wide areas on top of very much "older" beds, the two being lithically identical, so much so that "were it not for fossil evidence," as the Government geologist expressed it, "one would naturally suppose that a single formation was being dealt with." To me it seemed self-evident that no great interval of time could possibly have elapsed *between* the deposition of these two successive beds, which are so nearly identical in appearance; whereas the common interpretation of evolutionary geology said that a vast interval of time, represented by many millions of years, is here represented by this insignificant line between two strata which look perfectly conformable, with nothing to show for this long interval, either in the way of erosion or deposition.

Soon afterwards I ran across something even more significant. [In Sir Archibald Geikie's *Textbook of Geology* (1903)] I found some examples of an exactly similar conformity, but in the reverse order, with the alleged "old" rocks on top, this time, and the "young" strata underneath, but with every physical appearance of having been *actually deposited* in this order of sequence.

As the meaning of these well-known phenomena, the latter called overthrusts or thrust faults by geologists, penetrated his "young and unsophisticated" mind, he imagined that he had made "a very important discovery," namely, that it was impossible to prove that one fossil was older or younger than any other fossil and, therefore, that the fossil record lent no support to theories of evolution. Emboldened by what he had found, the impecunious author confidently offered readers of *Illogical Geology* "a thousand dollars to any one who will, in the face of the facts here presented, show me how to prove that one kind of fossil is older than another."[17]

In his little book Price drew attention to an instance of supposedly "upside-down" strata covering at least fourteen square miles in Alberta, just east of the Rockies, where "older" Cambrian fossiliferous rocks overlay "younger" Cretaceous rocks. Although a report by the Canadian Geological Survey described these formations as appearing "to succeed one another conformably," geologists, largely on the basis of the fossil content of the rocks, argued that they had resulted from overthrusting. In other words, Price concluded, "our eyesight and common sense is to be denied, in order to save the theory."[18] Shortly after publishing *Illogical Geology*, Price learned that the faulted area in Alberta, known as the Lewis overthrust, actually extended into Montana and covered an area of several thousand square miles. He conceded that some rocks, particularly in the Alps, had indeed been tilted up and overturned, but such formations could all be measured "in feet and yards, not miles." It seemed preposterous to him to suggest that a rocky region covering thousands of square miles had moved across the landscape. He felt certain he had finally found the evidence that would falsify the claims of evolutionary geologists. In an updated and expanded version of *Illogical Geology* published in 1913 as *The Fundamentals of Geology*, where he announced his new "discovery," he speculated that the Alberta-Montana overthrust would alone "do more than the hundreds of quite similar examples elsewhere have hitherto been able to accomplish in compelling complete reform in geological theory." Language seemed inadequate to express the compelling nature of this evidence: "I don't know what would convince the world, if this evidence here in Alberta and Montana is insufficient."[19]

His case thus apparently proved, Price announced a new geological principle, the *Law of Conformable Stratigraphical Sequence:* "ANY KIND OF FOSSILIFEROUS ROCK MAY OCCUR CONFORMABLY ON ANY OTHER KIND OF FOSSILIFEROUS ROCK, OLD OR YOUNG." This law, he said without a trace of false modesty, "is by all odds the most important law that has yet been discovered in connection with this whole subject of stratigraphical geology." Throughout the remainder of his long life, this principle served as the bedrock of his scientific opposition to evolution; for, he reasoned, if the fossils could not be dated sequentially, then the strongest evidence for evolution simply disappeared, leaving the field clear for special creation. And if deceptive conformities and thrust faults failed to support an evolutionary history of life on earth, they seemed to fit exactly with a catastrophic model that viewed the fossils as representing contemporary, rather than consecutive, populations buried by a single, universal deluge. "Flood geology," as he called his scheme, thus cleared up "beautifully every major problem in the supposed conflict between modern science and modern Christianity."[20]

CREATIONIST AT WORK

During more than half a century of arguing his case, Price occasionally revised the details, but the thrust of his attack remained steadfastly focused on the notion of successive geological ages, "the devil's counterfeit of the six days of creation as recorded in the first chapter of Genesis." In hundreds of articles and more than two dozen books—monographs and textbooks, debates and answers to questions, "sugar-coated" stories and allegories—he promoted his alternative flood geology and critically examined evolution from its scientific, theological, historical, and social angles.[21] He maintained his prolific output despite holding down a succession of time-consuming teaching positions in Adventist educational institutions. He began this phase of his life in 1907, when the newly organized Loma Linda College of Evangelists (now Loma Linda University) hired him to teach student nurses; it ended in 1938, when ill health—aggravated by a theological scandal at Walla Walla College, Washington, in which he was only peripherally involved—forced him to retire from his professorship of philosophy and geology. He had written his many books, he once explained, "chiefly while other people were sleeping or recreating; for all the best part of my life I have worked 12 or 15 or sometimes even 18 hours a day. . . . My Sundays and all holidays, together with the summer vacations have mostly been periods of sheer drudgery at the typewriter." His only freedom from work came during his years of retirement, between ages sixty-seven and ninety-two, which he spent in southern California.[22]

The book Price regarded as his *magnum opus*, a 726-page college textbook titled *The New Geology*, first appeared in 1923 and eventually sold over fifteen thousand copies. By the time of its writing he had come to regard his previous publications as "mere trial efforts" designed to elicit criticisms and corrections from his opponents. After honing his arguments for two decades, he finally felt ready to offer the world a systematic treatise that would serve as the basis for reforming the science of geology. Although he devoted some space in *The New Geology* to pointing out alleged logical and methodological fallacies in orthodox geology, overall he assumed a more positive stance than usual, hoping thereby to relieve himself of the charge that his work was "wholly iconoclastic."[23]

The New Geology, like so many of Price's previous works, featured the Genesis flood as the central geological event in the history of earth. It also highlighted what the author saw as his most significant contribution to geology: his "great" law of conformable stratigraphic sequence, which "alone is quite sufficient to relegate the whole theory of organic evolution to the lumber room of science, there to become the amusement of the future students of the history of cosmological speculation." He offered as an alternative to uniformitarian geology what he called the "new catastrophism," which was based on a single universal deluge as opposed to the old catastrophism of the French zoologist Georges Cuvier (1769–1832) and his disciple Agassiz, which entailed a series of catastrophes spread over immense periods of time. The evidence for such an event—particularly the mummified "Siberian elephants" that had frozen so suddenly their flesh could still be eaten—proved, he argued, "that a once beautiful world, well stocked with an amazing variety of plants and animals, was at some time in the long ago overtaken by a sudden and horrible world convulsion, the results of which we now have spread out over all the continents, in the form of the major stratified deposits."[24]

Price firmly believed that the evidence for a worldwide flood was "as well established as is the destruction of Carthage or the burning of Moscow," but because the specific cause of this catastrophe could only be surmised, he devoted little attention to the subject. He did, nevertheless, speculate in *The New Geology* and elsewhere that the cause of the deluge might have been "something of the nature of *a jar or a shock from the outside,* which would produce an abnormal tidal action, resulting in great tidal waves sweeping twice daily around the earth from east to west, this wave traveling 1,000 miles an hour at the equator." If this unidentified "external force" had knocked the earth's axis 23½ degrees from its original position perpendicular to the plane of its orbit, as he suspected, then "there would be forces let loose on the earth's surface sufficient to do an inconceivable amount of geological work."[25]

Because Price never swayed from his conviction that the idea of successive geological ages represented the crux of the evolution problem, he devoted comparatively little attention to biological questions. "What is the use of talking about the origin of species," he asked as early as 1902, "if geology can not prove that there has actually been a succession and general progress in the life upon the globe?" Besides, the development of Mendelian genetics, which seemingly allowed for only "definite and predictable" variations, had so eroded confidence in the efficacy of natural selection, Price felt he needed only to write the "funeral oration" for Darwinism. As he observed in *The Phantom of Organic Evolution* (1924), his most extensive critique of the *biological* arguments for evolution, "A dead lion needs no bullets." Price sat in the audience in 1921 when the English biologist William Bateson confessed to the American Association for the Advancement of Science:

> We cannot see how the differentiation into species came about. Variation of many kinds, often considerable, we daily witness, but not origin of species. . . . Meanwhile, though our faith in evolution stands unshaken, we have no acceptable account of the origin of "species."

Convinced that he had just heard "the swan song of Darwinism," Price, like other antievolutionists, immediately put Bateson's words to effective polemical use.[26]

Price had little quarrel with evolutionists regarding the origin of *species,* as the term was commonly used by biologists. Although he preferred to equate "species" with the originally created Genesis "kinds," and in later life regretted that he had conceded so much to evolutionists, he at times freely admitted that new species, narrowly defined, had evolved from "the great stocks, or families," created by God—and at a rate far more rapid than most evolutionists demanded. The more variation he allowed, the easier it was for him to sidestep "a real difficulty," namely, explaining "how the great diversity of our modern world may have come about after the world disaster of the Deluge, from a comparatively few kinds which were salvaged from that great cataclysm." In 1925 he justified his views in a revealing open letter to Adventist science teachers:

> Personally, I believe that these great family types are the ones that were originally created, and that a false issue has been raised over the "origin of species." . . . I think it is quite reasonable to suppose that all our cats are of one stock, that all our cattle are of a common origin, and that all the dogs and wolves may be of a common descent. To suppose this is only to suppose something which helps us to see how the great diversity around us may have come about from a comparatively few original stocks which sur-

vived the great world disaster which the Bible and a rational geology alike declare has actually taken place.

The point at issue between creationists such as Price and evolutionists was not variation but its extent and direction: "whether the general run of these changes have not all been in the direction of degeneration, not development."[27]

As so often was the case, Price found guidance in dealing with the species question in the writings of White, who in 1864 had written:

Every species of animal which God had created were preserved in the ark [of Noah]. The confused species which God did not create, which were the result of amalgamation, were destroyed by the flood. Since the flood there has been amalgamation of man and beast, as may be seen in the almost endless varieties of species of animals, and in certain races of men.

Unlike most early Adventist apologists, who interpreted the ambiguous "amalgamation of man and beast" to mean interbreeding *between* man and beast, Price insisted that White really meant "the mixing of races of mankind and the crossing or hybridizing of races of animals which God never meant to mix or cross." Though understandably reticent to discuss White's views in his published works, especially those intended for a non-Adventist audience, he privately subscribed to the common Adventist belief that Satan himself, "the great primal hybridizer," was "the real instigator of all the mixing and crossing of the races of mankind, and also the mixer of thousands of kinds of plants and animals which God designed should remain separate." Thus Price substituted demonic manipulation for natural selection in order to explain the origin of species.[28]

Publicly, Price tended to attribute the appearance of new species to environmental influences and *divine* intervention. In the chapter "Species and Their Origin" in *The Phantom of Organic Evolution*, he outlined a theory of common descent with modifications characterized by a process of "degeneration downward" rather than "development upward" and by quick rather than slow change. The most rapid change had probably occurred in the immediate post-flood period, when both humans and animals had encountered a novel environment. In a textbook for high-school students, he explained the process:

Very radical changes in the environment of plants or of animals tend to make the species vary; and if they survive in such new environments, their size or color or habits, or other physical "characters," will be different. This is the reason why the living forms are so different, in some instances, from their ancestral forms found as fossils in the rocks; for there has been a very radical change in their environment in passing from the antediluvian world to the modern one. Some species have changed so considerably that scientists do not recognize them as the same, but give them different specific or even different generic names, calling the older form "extinct," and the modern form a "new" species.

New postdiluvian species arose not so much by means of Darwinian natural selection or by a Lamarckian inheritance of acquired characteristics but because "the great superintending Power which is over nature, adapted these men and these animals and plants to their strange world." In accounting for the geographical distribution of plants and animals, which he admitted was a "very difficult" problem for creationists because nearly identical environments did not always possess the same flora and fauna, he similarly appealed to divine intervention, speculating that after the deluge animals spread out from Mount Ararat "under the direct guidance of the Creator."[29]

In explaining the origin of human races, Price variously cited the influence of the post-flood environment, the divine confusion of tongues associated with the Tower of Babel, and the negative results of interbreeding. In one of his earliest musings on the topic, he poetically traced the human family tree, emphasizing in one verse the climatic conditions that gave rise to the Negro race:

> The poor little fellow who went to the south
>   Got lost in the forests dank;
> His skin grew black, as the fierce sun beat
> And scorched his hair with its tropic heat,
>   And his mind became a blank.

In *The Phantom of Organic Evolution,* he observed that racial mixing in violation of God's intentions had led to rapid degeneration after Babel, producing not only Negroes and Mongolians but perhaps apes as well, which he thought might be "degenerate or hybridized men." The Heidelberg, Neanderthal, and Piltdown fossils, which physical anthropologists identified as human ancestors, Price saw as "degenerate offshoots which had separated from the main stock."[30]

Because of Price's unorthodox views and abrasive personality, it would be tempting to dismiss him as a mere charlatan who sought fame and fortune. That would be a mistake. True, he was at times arrogant, opinionated, mean-spirited, and prone to make extravagant claims. Perhaps he was even delusional. Certainly he understood the condition well, as he revealed in a condescending letter to a critic that unintentionally described his own personality:

> I have been quite a student of the psychology of controversial tactics. I know very well the facility with which the human mind tricks itself into believing what it wants to believe, until black becomes white and white becomes black. And persons who have never indulged very much in introspective analysis of their own motives, who have large self-confidence and have never had any doubts about their own correctness and their own fallibility, are very susceptible to self-delusion when it comes to dealing with the supposed "facts" of either history or science.[31]

Price may have been a fanatic, but he was no fool. Even the severest critics among his personal acquaintances never questioned his intelligence and integrity. If we hope to understand why he fought so tirelessly against evolution, often at great personal sacrifice and against tremendous odds, we must try to see evolution—and the world—as he saw them.

Price sincerely believed that evolution was scientifically unsound, but he admitted from the beginning that his primary objections to the theory were "philosophical and moral." Like other fundamentalists, he repeatedly warned that evolution not only made a shambles of Christian theology and ethics but threatened political freedom as well. The scope of the threat is conveyed by a query with which he began one of his antievolution tracts: "Do you know that the theory of evolution absolutely does away with God and with His Son Jesus Christ, and with His revealed Word, the Bible, and is largely responsible for the class struggle now endangering the world?" In two of his books, *Poisoning Democracy* (1921) and *Socialism in the Test Tube* (1921), he explicitly linked evolution and "Marxian" socialism. With Bryan, he thought that World War I, during which Germany put "the ruthless ethics of Darwinism . . . into actual practice," provided ample evidence of the threat evolution posed to human freedom.[32]

If Price shared the concerns of many fundamentalists regarding evolution, he also, by virtue of being a Seventh-day Adventist, had distinctive reasons for defending creationism. As a committed Adventist, he believed that he was living literally in the last days of earth's history, that in all likelihood he would personally witness Christ's Second Coming. His belief in the timing and reality of this event rested in part on a concrete reading of the prophecies found in the last book of the Bible, the Revelation of John. And he professed not to see how a person could believe in "the imminence of the events predicted in the last chapters of the Bible, when he has lost all confidence in the history recorded in its first chapters."[33]

Price attached special significance to a prophecy found in 2 Peter 3:3–7 alluding to events in "the last days." Peter predicted that there would be "scoffers," ignorant of Noah's flood, arguing that "all things continue as they were from the beginning of the creation." To Price, the language unambiguously identified uniformitarian geologists as the culprits. "Personally, I do not know of any prophecy in the Bible that impresses me more strongly as an accurate description of the present situation," he wrote. "This is a picture of *the scientific situation* just before the second coming of Christ, and it plainly tells us that false ideas about the past history of the world would be used as an excuse for disbelieving the doctrine of the second coming."[34]

As previously mentioned, the very identity of Seventh-day Adventists hinged on the conviction that the saints who greeted Christ at his Second Coming would be observing the seventh-day Sabbath in harmony with the fourth commandment found in Exodus 20:8–11:

> Remember the sabbath day, to keep it holy. Six days shalt thou labor, and do all thy work: But the seventh day is the sabbath of the Lord thy God: in it thou shalt not do any work. . . . For in six days the Lord made heaven and earth, the sea, and all that in them is, and rested the seventh day: wherefore the Lord blessed the sabbath day, and hallowed it.

The Sabbath doctrine seemed to demand a literal creation week, for, as Price cogently argued, if a person "does not believe that there ever was a real Creation at some definite time in the past, how can we expect him to observe the Sabbath as a memorial of that event, which in his view never occurred?" Thus, in his opinion, "the very logic of their belief in the Sabbath as the divine memorial of a literal creation" saved Adventists from falling victim to the evolution heresy. It also ruled out the popular notion of "creation on the installment plan," that is, creative acts interspersed over millions of years, which Price regarded as a "burlesque" of creation. For him, true creationism involved only "*one act of creation,* which may easily be supposed to have included all of those ancestral types from which our modern varieties of plants and animals have been derived."[35]

Confidence in the inspiration and authority of White, who warned of the insidious effects of evolution, played a key role in convincing Price of the correctness of his views. Despite occasional difficulties understanding her scientific statements, particularly those dealing with "amalgamation," he seems never to have doubted her word. In 1905, when his younger brother, Charles, questioned White's claims to inspiration, George advised:

> It is well to remember, Charlie, that Abraham is commended, not because he had things demonstrated for him, but because he "believed God." And if we withhold our assent to the truth of God [as revealed in White's writings] until every possible objection is mastered the plagues will fall upon us unprepared. We have a solemn and a glorious truth for these days which is either the *only* thing worth knowing or living for, or the whole thing is just a piece of clap-trap and old women's yarns.

George himself freely credited White with saving him from succumbing to evolution. In a typical statement addressed to an Adventist audience in 1927, he said: "Every thinking man among us must acknowledge that our safety and immunity in this respect is due to our faith in the simple Bible narrative, supplemented by the writings of Mrs. E. G. White."[36]

Price's success as an internationally known spokesman for creationism unquestionably fulfilled a craving for public recognition, but if he harbored any illusions of growing rich through his writings, they were quickly dispelled. To publish *Illogical Geology* in 1906, he incurred such considerable debt that it remained unpaid until 1918, when a Lutheran educational organization offered him the largest paycheck of his life: $1,200 to write on geology and astronomy for a series of home-study texts. All of the proceeds from *The New Geology* went directly to the Adventist church as reimbursement for his salary while writing it. He rarely made more than $50 in royalties a year, and much of this income went to pay for his extensive correspondence. *The Phantom of Organic Evolution,* which appeared shortly before the Scopes trial and benefited from the excitement surrounding it, brought him $400, but that was a fluke. Price wrote not for gold but for glory and God, believing that he had been divinely chosen for a special work. Describing his experience in Tracadie when he first resolved to devote his life to combating evolution, he wrote: "I felt a real call to do the job, as though someone had tapped me on the shoulder and told me to do it." In his view, rejecting such an entreaty would have jeopardized his chances for eternal life.[37]

Though hypersensitive to slights, Price made little effort to shield himself from scientific criticism. On the contrary, at times he eagerly solicited it, preferring to be insulted rather than ignored. Through the years he corresponded with a number of orthodox scientists and "buttonholed dozens of 'big' geologists" when opportunities arose. In 1906 he distributed hundreds of complimentary copies of *Illogical Geology*, many to scientists, requesting their comments and criticisms. "By far the most candid and thoughtful" response came from David Starr Jordan (1851–1931), president of Stanford University and the leading American authority on fossil fishes. Price, titillated by the risk of sparring with a superior opponent and grateful to be taken seriously, promised on one occasion "to become an evolutionist within twenty-four hours" if "the foremost ichthyologist in the world" could prove to him that one fossil was older than another. Off and on for nearly twenty years Jordan labored with Price to help him see that his case against geology was "based on scattering mistakes, omissions and exceptions against general truths that anybody familiar with the facts in a general way can not possibly dispute." Impressed by his correspondent's obvious intelligence and lawyerlike mind, as well as by his ignorance of geology, Jordan repeatedly but unsuccessfully urged him to "undertake some constructive work in Paleontology in the field and in laboratories."[38]

Price, an armchair scientist, understandably felt insecure about his lack of formal scientific training and his limited familiarity with the evidence he was disputing. Though he never claimed "to be an expert field geologist or a scientific collector," he nevertheless made the most of the little direct knowledge he did possess. He liked, for example, to claim that his experience selling books in eastern Canada and in southern Colorado, where he had spent one summer while in college, gave him invaluable "firsthand knowledge of field geology." On more than one occasion he actually visited Alberta to view the famous overthrust near Banff and found that it looked "perfectly natural and show[ed] no physical signs of abnormal conditions." In 1926, while he was on appointment to England, the Adventist church paid for him to spend a summer in the Swiss Alps investigating examples of "upside-down" strata. "Naturally you do not wish to be caught in any statements that you cannot hold to in making declarations regarding special geological formations," wrote the empathic president of the church. Still, Price's familiarity with the outdoor world remained rudimentary at best. In 1937, shortly before his retirement, some of his students at Walla Walla College invited him to accompany them on a fossil-hunting trip to the John Day country in Oregon. To their amazement, they discovered that the famed creationist could scarcely tell one fossil from another.[39]

Price's insecurity about his inadequate education led him to downplay the importance of formal training and advanced degrees. He argued defensively that "even a study of elementary logic, and of logical processes of thinking, is of far more importance than is a mere training in expert experimentation or in field observation." In view of the error that permeated modern science, he considered it a virtue that he had never been infected with the disease of "university-itis." As for graduate training, he professed to have more admiration for persons who "felt a divine call" than for ones who subjected themselves for years to a stultifying experience in order to gain that glittering prize, a Ph.D. degree. In truth, Price greatly valued academic credentials. Well into his fifties he toyed with the idea of enrolling in a university, and to the end of his life he wondered whether he had made the right decision in not obtaining an M.D. while teaching at the College of Medical Evangelists. When the medical school awarded him a B.A. in 1912, based in part on his "Authorship" and independent study, he proudly displayed the letters on the title page of his next book, *The Fundamentals of Geology*, which identified him as "Professor of Geology." And during the last forty-five years of his life, he often tacked an M.A. onto his name, a gift from the Adventist Pacific Union College.[40]

More than any other creationist of his time, Price concerned himself with questions of scientific method. In his early writings he expressed a particular fondness for the empiricism of Bacon and Newton, to whom he dedicated *The Fundamentals of Geology*. To bring about the desired reforms in earth history, he proposed a "Baconian study of geology" that would start with knowledge of the surface rocks and work downward. Although convinced that "inductive geology" pointed to a literal recent creation, he recognized that discussions of the Creation properly lay outside of science, "across the boundary-line in the domain of philosophy and theology." In harmony with his populist view of science, Price liked to point out that the vaunted "scientific method" involved nothing more than the application of "enlightened common sense." He argued in Baconian fashion that true science began with the "patient gathering of facts," not "shameful speculation." But he was sophisticated enough from having read Alfred North Whitehead (1861–1947) and other philosophers to realize that there were no pure "facts" unadulterated by interpretation. And he was clever enough to use this knowledge to attempt to place creation on an equal footing with evolution. Just as "geologists and paleontologists [looked] at their facts through the colored spectacles of Darwin and Lyell," so, too, did creationists view the natural world through the lens of Scripture. Because the value of a theory depended on its ability to give the most "reasonable explanation of the facts of nature" rather than on its pedigree, Price "cheerfully" admitted that the creationist explanation of origins was "suggested by our religion, and that it probably could never have been wholly worked out even as a hypothesis by the study of nature alone." In choosing between "the two alternatives now before the world," evolutionary geology versus world-catastrophe, there was only one legitimate question to ask: *"Will it work?"*[41]

Price's most acrimonious encounters with the scientific establishment occurred in the 1920s, when he twice became embroiled in battles with James McKeen Cattell (1860–1944), editor of *Science*. The first episode began in 1922, when Arthur M. Miller (1861–1929), a geologist at the University of Kentucky, published an article in *Science* titled "The New Catastrophism and Its Defender." In it he took Price to task not only for contributing to the recent attacks on evolution but for masquerading as a geologist even though he published primarily in the religious press and was "a member of no scientific body and absolutely unknown in scientific circles." Upon reading these disparaging remarks, Price dashed off an angry letter to the editor threatening "action for libel" but expressing a willingness to settle for the opportunity to rebut Miller. Cattell offered to correct any errors of fact but declined to print Price's geological views on the grounds that they "would not be of interest to scientific men." Denied access to the pages of *Science*, Price turned to the more sympathetic *Sunday School Times*, which, he reminded Cattell, had 200,000 intelligent readers. Writing with scarcely concealed fury, Price accused the editor of *Science* of being animated by "pseudo-scientific bigotry" and defended his own honor by pointing out that he carried membership in both the American Association for the Advancement of Science and the California Academy of Sciences. "The secret of the whole matter," he explained to his nonscientific audience, "of course, is that the Evolution doctrine has become a religion with these men, a violent, anti-Christian religion."[42]

No sooner had this quarrel died down than Cattell received an unsolicited review of Price's *New Geology* written by Charles Schuchert (1858–1942), a Yale geologist upset by the appearance in Price's book of thirty-two illustrations apparently taken from a text Schuchert had coauthored—and after Schuchert's publisher had pointedly refused Price permission to borrow so many pictures. "The man needs showing up and as widely as possible among lovers of science," wrote Schuchert, who in his review portrayed Price as "a fundamentalist harboring a geological nightmare." Cattell, still skittish from his recent fight with Price, begged off, later explaining that he did not want to criticize Price publicly because it was "undesirable either to print or to decline to print any reply that he may make." But Schuchert persisted, saying that he had talked the matter over with David Starr Jordan and others and they all agreed that the nonscientists fighting the fundamentalists needed "to have ammunition supplied them by us who know the facts." Under such peer pressure Cattell relented—only to face, as he had feared, another barrage of angry letters from Price, who this time was in no mood for compromise. "An apology and retraction now would do no good; I am through with this sort of foolery," Price fumed. "I will see what the law can do for me." There is no evidence that he followed through on his threat to take legal action, but for decades thereafter he protested his innocence against the charge of photographic plagiarism that blotched his otherwise spotless reputation for honesty.[43]

What actually happened remains a mystery. We know that Price bought a copy of Louis V. Pirsson and Charles Schuchert's *A Text-Book of Geology* (1920) shortly after its publication and finished "it all in 9 days consecutive reading—except Sabbath." Years after the event Price claimed that he had made a special transcontinental trip to obtain his illustrations from the same original sources that Schuchert had used. But then why did Price even bother to ask Schuchert's publisher for permission to use the illustrations if he had obtained them independently? And why did Price's publisher, the Adventist-owned Pacific Press, tell an inquiring pastor, "We have permission for every illustration in 'New Geology' from the publishers of these books, and we have it in written form in our vault at the present time. . . . The Pacific Press did not go to Schuchert for permission on these illustrations, because he did not own the copyright. They were owned by the publishing concern."[44]

Early in his battles with *Science* Price justified his practice of publishing in religious magazines by alleging that "strictly scientific journals" had not been open to him. He agreed that his critiques of thrust faults and deceptive conformities "ought to be discussed in a purely scientific way" but went on to note that "if this cannot be done through the orthodox scientific channels, I do not think that I ought to be blamed for allowing other editors to publish what I have to say." Price's language clearly implied that scientific journals had unfairly ostracized him, but at least some evidence suggests that he himself sought to circumvent the scientific community by taking his case directly to the people. Except for one incident in 1919, when the editor of *Scientific American* graciously declined to publish a lengthy manuscript, "The Largest Problem in Geology," there is no evidence before the late 1930s that Price even submitted his work to orthodox scientific publications. In fact, until the *Science* imbroglio he never complained of censorship. On the contrary, in 1917 he boasted of sharing with the public results "most scientists would feel obliged to reserve for the select few of some learned society, to be published subsequently in the Reports of its 'Transactions,' and to find their way after years of delay into the main currents of human thought." As late as 1924 he gave similar reasons for bypassing the scientific establishment:

> Over twenty years ago, when first making some of my geological researches, I came face to face with the question: How shall I publish my discoveries? If I wait for the dilatory methods of "orthodox" scientific pedantry, through the media of papers before scientific societies, etc., I will be grayheaded before I can force recognition of these discoveries. On the other hand, if I publish in some "unorthodox" (scientifically) way, such as through the popular or religious journals, I will very likely be boycotted by the standpat scientists, and almost turned out of the scientific synagogue.
>
> I chose the latter method; and the result has been as might have been expected. But I have never regretted my choice; for I believe a much greater number of people have been benefited by my publications.[45]

It was not until after the rebuff from Cattell that Price began crying foul, claiming (without documentation) that he had been "blackballed" by scientific societies and denied access to scientific journals by the "self-appointed censors of modern scientific discussion." Only then did he begin to say that, as a young man, he had first tried scientific journals before turning to the religious press. He blamed his exclusion in part on the "overspecialization" of modern science, which had created a situation in which "each specialist knows that in his own department the old ideas have broken down; but he still supposes that in the other departments the general theory is still being confirmed and fortified by modern researches." This arrangement allowed "a power-ful clique of standpat reactionaries" to control all of orthodox science and repress "anything but the most approved conventionality." The refusal of the Zoological Society of London to publish a scientific article by the English creationist Douglas Dewar (1875–1957) fueled Price's suspicion that evolutionists were out to muzzle scientific heretics.[46]

Price's first—and only—article in a mainstream scientific journal, the *Pan-American Geologist,* did not appear until 1937, about the time of his retirement. Apparently encouraged by the success of another deluge geologist in placing articles in this journal, Price submitted "Methodology in Historical Geology," in which he once again exposed "the logical atrocities of deceptive conformities and thrust-faults" and pushed for the acceptance of his new catastrophism. His radical sug-gestions for restructuring the science of geology provoked little public reaction, except for a mild note of dissent from the editor, perhaps written to deflect anticipated criticism for providing Price with a sci-entific platform. But when the journal folded a few years later, some geologists privately attributed its demise to the editor's poor judgment in publishing Price.[47]

Especially in later life Price complained that scientists did not take him seriously enough even to respond to his challenge. "Why this long delay?" asked the frustrated crusader in his seventieth year. "Why has not some competent university teacher taken up the task?" Perhaps most scientists concurred with the opinion of the Bryn Mawr geologist-geographer Malcolm H. Bissell (1889–1975) that Price's arguments were "really not worth answering," and some may have shared Bissell's suspicion that there was "something wrong with [Price's] mental pro-cesses." But a surprising number of reputable scientists, including Jordan and Bissell himself, who corresponded with Price for months before withdrawing in disgust, did attempt to refute the new geology. In 1925, for example, the European-trained geologist-priest Stephen Richarz (1874–1934), upset by the publicity his fellow cleric George Barry O'Toole had given to Price, published a detailed rebuttal to Price's arguments. With regard to his interpretation of overthrusts, the main pillar supporting the new catastrophism, Richarz wrote:

> There is not a single instance of fossils in the "wrong order" which cannot be accounted for by overthrusts or overturned folds. . . . It is false to say that geologists postulate the great overthrusts in order "to explain away 'wrong sequences' of fossils." . . . Over-thrusts have been found quite independently from fossils; they can be seen even in Precambrian formations, in which there are

no fossils at all. Thus the formidable argument of Mr. Price against the sequence of fossils, as unanimously accepted by all geologists, breaks down completely. By such phrases as "recent discoveries," "quite new," Mr. Price can deceive only those who are strangers in the science of geology.[48]

Price also received criticism from various mainstream scientists whose opinions he solicited. For example, George H. Ashley (1866–1951), the state geologist of Pennsylvania, who estimated that he had seen as many normal sequences of strata as he had fence posts, assured Price that geologists had made no mistake. Because every departure from the normal sequence of fossils could be "easily and readily explained" by direct observation in the field, he dismissed Price's flood explanation as "all bunk." Even the zoologist Austin H. Clark, himself a critic of Darwinism, assured Price that the evidence for deceptive conformities and overthrusts was unimpeachable. Virtually the only positive scientific reaction to Price's views came from William Bowie (1872–1940), chief of the Division of Geodesy of the United States Coast and Geodetic Survey, who applauded Price's critique of overthrusting. "I have always felt that there might be some doubt as to the reliability of the geological sequence of strata," he admitted to Price, adding that his lack of expertise in paleontology had kept him silent in the past.[49]

From time to time Price's own followers raised perplexing questions about flood geology. The editor of *Bibliotheca Sacra*, Melvin G. Kyle (1858–1933), who credited *The New Geology* with throwing "a wrench into the smooth running machinery of the evolutionary theory of world making," wondered why, when fossils were found in the wrong order, they were always in exactly the *reverse* sequence of that predicted by geologists. He professed to have no trouble accepting the possibility that strata might be found out of order, "but that such variation would produce the whole series of life in regular reverse order is rather too great a strain upon credulity." Thus backed into a corner, Price admitted that "in the majority of cases the fossils do conform fairly well to the familiar scheme"—"a very vital admission," noted Price's Adventist protégé Francis D. Nichol (1897–1966), who himself had a question or two. "It has always been something of a perplexity to me," confided Nichol, "to visualize just *how* amid the turbulence of the Flood and the period of subsidence immediately following, there would be sufficient opportunity for one stratum to take definite and distinctive shape and mold before another was hurled upon it by the restless ebb and flow of the waters." Nichol particularly pressed Price to explain the formation of the Grand Canyon: "Let's have the worst before us when we're dealing with the enemy, and if we perish, we perish!"[50]

On occasion Price found himself defending not only the geological but the scriptural validity of his flood geology. The amateur historian of science and prep-school science teacher Edwin Tenney Brewster (1866–1960) especially enjoyed pestering Price with questions about the compatibility of the the new catastrophism with the Old Testament. How, wondered Brewster, could the author of Genesis, writing about the antediluvian world *after* the deluge, refer to familiar geographical landmarks if, as Price claimed, the present-day topographical features of the region had resulted from the flood? If the Holy Land rested on stratified rocks deposited during the flood, one would expect the pre- and post-flood geography of the region to differ markedly; but Genesis refers to "the same rivers and plains, the same wildernesses, the same 'Mountains of Ararat,' that Noah saw before the Great Flood." Seen in this light, wrote Brewster, "the Prician hypothesis flatly contradicts the Bible."[51]

Brewster also liked to twit Price for invoking miracles when stumped to explain such occurrences as the preservation of the ark during the flood and the destruction of human remains afterward. "In view of the fact that Professor Price offers his Flood Theory as a *scientific* hypothesis," asked Brewster, how could he justify any miraculous explanations? Price, who recognized the problem from a scientific point of view, nevertheless defended his appeal to miracles:

> If I didn't believe in a God Who is running this universe, and Who has always been running it in the past, and if I were out to defend through thick and thin the naturalistic doctrine of "uniformity," I suppose I never could be brought to admit the possibility of a great world Catastrophe, even though such a theory might bring in a great simplification into many of these problems. But you can't scare me with the bugbear of "miracle." You will have to think up something else.[52]

INFLUENCE

Despite Price's missionary zeal and prolific pen, it took over fifteen years for his peculiar interpretation of Genesis to attract much attention. Symbolic of the lack of interest in his work was the fact that no publisher wanted his second book, *Illogical Geology*, which he finally printed at his own expense. To fellow Adventists, who needed no convincing on the subject of evolution, his earnest crusade must have appeared quixotic. As late as 1916, by which time Price had already written five books on the subject, an Adventist editor, considering still another of Price's manuscripts for publication, told a colleague that until recently he had tended "to make more or less of a joke" of Price's work. That his own church, which operated a number of colleges, employed him largely for manual labor and high-school teaching until 1920, on the eve of his fiftieth birthday, and later had to bribe colleges to hire him, making him feel like a "charity" case, tells much about the prophet's lack of honor.[53]

Early fundamentalist leaders scarcely noticed Price's crusade, in part because issues other than evolution seemed more pressing during the first two decades of the century but also because of what many evangelical Christians regarded as Price's cultish Adventist theology. The complimentary copies of *Illogical Geology* that Price gave away to influential religious figures brought little public recognition. One of the recipients, A. C. Dixon, later became an admirer of Price's, but at the time he was lining up authors for *The Fundamentals*, he invited the better-known Wright, not Price, to write on science and religion. Wright, who had grown up around Adventists and for a time had lived in the Miller homestead while teaching William Miller's grandchildren, seems to have played a key role in denying Price an early hearing among fundamentalists. In 1910 the editor of the *Bible Student and Teacher*, published by the Bible League of North America, sent Wright galley proofs of an article by Price titled "Evolution Revolutionized; or, A New Line of Argument against the Evolution Doctrine of Origins," requesting a brief response to accompany the essay. The article, despite having already been set in type, never appeared—apparently because of the severity of Wright's criticisms.[54]

Price's stock among non-Adventist fundamentalists rose rapidly with the publication in 1917 of his *Q. E. D.; or, New Light on the Doctrine of Creation*. Brought out by the Fleming H. Revell Company, founded by Dwight L. Moody's brother-in-law, this book helped to break down sectarian barriers and prompted William Bell Riley to invite Price to attend a meeting being organized in Colorado Springs for fundamentalist leaders. There he met Charles G. Trumbull (1872–1941), editor of the *Sunday School Times*, and W. H. Griffith Thomas (1861–1924), a prominent Anglican evangelical from Toronto, both of whom subsequently promoted Price's new catastrophism. Although Riley himself objected to the traces of Adventism that sometimes crept into Price's works, he became one of Price's biggest boosters, publicizing his books from pulpit and press and introducing him to readers of his own magazine as "one of the real scientists of the day [whose] writings are destined to profoundly influence the thinking of the future."[55]

By the late teens Price's byline was appearing with some frequency in fundamentalist publications, and Price was starting to move freely in non-Adventist circles. However, it was not until the early 1920s, when the evolution question moved to the top of the fundamentalist agenda, that Price's reputation as a scientific celebrity began to soar. Virtually the entire directorship of the fundamentalist movement lauded his contributions. Fort Worth's fiery J. Frank Norris assured fellow creationists that Price's geological work "utterly demolishes Evolution." John Roach Straton, pastor of the Calvary Baptist Church in New York City, hailed *The New Geology* as a "great and monumental" contribution from "a thoroughly up-to-date scientist." Harry Rimmer, Price's chief scientific rival among fundamentalists, called it "the most remarkable and up-to-date book of Geology extant today . . . a masterpiece of REAL Science." Arthur I. Brown ranked Price among "the world's leading Geologists" and characterized *The New Geology* as "undoubtedly, the sanest, clearest and most irrefutable presentation of the Science of Geology from the standpoint of Creation and the Deluge, ever to see the light of day." S. James Bole, who used the book as a supplementary text in his geology course at Wheaton, found Price's logic particularly "clear and compelling." In view of the developing controversy over evolution, Leander S. Keyser (1856–1937), reputed to be the "chief exponent of Lutheran orthodoxy in America" during the 1920s, could not help thinking that Price, like the biblical Queen Esther, had "come to the kingdom for such a time as this."[56]

And so it seemed—especially in view of the upcoming Scopes trial, which pitted the fundamentalist Bryan against the "infidel" attorney Darrow. Price had first met Bryan after one of the Great Commoner's lectures in Lodi, California, where Price was teaching after World War I, and by the early 1920s the two were corresponding from time to time. As the date of the trial approached, Bryan's advisors recommended that Price be recruited as an expert witness. Bryan, addressing Price as "one of the outstanding scientists who reject evolution," invited the flood geologist to come to Dayton at Price's own expense. Price, then teaching in England, sent his regrets with advice for Bryan to avoid scientific issues. Like Bryan, Price supported the campaign to outlaw the teaching of organic evolution in public schools because he thought it un-American to compel a parent to pay taxes "to have his child taught something that he utterly repudiates and considers anti-Christian." But Price also regarded the teaching of creation as unconstitutional. "If Genesis were taught in the public schools," he reasoned, "it would be an infringement upon the cardinal American principle of separation of church and state."[57]

During the trial Bryan gave a vague and confusing account of how Price's catastrophism supported creationism. He insisted that Price possessed "very good evidence" to dispute what most geologists said about the age of the earth—but then went on to admit that he himself accepted the antiquity of the earth and the day-age interpretation of Genesis 1. Price never forgave Bryan for what he regarded as a treasonous performance. Eager to avoid any personal responsibility for the debacle at Dayton, Price in later years forgot that he had urged Bryan to avoid scientific topics and imagined that he had tutored the attorney on how to confute the evolutionists with flood geology. Instead, complained Price, Bryan "conceded the entire geological argument to the evolutionists, with the pitiful results now known to all the world. No wonder he died a sad and disappointed man. But his is a good example of the outcome of every form of argument which does not take the record of Genesis at its face value." The man Price had once praised as having "mastered the essential scientific aspects" of the evolution question better than any other layman he knew, he now dismissed as knowing "almost nothing about the scientific aspects of geology, and the little he thought he knew was wrong."[58]

In failing to see the incompatibility between his own and Price's notions of earth history, Bryan had lots of company. Not one of the fundamentalist leaders who praised Price's attacks on evolution seemed to recognize that flood geology wiped out the scientific rationale for the traditional day-age and gap interpretations of Genesis. What Price offered as a mutually exclusive alternative to existing fundamentalist views, they unthinkingly accepted as simply another defense of the biblical narrative that rejected modern evolutionary science. Abundant evidence supports this claim. In 1924 Price complained to Straton that it gave him "a pain, as the boys say," to hear his "good friend" Riley, a day-age partisan, concede the validity of the geological column, the truth of which was "by all means the strongest argument in favor of the evolution doctrine." Ironically, Straton himself had just publicly equated Price's position on the age of the earth with that of Dawson, a leading nineteenth-century promoter of the day-age view. Rimmer, Bole, and Brown, all gap men, continued to talk about ruin and restoration before Adam while quoting Price against the evolutionists; and at least the first conceded the local extent of Noah's flood, hardly the basis for constructing a new geology. In trying to explain how rocks "ONE HUNDRED MILLION YEARS OLD (?) ARE LAID DOWN ON TOP OF ROCKS ONLY ONE MILLION YEARS OLD," Rimmer, thinking he was following Price's explanation of thrust faults, suggested that they were overturned during the deluge. Had he read Price more carefully, he would have learned that the stratified rocks had all been *formed*, not merely *overturned*, during the flood. And so it went, with Price becoming increasingly famous while his flood geology was misunderstood.[59]

In the years after the Scopes trial and especially after Price's return to America from England in 1929, he emerged as one of the two most popular scientific authors in fundamentalist circles, the other being Rimmer. In addition to appearing regularly in Adventist magazines, his prose frequently graced the pages of the most widely read fundamentalist periodicals, such as *Moody Monthly, Sunday School Times, Defender, Bible Champion,* and *Christian Faith and Life;* and it occasionally ran in such places as the *Catholic World, Princeton Theological Review,* and *Bibliotheca Sacra.* In 1928, "after several earnest solicitations," Price agreed to serve as a contributing editor to the *Bible Champion;* three years later, when it merged with another journal to form *Christian Faith and Life,* he became editor of a department called "Current Scientific Discoveries." In announcing the latter appointment, the journal introduced Price as an "authority" in geology, the "chief exponent" of flood geology, which "many students of both the Bible and geology feel convinced . . . offers a more adequate explanation of the puzzles of science and religion than the former methods of trying to harmonize the Bible and science."[60]

Price, who came to see himself as one of the chief architects of fundamentalism, credited three authors in particular with helping popularize his flood geology: O'Toole among Roman Catholics, Floyd E. Hamilton (1890–1969) among Presbyterians, and Byron C. Nelson (1893–1972) among Lutherans. In *The Case against Evolution,* which Price regarded as "the authoritative Roman Catholic work against the evolution theory," O'Toole had indeed cited Price to illustrate the difficulties of using fossil evidence to determine the relative age of strata; but being more concerned with the biological than the geological side of the argument, O'Toole said little about the deluge and its supposed effects. Hamilton, a popular Presbyterian apologist and missionary to Korea, featured Price's theories in two of his books, *The Basis of Christian Faith* (1927) and *The Basis of Evolutionary Faith* (1931). In both works he discussed flood geology in detail but refused to say whether special creation took place in six twenty-four-hour days or over "six indefinite periods." Of the three, the only true disciple was the Lutheran pastor Nelson (whose views will be discussed in the next chapter). His *Deluge Story in Stone* (1931), a history of flood geology, recognized Price as the "one very outstanding advocate of the Flood" in the twentieth century.[61]

After the passing of a euphoric period following the discovery of the new catastrophism, Price never "kidded" himself with the "fantastic notion" that his system of flood geology would supersede evolution in the schools, but for years he did expect to convert evangelical Christians to his way of thinking. After all, as he liked to point out, it did "not take a Solomon to see that the theory of organic evolution becomes nonsense if the Flood theory is regarded as true." By the mid-1930s he was happily claiming victory, confident that his new geology was "the leading view now held by Bible believers in all parts of the world." As we shall see, this was hardly an accurate assessment, as even Price later admitted, but he was not alone in sensing a shift toward his way of thinking. One knowledgeable evangelical noted that in the late 1940s Price's flood geology seemed to have "infiltrated the greater portion of fundamental Christianity in America," and a few years later the Baptist theologian Bernard Ramm (b. 1916) observed with alarm the "staggering" extent of Price's influence. Flood geology, Ramm said, formed "the backbone of much of Fundamentalist thought about geology, creation, and the flood." Still, most fundamentalists at mid-century, like those during the previous twenty-five years, failed to see the implications of flood geology for life on earth before Eden.[62]

SIX

The Religion and Science Association

In the mid-1930s Dudley Joseph Whitney (1883–1964), one of the few non-Adventist advocates of Price's new catastrophism, observed that fundamentalists were "all mixed up between geological ages, Flood geology and ruin, believing all at once, endorsing all at once."[1] How, he wondered, could evangelical Christians possibly convert the world to creationism if they themselves could not even agree on the meaning of Genesis 1? To bring order to this chaotic state of affairs, he pushed for the establishment of a society to create a united fundamentalist front against evolution. The resulting Religion and Science Association (RSA), though short-lived, represented the beginning of a transition from the one-man crusades of people like George McCready Price, Harry Rimmer, and Arthur I. Brown to small-group efforts that sought consensus and satisfied a desire for community.

The appearance of the first creationist societies in the 1930s followed a pattern typical of fundamentalism generally. Until recently most historians have alleged that fundamentalism lost its vitality after the embarrassing Scopes trial in 1925. However, as Joel A. Carpenter and others have shown, the absence of media attention can be deceiving. After the 1920s, fundamentalists did indeed tend to abandon their efforts to banish modernism from the established churches and to outlaw evolution in public schools. But rather than surrendering, they turned their energies toward developing a separate institutional base from which to evangelize the world: radio ministries, colleges, and the all-important Bible institutes, the greatest of which was the Moody Bible Institute of Chicago.[2] With respect to evolution, they turned from lobbying state legislatures to pressuring local school boards; and, despairing of ever converting the scientific community to their way of thinking, they set about to create their own alternative societies and journals.

THE INNER CIRCLE

Although the idea for a creationist society originated with Price, its implementation resulted in large part from the "unlimited pep" of Whitney, a rancher and sometime farm-journal editor from the agriculturally rich San Joaquin Valley of California who raised figs, walnuts, and poultry. A Berkeley native, he attended the University of California and received a B.S. degree from its college of agriculture in 1907, having taken basic courses in chemistry, botany, and entomology. After college he briefly pursued a graduate degree in agriculture and taught school in Hawaii for several years before finally settling in 1915 on a small ranch in the Lindcove area, near the central California town of Exeter. His eccentric behavior and unconventional beliefs prompted locals to dub the lonely bachelor "the sage of Lindcove." A onetime disciple of Alexander Dowie (1847–1907), the turn-of-the-century faith healer who helped launch the Pentecostal movement in America, Whitney remained sympathetic to the teachings of Pentecostals though disdainful of their distinctive practices. He detested most fundamentalists. After trying various Methodist, Presbyterian, and Baptist churches, he eventually assisted in founding the independent Lindcove Community Bible Church, where for years he served as a deacon and Sunday-school superintendent. His curmudgeonly manner often masked his deep spirituality, reflected in yearly readings of the entire Bible and twice yearly readings of the New Testament.[3]

Like many other evangelicals, Whitney subscribed to the ruin-and-restoration interpretation of Genesis 1—until early in 1925, when an Adventist acquaintance loaned him a copy of Price's *Q. E. D.; or, New Light on the Doctrine of Creation*, which taught him for the first time "about upside down mountains and other geological nightmares." With the fervor that comes with conversion he took up Price's cause and threw himself into preaching the gospel of flood geology, particularly to unbelieving fundamentalists. He thought it preposterous that

William Bell Riley, a day-age man, was trying to succeed Bryan as "the anti evolutionist *champion* (Whew) of the U.S.A.," and he felt nothing but contempt for "the typical fundamentalist evolution fighter," who knew little of science and cared little about logical consistency. "I am completely disgusted with the ignorance and imbecility of the anti evolutionists with hardly an exception," he wrote to Price in 1927. "Is there anything we can do about it?"[4]

Indeed, there was. In the February 1928 issue of the *Bible Champion*, which Whitney regarded as "the leading (may be) anti evolution paper in this country," he called on fundamentalists to abandon the "old, discredited" interpretations of Genesis and embrace flood geology. He argued:

Unless this hypothesis is accepted by Fundamentalists, they might as well, in my opinion, drop all opposition to the theory of evolution. For, if they adopt any other scheme of geologic history, they either are helpless in explaining the facts of geology, or they inferentially accept a theory of biologic history which means nothing else but evolution.

The very next issue of the *Bible Champion*, published by the Bible League of North America, announced the appointment of Price as a contributing editor, thus giving him a prominent platform for witnessing to fundamentalists.[5]

Because of his exclusion, real and imaginary, from the mainstream scientific community, Price had for years longed to have his own creationist journal and society. About 1906 he seems to have briefly edited a monthly called the *Modern Heretic*, and following its failure, he tried unsuccessfully to convince Adventist leaders to back the publication of a magazine he proposed calling the *Creationist*. However, by the 1920s he had grown so "tired of ephemeral and abortive attempts to turn the world upside down," he rejected an invitation to serve as editor-in-chief of a proposed antievolution magazine. The cause, he argued, deserved nothing less than "a dignified, scholarly presentation of facts and arguments" and would only be discredited by something called the "Monkey Magazine" that featured the " 'Jocko-Homo,' 'Puddle-to-Paradise' style of argument."[6]

Whitney's enthusiasm for flood geology prompted Price in the mid-1930s to resurrect his proposal for the *Creationist*. When he broached the idea to Whitney, the irrepressible farmer countered with a suggestion that they first form a society to sponsor the magazine. Thus was born in 1935 the Religion and Science Association, apparently the first antievolution organization in America aimed at resolving scientific and hermeneutical problems rather than restricting the teaching of evolution. The primary purpose of the association, explained Price, was to find "a harmonious solution" to the exegetical discord within fundamentalism.[7] Both Whitney and Price, however, possessed ulterior motives for creating the RSA: Whitney viewed it as an opportunity to leave the isolation of his ranch for a career as a traveling lecturer, campaigning "the country for creation," while Price saw it as a means of promoting his forthcoming book, *The Modern Flood Theory of Geology* (1935). He went so far as to draft a publicity statement for release by his publisher, Fleming H. Revell, putting the RSA on record as recommending his book and endorsing flood geology as "sound scientifically and theologically, and by far the best and most reasonable explanation of the facts of the fossils and the rocks." To ensure that no one would misunderstand the implications of such an endorsement, he added: "The Science and Religion Association also place themselves on record as condemning and repudiating the only other possible alternatives about the fossils: (a) The Day-Age theory, which is false scientifically and cannot be made to harmonize with the record of Genesis I; and (b) The Pre-Adamic Ruin Theory, which makes nonsense of the scientific facts and is utterly fantastic theologically."[8]

Practically, the first task was to recruit a prestigious panel of officers. Because Price, as he confessed, stood "to gain more by such an organization than almost anybody else," he thought it best to maintain a low profile, at least at first. His preferred slate of officers called for L. Allen Higley (1871–1955) of Wheaton College as president; Byron C. Nelson, a Lutheran minister from Wisconsin as vice president; and Whitney, whose considerable energy would keep the organization going, as secretary-treasurer. All agreed to serve and promptly set about to appoint a five-man board of directors, with Price as chairman and Higley and Nelson as ex officio members. For the fourth and fifth slots the group initially selected J. Gresham Machen (1881–1937) and J. D. Eggleston (1867–1953), both distinguished fundamentalists; however, other commitments led both men to decline. Whitney expressed a readiness "to turn handsprings with delight," if a good Catholic scholar could be found, but the only known candidate, George Barry O'Toole, had left the country for China. Higley's suggestion to invite Jay Benton Kenyon (1885–1968), a biologist on the faculty of Asbury College, a Holiness school in Kentucky, generated little enthusiasm. Eventually the group enlisted the services of Theodore Graebner (1876–1950), a Lutheran theologian nominated by Nelson, and—to ensure that flood geologists constituted a majority—Harold W. Clark (1891–1986), a former student of Price's who taught biology at an Adventist college in northern California. This ecumenical body, the founders promised in an announcement in *Christian Faith and Life,* would "take a united stand against the theory of evolution all around us."[9]

The presence of two Adventists and two conservative Lutherans among the association's six officers in part reflects the commitment of these two religious groups to a hyperliteral reading of Scripture. If anything, Lutherans in the Missouri, Wisconsin, and Norwegian synods, on the right end of the Lutheran theological spectrum, surpassed even the concrete-minded Adventists in their devotion to the plain meaning of the Word. Well into the twentieth century, liberal Lutherans ridiculed the Missouri Synod for holding that the Bible taught Ptolemaic astronomy, which placed the earth at rest in the center of the solar system. Theodore Graebner's own father, August L. Graebner (1849–1904), a scientifically curious theologian, insisted that if the Copernican system disagreed with the Bible, "then the heliocentric system must fall." To the charge that "the Missouri Church unceremoniously casts all science to a side in so far as the latter is in conflict with what seems to 'Missouri' to be the teachings of the Scriptures," he offered a resounding "Yea and Amen."[10]

On the subject of heliocentrism, Theodore Graebner adopted a more tolerant attitude than his father, saying little more than that he did not believe "that the argument for the Ptolemaean system on the basis of certain Scripture texts is conclusive." While conceding that the Bible used "ordinary human language" in describing astronomical events, he was no more inclined than his father to admit biological evolution. Theodore's religious roots ran deep in three conservative traditions: He was reared in the Wisconsin Synod, ordained in the Norwegian Synod, and, after 1913, employed by the Missouri Synod at its Concordia Seminary in St. Louis, where for decades he also edited the synod's "official English organ," the *Lutheran Witness.* Although his training in Lutheran schools had been primarily theological, as a young man he had taught biology at the Lutheran Ladies' Seminary in Minnesota, and he maintained an active interest in biology throughout his life. At least once he described himself as a "real scientist" by virtue of having had "a college course in natural science, [having] taught biology for six years, [having] devoted my leisure time to it for 30 years, [having] read all the modern college texts in biology, [and having] made special studies in museums and private collections."[11]

As a precocious teenager he had read Darwin's *Origin of Species* and other evolutionary works, only to be revulsed by "the anti-religious bias of evolution." It not only contradicted the biblical account of creation but, most damning of all, denied the Fall, thereby eliminating the need for redemption or a Savior. He found the theory to be equally suspect scientifically. Early in the century a professor at Luther College had convinced him that evolutionists were "reasoning in a circle," dating fossils by their position in the rocks while simultaneously dating the layers of rocks by their fossil content. But it was not until the early 1920s that he discovered in Price's *Fundamentals of Geology* "new and unexpected confirmation" of his suspicions. In 1922 he personally entertained the Adventist geologist when Price passed through St. Louis. Though he came to have a high regard for Price as "a professional geologist" of intelligence and integrity, and accepted his critique of orthodox geology based on thrust faults and deceptive conformities, he remained skeptical about attributing most of the fossil-bearing rocks to the deluge. In Graebner's various books on creation, including *Evolution: An Investigation and a Criticism* (1921), which he erroneously regarded as "the first scientific work printed in America against the evolution theory"; *Essays on Evolution* (1925), a collection of his previously published articles; and *God and the Cosmos: A Critical Analysis of Atheism* (1932), his *magnum opus*, he pointedly refrained from attempting to reconcile Genesis and geology. "I carry the data of stratigraphy, mountain formation, erosion, and the immense areas of volcanic origin in one compartment of my thinking, the narrative in Genesis in another, with a water-tight bulkhead between," he explained to Nelson. "I cannot harmonize the two. But that does not make me reject one or the other." Time, he hoped, would resolve any difficulties.[12]

Nelson, in contrast, devoted much of his life to searching for an accommodation between science and Scripture. The athletic, artistic son of John M. Nelson, a progressive U.S. congressman from Wisconsin, Byron C. Nelson belonged to the Norwegian Lutheran Church of America. As an undergraduate philosophy student at the University of Wisconsin, he nearly drowned under "the wave of evolutionary philosophy and pseudo-science" that swept over America's educational institutions, but Arnold Guyot's day-age interpretation of Genesis 1 temporarily kept him spiritually afloat. Shortly after the publication of Price's *New Geology* in 1923, Nelson obtained a copy and devoured its contents. As he later recalled, "It gripped me so that I read it in three days, scarcely sleeping." Price's vision of earth history seemed so much more satisfying than Guyot's, Nelson scarcely hesitated in casting his lot with the flood geologist. As a master's student at Princeton Theological Seminary in the mid-1920s, he wrote his thesis on the scientific and biblical meaning of the Genesis "kinds," concluding that both

Mendel and Moses taught the fixity of species. When a revised version of this work appeared as his first book, under the title *"After Its Kind": The First and Last Word on Evolution* (1927), the geological portions closely followed Price. A few years later a second book, *The Deluge Story in Stone: A History of the Flood Theory of Geology* (1931), featured Price as the "one very outstanding advocate of the Flood" in the twentieth century. An excellent artist, Nelson illustrated his own books, which he typed at a large desk in his study, "a green celluloid visor shading his eyes from the glare of a single bulb hanging from the ceiling, his shirt sleeves pulled up over short arms by rubber bands, his collar band awaiting a starched button-on collar." Largely as a result of his own labors, Nelson was by the mid-1930s able to report to Price that Norwegian Lutherans were "very generally with you on Flood geology."[13]

In the late 1920s, while pastoring a Danish Lutheran church in Perth Amboy, New Jersey, Nelson renewed a friendship with a fellow Wisconsinite, Thurlow C. Nelson (1890–1960), who had become head of the zoology department at nearby Rutgers University. At first the professor urged his friend to stick to theology and leave science alone, but a reading of *"After Its Kind"* convinced him that the preacher might benefit from a greater exposure to the biological sciences. On Thurlow Nelson's recommendation Byron Nelson enrolled in a genetics course at Rutgers, taught by a professor reputed to be "death" on creationists. Nelson not only passed the course but struck up a friendship with the teacher, who invited him to take an advanced seminar in which, Nelson reported, he "bred fruit flies by the thousands, attempting to bring about mutations by feeding them arsenic." For a period he became so enamored of science, he toyed with the idea of giving up his ministerial career to teach biology. Instead, he returned to Wisconsin to study geology at Madison and, soon thereafter, to minister to two small churches in the Oconomowoc Lake region.[14]

Nelson first attracted national attention as a creationist in 1931, when his young daughter came home from school with a book that described prehistoric humans conversing in grunts, "unfa, unfa, glug, glug." Nelson's protests led to the book's being banned in the public schools of Perth Amboy—and to news stories describing the incident as the first "outbreak" of such antievolutionism in the North. By the time Price and Whitney invited him to assume the vice presidency of the RSA, Nelson had been in correspondence with both men for a number of years. Shortly after the publication of *"After Its Kind,"* which Price recommended "unqualifiedly," Whitney, in characteristically aggressive fashion, had challenged Nelson to a mail debate to settle their differences "regarding the biological problems of evolution." The two creationists agreed fully on the special creation of plants and animals and on the truth of flood geology, but whereas Nelson insisted on the fixity of species since Eden, Whitney eagerly welcomed "every bit of *modification during descent* that can reasonably be asserted."[15]

The underlying problem, as Whitney saw it, was the shortage of space on Noah's ark. "If we insist upon fixity of species we make the Ark more crowded than a sardine can," he argued. "If we agree to all [variation] that can be demanded, we simplify the Ark problem greatly," because all present-day species could have descended from relatively few passengers on the ark. Such concessions convinced Nelson that Whitney had "given up half of the battle" to the evolutionists—all because he failed to recognize the immense size of the ark and the small number of "large, bulky species like elephants, rhincerouses [sic], lions, horses, cattle etc." that had to be accommodated. All of the large animals, Nelson surmised, could be "put on one floor easily," and the smaller species, such as rabbits, squirrels, and birds, did not take much room. Price, though concerned that such petty arguments over biological matters only diverted attention from "the vastly more important point of whether there has been a world-catastrophe in the long ago," sided with Whitney. "For my part," he wrote Nelson,

> I cannot but think that Whitney is on safe ground when he admits that he is willing to accept as of common origin any group of species which common sense tells us may possibly have sprung from common ancestors under the very abnormal and yet providentially arranged conditions immediately following the Flood. By conceding this much to the evolutionist we shift the battle front from the "species question," where the fight has raged altogether too long, to the geological question, where it ought to be, but which the evolutionists always wish to avoid.[16]

Whitney, who craved the limelight that had bathed the antievolutionists in the early twenties, constantly plotted ways to recapture the headlines. As a journalist, he knew firsthand that "the more freakish the idea," the greater the public interest, and the prospect of publicly embarrassing some big-name geologist threw him "into a spasm of delight." However, his (and Price's) basic objective in founding the RSA was to convert *fundamentalists* to flood geology. A self-confessed misanthrope, Whitney questioned the ability of conservative Christians to reason things out for themselves. "Just to give the different positions would do little good," he believed. "Readers are too ignorant to reason or to weigh evidence." What fundamentalists needed were official "pronouncements" issued by a "Supreme Court" of theologically orthodox scholars telling them what to think. Thus Whitney proposed that RSA members debate a particular issue by correspondence before submitting their material for a judicial decision. The findings would then be published for all to read.[17]

Heading Whitney's agenda of issues to resolve was the ruin-and-reconstruction theory, which he regarded as "the greatest obstacle to a clear understanding of creation by Fundamentalists in general." As he explained to Price, "It is a *Fundamentalist* dogma; and when I say dogma in that connection I mean DOGMA and then some. . . . The thing is an *obsession*. I do not think that you, as an Adventist where that theory has no standing, can quite realize what a grip that has."[18] Unfortunately for Whitney and Price, they had, perhaps unwittingly, chosen as president of the association a man staunchly committed to that very interpretation of Genesis: L. Allen Higley.

Among the officers of the RSA, Higley claimed the most impressive scientific credentials. He had earned a Ph.D. in organic chemistry from the University of Chicago in 1907 and had worked in industry, higher education, and government for nearly two decades before joining the faculty of Wheaton College—"THE college for Fundamentalists," in Whitney's opinion—where he taught chemistry and geology. For a number of years he had relied exclusively on Price's texts in teaching geology, but student complaints about collapsing the geological column into a single year had lately caused him to adopt other works as well. Although he was only a year younger than Price, his robust constitution and zest for geological fieldwork aroused the envy of the senior creationist, who ruefully admitted Higley's superiority as a field man. Theologically, Higley's trajectory closely paralleled Whitney's, following a path from charismatic Dowieism to nondenominational evangelicalism. But there the similarity ended, because on the subject of Genesis 1, the two found themselves in a spirited disagreement that eventually engulfed the entire association.[19]

At first Price seemed optimistic about the prospect of converting Higley to flood geology. Following his first meeting with the Wheaton professor, in August 1935, Price reported to Nelson: "Hitherto he has not been an out-and-out Flood geologist; though he of course has believed in a real Flood, and has been strong against evolution." Price hoped that Higley's recent visit to the Canadian Rockies, where he had seen for himself the evidence of "upside-down" fossils, had "cured him entirely" of his predisposition to accept the antiquity of life on earth. On a personal level Price found Higley to be "a most kindly, gentlemanly fellow, a very sincere Christian, with a sweet though superbly strong character." The better Price got to know him, however, the less sanguine he became. When he learned that Higley attributed the fossils to one or more pre-Adamic cataclysms, which left, as Price complained, "nothing worth while for this Flood of Noah to do, and virtually explains it all away as an affair of no geological importance," his patience began to wear thin. To Price, such a position was not only "illogical and inconsistent" but heretical. Higley's scheme illustrated "the permanent evil of wholesale adding to the Word of God"; his exegesis represented "about as bold a piece of twisting the Holy Word of God as I can conceive of." Besides, acceptance of a pre-Adamic cataclysm would negate Price's entire life's work, which, Higley quickly discovered, "centered around a single idea—Flood Geology."[20]

Higley himself was no less single-minded, having, as he explained, dedicated "years of careful investigation" to constructing a distinctive version of the ruin hypothesis, or what he preferred to call the "pre-Adamic cataclysmic Judgment." Higley claimed to have found biblical evidence of *two* creations and *two* floods. The period between the creation "in the beginning" (Genesis 1:1) and the Edenic creation was a probationary time for Satan and his evil angels. In a conversation with fellow RSA director Clark, Higley explained that "Lucifer was the first being created on this world and that he was put here several million years ago and that his fall was the cause of several million years of chaos in which the earth was without form and void. That was the time when the stratified rocks were laid down." At the end of the period, God destroyed "all plant and animal life" in a pre-Edenic flood, after which he repopulated the earth in a literal six-day creation about four thousand years before the birth of Christ. When Lucifer caused Adam and Eve to sin in the Garden of Eden, the rebellion against heaven continued, provoking God into destroying the earth a second time, in the worldwide flood of Noah. It is unclear whether Higley dated most of the fossil-bearing rocks from the period between the first creation and the first flood, as Clark recalled, or from the first flood itself, as other accounts imply; whatever the case, Price rightly perceived that Higley attached little geological significance to the second flood, the deluge of Noah.[21]

On the issue of flood geology Price, Whitney, Nelson, and Clark stood united, while Graebner, to his fellow directors' surprise, declared that he could accept neither the ruin nor the flood hypotheses. The obvious continuity between living species and the fossils seemed to rule out a universal pre-Adamic cataclysm, while a host of evidence made him doubt flood geology. "In spite of all that I have read about the flood theory to account for stratification, erosion, and fossils, I cannot view the mountains without losing all faith in that solution of the problem," he confessed to Higley. "I cannot believe that 10,000 feet of limestone were first laid down, loaded with fossils, and then eroded like the flood bed of the Mississippi, of the Colorado, and Columbia or even the Wisconsin, not to speak of the canyons and glens which owe their origin so plainly to glacial movements, and accept the Flood of Genesis as the explanation of these highly complex and really tremendous phenomena." As for how and when the fossil-bearing strata were formed, he professed agnosticism: "I do not know how long it took to lay down the strata, nor does anyone KNOW."[22]

IRRECONCILABLE DIFFERENCES

The deep fractures running through the creationist community became even more visible in connection with the RSA's first public venture, a two-day conference on creationism held at the Moody Church in Chicago, March 27–28, 1936. In advance of the meetings Graebner pleaded with Higley to keep the proceedings strictly scientific and especially to avoid opening and closing sessions with prayer. Missouri Lutherans, he explained, refused to "practise joint prayer with those not on the same confessional basis," and besides, the "greatly differing doctrinal platforms" of Price and himself, for example, made any religious interaction risky. When Higley replied that "prayer is not only proper but necessary," Graebner threatened to resign "unless religious exercises are eliminated from the public or private sessions."[23]

In harmony with the RSA's objective of establishing a united creationist front, Higley, the primary conference organizer, invited papers from a wide range of antievolutionists, including the three members of the association's book committee—Graebner, Leander S. Keyser of Wittenberg College, and Clarence H. Benson (1879–1954) of the Moody Bible Institute—and prominent representatives of the major hermeneutical factions within creationism: Price, Whitney, Nelson, and Clark for flood geology; Rimmer for the gap theory; and W. Bell Dawson (1854–1944), John William Dawson's son and a prominent Canadian engineer, for the day-age view. Though the RSA officers generally had little regard for Rimmer as a scientist and all disagreed with Dawson on his liberal reading of Genesis 1, Higley thought it important to cultivate their "good will" as fellow antievolutionists. He also took it upon himself to invite William H. Haas (1872–1960), a geologist and geographer from nearby Northwestern University, to give an illustrated lecture on "The Geologist and Time" that presented the scientifically orthodox view of an ancient earth. In evaluating the success of the conference, Higley ranked Haas's performance as "by far the most outstanding presentation of the entire convention," while Whitney fumed at the implied RSA endorsement of "evolutionary geology."[24]

Whitney, who along with Price, Dawson, and Rimmer had been unable to participate personally in the conference, judged the entire affair a disaster. He especially disagreed with Higley's decision to focus on arguments against evolution rather than for a particular view of creation, which Whitney regarded as "the first and most essential step" toward creating a fundamentalist consensus. And he was bitterly disappointed that just as creationists had finally gotten their little association off the ground, "Ruinites and evolutionary geologists" had captured it. In the beginning Whitney had targeted the "Wheaton crowd" as key allies in the battle for flood geology, believing that "the more Wheaton can be played up in our association, the more influence it will have with the Fundamentalist, which is what we want." But now that the conference had actually brought in seven new members from the Wheaton College faculty, presumably "ruinists," he feared that "at the next election the Flood geologists will find themselves in a minority and helpless."[25]

In planning and running the convention, Higley had consciously tried to avoid public conflict over differing views of Genesis, preferring to settle such matters privately, perhaps informally at a campground. Nelson proposed a field trip to the Rockies, where the upside-down strata could be examined "with our own eyes" and evenings could be spent sitting around a camp fire deciding amicably "whether we were flood geologists or long age geologists or pre-Adamic ruin geologists or what." Whitney, hoping for "a big gang and a wonderful time," pushed for a California "auto tour," with each party bringing its own camping gear. All plans fizzled, however, when both Nelson and Higley were forced to back out; as Whitney observed, meeting without Higley to debate interpretations of Genesis 1 would be like "playing Hamlet with Hamlet missing." Whitney himself did get a chance to meet Higley in late summer, when the RSA president stopped by the ranch on his way to visit the giant redwood forests. Whitney found Higley "a very pleasant man," but one who harbored disturbing views about pre-Adamic catastrophes. Surely, he thought, Higley's acceptance of "evolutionary geology was something to make the angels of darkness rejoice."[26]

The epistolary warfare between Higley and the flood geologists dragged on into early 1937, by which time even the noncombatant Graebner, who declared a plague on both sides, was feeling "overwhelmed by the mass of carbon copies to which this controversy in our Directorate has given birth." Whitney, worn down by Higley's intransigence, finally concluded that the professor would never abandon the pre-Adamic-ruin hypothesis because the issue for him was not "a matter of reason but of conscience." And, he added sadly, "a conscience is a terrible thing when it is binding." To break the stalemate, Whitney urged bypassing the president and taking the matter directly to the public. Nelson had for some time favored simply dumping Higley and installing a flood geologist, but Price, fearing that he himself would have to step in, urged restraint. Higley "is a good fellow in many ways, and he is honestly opposed to the entire evolution scheme,—as he knows it," Price reminded Nelson. "So let us not swap horses until we get across the river."[27]

But crossing the river proved to be no easy task. Despite the founders' goal of establishing a creationist consensus on Genesis 1, the RSA remained as divided as fundamentalists generally. An informal and incomplete poll of members' interpretations of Genesis revealed "about an even break between Flood geology, age days and ruin and reconstruction." "A swell gang we are," Whitney noted with irony, "trying to fight evolution when we can agree on nothing among ourselves except that evolution is wrong." Even the flood geologists in the group split over the extent to which plants and animals had changed—or evolved—since the Edenic creation. In a continuation of the earlier Whitney-Nelson debate, Whitney, backed by Clark, insisted that there had been moderate speciation *"within created types,"* that morphology, not fertility, identified the original "kinds." Nelson, on the contrary, believed that such concessions opened "the door of evolution so wide that I, for one, don't see a place to shut it." Surprisingly, Price, who had formerly sided with Whitney in allowing evolution within originally created families, now joined Higley in supporting Nelson. Unlike the rancorous dispute over a pre-Adamic cataclysm, this exchange remained on friendly terms, largely because Whitney admired Nelson for his scientific mind, tolerant spirit, and Nordic ancestry. Scandinavian peoples, wrote the Celtic Whitney, "must be admitted to be the cleanest and finest among the white races, unless the Highland Scots can rival them."[28]

To bluff outsiders into thinking that the association was still alive, Whitney urged that the group issue an official pronouncement on speciation. But rigor mortis was already setting in. Since the Chicago conference in the spring of 1936, the RSA had done little besides debate the twin issues of pre-Adamic geology and post-Adamic biology. When talk of a second national meeting began, Higley quashed it by observing that a small turnout would only embarrass the group, which at its peak had claimed a membership of fifty-six: thirty-one "active" members, who boasted some scientific competence, and twenty-five "associate" members, who paid a dollar a year for the privilege of underwriting their more accomplished brethren. Though a precise breakdown of denominational affiliations never appeared, Whitney at one point estimated that "over half" of the inquiries he received as secretary came from Lutherans, undoubtedly attracted by Graebner's and Nelson's participation.[29]

Among the RSA's officers, only the indefatigable Whitney refused to concede failure. As the association drew its last breaths in 1937, he revived Price's original idea of publishing the *Creationist,* launching the venture single-handedly as a mimeographed sheet nominally associated with the RSA. By this means he hoped to lay "the ground work for something more elaborate" in the future. Except for letters to the editor, Whitney wrote all copy himself, mostly defending flood geology against rival interpretations of earth history. With the start of the second volume in 1938, he dropped all mention of the RSA, and by spring of that year he was referring to "the late Religion and Science Association."[30]

In view of later developments, the association may have died at a propitious time. Within a few years Price and Nelson were feuding over the age of humans and the extent of glaciation, with Nelson, once a devoted disciple, now pushing creation back 100,000 years or more and warning his former mentor not to "make an ass" of himself by continuing to teach "absurd" ideas. Further geological study, particularly of the annual layers of sediment (varves) deposited by retreating glaciers, had so convinced Nelson of the antiquity of the earth and the geological significance of ice that he declared that "if the Bible said otherwise I would not believe the Bible." Fortunately, like George Frederick Wright decades earlier, he had discovered William Henry Green's influential writings on biblical chronology that allowed for a theologically orthodox break with Ussher. In his last book on science and the Bible, *Before Abraham: Prehistoric Men in Biblical Light* (1948), Nelson concluded that "the Bible, when properly interpreted, leaves men free to agree to any antiquity of the human race, however great that may be, which a genuine science makes it necessary to grant." By this time he was entertaining the possibility of "the creation

being back maybe a million years—a literal 7 day week creation—and
a Deluge which changed the face of the whole earth correspondingly
ancient."[31]

Graebner, never much impressed by the arguments for flood geology, grew ever more skeptical with the passage of time and by the end
of his life even conceded that theistic evolutionists might make good
Lutherans. When a colleague at Concordia Seminary, Alfred M. Rehwinkel (1887–1979), sought in the 1940s to publish a defense of Price's
flood geology aimed at Missouri Lutherans, Graebner tried to block
publication on the grounds that the book "would affect adversely if not
disastrously the work of our university pastors." As a young frontier
circuit rider, Rehwinkel had lived in the shadow of the famed "upside-
down" Chief Mountain; later he had studied geology at the University
of Alberta and briefly taught physical geography at a Lutheran college
in Canada before joining the theology faculty of Concordia in 1936. His
book *The Flood: In the Light of the Bible, Geology, and Archaeology*
(1951), which praised Price as the foremost defender of Genesis, "an
able geologist and a brilliant writer," finally appeared the year after
Graebner's death.[32]

Higley left Wheaton College in disgrace in 1939 after a handwriting
expert identified him as the author of scandalous letters to the trustees,
deceptively signed "An Alumna," charging the president, J. Oliver
Buswell, Jr. (1895–1977), with heresy. In these anonymous attacks
Higley accused Buswell, a day-age advocate, of teaching "the very
foundation of evolution and Modernism" and of trying to "make a
Modernistic College of Wheaton." The year after departing Wheaton,
Higley defended what Nelson called "his crazy Pre-Adamic Ruin business" in a book titled *Science and Truth* (1940). In describing the work
to Price, Nelson maliciously hinted that their former president had
gone east for reasons of mental health, a suspicion based on the mere
coincidence of Higley's living in a town with a mental institution.
Higley finished his academic career teaching science at The King's
College, an evangelical school then located in Delaware, but for the
rest of his life he seems to have remained aloof from organized creationism, thus depriving the movement of one of its scarcest resources,
trained scientists.[33]

Whitney, though never again on center stage, remained on the
fringes of creationism for years, a victim of his own eccentric personality. Since becoming a flood geologist, he had yearned to publish a
major book, and at times the goal seemed within reach. Harper reportedly expressed considerable interest in one of his manuscripts in
the late 1920s, and Eerdmans promised to bring out his debate with

the atheist Edwin Tenney Brewster in the mid-1930s. But plans always seemed to fall through. Finally in 1946 the obscure Christian Evidence League published his five-part *Case for Creation,* in which he argued for a post-flood re-creation of plants and animals; and nearly a decade later a New York vanity press issued *The Face of the Deep: A Defense of Divine Creation* (1955), dedicated to Price. By this time Whitney's optimism was waning. "I still do not know whether my book will be a complete flop or finally wake up the outside public (part of it) and then break over into the Fundamentalists and Conservatives," he wrote Price late in 1955. "At least I will have tried and will not mourn over my failure." Judged by the difficulty of finding a surviving copy today, his book failed to awaken anybody. He died in 1964, at age eighty-two, when he absentmindedly walked into a huge dump truck.[34]

Clark, the youngest and least influential of the RSA directors, went on to become a major creationist theorist in the middle decades of the century, but in so doing, as we shall see, he incurred the wrath of Price, who charged his former student and fellow Adventist with heresy. Price himself retired to southern California, where, assisted by a new circle of creationist friends, he continued his quixotic crusade to transmute flood geology into theological and scientific orthodoxy.

From one perspective, the RSA has all the appearance of a creationist fiasco, a grandly conceived scheme that merely highlighted the lack of theological agreement and scientific expertise among antievolutionists. Yet, seen from a different light, the association is not without historical significance. As the first such organization in America, it brought the leading creationists into contact with one another, in some cases for the first time; it gave birth to the first creationist periodical of which I am aware; and it prodded fundamentalists to select among the various interpretations of Genesis 1 rather than "believing all at once, endorsing all at once." Reflecting on the experience, Price later recalled that "nothing really came of the attempt; but I am not at all sorry that the attempt was made. Many subjects were cleared up as the result of the very extensive correspondence which we carried on in attempting to get organized."[35] Perhaps most important of all, the failure of the RSA led directly to the formation of a more homogeneous and influential creationist organization: the so-called Deluge Geology Society.

SEVEN

The Deluge Geology Society

As the Religion and Science Association self-destructed in the summer and fall of 1937, Harold W. Clark contemplated the future. "If we ever expect to do anything for the sake of creationism," he advised George McCready Price, "we must organize for ourselves and hold to our own particular beliefs and work it out in our own way." But because he could think of only "four or five who are doing anything along this line," he urged postponing the start of a successor society until more like-minded recruits could be found. Above all, he did not want "to start out with a big flare of trumpet and drum and fizzle out" the way the RSA had done. The aging Price, however, was impatient to start rebuilding, and in 1938 he and a nucleus of Adventist associates in the Los Angeles area formed the Society for the Study of Creation, the Deluge, and Related Science, better known as the Deluge Geology Society (DGS). To avoid the dissension that had rent the RSA, the founders of this professedly nonsectarian organization limited membership to persons who believed that the week of creation lasted no more than "six literal days, and that the Deluge should be studied as the cause of the major geological changes since creation." In other words, ruinites and day-agers were not welcome.[1]

CAPTAIN ALLEN TAKES COMMAND

The spark plug of the new society was a Whitney-like refugee from the RSA named Benjamin Franklin Allen (1885–1960), a lifelong bachelor who had migrated from the Ozarks to California with the onset of the Great Depression. A 1912 graduate of the University of Arkansas, where he had taken a number of courses in engineering and science, he went on to a miscellany of careers, including high-school principal, lawyer, legislator, infantry captain, and deputy secretary of state in Arkansas. In the late 1920s he served as state lecturer for the Arkansas Anti-Evolution League, helping to secure the passage of a popular referendum banning the teaching of evolution in public schools. Then for five years he assisted the Adventist evangelist H. M. S. Richards (1894–1985) in tent and tabernacle work before devoting himself full time to the promotion of flood geology, especially to the writing of a major book tentatively called "Noah's Flood." Without a steady income, he lived for years in destitution, crowded into his tiny trailer and supported largely by public funds and the charity of friends.[2]

A zealous Seventh-day Adventist, Allen proudly credited all of his scientific work on "coal, oil, volcanoes, earthquakes, mineral deposits, and mountain making" to the divinely inspired writings of Ellen G. White. Price's works he held in only slightly less esteem, believing them to be written "under the guidance of the Holy Spirit" and comparable in "style and forcefulness and logic" with those of Sister White. His only complaint about Price's writings was their technical language, which made them unsuitable for mass consumption. What Price needed, Allen concluded, was a scientifically qualified popularizer: "That's where such as *I* come in, who years ago studied all the main branches of science in the University, and *kept it up since* (which amounts to ten times more) in my reading and attention and thought. *Somebody* must digest, reinterpret, and *retail* this great material, this precious truth for which so much of the world is dying." Allen, described by Price as "a hardy out-of-door man," also felt that creationists needed to spend more time in the field, which Price, "the prince of Library research men," had been unable to do.[3]

Price, somewhat uncharacteristically, took Allen's veiled criticism without offense. Though many who met Allen immediately dismissed him as a crank, Price personally identified with the impoverished outsider. "Many used to call me a crank thirty or forty years ago," he wrote in 1939, when Allen's fortunes were at their lowest; "and I have often walked the streets of New York and Los Angeles with nothing to eat. So I know how to understand Brother Allen's situation." Price, now visibly slowing down, especially admired Allen's abundant energy—and his nearly fanatical commitment to flood geology. "Allen seems to be getting the Adventists lined up for this Flood geology in a way that I never could accomplish," Price happily informed Dudley Joseph Whitney. "I am not a 'promoter,' while he is,—and a live one." Unfortunately for Captain Allen, as he liked to be called, his abrasive, judgmental manner tended to repel the very people he wished to attract. A fellow creationist, still smarting from a prick of Allen's intemperate pen, aptly characterized him as "the whip, the self-appointed smeller out of heresy, and castigator of the brethren." "You pose as a very humble man, one who is not a scholar," continued the critic, "but under this 'Uriah Heep' masquerade you surpass all in your love of your own opinion and your orthodox complex."[4]

The Deluge Geology Society began meeting in Los Angeles in 1938 at the close of a lecture series by Price, who provided more inspiration than active leadership. Relaxed and jovial among friends, Price amused his associates by cackling whenever he thought he had made a good point. Most of the organizational and promotional work fell to Allen and several Adventist physicians associated with the College of Medical Evangelists, where Price had once taught. Among the most active medics were the nationally known neurologist Cyril B. Courville (1900–1968) and the young, Dutch-born dermatologist Molleurus Couperus (b. 1908). In addition to holding monthly meetings, the members circulated a mimeographed newsletter and published a journal, the *Bulletin of Deluge Geology and Related Sciences,* that in both content and appearance put the *Creationist* to shame. By 1942 membership stood at about four hundred, and just three years later Allen was claiming "a band of more than 600 loyal and sacrificing" supporters. "In no other part of this round globe," bragged Price, "could anything like the number of scientifically educated believers in Creation and opponents of evolution be assembled, as here in Southern California."[5]

Though Adventists constituted by far the largest contingent within the DGS, Allen thought it wise to "avoid having so many Adventists that it could be called an Adventist group." Thus the society actively sought to attract persons of other faiths, especially individuals suspected of having sympathy for flood geology. Arthur I. Brown reportedly kept "in constant touch" with the society and lectured to the group in June 1943. William J. Tinkle (1892–1981), a biologist at La Verne College, a Church of the Brethren school just east of Los Angeles, attended meetings when possible and published in the *Bulletin,* as did Whitney. From distant Houston, Texas, a young civil engineer named Henry M. Morris (b. 1918) early signed on as a member and later, as a graduate student at the University of Minnesota, sought Allen's advice on choosing a dissertation topic that might further the interests of flood geology. But the most active non-Adventist of all, indeed a core member, was Walter E. Lammerts (b. 1904), a horti-culturalist with the University of California at Los Angeles who drummed up interest in the DGS among fellow Missouri Lutherans, assuring them that though most members were Adventists, they included "some fine students of nature." Like his predecessors in the RSA, Lammerts hoped in time that "a united front of anti-evolution students may become a force potent enough to cause some of the evolutionists to at least consider seriously the many arguments against the theory." However, not all members welcomed the ecumenical approach. Clark, for one, worried that the DGS included "among its prominent leaders some [presumably Lammerts in particular] who are entirely unsympathetic to Seventh-day Adventist aims and views."[6]

In contrast to the RSA, which squandered all of its energies on internecine disputes, the DGS undertook some of the earliest collective field research connected with creationism, including investigations of allegedly human fossil footprints and an abortive search for Noah's ark. Creationists' curiosity in reported humanlike tracks was awakened in the late 1930s, especially by the publication in 1939 of an illustrated article in *Natural History* that reported the discovery of "mysterious, 15-inch, man-like tracks," found with dinosaur prints along the Paluxy River near Glen Rose, Texas. The author, Roland T. Bird (1899–?), a paleontologist with the American Museum of Natural History, went out of his way to discount such claims, but some science writers could not resist the temptation to tease a gullible public. A sensationalist piece in *Scientific American,* titled "The Carboniferous Mystery," predicted that "all the geologists will resign their jobs and take up truck driving" if such prints were found to be human. The mere thought of evolutionary geologists hitting the road brought cheer to the deluge men, who gleefully (but inaccurately) spread the news in the inaugural issue of the *Bulletin of Deluge Geology.*[7]

Early on Byron C. Nelson, who dismissed the claims of human origin as "absurd," cautioned Price to bear in mind that if the tracks had indeed been made by humans, the fact would not only falsify the theory of human evolution but "knock the Flood theory all to pieces." If the fossil-bearing strata had been laid down rapidly during the relatively brief deluge, as Price claimed, then Nelson could "not see how any man's footprints could be made in that strata." Nevertheless, excitement among flood geologists continued to mount. In 1943 the DGS began soliciting funds for "ACTUAL EXCAVATION" of reported sites, and at a society-sponsored symposium in November Allen and the two other society members who formed the Footprint Research Committee—Everett E. Beddoe, (1889–1977), an Adventist minister, and Clifford L. Burdick (b. 1894), a consulting geologist—presented "an extensive field report on fossil human footprints," accompanied by casts and photographs. Burdick, a graduate of the Seventh Day Baptist Milton College in Wisconsin, had embraced flood geology in the early 1920s, when as a recent convert to Seventh-day Adventism he had enrolled in an Adventist college to prepare for mission service and had there met Price. Allen promoted Burdick in creationist circles as "the only S.D.A. (so far as I have heard) who ever studied geology and mineralogy and geochemistry in a standard scientific institution. And, what is more, he took a *graduate* degree in that work." Burdick did indeed claim a master's degree from the University of Wisconsin, where he had studied geology for a year as a postgraduate student, but he had left degreeless after failing his oral M.S. examination.[8]

From the beginning the DGS treated the footprint project with a mixture of grandiosity and paranoia, fearful lest competitors steal their thunder or enemies thwart their work. In a 1944 announcement Allen cryptically reported that the society's special committee was conducting research of such a "spectacular character" that it forbade "all publicity on that topic, *so delicate is the situation and so important the subject.*" Out in the field DGS researchers guarded against spies "from unfriendly scientists" and advised local farmers, whom they offered $100 for finding new prints, to cover any evidence carefully "so as to keep it a secret." At one point Price suggested that the footprint committee contact Science Service, a private organization that disseminated scientific news, to find out which museums might have information about the tracks. But, he warned Allen, "if they suspect a 'crank' is after the information,—then they shut up and give no information. They long ago spotted me; and I feel sure they have also black listed you and our Society." To avoid detection, he proposed that the little-known Couperus send off an innocent letter of inquiry.[9]

Monetary need, however, ultimately overcame the fear of harassment, and in 1945 Allen, in an effort to secure financial backing for the project, publicly revealed the discovery of gigantic fossil footprints of humans far older than allowed by evolution, "thus at a single stroke defeating that theory." It was, he announced excitedly, "*one of the most spectacular challenges ever to come to us. With the necessary funds we can astound the scientific world!*" Even after Beddoe's discovery that Native Americans had "without question" carved some of the humanlike footprints, and an unsuccessful field trip to Texas in 1945, enthusiasm for the project remained high. And though the DGS never made the dramatic discoveries it hoped for, the fossil footprints, like the "upside-down" Chief Mountain, long remained icons of flood geology.[10]

THEORIES OF SATANIC ORIGIN

For a society ostensibly bound together by a common commitment to flood geology, the DGS enjoyed a surprisingly short honeymoon. Within a few years, members were wrangling over everything from hermeneutics to tectonics, angrily denouncing one another with a freedom usually reserved for family disagreements. The divisiveness stemmed in part from the members' ultra-Protestant insistence on the right of private interpretation, but a new factor had also come into play: the presence of younger university-trained scientists unwilling to accept the dictatorial pronouncements of armchair savants like Price. Perhaps the most extreme example of such rebellion was Clark's traumatic break with Price, his former teacher and fellow RSA director.[11]

Clark, an Adventist farm boy from New England, had first turned to the study of science and religion as a result of reading Price's *Back to the Bible* (1916). Later he had the good fortune of studying with the master when, as an older student returning to college after years of church-school teaching, he enrolled in Pacific Union College in 1920, the very year Price arrived. Upon graduating two years later, Clark stayed on to replace Price, who left for a position at Union College in Nebraska. During the 1932–1933 academic year Clark studied for a master's degree in biology at the University of California, becoming the first Adventist to obtain a graduate degree in that discipline. Though his major professor at Berkeley was an "ardent Darwinist" who knew about his student's fundamentalist beliefs, Clark detected none of the hostility creationists expected to encounter in such situations. In fact, he and his "open-minded" professor became personal friends.[12]

For nearly two decades Clark also maintained a cordial, if geographically distant, relationship with Price, who had come to trust his junior colleague "implicitly" in matters of science and Scripture. In 1929 Clark dedicated his first book, *Back to Creationism,* to Price—"Teacher, Friend, Fellow-warrior and Prophet of the New Catastrophism"—and he seldom strayed from his mentor's critique of orthodox geology. That summer, however, Clark spent the first of several vacations studying glaciation in the mountains of the West, and the evidence he saw convinced him that ice had once covered large portions of North America, perhaps for as long as fifteen hundred years after the flood. Such ideas were anathema to Price, who believed that the theory of continental ice sheets *"was invented as a sort of mental buffer, an intellectual shock absorber, to make the transition from the geological world to the present less violent and catastrophic."* He suspected that the creationist Louis Agassiz had formulated his glacial theory "under the hypnotic suggestion of the devil." After returning from Berkeley, Clark updated and enlarged his book, trying to make it as scientifically respectable as possible within the constraints of strict creationism. In addition to parting with Price on glaciation, he challenged the common Adventist notion, often associated with Price, that there had been "no change in species since the beginning." Hybridization, maintained Clark, had been "a very potent factor in the formation of new species."[13]

Upon receiving Clark's manuscript, Price raced through "every word" of it in forty-eight hours. He then dashed off a proud letter to Adventist church leaders praising his former student for his "fresh and interesting" treatment of geology, which added "very greatly and materially to the work that I have tried to do along this line." As for the biological part of the treatise, he wholeheartedly endorsed Clark's acceptance of "species-making by means of hybridization and otherwise," noting that these suggestions corroborated White's puzzling statement about amalgamation between humans and animals—an observation so controversial that "panicky" publishers had subsequently deleted it from her works. "If the Seventh-day Adventist people will all get behind these two ideas, Flood geology and plenty of species-making since the Flood, and if these two ideas can become widely known as the Adventist official teaching on these subjects, I believe that it would not be long before the scientific world would 'sit up and take notice,' " Price concluded, adding generously that he considered Clark's manuscript "better as a general survey of the subject than any single book of mine."[14]

Price might have written less effusively had he realized the direction in which Clark was heading. For some time Clark had been moving toward the conclusion, adamantly denied by Price, that the fossiliferous rocks had been deposited in the very order assigned them by geologists. But instead of viewing the fossils as representing successive populations of plants and animals spanning immense periods of time, as evolutionists did, Clark increasingly leaned toward interpreting them as marking the ecological zones—from ocean depths to mountain tops—of the antediluvian world. Thus the progressive sequence of the geological column merely reflected the order in which plants and animals had succumbed to the rising floodwaters. Any lingering doubts about the validity of the geological column were dispelled in the summer of 1938, when Clark, at the invitation of one of his students, visited the oil fields of Oklahoma and northern Texas and saw with his own eyes why geologists believed as they did. Observations of deep drilling and conversations with practical geologists gave him a "real shock" that permanently erased any residual confidence in Price's vision of a topsy-turvy fossil record. Before long he even came to see the evidence for overthrusting, including that associated with Chief Mountain, as "almost incontrovertible."[15]

On returning home to California, Clark mustered the courage to share his exhilarating yet disturbing thoughts with Price. "I am going to startle you," he announced—and then proceeded to tell Price "the facts":

> The rocks do lie in a much more definite sequence than we have ever allowed. The statements made in the New Geology do not harmonize with the conditions in the field. . . . All over the Middle West the rocks lie in great sheets extending over hundreds of miles, in regular order. Thousands of well cores prove this. In East Texas alone are 25,000 deep wells. Probably well over 100,000 wells in the Midwest give data that has been studied and correlated. The science has become a very exact one, and millions of dollars are spent in drilling, with the paleontological findings of the company geologists taken as the basis for the work. The sequence of the microscopic fossils in the strata is very remarkably uniform. . . . The same sequence is found in America, Europe, and anywhere that detailed studies had been made. This oil geology has opened up the depths of the earth in a way that we never dreamed of twenty years ago.

To harmonize these findings with a literal reading of Genesis, all one had to do was attribute the fossil record to the ecological zones of the Noachian world. "On this clue," he wrote, "I intend to reconstruct our geological arguments." He found it amazing how many geological facts fit into his new scheme: "The geosynclines, stratigraphic sequence, overthrusts, and glacial evidences all fall into line beautifully." Fearing that he might be upsetting Price, he ended with an expression of hope that his esteemed teacher would "not feel disturbed with the fact that

I am carrying some of these points beyond where you have left them in your books."[16]

Disturbed? After devoting his very life to demonstrating "that the fossils do *not* occur in any invariable order of succession"? Price could scarcely contain his fury, especially after learning that Clark was using an evolutionary text instead of *The New Geology* in his classes at Pacific Union College and was dismissing Price's self-described "*magnum opus*" as "entirely out of date and inadequate." Price angrily diagnosed Clark as suffering from "the modern mental disease of university-itis" and of currying the favor of "tobacco-smoking, Sabbath-breaking, God-defying" evolutionary geologists. Itching to call Clark a charlatan, but concerned lest he overstep the bounds of Christian propriety, Price prudishly alluded to "an ugly word which people use to signify a man who claims to have scientific knowledge which he does not possess. It is from the Italian, through the French; it begins with a *c* and it ends with the letter *n*. But I am not going to use it; for it is a sort of word which no Adventist ought to use toward another Adventist, even if he should think it appropriate."[17]

Clark's offhand remark in a letter to Allen that "even Price himself has changed many of his views, or he would if he could get the material that is available at the present time," particularly incensed Price. For this indiscretion he accused Clark of "actionable libel or legal slander" and demanded that church officials summon Clark before "an Advent-ist tribunal" to face charges. Failing that, he threatened to seek redress in a civil court. The mild-mannered Clark, chagrined by this unfortu-nate turn of events, repeatedly tried to placate Price, reminding him that the phrase *out-of-date and inadequate* had come from Price him-self and that he owed "a great debt of gratitude for the kindly sympathy you have always manifested toward me and my work." "Let us have peace," he pleaded. "I have no argument with you, if you would be reasonable, and would discuss points of difference in a sane manner, and would not make them a reason for accusations of heresy. You are forcing the issue, and will either force me off the map, or force me to fight back, neither of which I desire." Clark, who never wavered from strict creationism or even flood geology, reminded Price of the many ideas they continued to share:

I believe that the world was actually brought into existence on the first day of creation, about six thousand years ago; that it was organized during the creation week and clothed with plant life; and that all the major types of animals were formed at that time. The great geological changes I attribute to the flood, and, in a minor degree, to changes occurring since the flood. In that I think you and I stand together.[18]

But such words of reconciliation fell on stone-deaf ears. Price increasingly took to mocking Clark for his advanced training and for his ability to handle "the many trivialities of ecology" while fumbling "the larger problems of science." "It is notorious that the men who acquire skill in the trivial minutiae of some of the sciences, such as counting the bristles on a certain part of an insect's anatomy, or the coloration pattern of the back of a bird, or the overlapping habitats of two closely allied 'species,' are usually mere children when they attempt to deal with such larger world-causes as make up the chief part of the 'problems' of geology," he sneered. "In other words, I think you have just got into water too deep for you, when you say that this textbook we are dealing with is so entirely inadequate in its handling of its problems."[19]

In the spring of 1941 Price finally got his wish: a formal hearing with his adversary before a special committee of preachers appointed by the Adventist church. The proceedings, however, "turned out a fizzle" when the somewhat baffled ministers refused to condemn Clark and merely recommended that the two flood geologists get together to work out their differences. Though in no mood to compromise, Price agreed to a face-to-face meeting. "On some points," Clark recalled, Price "would say 'I settled that point fifty years ago, and there is nothing more to say.' On others he would say 'There are some things the Lord does not expect us to understand, and there is no use in trying to explain them.'" Given Price's sensitivity about implications that he had changed his mind, there was little reason to believe that he would now alter his thinking.[20]

Ironically, the recent publication of the biological part of Clark's manuscript under the title *Genes and Genesis* (1940) had pushed Price into doing just that. In his book Clark defended limited Darwinian natural selection—within genera, families, and even orders—against the "extreme creationism" of those who insisted that God had created every species. As we have seen, Price only a short time earlier had enthusiastically seconded Clark's call for limited evolution, saying that creationists should welcome all the species-making they could get. Now, however, his animus toward Clark caused him to reexamine his own biological beliefs, only to discover that he himself had conceded too much to the evolutionists. Thus in a humiliating recantation of his views of speciation, he expressed shame for having accepted so much crossbreeding between species and especially for having suggested that the prophet White had believed in the interbreeding of humans and animals. He reproached Clark for teaching the "wretched theory that there are some races of men 'still in existence' who are only sub-men, mere hybrids, half-human, but half-something else," a theory Clark had learned from Price himself. "Why," asked Price plaintively, "should you follow the worst that I have been writing instead of the best?"[21]

Through international war and peace Price kept up his petulant attack on the traitorous "semi-evolutionist of P.U.C.," calling him "an appeaser for the evolution theory" and an "intellectual Quisling." In 1946 Clark retaliated with *The New Diluvialism*, a detailed exposition of his ecological zonation theory sprinkled with jabs at Price. In this book, Clark explained artfully, he sought "to steer a middle course between the Scylla of the evolutionists and the Charybdis of many diluvialists whose zeal exceeds their information," an unmistakable reference to Price. In a passage calculated to drive the already unstable Price to distraction, Clark credited his teacher with having first pointed out the geological significance of pre-flood ecological districts. Indeed, such references litter Price's writings, but he understandably declined the honor of fathering a theory he now believed delivered creationists "hog-tied into the hands of the evolutionists."[22]

Clark's modified flood theory possessed the decided advantage of allowing adherents to retain a geologically significant deluge while simultaneously accepting the compelling evidence of order in the fossil record, of mass movements of the earth's crust, and of a series of ice ages—all within six thousand years. In the opinion of one grateful, if hyperbolic, flood geologist, *The New Diluvialism* marked "an epoch of importance in the history of science." Price, of course, viewed Clark's book not as a friendly amendment to flood geology but as an "attempt to save the most sacred cow of all the sacred cows of the evolutionists, namely, the alleged chronological sequence of the fossils." He accused Clark of deceptively featuring the data that favored the ecological-zonation theory while ignoring or downplaying contrary evidence. "In selling strawberries," he lectured, "it is not considered good ethics to put all the big ones on top of the basket, and thus cover up all the little scrawny ones at the bottom." He knew the practice all too well.[23]

Flouting good ethics himself, and frustrated by the failure of Adventist leaders to throttle Clark, Price launched his ultimate weapon, a jeremiad called *Theories of Satanic Origin*. The not-so-subtle title, Price explained, came from White, who had introduced the phrase "less than ten years after the Lord providentially led me to work out a scientific demonstration that these geological 'ages' based on the fossils are artificial and untrue." In an unrelenting denunciation of the author of *The New Diluvialism,* left nameless but not unidentified, Price called for a veritable return to the days of witch-burning. It was, complained Clark, "the most unjust and scandalous attack that I have ever known from anyone who professed to be a friend and a brother." Though injured, Clark resisted the impulse to strike back. And just twenty years later, in a classic example of turning the other cheek, he wrote a sympathetic biography of Price, whose "bitter lashing and castigation" of Clark had cost Price the respect of less forgiving co-workers.[24]

Clark was not alone among DGS members in pushing for greater acceptance of microevolution within the originally created kinds. In the early 1940s another university-trained Price protégé, Frank Lewis Marsh (1899–1992), joined Clark in advocating post-Edenic speciation. As a young man Marsh had aspired to become a physician, but poverty forced him to settle instead for a career first in nursing and later in teaching. As a student at Emmanuel Missionary College in the late 1920s he studied geology under Price and came to idolize the "old warhorse." While teaching at an Adventist school in the Chicago area, Marsh took advanced work in biology at the University of Chicago and obtained an M.S. in zoology from Northwestern University in 1935, specializing in animal ecology. Later, after joining the faculty of Union College in Lincoln, he completed a Ph.D. in botany at the University of Nebraska in 1940, where he wrote his dissertation on plant ecology and became the first Adventist to earn a doctoral degree in biology. While attending these secular universities, he resisted the impulse to challenge his professors on the issue of evolution, telling himself that he "was there to learn what they had to offer," not to convert them to his way of thinking.[25]

Like Clark, Marsh never deviated from a literal, recent creation and universal flood, but the more he learned, the more he questioned the notion that all species had originated by separate creative acts. Zoologists, he noted, had identified thousands of species of dry-land animals alone, yet Adam had been able to name all of them in a single day. Thus it seemed unreasonable to equate the Genesis kinds with the multitudinous species of the twentieth century. Besides, as he once explained to Price, his close association with evolutionists over the years had given him "an understanding of their way of thinking" and a confidence in their taxonomic work that Price could never appreciate. "You have never rolled up your sleeves and worked as one of their crowd on various research projects as I have," he reminded the self-taught geologist.[26]

In his first book, *Fundamental Biology* (1941), written from the point of view of "the Seventh-day Adventist fundamentalist," Marsh portrayed the living world as the scene of a cosmic struggle "between the Creator and Satan." Taking his cues from White, he speculated that amalgamation or hybridization had been "the principal tool used by Satan in destroying the original perfection and harmony among living things." The black skin of Negroes was only one of many "abnormalities" engineered in this diabolical way. Despite generally agreeing with Clark on the natural origin of species, he thought his California friend had gone too far in allowing for the crossbreeding of "kinds." In contrast to Clark, who believed that the biblical reference to plants and animals producing "after their kind" was merely a moral principle, Marsh regarded the Genesis statement as a biological law that forever separated the different "kinds." The examples of interbreeding between "kinds" that Clark gave in *Genes and Genesis* struck Marsh as being more appropriate to *Ripley's Believe It or Not* than to a scientific treatise. And he was sure that Clark's proof of human-beast crossings, cited as confirmation of White, "would not last five minutes under a scientific cross examination." Such lapses he attributed to the fact that Clark had studied only at Berkeley, while he himself had been privileged to take advanced work at *three* institutions of higher learning: Chicago, Northwestern, and Nebraska.[27]

Aware that Price tended to turn every difference of opinion into "a more or less knock-down-drag-out debate," Marsh anticipated the worst from his hero—and he was not disappointed. The younger man took special pride in having coined a new term, "baramins" (from the Hebrew words *bara*, "created," and *min*, "kind") to denote the original units of creation, but Price coolly dismissed this innovation as a cheap grab for public attention. And he laid his former student out cold, as Marsh described it, for "having fallen for evolutionary propaganda." But needing all the allies he could recruit for his fight with Clark and taking advantage of a growing rift between Marsh and Clark over amalgamation between humans and beasts, Price soon backed off and began supporting Marsh. Upon reading Marsh's *Evolution, Creation, and Science* (1944), Price could not resist taunting Clark with how much Price appreciated Marsh for "leading us out of the fog in the vexed matter of the 'species' question." Before long he had anointed Marsh as his successor, telling him, "When I retire from the battle my mantle shall fall upon you."[28]

"No one but a fool can claim that there is any evolution heresy" in Marsh's writings, declared Price after his change of heart. But many of his fellow DGS members felt otherwise. On circulating the manuscript of *Evolution, Creation, and Science* to society members, Marsh soon found himself having to squelch "the wild rumor which comes to me that some of you men think I am an evolutionist." He begged his "brother fundamentalists" not to equate limited variation with evolution. "It is the constant refusal on the part of many special creationists to recognize the change in nature which *actually does occur* which has led [H. H.] Newman [one of Marsh's teachers at the University of Chicago] to dub us 'ignorant, dogmatic, or prejudiced'—and I think he is justified in saying just that!" When the book finally appeared, Lammerts accused him of being "much more of an evolutionist than any man I have ever known." In reviewing the work at a society meeting described by Allen as "*a most remarkable event*," the Lutheran botanist portrayed Marsh as a man "willing to grant everything that evolutionists ask for and boil it down to just about five thousand years, on a scale that they don't even dream of in hundreds of thousands, or even millions, of years."[29]

Beginning with *Evolution, Creation, and Science*, Marsh sanitized his major published works to avoid mentioning "the Spirit of Prophecy" (that is, the writings of White), which he feared would only repel non-Adventist readers. Hoping that a defense of creationism written by a credentialed biologist might cause the scientific world to take note, he had his publisher send out complimentary copies to prominent evolutionists, including the Harvard zoologist Ernst Mayr (b. 1904) and the Russian-born Columbia geneticist Theodosius Dobzhansky (1900–1975). The former declined to comment, preferring to take the advice of "the rabbit in Walt Disney's film *Bambi*: 'Don't say anything if you don't have anything nice to say.' " Dobzhansky, however, believing that "the majority should at least consider the minority view and subject it to criticism," engaged in an extended correspondence with Marsh that vividly reveals the issues dividing creationists and evolutionists.[30]

Shortly before the appearance of Marsh's book, the Orthodox Russian émigré had remarked in his *Genetics and the Origin of Species* (1937) that "among the present generation no informed person entertains any doubt of the validity of the evolution theory in the sense that evolution has occurred." But reading Marsh's book convinced him otherwise. In reviewing it for the *American Naturalist*, Dobzhansky announced that Marsh had written what he had previously thought to be impossible: a sensibly argued defense of special creation. Dobzhansky expressed particular surprise at discovering how much evolution (within "kinds") a creationist such as Marsh was willing to grant: "He outbids evolutionists on the score of the speed of the changes, for he assumes that all dogs, foxes, and hyenas are members of a single 'kind,' and, therefore, must have descended from a common ancestor in any event less than 6000 years ago. In rejecting macroevolution, Marsh's book taught the valuable lesson that "no evidence is powerful enough to force acceptance of a conclusion that is emotionally distasteful."[31]

Although Dobzhansky found Marsh's ideas scientifically invalid and religiously subversive, he respected the church-college biologist for being "the only living scientific anti-evolutionist." The obscure creationist, for his part, could scarcely conceal his delight at having the unexpected chance to argue his case before one of the leading evolutionists in the world. In response to Dobzhansky's comment about outbidding the evolutionist, Marsh amended and clarified his position. He explained that God in originally stocking the world with plants and animals had created not only "kinds" but "varieties" within those kinds, capable of crossbreeding. Thus he professed not to claim, for example, that all dogs and foxes had descended from a single ancestor—ignoring for the moment the problem of overcrowding on Noah's ark. And though he had said virtually nothing about the mechanism of change in his book, he now admitted that the concepts of natural selection and survival of the fittest were "really extremely important" in explaining the present distribution of species.[32]

The central issue separating the two biologists hinged on the nature of scientific proof. Marsh, who assimilated all the evidence of microevolution into his creationist paradigm of changes within "kinds," demanded nothing less than laboratory-based demonstrations of macroevolution. But, as Dobzhansky pointed out, the evidence for such large-scale evolution rested on inference, not direct observation. Because macroevolution took place in geological time, he patiently explained, it could "be proven or disproven only by inference from the available evidence." Marsh, predictably, found this argument unconvincing. "Alas! Inferential evidence again!" he exclaimed. "Is there no *real* proof for this theory of evolution which we may grasp in our hands?" Eventually, explanation gave way to frustration, with Dobzhansky finally brushing Marsh's concerns aside with the quip "If you demand that biologists would demonstrate the origin of a horse from a mouse in the laboratory then you just can not be convinced."[33]

Marsh rejected the inferential evidence for macroevolution primarily because of his prior commitment to the scientific and historical veracity of the Bible. In justifying such an allegiance, he maintained that "in not one single instance" had the Bible been proven false. "That very real fact should mean something to us as scientists," he argued. "In the light of that fact, the Genesis statement regarding the origin of living things must likewise be tested if we are to make wise use of the sources of truth at our disposal." Dobzhansky found this line of reasoning unpersuasive, but he credited Marsh for at least stating candidly in his book that "the account given by the Bible is settled for you before you begin to consider the biological evidence."[34]

After more than two months of nearly weekly exchanges, Marsh trusted Dobzhansky sufficiently to expose some of his innermost fears and feelings. In what he suspected would be his final letter, he assured the geneticist that he was not "a chronic grouch who goes about looking for something to argue about." Nor was he "looking for ease, comfort, and a good name." Though he disliked being at odds with his scientific brethren, he was "willing to take it on the chin" if principle required it and if he could only get mainstream scientists to accept special creation as a legitimate alternative to evolution. In closing he expressed the hope that Dobzhansky would find "some diversion in these letters, some pleasant mental gymnastics, and possibly experience a broader acquaintance with unusual ideas so that the benefits will not all be going one way." Six years later, in the third edition of his *Genetics and the Origin of Species*, Dobzhansky cited Marsh as the exception to the rule that "an informed and reasonable person can hardly doubt the validity of the evolution theory, in the sense that evolution has occurred." The creationist biologist proved "only that some people have emotional biases and preconception[s] strong enough to make them reject even completely established scientific findings." It wasn't much of an acknowledgment, but Marsh appreciated the recognition.[35]

THE ROCK OF PRE-GENESIS TIME

The issues raised by Clark and Marsh may have buffeted the DGS's fragile vessel, but they did relatively little damage, in part because the two dissenting biologists lived so far from the core group. A much graver danger closer to home threatened, in Allen's words, to destroy the society on "the same rock, *pre-Genesis time for the earth*," that had wrecked the old RSA. As early as 1943 Allen was complaining of disruptions resulting from "the theory of radioactive time which would place 1 billion 600 million years between verses 1 and 2 of Genesis I." Largely to accommodate new evidence from radioisotope dating, some of the younger, better-educated members of the society were pushing for acceptance of an ancient earth and solar system, with most denying the existence of life on earth until the Garden of Eden. In this way they could salvage a literal six-day creation and fossil-burying flood while accepting the latest scientific findings about the age of the solar system.[36]

Leading the insurgents was the youthful dermatologist with the mel-
lifluous name, Molleurus Couperus. Reared as an Adventist in the
Netherlands, where White had only an attenuated influence, Couperus first came to the United States to study in denominational schools.
After graduating with an M.D. from the College of Medical Evangelists, he moved to New York City for advanced work in dermatology at
Columbia University. There he sat in on Dobzhansky's lectures and
developed a special interest in radioactive dating, based on the discovery that the ages of rocks could be determined approximately by
measuring the extent to which a particular radioactive element, say
uranium-238, had decayed into its daughter-product, lead-206. By the
time he settled in southern California in the early 1940s and became
active in the DGS, he had already reached the conclusion that the
entire hypothesis of evolution stood or fell on one point: "the determination of the age of the various layers and the age of the contained
fossils by their radioactivity and helium content."[37]

For a while Couperus remained undecided about the implications
for a recent creation. Perhaps, he explained to Price, God had created
the rocks with the appearance of age, just as he had created mature
trees in Eden. "I think that we cannot know just which way the Lord
formed the material of this earth, or of the rest of the universe." But if
the evidence indicated an old earth, he felt confident the needed time
could be accommodated between the first two verses of Genesis. Allen
and other young-earth members of the DGS, who were counting on
Couperus to refute the time claims of the geochemists, soon learned to
their dismay that he was giving medical students an age of at least 1.6
billion years for the earth. The doctor tried to put their minds at ease
by insisting that his conclusions had "nothing to do with either the
problem of creation as such or the problem of evolution, but only with
the problem of the time of the initial creation of the earth." But both
he and they knew better.[38]

For decades Price himself had waffled on the question of an ancient
earth. In his early years as an antievolutionist, he had freely admitted
that the solar system might have been created "any number of millions
of years in the long ago." And though he had abandoned that view in
the 1920s, he never made a fuss over the issue of pre-Genesis time—so
long as no pre-Adamic life was involved. During the RSA controversies
he had told his fellow directors that Higley could "make the mass of the
earth just as old as he sees fit. All I am concerned about is the part of
the earth which was disturbed at the time of the Flood." In his book
Genesis Vindicated (1941), Price went out of his way to distinguish
between an ancient universe and a young solar system. Because Gen-

esis said "absolutely nothing at all" about the age of the universe outside the solar system (except for a passing reference in Genesis 1:16 to also making the stars), he saw no need to argue that "the age of the whole sidereal universe must be 'cabin'd, cribb'd, confined' within the compass of the creation week of the first chapter of Genesis." But since Exodus 20:11 declared that "in six days the Lord made heaven and earth, the sea, and all that in them is," he favored "an absolute creation, *ex nihilo*, of the material of our earth [and solar system] at this beginning of the six days." In a note at the end of his book, he dismissed the new radioactive method of dating rocks as being "full of fallacies, of slipshod methods, and of sheer charlatanry."[39]

Believing that he had the solid backing of Price on this matter, Allen attacked Couperus for violating "the integrity of the Creation Week" and for following "the facts of nature" rather than the Word of God. At the same time Allen encouraged his friend Burdick to write an exposé of the "Gross Fallacies in the Radioactive Time Theory," grandly described by Allen as *"one of the most important papers ever produced on the relation of science and the Bible."* According to Burdick, radioisotope dating suffered from the same "glaring weakness" that undermined calculations based on sedimentation: the assumption of uniformity. As corroborating testimony he cited the prewar warning of Watson Davis (1896–1967) in the *Science News Letter* that "the radioactive 'time clock' method of determining the age of the earth may be proved wrong if uranium can be split up in the strange new manner" conceived by physicists—an eventuality subsequently demonstrated in dramatic fashion by the detonation of the first atomic bombs.[40]

By June 1945 Allen had grown so obstreperous over the age of the earth that the society's board of directors removed him as secretary and installed Couperus in his place. That action only enraged the combative captain, who made life miserable for his perceived enemies. By October Courville and Couperus were so fed up with his abuse they submitted their resignations from the board. When their colleagues refused to accept their offers, Allen turned in *his* resignation, reportedly saying that "it had become clear to him finally that there had been something very much wrong with his attitude and conduct, and that he had been very selfish." The board declined his offer, too, but on condition that he attend no board meetings for six months and that he "immediately stop all letter writing and activity regarding the organization." A week later the board voted to cut off his monthly allowance and to change the name of the organization to the Natural Science Society. Before the end of the year the officers had transferred all of the assets of the DGS to the new entity and had decided to discontinue the old *Bulletin of Deluge Geology* and begin a new journal, the *Forum for the Correlation of Science with the Bible,* edited by Couperus. To make it palatable to "outsiders," Couperus dropped all references to "the Spirit of Prophecy," the code phrase for Ellen White.[41]

Allen, needless to say, did not take these changes silently. In a document headed "The Original Society Illegally Supplanted and All Scriptural Standards Abandoned" he lashed out at the *"small group of illegal usurpers"* who were introducing heretical notions that undermined the most central belief of Seventh-day Adventism: Sabbath-keeping as a memorial of a six-day creation. Because he had refused to join the apostates, he complained with some justification, he had been *"held up to scorn everywhere* as 'hard to get along with,' 'uncooperative,' 'ignorant of science,' and still worse."[42]

Allen felt especially aggrieved by Price's role in the affair. For years Allen had served alongside Whitney as one of Price's most loyal lieutenants, but now he found himself pushed aside by a smooth-talking skin specialist who questioned the very idea of a universal flood. Remembering his own flirtations with an old earth, and perhaps swayed by Couperus's urbanity and intelligence, Price refused to chide him for his divisive views, urging him only to think over the declaration of the fourth commandment that "in six days, God made . . . A L L that in them is." Such inclusive language made it difficult for Price not to think that the Edenic creation included "the granite and all the primitive or azoic rocks, or in other words, the body of the earth." As for the radiometric findings that so impressed his younger colleague, Price wrote: "Too many blunders concerning time-estimates during the past century or so arise in my mind; and I have to say with the Scotsman: 'I ha' m' doots.' " By the time of the big breakup he did not know where he stood on the matter, except he was sure that whenever God created the solar system "He was not indebted to preexisting matter," a White shibboleth.[43]

The first volume of the *Forum,* under Couperus's editorship, carried a cluster of articles on the age of the earth and the meaning of the first two verses of Genesis. Burdick and Clark defended what the latter called "the ultra-literal view," that God spoke the matter of the earth into existence at the beginning of the six-day creation, while John Lowell Butler (1896–1977), an ex-Adventist singing evangelist who had sat beside Clark in Price's classroom in the early 1920s, argued for an ancient earth. Price, by this time unwilling to side with Clark on any matter, finally declared himself unequivocally for an old earth. He announced that because he no longer believed that "the fourth commandment was decisive in favor of the short chronology for the age of the earth," he felt free "to let the scientific evidence decide the matter." And the scientific evidence, as he now saw it under Couperus's tutelage, overwhelmingly favored "the view that the body of the earth had existed long before Creation Week, though of course it was not then stocked with any plants or animals." He estimated that the physical universe must have started "about 2,000 million years ago."[44]

Price's about-face on so vital an issue created a sensation among fellow believers. "Since Price came out for the new theory," Allen wrote excitedly in 1948, "many of our S.D.A. leaders, and even college professors, seem to think that 'if Price favors the new theory SURELY IT CAN'T be at all evolutionary, or contrary to Deluge Geology,' etc." Ironically, by the time Allen penned these words, Price had already abandoned what he called "uniformitarian" old-earth creationism for "supernaturalist" young-earth creationism. Apparently a rebuke from his old friend Whitney proved decisive in convincing him that the arguments for a recently created earth were "rock-bottom truth." Price now saw clearly that "this radioactivity business . . . is full of uncertain assumptions and fallacious arguments. I cannot have any more scientific respect for it than for the argument about 'varves,' or about 'index fossils.' They all belong together,—just some of the tricky methods used by the Great Deceiver to befuddle the people of the last days." He acknowledged his inability to answer all the arguments of "the radioactivity men," but comforted himself with the realization that "an old man with one foot in the grave and the other on a slippery banana peel cannot hope to clean up such a confused and confusing subject."[45]

Allen, still smarting from Price's betrayal in siding with Couperus, adopted a wait-and-see attitude. In an emotional letter filled with underlining, circled words, and capital letters he informed a young-earth colleague that "Price's 'quickie' doctrinal 'recantation' " left him unimpressed. "It mostly goes only to prove how hot the spot was in the house he himself had set on fire. But probably even he knows that this alone will by no means put out the fire." Allen feared that "great harm" had already been done by Price and his satanically inspired pals. Thus he continued with his plan, worked out in "the utmost secrecy," to publish a condemnation of creationists who adopted "the ancient heathen conception" of an old earth. In it he denounced radioisotope dating as merely "a new modification in the evolutionary geologic age system," incompatible "with every principle of Deluge Geology."[46] But the "harm" had indeed been done. From the mid-1940s on, flood geologists, especially within the Adventist tradition, enjoyed the freedom to choose either an old or young earth—so long as they admitted no earthly life before Eden.

Old-earth advocates were, understandably, even more upset than Allen with Price's recantation. And the reaction of some of them showed that the conservatives held no monopoly on tactlessness and paranoia. Butler, fearing White's predicted persecution at the hands of Catholics, saw a papist plot behind it all. The attitude of six-thousand-year men like Allen and Marsh seemed strikingly similar to that of "Roman Catholic spies and agents" then infiltrating the *"conservative"* ranks of American Protestantism. "Catholics are very clever and very patient through the years in disguising themselves and in getting into every Protestant and anti-Catholic organization," he explained to Couperus. Though Butler stopped short of accusing Price of being in the service of Rome, he wrote his former teacher a sarcastic letter attributing his fickleness to hysteria. "I hope your inner sight is not failing with your eyesight," he told the nearly eighty-year-old man.[47]

By 1947 "nearly everyone" had come to the conclusion that the Deluge Geology Society/Natural Science Society was dead. Marsh, having waited eagerly for the promised reports on antediluvian humans, finally concluded in disgust that Allen and his associates were "either a bunch of fakers . . . or just misguided well-intentioneds who actually have some valuable scientific data that they are trying to keep hid." Even Couperus, the society's secretary and editor, conceded that the group had "done practically nothing in an organized way since last December." Allen may have been a pest, but at least he prodded people into action. The new leaders quickly learned that it was "impossible to get anyone to do anything that takes more than just a minor effort." The publication of the 1947–1948 volume of the *Forum* sapped the society of its last breath, though some former members in Loma Linda kept a Seminar on Science and Religion going through the 1950s.[48]

After the breakup in 1945, the ostracized Allen had teamed up with Burdick to form Amazing Discoveries, Inc., with Allen assuming primary responsibility for library research and Burdick for fieldwork, such as hunting for fossil footprints in the Southwest. The two also collaborated on a top-secret project they called the Sacred History Research Expedition, dedicated to finding Noah's ark, a dream of Allen's since the 1920s. The conspiratorial Allen planned to keep news of the anticipated discovery secret until "M-Day" (Message Day), when the find would be announced to the world in repeated international broadcasts followed by sound newsreels. "Only a *Sudden, Pent-up,* and *Spectacular Announcement,* fully prepared in secret, can gain the full attention of the whole world to our *Message,*" he explained in urging strict confidentiality. Burdick would later ascend Mount Ararat, but Allen's dream never materialized.[49]

Increasingly isolated and cantankerous, Allen continued until the end of his life to pursue vindictive and grandiose plans. In the early 1950s he sought out the new president of the Seventh-day Adventist church, W. H. Branson (1887–1961), to offer his services in rooting out members who followed the dangerous "new fad" of radioactive dating. The cagey executive, having no desire to play Allen's little game, would say only that "the Biblical record of Genesis is still our official viewpoint as a people and I hope it shall ever be so." A few years later Allen claimed to have invented a new method for discovering uranium ore that had won the backing of a prominent Harvard geochemist. "I AM SURE," he wrote to a friend, "it is a *miracle* of God, that I, even I, should be enabled to arouse and CHALLENGE the world's (best) geochemist." And before his death in 1960 he was badgering Cecil B. De Mille, the famous Hollywood producer, to follow up *The Ten Commandments* with a screen spectacular depicting the story of Noah and the flood.[50]

Couperus, having committed himself to science, slowly metamorphosed into a theistic evolutionist. In the mid-1950s he studied physical anthropology at UCLA and began attending professional anthropological meetings. Later he struck up a friendship with Louis B. Leakey (1903–1972), codiscoverer of the Zinjanthropus skull at Olduvai Gorge, and on three occasions in the 1960s he spent extended periods with Leakey in East Africa studying the remains of hominid fossils. These experiences convinced him not only of the great antiquity of the human race but of its evolution as well. Though he liked to say that he remained "a creationist," his minimalist version of creationism required little more than the supernatural origin of a single cell. Allen himself could scarcely have scripted a more fitting end to the DGS story.[51]

EIGHT

Evangelicals and Evolution in Great Britain

During the decades immediately following the fundamentalist crusade of the 1920s, reports of a creationist revival in Europe, especially in Great Britain, buoyed the hopes of American antievolutionists. For years they had dreamed of reinforcements from abroad, but even conservative Christians had refused to come to their aid during the battles that followed World War I. As historian George M. Marsden has explained, British evangelicals, always a minority, had developed a stronger tradition of theological toleration than their revivalist brothers and sisters in America, who until the twentieth century had never experienced minority status. Thus, while displaced American evangelicals fought to recover their lost position in society, British evangelicals adopted a nonmilitant live-and-let-live philosophy that stressed personal piety over doctrinal purity.[1] Even the venerable Victoria Institute, long a haven for British creationists, had by the 1920s become a stronghold of liberal evangelicalism and theistic evolutionism and a shadow of its former self. Not until the 1930s did Britain's handful of creationists muster the energy to organize themselves as the Evolution Protest Movement.

CREATIONISM AT THE VICTORIA INSTITUTE

The near-simultaneous appearance of such works as Darwin's *Origin of Species* and the notorious *Essays and Reviews* (1860), which exposed the Bible's "erroneous views of nature," prompted a group of concerned Britons in 1865 to organize the Victoria Institute, or Philosophical Society of Great Britain. The institute's stated objective was to defend "the great truths revealed in Holy Scripture . . . against the oppositions of Science, falsely so called." Though not officially opposed to evolution, during the last decades of the nineteenth century it attracted a number of scientists skeptical of Darwinism, including the North Americans John William Dawson and Arnold Guyot, the last of the nineteenth-century scientists to cling to creation. Its principal founder, James Reddie (d. 1871), regarded Darwin's theory as "*inharmonious, inadequate, inconsistent,* and utterly *incredible,*" and one of the three original vice presidents, Philip Henry Gosse (1810–1888), represented the most thoroughgoing creationism found in post-Darwinian British science. In addition to preaching for the millenarian Plymouth Brethren, he worked as a marine biologist. Dubbed "an honest hodman of science" by Thomas H. Huxley, Gosse suffered the misfortune of being immortalized by his disillusioned son, Edmund, in the bittersweet autobiography *Father and Son.* In response to the assertions of geologists that fossils chronicled the history of life on earth, Gosse tried, according to his son, "to justify himself as a strictly scientific observer who was also a humble slave of revelation" by suggesting that God had created the earth with the appearance of age. Thus Adam, for example, though never attached to a mother by means of an umbilical cord, possessed a navel, or *omphalos,* the title Gosse gave to his quaint book on the subject, published in 1857.[2]

During its heyday in the late nineteenth century, the Victoria Institute enjoyed considerable prosperity, with membership, which included a number of Fellows of the Royal Society, reaching a high of 1,246 in 1897. Within less than twenty years, however, enrollment had plummeted to under one-third that number, and apathy had set in. Between 1924 and 1928, when George McCready Price attended virtually every meeting while living near London, he found the institute wallowing in "a rut." And though a British subject by virtue of his Canadian birth, he also found little sympathy for what was widely regarded as his peculiarly American brand of antievolutionism. Shortly after his arrival in England, the officers of the institute, of which he had been a member for several years, graciously invited him to share his views. Thrilled by the honor, he gave a synopsis of the new catastrophism, accompanied by a guarantee that it would provide a "final solution" to the problems raised by evolution. But most institute members remained unpersuaded. One cleric in the audience likened Price to "the proverbial bull in the china shop. With one horn he has destroyed the science of Geology and with the other that of Paleontology, and then, standing among the ruins, he tells us to put the fragments together and construct a new system of Geology." A few examples of upside-down strata seemed insufficient to warrant "rejecting the whole system of stratigraphical geology built up after more than a century's painstaking observation and collection of facts by thousands of competent observers and workers in all parts of the world." A disappointed scientific member of the institute rudely warned his fellows against accepting "blindly the enthusiastic material of some newly-inspired amateur."[3]

The next year Price won the institute's triennial Langhorne Orchard Prize—and twenty guineas—for an essay on "Revelation and Evolution." But he still failed to win the respect he coveted. The longtime editor of the institute's journal, which published the essay, attached a disclaimer expressing his misgivings about Price's attempt to foment "a new crusade against Evolution" in the United Kingdom. The editor seconded another member's condemnation of Price's effort "to drive a wedge between Christians and scientists" in Britain, as he had done across the Atlantic.[4]

Price suffered even greater humiliation when the Rationalist Press Association lured him into a highly publicized debate in Queen's Hall, London, with the ex-Franciscan priest turned rationalist philosopher Joseph McCabe (1867–1955). Price, who until this time had avoided face-to-face confrontations, found himself before a hostile audience of about three thousand, matched against a skilled and experienced debater. In contrast to McCabe, who entertained the crowd with pictures on a screen, Price took the professorial approach, standing by a stack of books from which he quoted at length. During Price's last fifteen minutes hecklers interrupted him so often he could not finish his rebuttal. "At one time," he wrote home to a friend in America, "I suppose a thousand people were on their feet at once, yelling and arguing with me or with their next-seat neighbours. It was a lively time."[5]

Generally, Price received anything but a lively reception during his four years in Great Britain. "I have been somewhat disappointed in the apparent lethargy of the friends of the Bible over here, regarding the subject of evolution," he wrote to the American fundamentalist John

Roach Straton. The common people learned little but evolution in their churches and schools, and the British "clergy were so unmercifully castigated and browbeaten by that intellectual snob and bully, T. H. Huxley, that they have never dared to call their souls their own since then." Considerable testimony supports Price's assessment. When the journal *Nature* solicited the reactions of leading British clerics, scientists, and educators to the antievolution turmoil in America in the mid-1920s, no one seemed to fear a similar outbreak of "intellectual terrorism" in Great Britain. "The normal educated Christian in Great Britain regards the process of evolution as the machinery by which God has created man," observed E. W. Barnes (1874–1953), the liberal bishop of Birmingham, in a typical statement. "Every divine of any eminence among us accepts this point of view." Indeed, Price knew only two prominent English divines, one a Presbyterian, the other a Wesleyan, who were actively fighting evolution.[6]

The fortunes of British creationism improved slightly in 1927, when the Victoria Institute installed as its president the aging John Ambrose Fleming (1849–1945), a distinguished electrical engineer and inventor who two years earlier had been nominated by the pioneer of wireless telegraphy Guglielmo Marconi (1874–1937) and other scientific luminaries for a Nobel Prize for his contributions to the telephone, telegraph, radio, and electric light. Son of a Congregational minister, he had moved as an adult to the evangelical wing of the Church of England. He unquestioningly accepted the miracles and prophecies of the Bible, including the imminent Second Coming of Christ, and resolutely defended its teachings against higher critics and scientific materialists. His opposition to evolution, such as it was, stemmed from his growing conviction in the years after World War I that Darwin's theory was doing more than anything else to weaken public faith in the Scriptures.[7]

Although Fleming fancied himself a creationist and insisted on the special creation of the soul, his toleration of divinely guided development and his acceptance of pre-Adamic humans often led others to mistake him for a theistic evolutionist. In 1927, when he addressed the Victoria Institute on the relationship between evolution and revelation, one member congratulated him for presenting "an able, well-reasoned case for Evolution." Another contrasted the views of Fleming with those of Price, noting that the former, "with better knowledge of the present attitude of Evolutionists, finds no such *impasse* between Scripture and the latest scientific account of Evolution." Fleming, taken aback by this reading of his stand, hastened to clarify his position. If evolution means "a self-acting impersonal agency, bringing about the development of the Universe, then my object was not to make out a case *for* it, but *against* it, and to show that there are discontinuities in Nature which cannot be bridged by any of the physical or natural agencies with which we are acquainted at present," he explained. "Even if the term 'Evolution' is restricted to denote the slow changes from the simple to the complex, then I hoped I had shown that, nevertheless, all such processes require guidance." In short, he was willing to accept development so long as it did not preclude design and occasional divine interventions. Certainly, he had no dispute with geologists who stretched the history of the earth over millions of years.[8]

Human history, however, concerned him greatly. By the mid-1930s Fleming had reached the following conclusions: First, pre-Adamic humans, clearly distinguishable from animals, had inhabited the earth prior to "the arrival of the Adamic man at the approximate date 5500 B.C." Second, the creation of the Adamic man, who was to convert existing humanity to a belief in the Creator, produced "a being more eminently endowed with psychical faculties of initiative, authority, and powers of intercommunication than before." Third, because Adam and Eve were not the parents of the whole human race, the Bible revealed only a small, but central, segment of human history. For his presidential address to the Victoria Institute in 1935, Fleming summarized his views on anthropology and the Bible—and caught the attention of the leading London newspapers. Despite his acceptance of planetary and animal evolution and the antiquity of the human race, the papers portrayed him as having "delivered a fierce and heavy assault on the centre of modern scientific doctrine, the principle of the evolution of the universe, of the animals and man." The eminent scientist, they said, had gone further in opposing evolution than even most churchmen. The flap generated by Fleming's address moved the anthropologist Sir Arthur Keith (1866–1955) to write a book-length reply to the new critics of Darwinism in which he warned darkly that "Fundamentalism is certainly much more prevalent in England and in Scotland than educated and liberally-minded people have hitherto believed." Perhaps so, but it was a fundamentalism few Americans would have recognized.[9]

THE ACWORTH CIRCLE

Before returning to North America in the late 1920s, Price, noting that British creationists were "scattered and divided into a number of small, weak, and insignificant groups or societies," called for a unification of effort that would, he predicted, "bring out a surprising manifestation of strength in defense of the Bible." In the early 1930s Sir Ambrose, as he was now called, joined Bernard Acworth (1885–1963) and Douglas Dewar in forming Britain's first creationist society, the Evolution Protest Movement. Acworth, the son and grandson of Anglican clergymen, had trained at the Royal Naval College before embarking on an illustrious career as a submariner, which included winning the Distinguished Service Order during World War I and later becoming one of the pioneer advocates of sonar. After his retirement from the Royal Navy about 1930, he turned to free-lance journalism, serving as naval correspondent for such newspapers as the *Morning Post* of London and the *Yorkshire Post*. A staunch opponent of socialism, air power, and imported oil, he twice stood unsuccessfully for Parliament, in 1931 and again in 1942. In 1929 he published the first of over a dozen books, *This Bondage*, an eccentric critique of evolution, relativity theory, and air power, provoked in part by the military's growing infatuation with airplanes. Convinced (for reasons beyond my comprehension) that biologists derived one of their most conclusive proofs for organic evolution from the "mysterious and wonderful" migratory habits of birds, he sought to demonstrate that "the *scientific* treatment of birds in flight"— especially the observation that such birds do not feel the wind—gave no sustenance to the theory of evolution.[10]

Although the scientific community paid Acworth's unconventional ideas little notice, his book caught the attention of Douglas Dewar, barrister and amateur ornithologist, who was himself beginning to doubt the validity of organic evolution. As a teenager he had enrolled as a scholarship student in Jesus College, Cambridge, and in 1895 had taken a first-class honors degree in natural science. He then studied Hindustani and law before entering the Indian Civil Service, in which he eventually rose to the position of auditor general. While stationed in India, he took up zoology as a hobby and became an acknowledged expert on Indian birds. In 1909 he collaborated with the deputy superintendent of the National History Museum in Calcutta on a book called *The Making of Species*, a proevolutionary work in which the authors sought to steer "a course between the Scylla of use-inheritance and the Charybdis of the all-sufficiency of natural selection." Because of the "ghastly mess" the British were making of India, he took early retirement in 1924, returned to England, and, living off his pension, turned his attention to a critical evaluation of evolution.[11]

Because all of his teachers and textbooks at Cambridge had taught evolution, Dewar had left school "a rather half-hearted" Darwinist.

But, as he later explained, "subsequent study and field work led me first to reject Darwinism and finally evolution." He also grew increasingly concerned about the great harm evolution was doing "to the morality of the white races." By 1931 he had completed the treatise *Difficulties of the Evolution Theory*, a catalog of alleged problems ranging from geographical distribution to blood reactions. Swayed in particular by the scarcity of transitional forms in the fossil record and by his inability to identify natural forces capable of originating new organic types, he recommended adopting "a provisional hypothesis of special creation . . . supplemented by a theory of evolution." The biologist, he argued, had only two choices: "Every kind of living organism must have originated by one or the other of two processes— evolution or special creation."[12]

After finishing the first draft of his manuscript, Dewar stumbled across the work of the French zoologist Louis Vialleton (1859–1929), late professor of comparative anatomy at the University of Montpellier. Vialleton had attracted considerable attention in the 1920s for suggesting a theory of discontinuous evolution roughly analogous to the quantum theory of light. This sometimes led antievolutionists erroneously to equate his position with special creationism and to hail him as "the greatest comparative anatomist of the twentieth century." Dewar, hoping to make his own views more palatable to skeptical readers, liberally sprinkled his text with references to the Frenchman before sending it off to the printer.[13]

For the same reason he also carefully sifted out all mention of religion. Although a committed Christian who believed in the plenary inspiration of the Bible, he wanted potential readers to say "Here is a book, written without theological bias, which combats the doctrine of evolution on purely scientific grounds." In private he readily confessed his fundamentalist leanings. "I say, the Bible cannot contain false statements, and so if its statements undoubtedly conflict with the views of geologists, these latter are wrong," he wrote one correspondent in 1931. To accommodate the geological evidence of an ancient world, he at first proposed an idiosyncratic gap theory of his own making. Unlike the ruin-and-restorationists, he placed an interval between the first two *chapters* of Genesis rather than between the first two *verses*. The fossils, he believed, were relics of life on earth between the six-day creation of Genesis 1 and the much later Edenic creation described in Genesis 2, which introduced a "fully spiritual man." By the 1940s he had come to favor a single, ancient creation in which "all the main types of living beings were brought into existence by one creative act in considerable numbers, each type in the parts of the earth that were best suited to its habits."[14]

Dewar's book, printed at his own expense because no publisher would touch it, generated about as much public interest as the average Sunday sermon. It did, however, catch the eye of Price, who brought the book to the attention of antievolutionists in America—and brought his own works on flood geology to the attention of Dewar. At first Dewar, pleading ignorance in geological matters, refused to comment on Price's geology, but within a few months he was telling Price that "the more I think over your views the better I like them." Dewar especially empathized with Price's lonely struggle to win acceptance for flood geology, and he hoped that with the help of supporters such as Byron C. Nelson, Dudley Joseph Whitney, and himself, they might "yet be able to establish [Price's] theory, or one very like it." But despite his professed fondness for flood geology, with its common-sense reading of Scripture, Dewar found several key aspects of Price's new catastrophism difficult to accept. He failed to see, for example, how a fossil bed even a thousand feet thick could have been laid down in a few months, and, like Harold W. Clark, he regarded the evidence for a predictable geological column as too great to dismiss. For a while he entertained the notion that during the deluge animals had been sorted according to their specific gravity, which would explain, among other things, why vertebrates are not usually found in the lowest deposits. "Until their flesh had rotted they would float, whereas the shelled invertebrates would begin to sink as soon as they were dead." Similarly, he suspected that "the comparative scarcity of bird fossils would be explained by the fact that as their bones are aerated, their skeletons float." The thought prompted a quick experiment: "I have just verified," he wrote Price gleefully, "the fact that they do float by dropping the skull of an Indian skimmer into a basin of water!"[15]

Dewar's reservations about flood geology multiplied as he got to know Lt.-Col. Lewis Merson Davies (1882–1955), already by the early 1930s the most knowledgeable geologist and paleontologist in the world of creationism. Though both men had lived in India at the same time, they apparently had remained unaware of each other's existence until the publication of Dewar's *Difficulties of the Evolution Theory* brought them together. Davies, born in central India, spent his boyhood in Scotland before joining the Royal Artillery in 1900. Like the American Harry Rimmer, he took up boxing in the military, eventually winning the middleweight championship of Officers of the Army and Navy. For nearly two decades, from 1907 to 1926, he served in the Indian Artillery. While stationed on the northwest frontier of India, he began studying the geology of the area and subsequently published several papers on the topic. Upon returning to Scotland to complete his military service, he settled in Edinburgh and began reading at the university for an M.A. in geology, which he finally received in 1934. To fend off critics who might dismiss him as just "an old soldier" and not take him as a serious geologist, he continued his graduate studies at the University of Edinburgh, obtaining a Ph.D. in 1938 and a D.Sc. three years later. Thereafter he continued his geological investigations at the Grant Institute of Geology and came to be recognized as "a leading authority on the determination and distribution of Foraminifera— particularly those found in late Cretaceous and early Tertiary strata." Associates generally found him a likeable and well-informed colleague, perhaps better known for his work with the Boys Brigade movement in Scotland than for his stand against evolution.[16]

Religiously, Davies was what Dewar called a "Fundamentalist." In one of his earliest publications on science and religion, a tract called *The Significance of Modernism* (1927), Davies defended both a *"literal Bible"* and a literal Second Coming. Like Price, whom he met briefly during the flood geologist's sojourn in England, he attached great significance to the prophecy of 2 Peter 3:3–7—that scoffers in the last days would deny Noah's flood, declaring that "all things continue as they were from the beginning of the creation"—which he took to be an indictment of uniformitarian geology. As Davies read the passage, Peter was saying "that belief in uniformity is to lead to rejection of belief in the Flood, to rejection of belief in the Second Advent, and the rejection of belief in the future judgment and everlasting perdition of sinners." Also like Price, he gave geology a privileged position in any discussion of organic evolution. "You see," he wrote, "monkeys are not turning into men to-day, nor are lizards turning into birds. If such things ever happened, it was in the remote past; and only the geologist—studying fossils—deals with that past."[17]

In his most widely read work, *The Bible and Modern Science*, which began in 1923–1925 as a series of articles in the *Indian Christian* and went through four book editions by 1953, Davies wrote approvingly of Price as a fellow creationist and student of Bible prophecy. But he thought it impossible to assign, as Price did, "practically all geological phenomena to the effects of one brief Deluge. The very nature of *local* successions, quite apart from any 'onion coat' theory, makes that idea, to my mind, impossible." Privately he cited additional evidence against flood geology: the existence of multiple fossil forests, "regularly rooted and *in situ*, one above the other," which ruled out a single burial;

"fossils within fossils" found in "conglomerates with a matrix containing quite a different fauna, belonging to a higher level in that sequence"; and the fact that Ararat, being formed mainly of sedimentary deposits, would have been "of pea-soup consistency" at the end of the flood and thus incapable of supporting the loaded ark. It was "inconceivable," insisted Davies, "that a Flood—a single short-lived Flood—could have produced these results." He also faulted Price for giving insufficient weight to the scriptural testimony in favor of "*successive* creations." The indefinite gap between the first two verses of Genesis provided ample time, he argued, for at least one creation, and maybe several, before Eden. In terms of adhering to the literal Word of God, he yielded nothing to Price. "Remember," he wrote, "that I have the Bible very much behind me in holding that the earth is very old, with a history going far back beyond the Six Days (which, incidentally, I take as literally as you do)."[18]

THE EVOLUTION PROTEST MOVEMENT

Dewar's failure to find a publisher for *Difficulties of the Evolution Theory* tipped him off to the depth of prejudice against creationism. But it was his experience trying to publish work critical of evolution in a mainstream scientific journal that seems to have convinced him that creationists should start their own organization. The incident began when Dewar and a collaborator submitted a short paper on the distribution of mammalian fossils for publication in the proceedings of the Zoological Society of London, of which both authors were fellows. Their statistics, showing a greater number of mammalian genera in the lower Oligocene than at present, appeared to undermine evolutionists' claims about the incompleteness of the fossil record and thus "to shatter the whole case for evolution." The society, however, declined to publish it. "We got the opinion of a first-rate paleontologist and geologist about it," explained the secretary, "and he told us that although it must have taken a very long time to compile it, he thought this kind of evidence led to no valuable conclusion." The distinguished geneticist J. B. S. Haldane (1892–1964), who questioned the suitability of the proceedings for such a paper, offered a simple explanation of Dewar's findings: The fossil record represented many successive faunas over millions of years, while the comparatively few existing mammalian fauna reflected "a single moment of evolution." Dewar, nevertheless, interpreted the rejection as an attempt by those who had made evolution "a scientific creed" to muzzle independent-thinking creationists.

Articulating what would soon become a *creationist* dogma, he argued that "Those who do not accept this creed, are deemed unfit to hold scientific offices; their articles are rejected by newspapers or journals; their contributions are refused by scientific societies; and publishers decline to publish their books except at the author's expense."[19]

Under these circumstances, creationists clearly needed a home of their own. The liberal evangelicals who controlled the Victoria Institute, though willing to share their platform, refused to lead a crusade against evolution. Thus when Captain Acworth in 1932 proposed to form an antievolution society that would confine itself, "as far as that might be possible, to the scientific rather than to the philosophical and religious plane," Dewar enthusiastically seconded the proposal. Meeting with five like-minded conservative evangelicals in Acworth's London office, they organized the grandly named Evolution Protest Movement (EPM) to halt the evil threatening to bankrupt modern civilization. "The *goal* of evolution," Acworth declared, "through psycho-analysis, is *moral degradation;* through organised mass birth-control, and sterilisation, *extinction;* and through its social creed of communism, *revolution.*" Only by overthrowing evolution could England once again be prosperous, merry, and free. To lead the counter-revolution, the founders chose Fleming as president, Acworth as chairman, and Dewar as secretary-treasurer.[20]

For its first few years the EPM seems to have existed as a paper organization; its public launching did not occur until February 1935, when Fleming introduced the movement at a meeting at Essex Hall in London, attended by over six hundred. After listening to addresses by Acworth, Dewar, and others, the audience voted overwhelmingly to protest "against the continued teaching of the theory of evolution in the schools and colleges and pulpits of Great Britain and the Dominions on the ground that it was not established as a scientific truth." Though the founders emphasized their scientific concerns, they hoped ultimately "to bring the people back to the Bible, and to the saving faith of our fathers."[21]

During the early years of the EPM, Dewar, "an energetic little man," carried the movement almost single-handedly, serving first as secretary and after 1946 as president. In 1936 he brought out a booklet titled *Man: A Special Creation,* in which he marshaled scientific and biblical evidence against human evolution. He took advantage of the opportunity to announce the recent organization of the EPM and to warn evolutionists that they were "no longer going to have things all their own way." Writing to Price a short time later, he optimistically reported "making some headway in England in the campaign against evolution," so much, in fact, that he anticipated "a rather sudden change in opinion" before long. Although "the rationalists and orthodox biologists" were stepping up their boycott of creationism, he was sure the protesters had them on the run. At a personal level, he expressed the hope that his booklet would sell enough copies to get him "fairly well-known" and thus allow him to take "the bigger risk" of publishing another work, explaining the entire fossil record along creationist lines.[22]

Welcome evidence that he was not being ignored appeared the next year in the form of a book by the paleontologist Arthur Morley Davies (1869–1959), written as a reply to *Difficulties of the Evolution Theory*. Davies attributed the recent assertiveness of "the literary disbelievers" in evolution in part to the influence of Dewar, one of the few "qualified biologists" to challenge evolution. But unfortunately for the cause of creationism, Dewar was, in Davies's opinion, "too honest a man of science to make a good advocate." In evaluating Dewar's critique of evolution, Davies noted with apparent surprise that the new creationists had "retired from the position held by Linnaeus, Cuvier and Agassiz, abandoning species and genera to the evolutionist, and making the Family or some higher category their line of defence." Davies applauded this shift but deemed it insufficient to accommodate all the evidence of evolution. By this time Dewar was moving in the opposite direction. In *More Difficulties of the Evolution Theory* (1938), a reply to Davies dedicated to Albert Fleischmann, Dewar confessed that since writing his earlier book he had "begun to suspect that the units of creation may be more restricted than those suggested by Vialleton." He had also come to suspect, partially as a result of Whitney's influence, that the ages of rocks given by radioisotope dating were unsound.[23]

In the spring of 1933 Dewar lectured to a crowd of about a thousand at the Midland Institute and was "received with cheers." In the euphoria of the moment he volunteered, as a spokesman for the Evolution Protest Movement, to debate any evolutionist who dared to face him. At first there were no takers. Even McCabe, Price's old nemesis, declined on the grounds that the amateur ornithologist lacked "firsthand familiarity with the material" as well as "the preliminary training which is necessary for its interpretation"—limitations, Dewar pointed out, that would also have disqualified Charles Darwin and Alfred Russel Wallace (1823–1913). In 1937 McCabe finally relented and agreed, as a representative of the Rationalist Press Association, to meet Dewar in London. As Frank Lewis Marsh would later argue in his epistolary debate with Theodosius Dobzhansky, Dewar insisted on experimental, laboratory evidence of the transformation of one plant or animal into another. "If we define a species as an interbreeding community," he declared victoriously, "*no new animal species has been produced experimentally.*" In the 1940s Dewar teamed up with Merson Davies to debate another representative of the Rationalist Press Association, the geneticist Haldane. In Haldane's opinion he successfully answered every argument against evolution except one: "that it is contrary to the doctrine of infallible Scriptures."[24]

Unlike their North American brethren, who fought for decades to control the classroom, the British creationists, unfettered by constitutional prohibitions against the union of church and state, focused their greatest efforts on breaking the evolutionists' monopoly of the airwaves. "Were it not for the monopoly enjoyed by the BBC, such unblushing suppression of the truth would not be possible," Dewar and Davies alleged. "One company would see that another did not disseminate too great nonsense. But as things are, all Britons must listen to what the B.B.C. chooses that they shall hear; and they have no means of applying criticism or test if they disagree with it." Decades would pass, however, before their dream of broadcasting creationism would be realized.[25]

The British leaders of the EPM also dissented vigorously from the American flood geologists' call for "complete agreement among creationists," that is, the acceptance of deluge geology. "The present state of scientific knowledge is utterly inadequate for any one to determine which of some twenty different interpretations of the first two chapters of Genesis is the correct one," Dewar insisted in a letter to Whitney. In the absence of sufficient knowledge even to determine the original units of creation, the number of creations, or the antiquity of human life, Dewar thought it preferable from a tactical point of view to "concentrate on exposing the fallacies of evolutionism." Thus the EPM imposed no tests on its members beyond a general commitment to the proposition that evolution was scientifically and biblically untenable.[26]

During his quarter-century association with the EPM, which corresponded with the nadir of conservative evangelicalism in Britain, Dewar saw the movement slowly recruit about two hundred members and establish small branches in Australia and New Zealand. An attempt to colonize North America met with little success. In the late 1940s James D. Bales (b. 1915), a young Church of Christ minister on the faculty of Harding College in Arkansas, inspired by the success of the EPM, unilaterally launched the Evolution Unscientific Movement together with a journal called *The Thinking Christian*. On seeing a copy of the first issue, Dewar invited the enterprising preacher, who also possessed a Ph.D. in the history and philosophy of education from the University of California, to merge his infant organization with the EPM and become the American secretary of the parent society. Bales leaped at this opportunity and announced the merger in the second issue of his quarterly. But unfortunately for the EPM, the journal died after only one more issue, and thereafter Bales increasingly turned his considerable energies to fighting communism in the right-wing Christian Crusade of Billy James Hargis (b. 1925). On occasion he took time to blast the evolutionary foundations of communism, but he did little to further the interests of the EPM in North America besides distributing some of its literature.[27]

The EPM suffered an even greater setback when it failed to win the public endorsement of C. S. Lewis (1898–1963), perhaps the best-known Christian apologist of his day and a personal friend of Captain Acworth's. In the 1940s and early 1950s Acworth tried repeatedly to cajole the Oxford don into joining the protest against evolution, but Lewis, believing that nonmaterialistic evolution posed little threat to Christianity, refused to take sides or even to contribute a preface to one of Acworth's books. He feared that among his growing band of disciples some might take umbrage at his association with anti-Darwinists. "When a man has become a popular Apologist," he explained, "he must watch his step. Everyone is on the look out for things that might discredit him." Privately, however, he found Acworth's arguments against evolution increasingly compelling—and the pretensions of many biologists repellent. In 1951 he confessed that his belief in the unimportance of evolution had been shaken while reading one of his friend's manuscripts. "I wish I were younger," he confided to Acworth. "What inclines me now to think that you may be right in regarding it [evolution] as *the* central and radical lie in the whole web of falsehood that now governs our lives is not so much your arguments against it as the fanatical and twisted attitudes of its defenders." Such words no doubt gladdened Acworth's heart, but they did nothing to enhance the EPM's limited credibility among evangelicals.[28]

ROBERT E. D. CLARK AND THE VICTORIA INSTITUTE

When British creationists founded the Evolution Protest Movement, they did not simply abandon the Victoria Institute to evolutionists. Several of them remained active in the institute for years, with Fleming and Dewar serving, respectively, as the institute's president and vice president. But—with one notable exception—the creationists

shifted their primary allegiance to the EPM. The exception was Robert
E. D. Clark (1906–1984), who emerged in the years after World War
II as Britain's foremost creationist but who found the theological dogmatism and scientific ignorance of certain EPM leaders too much to
stomach. The son of a lawyer father and a missionary mother, Clark
was born in present-day Pakistan, in an area opened up to Christianity
by his paternal grandfather. As a student at St. John's College, Cambridge, he took a double first in the natural sciences tripos in 1928 and
four years later won a Ph.D. in organic chemistry from the same institution. Until World War II he worked as an industrial chemist, but
quit to become a schoolmaster when he was assigned to military work.
During the war he coauthored a popular survey of modern chemistry,
Order and Chaos in the World of Atoms (1942), his first book. He
ended his career as a senior lecturer in chemistry at Cambridgeshire
College of Arts and Technology.[29]

Religiously, Clark considered himself a "catholic" evangelical who
abhorred sectarianism and dogmatism of any kind. Though reared in
the Church of England, he eventually walked out of the established
church because as a strong pacifist he could no longer accept its positions on such issues as military service, oath-taking, and recognition of
the British sovereign as the head of the church of God. Through the
years he worshiped at will with Baptists, Congregationalists, Brethren,
and other evangelical groups, frequently filling their pulpits as a lay
preacher. The Inter-Varsity Fellowship, the most dynamic of the conservative evangelical organizations in Britain, brought out his first book
on science and religion, *Scientific Rationalism and Christian Faith*
(1945), and from 1943 to 1947 published *Current Notes and Abstracts
on Science and Theology*, which he edited. The Brethren's Paternoster
Press, under whose imprint several other of his books appeared, hired
him in the late 1940s as a scientific editor and for three years, beginning in 1948, supported the quarterly *Science and Religion: A Review
of Current Literature*, the successor to *Current Notes*. Although Clark
stopped short of advocating the inerrancy of the Bible, he believed
fervently in the literal Second Coming of Christ and devoted his last
years, like Price, to writing a book on the eschatological prophecies of
Revelation.[30]

Even as a boy the sweet-tempered but sharp-tongued Clark found it
difficult to take the claims of evolutionists seriously, and as a student at
Cambridge he concluded that their attitude was "completely stupid."
He suspected that many of them regarded evolution as little more than
"shop window stuff." Nevertheless, he professed a willingness to accept evolution if the scientific evidence demanded it, confident that he

could ultimately reconcile it with the biblical record. In interpreting Genesis, he followed the suggestion of fellow Victoria Institute member P. J. Wiseman (1888–1948) that the Mosaic creation story had not *occurred* in a single week but had merely been *revealed* in six literal days. This scheme left Clark free to follow science in accepting the antiquity of the earth, a local flood, and even the common ancestry of humans and apes—by translating Genesis 1:26 to read, "Let us make the already existing species man into a being with our image." His commitment to special creation, both divine and diabolical, rested almost exclusively on the evidence of design in the universe, which he thought could only be explained by supernatural intervention.[31]

Flood geology struck Clark as being no more plausible, and far less scientific, than evolution. As a young man he had read *The New Geology* and found it "stimulating and refreshing," but his respect for Price diminished with the years. As editor of *Science and Religion* in the late 1940s, he not only rejected a manuscript Price submitted for publication but refused even to print a brief notice announcing the flood geologist's latest pamphlet, *Feet of Clay: The Unscientific Nonsense of Historical Geology* (1949). Instead, he rebuked the senior creationist for adopting an "unChristian" tone in railing against evolutionists. Disgusted in particular by the subtitle and a subheading, "The Sacred Cow of Geology," Clark lectured Price on the principles of Christian behavior: "Should we like it if non-Christians talked about the sacred cow of revelation or the unscientific nonsense of Christianity? Of course not. Then we should do to others, said our Lord, what we should like them to do to us!" Clark reluctantly publicized Whitney's offer of a hundred pounds to any reader of *Science and Religion* willing to debate the age of the earth, with the American arguing for 4004 B.C., "within a few hundred years." But the editor added a note expressing his surprise that anybody would find such a thesis worthy of discussion: "It had never dawned upon our (perhaps prejudiced) mind that this view received much support from any line of enquiry whatsoever—Biblical or scientific." In a personal letter he begged Whitney to take the high road in opposing evolution rather than "pouring out arguments which at best will only make the Lord's enemies to blaspheme." At first Whitney savored the prospect of taking on Clark himself, the only scientifically competent creationist he knew who rejected flood geology, but when Clark spurned his offer, the jilted rancher mocked the Englishman's credentials and questioned his common sense.[32]

Traveling on the train from Kent to Cambridge one day in the mid-1930s, the young chemist turned his thoughts to the relationship between evolution and his own discipline. Just as the geologist Merson Davies saw evolution primarily as a geological problem and the biologist Dewar viewed it largely as a biological problem, Clark visualized evolution "first and foremost [as] a *chemical* problem," which at its most basic level required explaining "how chemical molecules of gigantic complexity came into existence and have been able to arrange themselves in increasingly complicated ways." Since the late nineteenth century, various scientific writers had alluded to the apparent incompatibility between evolution and entropy or the second law of thermodynamics, a mathematical formula according to which the amount of energy available for work in an isolated system could not increase. A more general expression of the law held that "disorder will tend to increase, but that order can never arise spontaneously from chaos." As Clark rode through the English countryside, the implications of the generalized version for the theory of evolution suddenly struck him: Evolution could "never involve any real rise in the degree of organisation of an organism." "If in past ages complex organisms ever did evolve from simpler ones, the process took place contrary to the laws of nature, and must have involved what may rightly be termed the miraculous," he concluded. "For this reason the doctrine of evolution can never legitimately form a part of naturalistic philosophical or sociological thought, nor can it ever be rightly used to support such dogmas as the inevitability of progress." Clark was sufficiently well versed in the scientific literature to know that most commentators dismissed the conflict between evolution and entropy as illusory, arguing that "just as a part of the energy of hot steam may be converted into highly ordered work at the expense of the remainder so, during the course of evolution, animals may in the last resort have obtained their organisation at the expense of the sun's energy which has degraded on the earth's surface." Nevertheless, Clark insisted that the sun's energy by itself could never give rise to *new* types of organization.[33]

Clark worked up his ideas for publication in the prestigious journal *Nature*, but the editors rejected his paper because it lacked mathematical sophistication. Not having the mathematical skill to develop his argument properly, he settled for publication by the Victoria Institute, which awarded him the Langhorne Orchard Prize, previously given to Price. Fleming, Dewar, and Merson Davies all recognized immediately that Clark had contributed a major new weapon to the arsenal of creationism, and the publication of a revised version of the paper in his widely read book *Darwin: Before and After* (1948) put the entropy argument in the hands of a generation of antievolution warriors who used it with abandon—and usually without Clark's sensitivity to the scientific issues involved.[34]

Clark, perhaps the most visible and articulate creationist of his generation, tended to keep his distance from other antievolutionists. Though he labored for years as an officer of the Victoria Institute and edited its journal, he refused until shortly before his death even to join the EPM, much less to to serve as its president, which he was requested to do. He had joined the institute in his youth because of family connections and a shared commitment to Christian evidences, and throughout his life he preferred the company of the relatively open-minded theistic evolutionists in the Victoria Institute to that of the more dogmatic creationists in the Evolution Protest Movement. Even at the institute he was frequently embarrassed by the "very strange" and sometimes nonsensical utterances of the creationist president, Fleming. Clark, who believed in special creation primarily for religious, not scientific, reasons, functioned for decades as the conscience of modern creationism.[35]

NINE

Evangelicals and Evolution in North America

E xcept for occasional spasms from small bodies such as the Religion and Science Association (RSA) and the Deluge Geology Society (DGS), organized creationism in North America appeared to be all but dead during the second quarter or so of the twentieth century. Even conservative evangelicals who instinctively rejected organic evolution rarely roused themselves to take action against it. In typical fashion the leaders of the American Scientific Affiliation (ASA), founded in 1941 by evangelical scientists concerned about the quality of the Christian witness on science and religion, devoted more energy to appraising than to opposing evolution. In marked contrast to the firebrands in the DGS, the ASA evangelicals preferred to avoid confrontations with evolutionists. Writing to Dudley Joseph Whitney at mid-century, Robert E. D. Clark chastised the obstreperous creationist for always "putting a spoke in the wheels" of God's work rather than pushing it forward like the positive young men in the American Scientific Affiliation.[1] During the 1940s and 1950s the ASA served as the principal evangelical forum for discussing the pros and cons of evolution and for evaluating the critiques of George McCready Price and Harry Rimmer. Under the influence of an inner circle closely connected with Wheaton College, the prevailing sentiment in the ASA shifted away from strict creationism to progressive creationism and even theistic evolutionism.

Eventually, the ASA liberals proved so successful in nudging evangelicals toward evolution, they helped to spark a conservative creationist counteroffensive in the 1960s.

THE AMERICAN SCIENTIFIC AFFILIATION

The idea for an American association of evangelical scientists dedicated to correlating science with the Bible came from Irwin A. Moon (1907–1986), an associate of the Moody Bible Institute who toured the country giving illustrated "Sermons from Science." Troubled by the many students he encountered who were learning un-Christian science—and the many Christian workers who were teaching unscientific Christianity—in 1941 he talked William H. Houghton (1887–1947), president of the institute, into inviting a number of scientists of known orthodoxy to Chicago to discuss the need for a new organization. Five showed up: John P. Van Haitsma (1884–1965), a biologist from the Christian Reformed Calvin College; Peter W. Stoner (1888–1980), a Congregational mathematician from Pasadena City College; Irving A. Cowperthwaite (b. 1904), a Baptist industrial chemist from Boston; Russell D. Sturgis (1897–1969), a Baptist chemist from Ursinus College; and F. Alton Everest (b. 1909), a Baptist electrical engineer on the faculty of Oregon State College in Corvallis. From this meeting emerged the ASA, dedicated to producing and disseminating "accurate" information on the relationship between science and religion. Each new member had to agree to the following creedal statement:

> I believe in the whole Bible as originally given, to be the inspired word of God, the only unerring guide of faith and conduct. Since God is the Author of this Book, as well as the Creator and Sustainer of the physical world about us, I cannot conceive of discrepancies between statements in the Bible and the real facts of science.

Thus was launched what one enthusiast called "the most important movement since the Reformation."[2]

The personable Everest quickly emerged as the organizational leader of the group, serving as president of the ASA through its crucial first decade of existence. A native Oregonian, he earned degrees in electrical engineering from both Oregon State College and Stanford University before returning to Corvallis in 1936 as a specialist in radio and television. At the end of World War II, during which he conducted underwater sound research for the National Defense Research Committee, he joined Moon in founding the Moody Institute of Science, with which he remained affiliated until 1970. Everest grew up as a conservative Baptist, partial to the Scofield Bible. As a young man he read both Rimmer and Price, but his favorite science-and-religion celebrity was the physician Arthur I. Brown, whose dignified style and scientific manner he determined to emulate. Thanks largely to Everest's diplomacy and persistence, the ASA during its first decade grew from 5 to 220, over half of whom came from Mennonite, Baptist, or Presbyterian churches and resided in the middle Atlantic or midwestern states. By 1961 membership had soared to 860.[3]

The ASA had barely gotten off the ground when Everest faced his first critical decision: how to relate to the Deluge Geology Society. One of the ASA's first recruits was the Lutheran biologist Walter E. Lammerts, who happened to be a neighbor of Moon's as well as a disciple of Price's. Through Lammerts, it seems, Everest and Benjamin Franklin Allen, secretary of the DGS, learned of each other's existence. Within months after returning from the ASA's organizational meeting in Chicago, Everest was eagerly telling his fellow ASA officers about the "large crowds of non-scientific folk" that the DGS drew to its meetings, and Allen was inviting the ASA to publish its material in the *Bulletin of Deluge Geology* until the fledgling society could start a journal of its own. Although Everest appreciated Allen's "friendly overtures" and professed to like the tone of the *Bulletin*— "dignified but definite in its presentation of its views"—he questioned the wisdom of becoming too entangled with the DGS, perhaps because of its "strong Seventh-Day Adventist flavor."[4]

The deluge geologists pushed for even closer cooperation: the joint publication of "a truly interdenominational" magazine devoted to science and religion. In supporting this proposal for pooling editorial resources, Lammerts urged Everest not to let differences over details, such as orthodox geology versus flood geology, get in the way. When Everest, who was spending the war years in southern California, called a meeting to discuss the proposal, three of the eight persons who showed up at Stoner's Pasadena home had close ties with the DGS: William J. Tinkle, Lammerts, and, to Everest's great surprise, Price himself, who came as Tinkle's guest. In describing the meeting to the ASA executive council, Everest wrote that he did not relish "the idea of becoming affiliated with a deluge society right off. . . . We would never hope to gain even the Christian geologists if we espoused [Price's] cause. I am convinced that we will have to do our job the slow, hard way and with men who are not sold out to some idea." In informing the flood geologists of the decision to remain unattached, he expressed his personal hope that "the various 'theistic scientists' can get together and present a united front to the world." He himself had already joined the DGS, though the next year he pointedly instructed Allen to list him as a subscriber, not a member.[5]

Despite Everest's gentle rebuff, the pesky flood geologists refused to go away. Early in 1944 Whitney approached the Moody people about publicizing his offer of $1,000 in war bonds to any geologist who could refute his critique of geological ages. Houghton, who had fostered the ASA to vet such proposals, passed Whitney's materials on to Everest for evaluation. Rather than dismissing the idea out of hand, the ASA president grabbed it as "a prod" to get the affiliation moving and to determine where it stood with respect to the various interpretations of Genesis. Stoner, who believed that the days of Genesis 1 represented "breaks between the great geological ages," advised him to have nothing to do with the flood geologists. "I consider this whole deluge geology idea to be extremely unscientific," he wrote Everest, adding that its advocates seemed to have "a chip on their shoulders." An earth scientist at his college had told him that no reputable geologist accepted Price's theory, though some required their geology students to refute it before giving them credit for their work.[6]

Everest nevertheless wrote Whitney an encouraging letter, inviting him to "feel free to call us at any time" if there was anything the ASA could do to help. By return mail Whitney, mistaking politeness for a genuine offer of assistance, proposed that the ASA "initiate a formal discussion of the question whether ages geology should be believed, or Deluge geology, or some third kind of earth history, if there is any such"—with the $1,000 in war bonds going to the scientist who successfully discredited flood geology. Distressed by the predicament his kindness had gotten him into, Everest begged the ASA officers for help. "I have a bomb in my lap which I would like to pass on to the rest of you," he wrote with only slight overstatement. The consensus of the council was to let the matter drop and avoid publicizing "the differences of opinion within our own group." Besides, as Stoner pointed out, no self-respecting geologist could afford to take Whitney's money. "It is something like the position I would be in if I claimed 2 × 3 was 7 and challenged the Mathematical world to debate the subject," he explained. "No Mathematician of standing would accept such a debate." Clearly, concluded Everest with newfound wisdom, the flood geologists presented the ASA with an "insurmountable problem."[7]

The early flap over flood geology tended to cast a pall over any Seventh-day Adventist who wanted to join the ASA, especially because the disruptive Whitney was often mistaken for being a member of the sect. In 1946 the executive council debated whether Adventists should even be accepted into membership, presumably as much because of their fidelity to Ellen G. White as to flood geology. Tinkle, who knew Adventists well because of his association with the Deluge Geology Society, urged that they be admitted "if good scientists and men of good judgment." Though his advice prevailed, only two Seventh-day Adventists elected to join the ASA during its first decade.[8]

J. LAURENCE KULP AND THE CRITIQUE OF FLOOD GEOLOGY

Even without many Adventists physically present, the ASA continued through the late 1940s to wrestle with the incubus of flood geology. The member who finally exorcised it was J. Laurence Kulp (b. 1921), a geochemist who joined the ASA in 1945. As one of the first American fundamentalists trained in geology, he contributed more than any other scientist to splitting conservative Protestants into self-consciously separate camps of "evangelicals" and "fundamentalists." Though brought up as a practically atheistic Episcopalian, he converted to fundamentalism at age sixteen and began attending an exclusive Plymouth Brethren assembly. At first he subscribed to a literal six-day creation and universal flood, but he soon learned from his Scofield Bible how to use the gap theory to reconcile the claims of historical geology with biblical inerrancy. By the time he entered Wheaton College as a junior transfer student from Drew University, he had moved on to the day-age interpretation of Genesis while clinging to the special creation of biological "kinds," especially humankind. After several years with the uncompromising Brethren, he found Wheaton surprisingly progressive, and under its liberalizing influence he came to understand that the Bible needed "correct interpretation." At the time he received his bachelor's degree in chemistry in 1942, he was living on "the liberal fringe" of the Wheaton community and drifting rapidly from his fundamentalist moorings.[9]

After a year in graduate school at Ohio State he moved on to Princeton University, where he obtained a Ph.D. in physical chemistry in 1945. While living in New Jersey, he continued worshiping with the Brethren and became active in the Inter-Varsity Christian Fellowship, giving talks to college students on how to make scientific sense of the Bible. Questions raised during these encounters, together with a desire to spend more time outdoors, prompted him to join a friend in auditing a freshman geology course while completing his graduate work at Princeton. The professor, Richard M. Field (1885–1916), eager to recruit chemists to geology, invited the two friends to accompany him on field trips and for ten weeks volunteered every Monday evening to tutoring them in his home. The more Kulp learned about earth history, the more curious he became; so when an opportunity arose to take a two-year postdoctoral fellowship in geology at Columbia University, he jumped at the chance, which allowed him to complete all the course work and some of the research required for a Ph. D. in geology. When his fellowship ended, the geology department at Columbia hired him to develop geochemistry as a new field. In part because of lingering questions about the harmony of Genesis and geology, he decided to focus his research on the exciting new area of radioisotope dating, which was rapidly transforming geology from an inexact, descriptive science into a rigorous, quantitative discipline. To learn the latest techniques, he traveled to the University of Chicago, where Willard F. Libby (1908–1980), the discoverer of carbon-14 dating, was applying his methods to archaeology. In seminars at Chicago's Institute for Nuclear Studies, where Kulp rubbed shoulders with the Nobel Prize–winners Enrico Fermi (1901–1954) and Harold C. Urey (1893–1981), he mastered the new technology. He returned to Columbia in 1950 to set up the second carbon-14 laboratory in the country and to pioneer in the application of radiocarbon dating to geological problems.[10]

The ASA provided Kulp with his first forum for debating the religious implications of geology openly and scientifically—and gave the sharp-witted, self-possessed geochemist a platform from which to expose the errors of less sophisticated Christians. Though once enamored of Rimmer's and Price's apologetical writings, by the mid-1940s he had come to fear that such "pseudo-science" would only bring derision to orthodox Christianity, which he still fervently espoused. In commenting privately on a paper Henry M. Morris had submitted to the ASA, arguing that a Christian could not consistently believe in evolution, Kulp, who had only recently retreated from the same position himself, warned that such a logically flawed thesis would only bring harm to the cause of Christ. Morris's "fundamental error," he pointed out, was to confuse "geology with evolution which is so common among writers on 'Science and the Bible.'" By this time Kulp, as a result of broad reading and frequent discussions with other liberal evangelicals, was inclined to view the Genesis story of creation as analogous to a series of impressionistic oil paintings intended only to show the existence of a Creator, not the details of earth history.[11]

Early in 1948 Everest, still fretting about how to handle flood geology, approached Kulp with a proposal to explore the subject at the next annual convention, scheduled for late summer. Kulp replied that, though interested in the president's suggestion, he preferred to discuss a topic "of far more importance to the A.S.A. than the necessary destruction of flood geology": the antiquity of humans. As an ad hoc geological advisor to Marie Fetzer (b. 1925), a Wheaton alumna working on a master's degree in physical anthropology at Columbia, he had been scanning the scientific literature on human history. Having "only read the various pseudo-scientific statements in Christian apologetical literature which blindly asserted that there was no evidence for man or manlike creatures earlier than 10,000 years ago," he was shocked to discover the strength of the case for human antiquity. "A careful study of the tremendous number of geological facts concerning the chronology of the Pleistocene period make[s] it apparent that such creatures have been on the earth probably hundreds of thousands of years," he told Everest. "Although I am not at all convinced of the theory of evolution as it is usually held and certainly have no tolerance for the non-Christian philosophy that is built thereon, I have learned too many things about the history of the earth's crust to attempt a major *refutation* of, at least, certain aspects of the biological theory."[12]

The program for the third annual convention of the ASA announced that J. Laurence Kulp, lecturer in geology and chemistry at Columbia University, would be presenting a paper titled "Antiquity of Hominoid Fossils." The accompanying abstract promised both titillation and consolation: "Considerable scientific evidence is presented that man-like creatures have been on the earth for at least many tens of thousands of years. This requires a reexamination of our interpretations of Genesis but is not disastrous to a strong conservative apologetic." Needless to say, the public presentation, Kulp's first at an ASA convention, sparked lively—and concerned—discussion. Perhaps the most troubled response came from Edwin Y. Monsma (1894–1972), a botanist at Calvin College, which was hosting the convention. Though grateful for the presence of a competent geologist, Monsma questioned any "time reckonings" based on the assumption of uniformity. Besides, he could see no way to harmonize human antiquity with a literal reading of Scripture. Kulp tried to placate his host by pointing out that only *one* assumption—a uniform rate of radioactive disintegration—was necessary to prove a very old earth, that an ancient earth implied nothing about "evolution as such," and that the scriptural objections could be met by interpreting the days of Genesis as "phases of creative activity." Monsma, who favored a young earth, remained unpersuaded. Like many other evangelicals who took the Bible literally, he found the idea of death and destruction before the fall of Adam and Eve an insurmountable obstacle to accepting the conventional notion of long geological ages and particularly the antiquity of human fossils. For Kulp, the barrier was easily skirted by distinguishing between death in the animal kingdom and death in a spiritual sense. Before the discussion ended, Kulp was dismissing the idea of a six-thousand-year-old earth as "foolishness" and warning those present not to "laugh off this business of the antiquity of man."[13]

Later in the convention, when Monsma himself presented a critical examination, "Some Presuppositions in Evolutionary Thinking," including so-called uniformitarianism, Kulp responded with a personal testimony:

> Over the last fifty years there have been practically no Christians in the field of geology. I was trained as a chemist before I felt that the Lord wanted me to go into geology. I went into it very critically, and I am still overly critical of all information I receive. However, most of us do not understand enough geology to appreciate the geologist's method of securing geological data. He is not one millionth the philosopher that he is usually given credit for being.

To Kulp, the fossil record presented the Christian with only two choices, only one of which was acceptable: either the Creator had deceptively given the earth the appearance of great age, or the history of life on earth had really spanned vast amounts of time. In response to one person's suggestion that the ruin-and-restoration hypothesis might prove helpful in accommodating the geological evidence, Kulp pointed out that a catastrophe great enough to destroy Satan and his hosts would have done "awful things to the neatly placed geological strata and there isn't any evidence of such a thing in the geological record." As far as he was concerned, the idea of a pre-Adamic ruin was best "left in the 1500s when it was originated."[14]

By 1948 many evangelical scientists in the ASA were ready to follow Kulp in boldly shedding the trite fundamentalist apologetics of the past, especially when told that scientific honesty demanded it and assured that doing so would not affect the "inspiration of the Scriptures even a little bit." Indicative of the changing intellectual climate in the organization was Kulp's appointment that year to the executive-council seat vacated by Monsma, who in the opinion of at least one conventioneer had also lost his debate with Kulp. Everest, who feared internal disharmony even more than doctrinal heterodoxy, worried only that "this problem of the age of man may be resolved as quietly and softly as possible." Nevertheless, he proceeded with plans to have Kulp demolish flood geology at the next annual convention, to be held in Los Angeles, in Price's backyard.[15]

Although Everest by this time was relegating the *Bulletin of Deluge Geology* to the "screwey [sic] shelf" of his library and was growing increasingly impatient with Lammerts's "unreasonable" defenses of flood geology at local ASA meetings, he did not want Kulp unnecessarily to give offense in discussing "this critical subject." Because southern California was the "hotbed" of flood geology, Everest expected the audience to include not only Price and his most devoted disciples but also many local theologians, "and the presentation of a destruction of something they hold very close to Biblical truth itself, if not done very carefully and wisely, might turn them from the ASA." As the time of the anticipated showdown approached, Everest coached Kulp to stress the facts rather than "how absurd Prof. Price's ideas are. . . . I am hopefully looking forward to a cool, dispassionate, and factual presentation that will essentially close the subject." Kulp completed his paper just in time for the annual meeting, staying up most of three nights in a row to do it, but pressing professional obligations prevented him from going to Los Angeles in person. Thus at the last minute Everest had to recruit Kulp's friend Fetzer to present one of the most consequential and controversial papers in the brief history of the ASA.[16]

Kulp began his demolition job by inflating flood geology into a straw man of Bunyanesque proportions, claiming that it had "grown and infiltrated the greater portion of fundamental Christianity in America primarily due to the absence of trained Christian geologists." He then proceeded to show that the "major propositions of the theory are contraindicated by established physical and chemical laws." He focused in particular on "four basic errors" that flood geologists commonly made: equating historical geology with evolution; assuming "that life has been on the earth only a few thousand years, [and] therefore the flood *must* account for geological strata"; misunderstanding "the physical and chemical conditions under which rocks are formed"; and ignoring recent discoveries, such as those associated with radioactive dating, that undermined their assumptions. He accused Price himself of both ignorance and deception. The flood geologist, for example, liked to cite a nineteenth-century geological report that allegedly demonstrated the conformability of Precambrian limestone on top of Cretaceous shale in the Canadian Rockies to undermine confidence in thrust faulting. Price neglected, however, to tell his readers that the author of the report, though indeed mentioning that the two formations *appeared* to succeed one another conformably when viewed from a particular perspective, discussed at much greater length the unambiguous evidence of tectonic movement. Kulp concluded on an evangelistic note, emphasizing the compatibility between geology and the plenary inspiration of the Scriptures and justifying his negative tone on the grounds that "this unscientific theory of flood geology has done and will do considerable harm to the strong propagation of the gospel among educated people."[17]

The expected fireworks never ignited. Price sat silently in the front row through both presentation and discussion. Only when the chairman pointedly invited him to speak did Price say "something very brief which missed what everyone was waiting for," reported an obviously disappointed Everest to Kulp. Allen, whose pyrotechnics would no doubt have met everyone's expectations, unaccountably failed to hear about Kulp's attack until it was too late to attend. At the close of the session, Price greeted Everest cordially and expressed his intention to study Kulp's criticisms carefully. The next day he wrote Everest a letter congratulating the ASA on promoting "free and open discussions of mooted points of science, together with a frank and positive allegiance to the basic facts of the Christian religion." He professed to appreciate the critical evaluation of his views, because "only in this way can we hope to arrive at settled and lasting truth."[18]

Other flood geologists responded less genially. One unidentified ASA member dismissed Kulp's critique as the product of a mind under the influence of "the orthodox geological viewpoint." Alumni of the old Deluge Geology Society in the Los Angeles area brought in Clifford L. Burdick from Arizona, the best-trained flood *geologist*, to plot ways of countering Kulp. Though Burdick disagreed totally with Kulp's assessment of flood geology, he cautioned his friends that they could not "lightly laugh off his presentation of the case for orthodox geology." In awe of Kulp's "position and degree," he resolved (once again) to complete his own academic training in geology. In the meantime he wrote a detailed rebuttal, "Deluge Geology—Fact or Fiction?," which he tried unsuccessfully to get published in the *Journal of the American Scientific Affiliation*.[19]

Kulp's critique of deluge geology represented only the front of his attack in the late 1940s on the assumptions underlying flood geology. In a second paper read at the Los Angeles convention, on carbon-14 dating, he argued that "preliminary work indicates Neanderthal remains (the youngest stratigraphically of the prehistoric fossil men) are at least older than 25,000 years." And he lined up an ally, Cordelia Erdman (b. 1924), to defend the orderliness of the geological column, in the paper preceding his on flood geology. Erdman, a musician and Inter-Varsity activist, had been a classmate of Kulp's at Wheaton before going on to The King's College, where she had studied geology with L. Allen Higley and adopted his ruin-and-restorationist views. Despite a weak scientific background, at Kulp's urging she had enrolled in Columbia University as a graduate student in paleontology, working with the evolutionist Norman D. Newell (b. 1909). She earned her master's degree shortly before returning to Wheaton in the fall of 1949 as an instructor in geology. During her five-year tenure on the Wheaton faculty she helped to turn fundamentalist opinion against Price by assigning Kulp's "Deluge Geology" to her classes.[20]

Kulp was also working behind the lines to ensure that the ASA's publications gave neither aid nor comfort to flood geology. The ASA's first book, *Modern Science and Christian Faith* (1948), edited by Everest, included the chapter "Geology and the Bible" by Edwin K. Gedney (1904–1980), an Advent Christian science professor at Gordon College of Theology and Missions with master's degrees in geology from both Brown and Harvard. A partisan of the day-age interpretation of Genesis, Gedney refused even to mention the flood-geology alternative. In an effort to stave off conservative critics, Everest, at Lammerts's insistence, added a conciliatory note, erroneously conflating the views of Price and Harold W. Clark:

> In all fairness it should be mentioned that there are those who believe the various geological strata do not have any time significance, but merely indicate ecological zones. . . . Prof. George McCready Price . . . is the foremost champion of this thesis.
>
> The author of this chapter, however, believes that the above view is a result of an abbreviated experience with the practical use of fossils and a failure to comprehend the history of age correlation. . . .

The first printing of five thousand copies sold out quickly, and in planning a second edition, Everest asked Kulp to supervise the revision of the chapter on geology. Kulp urged Gedney not to pass up the opportunity to expose "the utter inadequacy of Flood Geology." Writing early in 1949, before completing his own critique, Kulp stressed the need for "a thorough refutation" of this "ludicrous theory." Gedney, however, resisted the suggestion, fearing that non-Christian readers would be put off by an extended discussion of Price's theory. "We agree," Kulp finally informed Everest, "to hit Flood Geology by omission rather than by attack for the purpose of the present volume."[21]

Largely because of Kulp's powers of persuasion, flood geologists found themselves increasingly isolated within the ASA. Even old allies such as Everest, who once cultivated their participation, now wanted nothing more to do with them or their disreputable theory. For both friends and foes alike, Kulp dictated the terms of debate. Perhaps more than any other member, recalled Everest, Kulp "made it clear that the true role of the ASA was studying the problem and not promulgating any particular answer to the problem." And, as we shall see, for years to come flood geologists lived uncomfortably under the shadow cast by Kulp's 1949 paper, distressed and provoked by the knowledge that the world's greatest evangelical geologist, who professed equal fidelity to the Bible, condemned their reconstruction of earth history as mere hokum.[22]

QUALITY CONTROL

At the height of the anti-Price offensive, one leader expressed the fear that the ASA would "become known primarily as an anti-Rimmer organization." Because Rimmer had the ear of so many fundamentalists, some ASA members deemed him a greater obstacle to progress than Price; but as an evangelical insider with honorary doctorates from Wheaton and other fundamentalist colleges, he largely escaped the public derision heaped on the outsider Price. The move against Rimmer began when Everest learned that Van Kampen Press, the ASA's own publisher, was considering reprinting Rimmer's books on science and religion. Well aware of their scientific limitations, he asked Van Kampen to delay a commitment until the ASA could evaluate a representative sample. The reviews were predictably negative. Not one of the seven scientists who evaluated *The Theory of Evolution and the Facts of Science,* including the conservatives Lammerts, Tinkle, and Monsma, recommended ASA approval. Monsma, who thought the book should never have been published in the first place, offered a typical appraisal, faulting Rimmer for his "inaccuracies and overstatements" as well as for his reliance on ridicule. The consensus, as reported by the biologist J. Frank Cassel (b. 1916), whom Everest had selected to head up the review process, held "that the book in question is hopeless, and that it would be better to write a new book than try to rewrite this one." Because Rimmer presented such an easy target, Cassel warned that the ASA would need to "guard against derision, and the basing of our arguments upon false premises, misunderstanding, or misrepresentation of research—Rimmer's glaring faults." Rimmer's other books fared no better. As Bernard Ramm said of *The Harmony of Science and Scripture,* they were "filled with both sense and nonsense; the very good and the very bad."[23]

By the time the ASA finished reviewing the Rimmer corpus, Van Kampen Press had already reached a negative judgment, leaving the affiliation's leaders with the dilemma of deciding what to do with the potentially explosive information they had collected. As word of the project leaked, curious members began clamoring to read the documents. To protect the reviewers from embarrassment and possible reprisals, the ASA finally issued a "confidential" report for members only in which authors were identified solely by discipline. Even so, the new secretary-treasurer, Russell L. Mixter (b. 1906) of Wheaton, worried that Rimmer would discover who was mimeographing and distributing the reports. "Wouldn't he have fun if he knew just who was doing what?" he nervously asked Everest in a letter that playfully referred to Rimmer as "a joke(r)." Under the circumstances, publication seemed out of the question. "Rimmer isn't writing and speaking on these subjects now," noted Everest in justifying not going public, "and the man is a fine preacher." Besides, Rimmer still had a large and loyal following among the very people the ASA hoped to reach.[24]

As Everest quickly learned, not everyone appreciated the ASA's efforts to control the quality of evangelical publications on science and religion. A particularly upsetting episode centered on one of the affiliation's earliest recruits, a young Canadian engineer and linguist named Arthur C. Custance (1910–1985). English born, Custance had immigrated to Ontario in 1928. While farming and logging, he took a correspondence course in mechanical drafting; in the early 1930s he entered the University of Toronto, where he eventually earned an honors M.A. in biblical languages. During the 1940s he worked as a mechanical engineer in the Toronto area. In 1937, while still an undergraduate at Toronto, he took the lead in organizing a Canadian version of the Victoria Institute called the Kelvin Institute. Never more than a local discussion group, the institute at its peak numbered under twenty regular members, the most prominent being John R. Howitt (1892–1985), a psychiatrist and dedicated antievolutionist. Before the war forced a suspension of meetings in 1941, the institute issued three volumes of duplicated proceedings, including a paper by Custance exploring the connection between "Modern Geology and the Bible." To resolve apparent differences, he appealed indiscriminately to both the gap theory and Price's flood geology.[25]

By the time Custance joined the ASA in 1943, he had turned his attention to biblical anthropology and was well into a book on the subject. Everest, eager to establish a cooperative arrangement between the ASA and the Kelvin Institute, invited the unknown Anglican evangelical to contribute the chapter on anthropology to the ASA's *Modern Science and Christian Faith*. Custance promptly delivered the requested manuscript—only to be told a short time later that it was unsuitable for publication in both length and content. The referees, especially those from fields closely related to anthropology, had "strenuously" objected to its publication, apparently because of its dogmatic tone and its rejection of human antiquity. "Personally I accept the Word of God from cover to cover as Verbally inspired," Custance explained defensively. "To me it is the Touchstone of Truth. It *is* truth, and does not require to be proved, as the criticisms would suggest. I took it as a starting point, and then I tried to reconcile the findings and speculations of modern Anthropologists with the truth." He found the referees' implications of Jewish bias especially offensive. Insisting that he was no Jew, he traced his ancestry back to the Normans who came to England with William the Conquerer. Under the circumstances he felt he had no choice but to withdraw his paper and resign his membership in the ASA, which seemed to be "subordinating individual opinion after the manner of the Roman Catholic church." Everest, clearly distressed by this turn of events, pleaded with Custance to reconsider: "This is the first request of this nature we have received, and my sincere prayer is that it is not a shadow of collapse [that ended] several earlier attempts toward similar organizations. . . . But we have barely started! If everyone withdraws upon a difference of opinion, we most certainly will not last long." Custance refused to budge, and Everest soon turned his attention to a much greater threat: a difference of opinion over evolution.[26]

PROGRESSIVE CREATION OR THEISTIC EVOLUTION?

Having turned their backs on Price and Rimmer, the leaders of the ASA faced the task of formulating an alternative evangelical position on evolution. Although the affiliation had never taken an official stand on Genesis, besides requiring members to sign a doctrinal statement affirming their belief in the inspiration of the Bible, the founders seemed to have favored personally a literal reading of the creation story and in the beginning assured at least one prospective member that "the stance of the society would be anti-evolutionary." Before long, however, the society was moving rapidly in the direction of theistic evolution, with some members stopping off at what they called "progressive creation."

By the early 1950s Whitney could describe reading the ASA's journal as being similar to watching "a 'movie-talkie' of supposedly *very devout* Fundamentalists *filmed in the very act of becoming Modernists.*"[27]

Modern Science and Christian Faith, the ASA's 1948 handbook for students, reflected the affiliation's early creationist orientation—as well as the diversity of viewpoints within the organization. Along with Gedney's day-age interpretation of geology appeared "Biology and Creation" by Price's friends Tinkle and Lammerts. The two geneticists drew on illustrations from their field to argue against the evolutionary origin of species, but their uninspired, technical style resulted in numerous complaints to the editor that their chapter was "unintelligible." In informing the authors of the need for revision, Everest happily reported that his greatest fear, "being disowned by the evangelicals," had not materialized. In fact, the book had sold so well the publisher was planning a second run of ten thousand copies. To help reach this large audience, Everest urged Tinkle and Lammerts to spruce up their writing and tone down their attack on evolution. "Try," he advised, "to weed out little phrases which would classify the chapter as 'anti-' rather than 'non-evolutionary.' "[28]

Leading the ASA away from its antievolutionist stance were two shirttail relatives from Wheaton, Russell L. Mixter and J. Frank Cassel, who did for biology what Kulp was doing for geology. Mixter, who joined the ASA in 1943, had majored in literature and minored in biology at Wheaton in the 1920s, at a time when S. James Bole was "the whole Biology Department." In the one geology course then offered, Bole required Mixter and the other students to read Price's *New Geology*. As a student Mixter attended the First Baptist Church in Wheaton but left in protest when a "liberal" pastor assumed the pulpit; later he joined the United Gospel Tabernacle. After graduating in 1928, Mixter went on to Michigan State College for an M.S. in zoology and to the University of Illinois School of Medicine in Chicago for a Ph.D. in anatomy, which he obtained after returning to Wheaton to join the faculty. By the late 1930s he had moved from a brief flirtation with flood geology to a lifelong commitment to what he called progressive creation. This view accepted both the geological evidence of an ancient earth and the biological evidence of organic development. The writings of two evangelical theologians played a central role in facilitating this shift. Floyd E. Hamilton's *Basis of Evolutionary Faith* (1931)—ironically one of the works that helped to popularize Price's ideas among orthodox Presbyterians—taught him that the biblical phrase "after their kind" did "not rule out biological change which is clearly demonstrated or implied by a study of genetics and paleontol-

George Frederick Wright
in the early 1880s (Oberlin
College Archives)

S. James Bole as he appeared
in the 1922 Wheaton College
yearbook (courtesy of David C.
Lindberg)

Dudley Joseph Whitney about 1960
(courtesy of the Reverend Dayton
Kitterman)

L. Allen Higley, undated (*National
Cyclopaedia of American History*,
1961)

Harry Rimmer at the Central Baptist Church in New York City (*Sunday Mirror Magazine*, December 10, 1939)

George McCready Price in 1906 (courtesy of Molleurus Couperus)

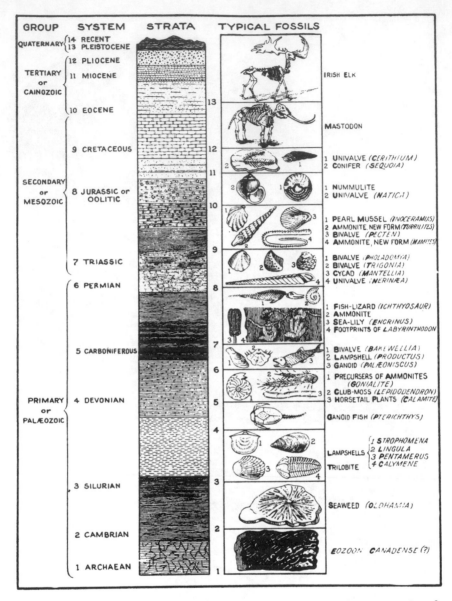

GROUP	SYSTEM	STRATA	TYPICAL FOSSILS

Quaternary: 14 RECENT / 13 PLEISTOCENE

TERTIARY or CAINOZOIC: 12 PLIOCENE / 11 MIOCENE / 10 EOCENE — IRISH ELK — MASTODON

SECONDARY or MESOZOIC: 9 CRETACEOUS / 8 JURASSIC or OOLITIC / 7 TRIASSIC / 6 PERMIAN

1 UNIVALVE (CERITHIUM)
2 CONIFER (SEQUOIA)

1 NUMMULITE
2 UNIVALVE (NATICA)

1 PEARL MUSSEL (INOCERAMUS)
2 AMMONITE, NEW FORM (TURRILITES)
3 BIVALVE (PECTEN)
4 AMMONITE, NEW FORM (NAMITES)

1 BIVALVE (PHOLADOMYA)
2 BIVALVE (TRIGONIA)
3 CYCAD (MANTELLIA)
4 UNIVALVE (NERINÆA)

1 FISH-LIZARD (ICHTHYOSAUR)
2 AMMONITE
3 SEA-LILY (ENCRINUS)
4 FOOTPRINTS OF LABYRINTHODON

PRIMARY or PALÆOZOIC: 5 CARBONIFEROUS / 4 DEVONIAN / 3 SILURIAN / 2 CAMBRIAN / 1 ARCHAEAN

1 BIVALVE (BAKEWELLIA)
2 LAMPSHELL (PRODUCTUS)
3 GANOID (PALÆONISCUS)

1 PRECURSERS OF AMMONITES (GONIALITE)
2 CLUB-MOSS (LEPIDODENDRON)
3 HORSETAIL PLANTS (CALAMITE)

GANOID FISH (PTERICHTHYS)

LAMPSHELLS:
1 STROPHOMENA
2 LINGULA
3 PENTAMERUS
TRILOBITE:
4 CALYMENE

SEAWEED (OLDHAMIA)

EOZOON CANADENSE (?)

ABOVE: The geological column explained by a flood geologist (George McCready Price, *The Predicament of Evolution*, 1925)

ABOVE, RIGHT: The celebrated Chief Mountain in Montana, where old Precambrian rocks rest on much younger Cretaceous rocks at about the tree line (the frontispiece in George McCready Price's *The New Geology*, 1923)

BELOW, RIGHT: Diagram showing how "evolutionists" use thrust faulting to explain the "upside-down" Chief Mountain (Byron C. Nelson, *"After Its Kind,"* 1927)

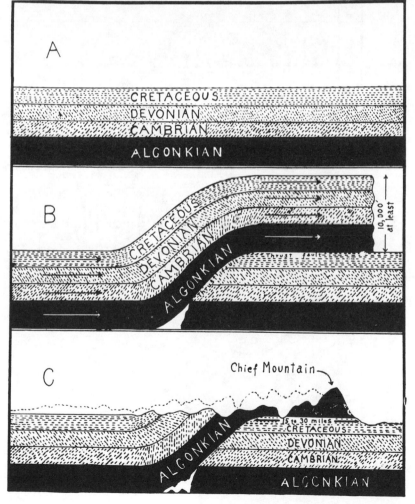

A

CRETACEOUS
DEVONIAN
CAMBRIAN
ALGONKIAN

B

CRETACEOUS
DEVONIAN
CAMBRIAN
ALGONKIAN
Thrust fault
10,000 at least

C

Chief Mountain
15 to 30 miles
CRETACEOUS
DEVONIAN
CAMBRIAN
ALGONKIAN
ALGONKIAN

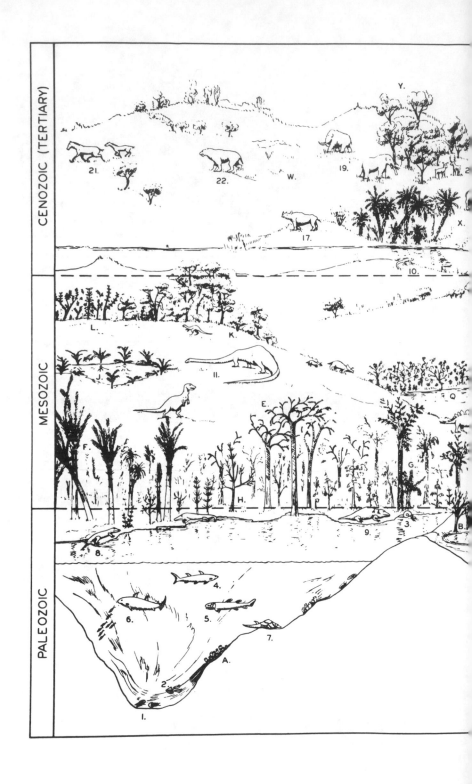

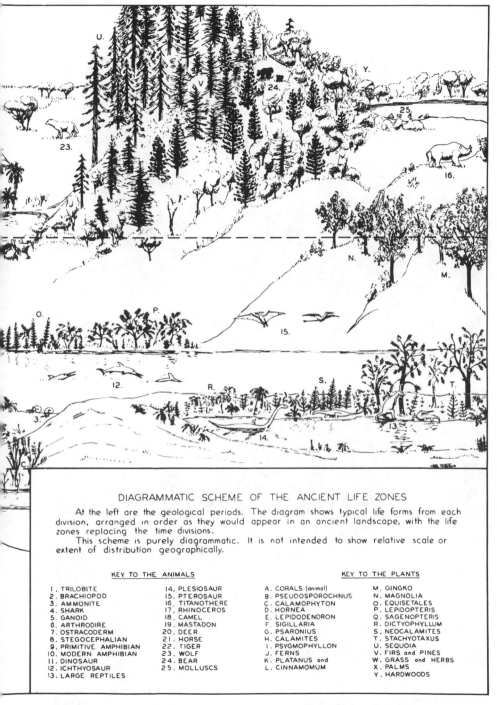

DIAGRAMMATIC SCHEME OF THE ANCIENT LIFE ZONES

At the left are the geological periods. The diagram shows typical life forms from each division, arranged in order as they would appear in an ancient landscape, with the life zones replacing the time divisions.

This scheme is purely diagrammatic. It is not intended to show relative scale or extent of distribution geographically.

KEY TO THE ANIMALS

1. TRILOBITE	14. PLESIOSAUR	A. CORALS (animal)	M. GINGKO
2. BRACHIOPOD	15. PTEROSAUR	B. PSEUDOSPOROCHNUS	N. MAGNOLIA
3. AMMONITE	16. TITANOTHERE	C. CALAMOPHYTON	O. EQUISETALES
4. SHARK	17. RHINOCEROS	D. HORNEA	P. LEPIDOPTERIS
5. GANOID	18. CAMEL	E. LEPIDODENDRON	Q. SAGENOPTERIS
6. ARTHRODIRE	19. MASTADON	F. SIGILLARIA	R. DICTYOPHYLLUM
7. OSTRACODERM	20. DEER	G. PSARONIUS	S. NEOCALAMITES
8. STEGOCEPHALIAN	21. HORSE	H. CALAMITES	T. STACHYOTAXUS
9. PRIMITIVE AMPHIBIAN	22. TIGER	I. PSYGMOPHYLLON	U. SEQUOIA
10. MODERN AMPHIBIAN	23. WOLF	J. FERNS	V. FIRS and PINES
11. DINOSAUR	24. BEAR	K. PLATANUS and	W. GRASS and HERBS
12. ICHTHYOSAUR	25. MOLLUSCS	L. CINNAMOMUM	X. PALMS
13. LARGE REPTILES			Y. HARDWOODS

KEY TO THE PLANTS

Diagram illustrating the ecological zonation theory of flood geology (Harold W. Clark, *The New Diluvialism*, 1946)

ABOVE: George McCready Price, left, with his disciple Frank
Lewis Marsh in southern California in 1960 (courtesy of Richard
M. Ritland)

ABOVE, LEFT: Byron C. Nelson, in dark coat and hat to the
left, with unidentified geologists in about the late 1920s examining fossil reptile tracks near Woodbridge, New Jersey (courtesy
of Paul Nelson)

BELOW, LEFT: Molleurus Couperus, right, with Louis B. Leakey
in 1967 at Olduvai Gorge, where the skull of Zinjanthropus
had been found eight years earlier (courtesy of Molleurus
Couperus)

ABOVE: Russell L. Mixter, J. Laurence Kulp, and F. Alton Everest, left to right, at the 1950 meeting of the American Scientific Affiliation (courtesy of F. Alton Everest)

ABOVE, LEFT: J. Laurence Kulp addressing the American Scientific Affiliation at Goshen College in 1950 (courtesy of F. Alton Everest)

BELOW LEFT: Walter R. Hearn in Berkeley, California, in the late 1970s (courtesy of Walter R. Hearn)

ABOVE: Richard M. Ritland, Frank Lewis Marsh, Harold G. Coffin, P. Edgar Hare, Ernest S. Booth, and an unidentified companion, left to right, on a 1960 field trip to Yellowstone National Park (courtesy of Richard M. Ritland)

BELOW: Clifford L. Burdick, Richard M. Ritland, and Frank Lewis Marsh, left to right, at the South Rim of the Grand Canyon during the summer of 1959, when Ritland introduced his Adventist colleagues to geological problems facing the church (courtesy of Richard M. Ritland)

Duane T. Gish debating at Iowa State University in January 1980 (courtesy of the Institute for Creation Research)

Henry M. Morris, left, and John C. Whitcomb, Jr., in 1984 (courtesy of John C. Whitcomb, Jr.)

Walter E. Lammerts with his prize roses in Livermore, California, in about the 1960s (courtesy of Walter E. Lammerts)

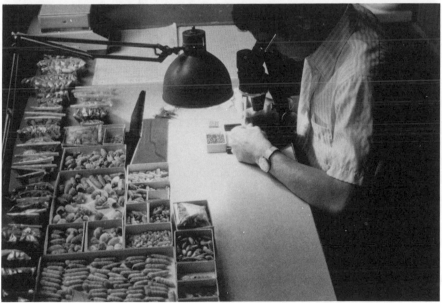

ABOVE: Nicolaas A. Rupke, as an undergraduate comparative anatomy student at the University of Groningen in 1964, introducing his classmates to George McCready Price's *Evolutionary Geology and the New Catastrophism* (courtesy of Nicolaas A. Rupke)

BELOW: Kurt P. Wise measuring Cerion for Stephen Jay Gould as a graduate research assistant at Harvard University in 1981 (courtesy of Kurt P. Wise)

EVOLUTION MODEL

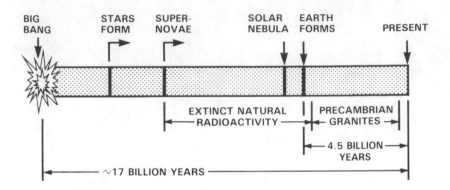

CREATION MODEL

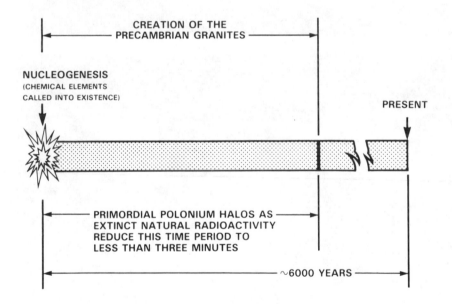

Figure 2.4 Models of Origins
(a) Based on Evolution (b) Based on Creation

Diagram from Robert V. Gentry's *Creation's Tiny Mystery*, rev. ed. (Knoxville, TN: Earth Science Associates, 1988) showing how about 4.5 billion years of geological history can be compressed into less than three minutes (courtesy of Robert V. Gentry) ogy," while an article by Wheaton president J. Oliver Buswell, Jr., "The Length of the Creative Days" (1935), showed him how to accommodate Genesis and geology by interpreting the Mosaic days figuratively.[29]

The year Mixter joined the ASA one of his wife's cousins married one of his former zoology students, J. Frank Cassel, who had graduated from Wheaton in the class of '38. After Wheaton, Cassel earned a master's degree in biology from Cornell University, served several years in the U.S. Army, and then taught at Colorado A & M (now Colorado State University) while working on a Ph.D. at the University of Colorado. In 1950 he took a job teaching biology at North Dakota Agricultural College (now North Dakota State University) in Fargo, where he remained until his retirement over three decades later. Though sporadically a Baptist, over the years Cassel maintained stronger ties to the Inter-Varsity Christian Fellowship than to any particular denomination. As an undergraduate he had learned the gap theory from Higley, his geology teacher, and the day-age theory from Buswell, the college president; and though he preferred the latter to the former, he grew increasingly skeptical of all efforts to force a fit between science and Scripture. At the same time he abandoned his original hope of proving the evolutionists wrong. Instead Cassel embraced what he called "evolutionary theism," an elusive position somewhere on the spectrum of belief between progressive creationism and theistic evolutionism.[30]

Together, Mixter and Cassel sought to bring evangelicals into harmony with modern biology, even if that meant creating temporary disharmony. Mixter, as a professor at Wheaton, had the greater influence; Cassel, as a teacher in secular schools, had the greater freedom. Mixter began undermining fundamentalist beliefs in a recent creation and the fixity of species in papers that he presented at ASA conventions in the late 1940s and subsequently published collectively as *Creation and Evolution* (1950). In this monograph, which enjoyed wide use in evangelical colleges and seminaries, he dismissed Noah's flood as an event of no biological consequence and, going one step beyond Hamilton, who had limited evolution to the family, advocated the acceptance of evolution "within the order." Creationists, he concluded, could "believe in the origin of species at different times, separated by millions of years, and in places continents apart." Meanwhile, Cassel, who signed on with the ASA in 1947, was urging the affiliation to tone down its antievolutionism and adopt a more positive attitude, approaching evolution "not with a chip on our shoulder but in the spirit of inquiry." Writing to a fellow biologist in 1948, he confessed that both Mixter and he had "felt for some time that many evangelical scholars such as those represented in the Affiliation take much too negative an attitude, namely that 'Evolution' is of the devil and it is our job to refute it from the word 'go.' " By this time it was an open secret that neither Mixter nor Cassel identified with fundamentalism any-more.[31]

In the summer of 1951, Cassel sent Everest the abstract of a paper on evolution he planned to read at the annual ASA convention later that year. His message could not have been blunter: "Evolution has been defined as 'the gradual or sudden change in animals and plants through successive generations.' . . . Such changes are demonstrable. Therefore, evolution is a fact." Not wishing to arouse "any sleeping dogs" he could not tame, he solicited advice from the politically astute founder on how best to deal with such a sensitive issue. Everest urged him to think carefully before handing critics a "cudgeal" [sic] with which to beat the ASA, adding, "We fall heir to plenty of antagonism without asking for it this way." Cassel's visceral response to this warning, addressed to both Everest and the anthropologist James O. Buswell III (b. 1922), who was hosting the convention, reveals the intense feelings that the subject of evolution still aroused in the ASA:

> . . . there is evolution and we believe in it. It's just against the fact that Russ, and both of you, and I feel that we can't call it by name for fear of its being used as a "cudgeal" as you so aptly state, Alton, that I am crying.
>
> . . . my point is, until we admit something exists, we can't study it—and evolution is in need of much thorogoing study by conservatives. At least among us girls, let's call it by name, talk it over, and then do something about it. Let's stop this wasting time saying, now this is evolution, and I believe in it, but it really isn't evolution—because something else is evolution.
>
> . . . I have no particular love for my statement—"Therefore, evolution is a fact." At least I didn't until you two reacted so beautifully and predictably to it. I tried to simply say that 2 plus 2 equals 4, rather than put it as we have been for so long in A.S.A.—"It is axiomatic what 2 plus 2 equals, but I can't say it cause it's a dirty word—and besides that's not what I'm talking about, anyhow." This time it happens to be what I'm talking about. But if you feel, and apparently both of you do, that it is a bit too strong for publication as yet—that it might not bring honor to the cause of Christ, then I have several possible suggestions [e.g., deleting or rewording the sentence].

. . . The really pertinent question now becomes—should I present the paper at all? Do you think A.S.A. is ready for it? . . . Do you think I can say this without getting tossed out on my ear, disbarred from A.S.A., and having all right to the title "Fundamentalist" even now and for evermore denied me?

I think we have pointed up here a very basic problem—basic I believe to the very existence—at least effective existence of A.S.A. We find ourselves in a philosophically untenable position, that of admitting that certain processes are functioning in the universe, but not being willing (and justifiably so) to carry our reasoning with these facts to their ad absurdum. Evolution has occurred, is occurring. But what are its limits? How far does it go? Frankly, I find myself shuddering and recoiling at some of the distances Russ seems willing to go. And yet why should God have created the first birds with tail vertebrae and teeth (Jurassic), and later some without a long tail but with teeth (Cretaceous), before He got around to modern birds without either? Does He like to experiment? Is it a sequence to confuse us? Are there really reptilian ancestors? Are the present fossils what they are claimed to be? . . .

Cassel ended his letter with a heartfelt cry for more scientific evidence—the criterion he had chosen for "rightly dividing the word of truth."[32]

Mixter and Cassel's sporadic push for greater openness toward evolution dovetailed perfectly with Kulp's agenda for the ASA, spelled out explicitly at an informal meeting of members in the New York City area in the summer of 1953. Out of a gathering of mostly graduate students from Columbia University and faculty members from Shelton College, held in Kulp's New Jersey home, issued a controversial report that called for *examining* rather than *defending* the faith, for developing a Christian philosophy of science rather than studying "the relationship of the facts of science and the Holy Scriptures." To "get away from the predominating flavor of an 'anti-evolutionary' society" and become a nondefensive "learned society" of evangelical scientists, the group recommended discarding the original doctrinal statement, which they saw as a "red flag" warning real scholars to stay away.[33]

When minutes of the meeting fell into the hands of conservatives, they reacted with predictable horror. Custance, who had just decided to rejoin the ASA, could scarcely believe what he was reading. In a seven-page outburst sent to his friend Howitt, he denounced "the Kulpians" for placing scientific respectability above religious commitment, for failing to realize that "in many cases theology must decide the issue, not scientific 'fact.' " The New York liberals, he charged, would "welcome all theistic believers in Evolution, but exclude all Christian men who believe in Flood Geology, and other such heresies. . . . Drop all anti-evolutionary emphasis and in effect concede the issue and be 'sensible' enough not to kick the tide. . . . Send a good will team to Moscow University!!?" Surprisingly, Howitt passed the letter on to ASA leaders for comment. Buswell, for one, could not recall ever reading anything so satanic and distorted. Fortunately for the Kulpians, Custance's days in the ASA were numbered. Shortly after rejoining he resigned for the second time—not so much because of ideological differences as because of the embarrassing breakdown of his marriage.[34]

Perhaps the best documentation for the ASA's abandonment of strict creationism comes from the events leading up to the publication of *Evolution and Christian Thought Today* (1959), edited by Mixter. The project took form in the late 1940s, when ASA members began discussing the possibility of writing "a scholarly, carefully prepared refutation of the evolutionary hypothesis" to appear on the centennial of Darwin's *Origin of Species.* In describing the proposed volume to ASA members, Everest explained that such a book was needed because "Darwin's doctrine engenders an atmosphere of materialism and presents very real intellectual barriers to the propagation of the Gospel." An accompanying prospectus unambiguously declared the creationist objective: "To show the inconsistency of evolution when the evidence is carefully evaluated, and to show that creation is a thoroughly sound explanation of the observations." Upon receiving this news from the president, Tinkle, an uncompromising creationist, congratulated him for his "broadside" against evolution and for launching a campaign against the insidious theory. His only reservation concerned heaping "more honors upon an undeserving man by publishing it on the centenary of Darwin's book."[35]

For years the book project elicited little comment, in part because the Darwin centennial lay so far in the future but also because the new generation of ASA leaders—especially Kulp, Mixter, and Cassel—wanted nothing to do with an old-fashioned creation-versus-evolution polemic. "I, for one," declared Kulp in a letter to Mixter, "will have no part of a straight 'anti-evolution' treatise which is the form in which it was originally proposed." By 1955 even Everest, writing as chairman of a newly constituted Darwin Centennial Book Committee, confessed that his views on evolution had "matured considerably during the course of the past ten years. . . . I cannot now recommend as a primary purpose of the proposed volume 'to show the inconsistency of evolution.' This is far too negative an approach." Though his faith in the inerrancy of the Bible remained unshaken, he no longer believed in holding "to grandfather's interpretations just because grandfather was a great and good man," and he refused to tickle "the average Christian's fancy by 'fighting evolution.'" Of the four other members of his committee, which included Mixter, only the conservative Monsma defended the originally stated objective.[36]

Despite occasional protests, the project moved forward, with Mixter in the lead as editor. For years Mixter had generally restricted evolution to biological orders, but in a paper read at the annual ASA convention in 1957 he pushed the limits of evolution at least as far as phyla, a much larger grouping of animal life, and expressed a reluctance to continue identifying the original units of creation on the basis of ever-narrowing gaps in the fossil record. It seemed to one commentator that Mixter was arguing for the acceptance of evolution "as a *tentative* working hypothesis until we can come up with something better." Although Mixter stopped short of an outright endorsement of theistic evolution, conservative creationists pounced on him for "completely reversing his position" and "capitulating to evolution," while evolutionists congratulated him for finally seeing the light. An admiring Cassel paraphrased Mixter's message in the following words:

> I, an evangelical Christian, can accept the basic concepts of evolution. Although not exclusively demanded by the data involved, it is certainly allowed, and in fact I can see no better or more logical way to handle the data. I believe in Creation, and simply affirm that in the light of the evidence now available, I think some evolution—this is, development of present-day forms by differentiation of previously existing forms—the most likely way God accomplished much of His Creation.

Such candid remarks prompted one troubled ASA member to lament the "growing conviction that inexorable pressure of expanding knowledge is about to force us to accept some formulation of the theory of evolution, including the evolutionary origin of man, and that we must adjust our thinking in accordance with this eventuality."[37]

Given his growing reputation, deserved or not, as a theistic evolutionist, Mixter recognized the need for extreme caution when discussing organic development. Adhering to the principle that "if you call something a rose you expect it to smell like a rose," the Wheaton biologist urged contributors to the volume to use the word *evolution* only when referring to "the total theory of derivation of all living things from the few, or one, original form of life," and to substitute inoffensive synonyms such as "development, derivation, [and] variations" when describing the formation of new life forms. One of the contributors, James O. Buswell III, cleverly suggested adopting the euphemism "scientific creationism" to cover views ranging from progressive creationism to theistic evolutionism, but the new label never caught on— until the conservatives later co-opted it for flood geology. In the end the stable of authors Mixter signed up to write the various chapters— including such familiar names as Buswell, Cassel, and Cordelia Erdman Barber, as well as the relatively unknown Walter R. Hearn— guaranteed that criticism would come no matter how sweet-smelling the language.[38]

In marked contrast to Cassel, Barber, Buswell, and several other contributors who followed Mixter in defending "descent with modification but always within predetermined limits," the brash young Hearn (b. 1926) refused to put any arbitrary limits on the extent of evolution. In an iconoclastic chapter, "The Origin of Life," coauthored with Richard A. Hendry (b. 1929), he deliberately stirred evangelicals with the declaration that Christians must be willing to accept "that life arose from inanimate matter through a series of physico-chemical processes no different from those we can observe today." Describing (some would say dismissing) Genesis as "a brief but beautifully poetic narrative," the two biochemists went on to argue that because "the Bible gives little specific information about the *ways* in which God has worked in nature," there seemed to be "little or no limitation on possible mechanisms" that Christian scientists could use in explaining life.[39]

Hearn's radicalism no doubt stemmed in part from the impatience of youth, but it also reflected a lack of socialization in evangelical schools. Unlike so many of the intellectual leaders in the ASA, which at times resembled a Wheaton alumni association, Hearn neither attended nor taught in a Christian college, with the exception of three years on the faculty of the nominally Baptist Baylor University College of Medicine. Reared as a Southern Baptist in Houston, Texas, Hearn earned a bachelor's degree in chemistry at the local Rice Institute. During his college years he actively participated in the Baptist Student Union and on occasion attended a Bible-study group led by Henry M. Morris, the only Christian he knew on the Rice faculty. Despite his affection for Morris, he never shared the flood geologist's commitment to hard-core creationism, preferring instead the less dogmatic views of R. E. D. Clark. After graduating from Rice, Hearn earned a Ph.D. in biochemistry from the University of Illinois and then spent brief periods on the faculties of the Yale and Baylor medical schools. During his time at

Baylor he collaborated with the physiologist Roger Guillemin (b. 1924), who went on to share the Nobel Prize in Physiology or Medicine with Hearn's successor, for work they had all done on the peptide hormones of the brain. In 1955 Hearn joined the chemistry faculty at Iowa State College in Ames, where he stayed until 1972.[40]

As one of the first biochemists to play an active role in the ASA, Hearn felt a God-given responsibility to inform members about the growing importance of biochemistry in theories of evolution, which had previously focused on evidence from paleontology and morphology. Because of his outspokenness, he often found himself the center of controversy within the ASA, but because of his unfailingly sweet temper, he seldom made enemies. Even the archconservative Howitt, who complained that Hearn "never yielded a millimetre" in debate and who resented the apparent arrogance of liberals such as Kulp and Buswell, treasured Hearn's friendship—even after Hearn, following a research leave in Berkeley in the late 1960s, let his hair grow long and tied it "at the back like that of a woman." During the 1950s ASA meetings often turned into what Hearn called "marvelous orgies of argument and prayer combined." Though he and Cassel shared much in common, the two liberals frequently fought over the proper relationship between science and religion, with Cassel defending a romantic, vitalistic view of biology in opposition to Hearn's coldly mechanistic, metaphysically barren approach. Cassel liked to say that Hearn left Christ at his laboratory door, while Hearn accused Cassel of seeing Christ in his microscope. In the end, however, it was Hearn who convinced Cassel to go all the way with evolution, from life to humans.[41]

Although some ASA members, such as Everest, found the slide toward evolution more exhilarating than painful, others, such as Cassel, experienced immense personal distress. The following autobiographical testimony poignantly captures some of the emotional turmoil of the 1950s:

> First to be overcome was the onus of dealing with a "verboten" term and in a "non-existent" area. Then, as each made an honest and objective consideration of the data, he was struck with the validity and undeniability of datum after datum. As he strove to incorporate each of these facts into his Biblico-scientific frame of reference, he found that—while the frame became more complete and satisfying—he began to question first the feasibility and then the desirability of an effort to refute the total evolutionary concept, and finally he became impressed by its impossibility on the basis of existing data. This has been a heart-rending, soul-

searching experience for the committed Christian as he has seen what he had long considered the *raison d'être* of God's call for his life endeavor fade away, and as he has struggled to release strongly held convictions as to the close limitations of Creationism.

The pain suffered by Cassel and his liberal friends elicited little sympathy from conservatives within the ASA, who thought the affiliation had gone "soft" on evolution, or, in the colorful phrase of one member, "had gone to the apes." In their opinion, the drift toward evolution was motivated not by intellectual honesty but by "the malignant influence of 'that old serpent, called the Devil, and Satan, which deceiveth the whole world' (Revelation 12:9)." With some justification, they regarded the term *progressive creation* as merely a euphemism for heterodox theistic evolution, the acceptance of which, declared one critic, was tantamount to placing "a kiss of death on [the] Master."[42]

The defection of some ASA leaders from the evangelical fold seemed to confirm conservative fears about the spiritual consequences of evolution. By far the most prominent defector was Kulp, who not only resigned from the ASA but also severed relations with the Plymouth Brethren. After years of involvement in the Inter-Varsity Christian Fellowship, the ASA, and the Young Life Institute, he found by the late 1950s that he no longer cared about liberalizing young fundamentalists. Although he served as an influential member of the Fuller Seminary board of trustees in the early 1960s, he had by this time moved both theologically and scientifically beyond the parochial concerns of his evangelical friends in the ASA. At least some attributed his departure to his not wishing "to be embarrassed by his association with us whilst working his way to the top." Rumor had it that once he achieved worldly success, he would consider "his return to the fold," but he merely drifted farther and farther from orthodox Christianity.[43]

The ASA liberals, for their part, displayed equally little charity toward the plight of the conservatives. Often self-conscious about their own fundamentalist heritage and increasingly embarrassed by the antievolution rhetoric of their creationist colleagues, they damned strict creationism as "pseudo-science" and a menace to the Christian faith. By the late 1950s strict creationists like Howitt, completing five years of service on the ASA council, no longer felt welcome at affiliation conferences. "I think that they will be very glad to get rid of me," he wrote shortly after attending his last council meeting. If he had become "a pain in the neck," as he suspected, it was only because he had defended the ASA's original principles. Under the circumstances a split between liberals and conservatives, mirroring the rift that had already divided the larger evangelical community into neo-evangelical and fundamentalist camps, appeared inevitable.[44]

CREATION AND EVOLUTION AT WHEATON COLLEGE

Because of the tight Wheaton-ASA connection, no institution suffered more fallout from the ASA's explosive affirmations of evolution than the acknowledged flagship of evangelical colleges. The furor generated by Mixter's paper at the 1957 ASA convention prompted the college president, V. R. Edman (1900–1967), to ask his besieged biologist to prepare a public statement for damage control. In "A Wheaton College View of Creation and Evolution," Mixter tried to finesse his way out of an increasingly uncomfortable situation while categorically denying that Wheaton professors believed that humans "evolved from ape-like creatures." Genesis, he wrote, identified God as the Creator of the world, but "just how much He created and how much He left to hereditary processes which He also started must be left to investigation." Fortunately for conservative Christians—and Mixter's job—he found the fossil record "too scanty and too broken to prove total evolution." Despite the ambiguity of his apology, it seemed to please the president, who sent it to the trustees as evidence that "we at Wheaton are avowed and committed creationists."[45]

The appearance of *Evolution and Christian Thought Today* scarcely enhanced Mixter's stock with Wheaton's right-wing constituents. Although book reviewers for *Eternity* magazine voted it the most significant book of the year in 1960, the conservative Christian press, reported Hearn, generally "felt obliged to put that word, 'evangelical,' in quotation marks when referring to authors of some of the chapters, or to prefix it with the adjective, 'so-called.'" Once again Mixter had to assure the anxious college president that he and his associates in the biology department were not teaching theistic evolution, defined as "all the evolution Darwin taught but evolution guided by God," but progressive creation. The latter, he explained benignly, "means that God created many species and after their creation they have varied as the result of mutation and selection so what was once one species has become a number of species, probably as many as are now found in an order or family."[46]

Just as the controversy over *Evolution and Christian Thought Today* began to wane, the science faculty at Wheaton arranged a college-sponsored science symposium, "Origins and Christian Thought," scheduled for February 17–18, 1961, in which many of the contributors to Mixter's edited book would participate. The event attracted not only President Edman but several local pastors, including Harold P. Warren (b. 1906) of the First Baptist Church in Wheaton, who sat in shocked silence while speaker after evangelical speaker challenged traditional interpretations of Genesis. Within days of the close of the symposium, Pastor Warren was circulating a statement detailing the heresy he had heard at Wheaton: Hearn insisting that the first three chapters of Genesis "MUST be taken figuratively," Cassel speculating that humans and apes shared a common ancestor, Buswell contending that humans and nonhumans could only be distinguished on spiritual grounds. Buswell's strained effort to baptize such views as "scientific creationism" Warren dismissed as a "smoke screen" thrown up to camouflage Buswell's own acceptance of evolution. "When you open the door on this 'loose figurative interpretation' approach," Warren asked, "what kind of a Supernatural Bible do you have left? What kind of a redemptive message do we have to preach?" Only two remedies, he concluded, could save Wheaton: "a REVIVAL, or a house cleaning!"[47]

Close on the heels of Warren's privately circulated document appeared a front-page story in the widely read fundamentalist *Sword of the Lord* proclaiming "Evolution at Wheaton College." In it the writer, another Baptist minister, reported that the symposium participants had taught "outright theistic evolution" while giving "no serious consideration . . . to what Genesis, literally understood, claims." "It is a sad, sad day," he observed, "when one of the best-known Christian colleges sponsors this kind of thing for the benefit of its students." Before long, officials of the General Association of Regular Baptist Churches (GARB) had joined the attack, warning Christian parents about the spiritual risk of sending their children to Wheaton. In response to the queries of one troubled mother whose son planned to enter Wheaton, Edman, declaring himself a fundamentalist, insisted that "no faculty member of Wheaton believes or teaches evolution." Stretching truth to the breaking point, he assured her that every one of the symposium speakers was "a convinced creationist."[48]

Under constant fire, Mixter shielded himself with the cloak of orthodoxy, protesting that he had gone no farther than many stalwart evangelicals of the past. In a groveling letter to Edman reprinted in the widely distributed pamphlet *Christianity and Science at Wheaton* (1962), he affirmed his belief in the first chapters of Genesis as "a factual account of what happened," in the special creation of "grass, fruit trees and herbs," and in the supernatural creation of "Adam as a real person and as the first man." On occasion his professions of orthodoxy necessitated rapid backpedaling from previously stated positions. In an essay for a youth magazine in 1959, he had written that "fossils indicate that modern man goes back finally to Australopithecus, the South African 'ape man' who walked upright, had a good sized brain, and made tools." When confronted with this statement in the early 1960s, he decided that Australopithecus was *not* "an ancestor of today's man but a race of creatures similar in anatomy to man but not his progenitor." An early statement that "one can believe in creation and evolution—provided that one defines the terms used" also came back to haunt him. "Exactly so!" exclaimed a gleeful critic in *Sword of the Lord* after catching Mixter at his own semantic game. Particularly galling to the strict creationists was the Christian public's inability "to discern the fact that the difference between [the Wheaton] concept of creation and the generally accepted concept of theistic evolution is so minuscule as to be quite negligible." But, as even Humpty Dumpty realized, words could mean what one wanted them to mean.[49]

TEN

John C. Whitcomb, Jr., Henry M. Morris, and *The Genesis Flood*

In 1954 Bernard Ramm, an evangelical philosopher and theologian close to the inner circle of the American Scientific Affiliation (ASA), published the most talked-about creationist book to appear in years, *The Christian View of Science and Scripture*. In it he tried to steer readers away from the "narrow bibliolatry" of the typical fiat creationist and guide them to what he called progressive creationism, which did away with the necessity for a young earth, a universal flood, and the recent appearance of humans. On the occasion of the book's twenty-fifth anniversary, the ASA celebrated the publication of the book as "a pivotal event" in the recent history of science and religion. [1]

In fundamentalist circles, says historian George M. Marsden, Ramm's work created "the largest stir" since the controversy a few years earlier over the Revised Standard Version of the Bible. By the time his book appeared, Ramm was emerging as a leader of the "new evangelicals," who distinguished themselves from the old fundamentalists by embracing "responsible" scholarship and shunning belligerency. Instead of separating themselves from the "world," they sought to engage modern culture and to promote social betterment. *The Christian View of Science and Scripture,* which "challenged the fundamentalist assumption that a high view of biblical inspiration implied that the Bible was a reliable source of scientific data," served as a model of neo-evangelical scholarship. Indeed, within months after it hit the bookstores, the evangelist Billy Graham (b. 1918) was calling for a view of biblical inspiration "along the line of the recent book by Bernard Ramm."[2]

As a lad growing up in the American Northwest, Ramm had fallen in love with science. After high school he enrolled in the University of Washington intending to major in chemistry, but he soon shifted to speech, a subject of keen interest because of a childhood stammer. By this time he had converted to evangelical Christianity, and before long he decided to become a Baptist minister. The years immediately following World War II found him teaching at the Bible Institute of Los Angeles while simultaneously pursuing a doctoral degree in philosophy at the University of Southern California.[3]

As a young Christian he had daily read his Scofield Bible and from it absorbed the gap theory of creation. He had also become enamored of Harry Rimmer's writings on science and religion, the most popular ones on the evangelical market at the time. The more Ramm learned about the philosophy of science, however, the more critical he became of the gap theory and Rimmer's "sloppy" attempts at harmonization. In Los Angeles he discovered the small but active ASA cell and formed a close friendship with F. Alton Everest. Although ineligible to become a regular ASA member because he lacked the required scientific credentials, he hosted the annual meeting in 1949, the occasion for J. Laurence Kulp's withering critique of deluge geology. At Biola (as the Bible Institute of Los Angeles was called) Ramm inherited a course on science and religion, for which he devoted considerable time and thought preparing lectures. Hating to let all of his effort go to waste after leaving southern California to teach at Bethel College and Seminary in St. Paul, Minnesota, he composed *The Christian View of Science and Scripture,* a polemic against "hyperorthodoxy" dedicated to Everest and checked by Kulp for "technical accuracy."[4]

Eschewing philosophical subtlety, Ramm divided evangelical writers on science and religion into two opposing camps: one representing the "ignoble tradition" of the fundamentalists, who adopted "a most unwholesome attitude toward science, and . . . used arguments and procedures not in the better traditions of established scholarship"; the other encompassing the "noble tradition" of such learned nineteenth-century Christians as James Dwight Dana and John William Dawson, "who have been patient, genuine, and kind and who have taken great care to learn the facts of science and Scripture." During the past half-century the narrow-minded enemies of science had "buried the noble tradition," which Ramm now sought to resurrect.[5]

In Ramm's view, no one typified the "ignoble tradition" more than Rimmer and George McCready Price, who had inflicted the disreputable gap and flood theories on unsophisticated Christians. In part because of Rimmer's long reign as "Fundamentalism's outstanding spokesman in matters of Bible and science," the gap theory had become "the standard interpretation throughout Fundamentalism . . . so sacrosanct with some that to question it is equivalent to tampering with Sacred Scripture or to manifest modernistic leanings." But in Ramm's opinion, neither biblical nor scientific scholarship supported it. Geologists, for example, had found no trace of the radical break in the geological record demanded by the ruin-and-restoration scheme. In view of the tenacious hold of the gap theory on the fundamentalist mind, Ramm regarded Price's influence as "one of the strangest developments of the early part of the twentieth century." After all, the flood and gap theories offered incompatible explanations of the evidence for geological ages—the former attributing it to Noah's deluge, the latter assigning it to the pre-Edenic period. Besides, as Ramm pointed out in a half-page footnote, Price lacked the competence, training, and integrity needed to be taken seriously. But despite these impediments and the opposition of the entire geological establishment, Price had exerted a "staggering" influence on American fundamentalism and his ideas had come to form "the backbone of much of Fundamentalist thought about geology, creation, and the flood."[6]

To accommodate both science and Scripture, Ramm proposed combining the pictorial-day theory of Genesis 1 with what he called progressive creationism. Though once fond of the day-age theory popularized by Dana and Dawson, and still convinced that the first chapter of the Bible recorded "the broad outline" of earth history, he had, like the English creationist R. E. D. Clark, recently adopted P. J. Wiseman's conclusion that "creation was *revealed* [pictorially] in six days, not *performed* in six days"—or even in six identifiable geological ages. The idea of a progressive creation allowed God to intervene on occasion to bridge the gaps in the geological record. On these occasions he created new "*root-species*," which then evolved, or "radiated," into the species known today. The whole process, spanning millions of years, was God's method of preparing a suitable home for humans, the crowning work of creation:

> The vast forests grew and decayed for his coal, that coal might appear a natural product and not an artificial insertion in Nature. The millions of sea life were born and perished for his oil. The surface of the earth was weathered for his forests and valleys.

From time to time the great creative acts, *de novo,* took place. The complexity of animal forms increased. Finally, when every river had cut its intended course, when every mountain was in its purposed place, when every animal was on the earth according to blueprint, then he whom all creation anticipated is made, MAN, in whom alone is the breath of God.

"This is not evolution," Ramm insisted. "It is progressive creationism."[7]

Ramm ambitiously hoped that his theory would "form the basis of a new biological synthesis" that would be to biology what Einstein's relativity theory was to physics. This never happened. In fact, his peculiar version of progressive creationism failed to win the support of even a majority of evangelical scientists. Nevertheless, *The Christian View of Science and Scripture* sold tens of thousands of copies and profoundly influenced the way in which many orthodox Christians answered the questions posed by creation and evolution. Restless evangelical scientists, long constricted by the narrow interpretations of Rimmer and Price, thanked Ramm for giving them the theological "breathing space" their research seemed to demand. As the biologist J. Frank Cassel testified, "Naturally when I see the need of more time, and some Biblical scholar says, 'Sure, you may have more time,' I tend to jump on his band wagon very quickly." Years after the appearance of Ramm's book another grateful ASA member credited it with having paved the way for "the majority of Christian biologists" to accept evolution. He found it curious, however, that Ramm had "opened the door for a Christian reconciliation with evolution but stopped short of going through that door" himself.[8]

JOHN C. WHITCOMB, JR.

If Ramm thought he was officiating at the funeral of flood geology, he was badly mistaken. His book may have opened the door to evolution for many evangelicals, but it provoked others into mounting a massive campaign in defense of young-earth creationism. The reaction began with an angry young Bible teacher and seminarian named John C. Whitcomb, Jr. (b. 1924). To Whitcomb's way of thinking, *The Christian View of Science and Scripture,* with its unbiblical notions of a geographically and anthropologically local flood, offered "final proof of the logical absurdities to which one is driven as an evangelical by following uniformitarian geology." Instead of easing his mind about the harmony of Genesis and geology, Ramm's book had heightened Whitcomb's perplexity about how to reconcile "the obvious teaching of Genesis concerning the magnitude and uniqueness of the Flood and the demands of science (that is, of Dr. Kulp, etc.) that geology can allow for no such Flood." And if death had resulted from Adam's sin, as the Bible taught, then how, he wondered, could fossils antedate the Edenic creation?[9]

The well-educated, cosmopolitan Whitcomb hardly fit the Menckenesque stereotype of the rustic, semiliterate fundamentalist preacher. The son of an army officer who served as chief of staff under General George Patton in the Third Army, Whitcomb lived for several years as a boy in northern China, where he learned to speak Chinese. When the time came to attend college, he matriculated in Princeton University, where he studied historical geology and paleontology for a year before graduating in 1948 with honors in ancient and European history. During his freshman year he had converted to evangelical Christianity as a result of hearing a former missionary to India, and following graduation he enrolled in Grace Theological Seminary, a fundamentalist institution established in Winona Lake, Indiana, by the Fellowship of Grace Brethren Churches. After earning a B.D. degree in 1951, he remained at the seminary teaching Old Testament and Hebrew (and the gap theory of creation) while continuing his graduate studies.[10]

In the summer of 1953 the ASA held its annual convention on the Grace campus. Whitcomb especially liked a presentation by Henry M. Morris, "The Biblical Evidence for a Recent Creation and Universal Deluge," which defended flood geology against the competing day-age, ruin-and-restoration, and pictorial-day theories. By this time most ASA members, having been instructed by Kulp, held the flood theory in low regard, and Whitcomb was distressed to hear Morris being "politely denounced" by those in attendance. For Morris, the highlight of the convention was discovering another Christian scholar who shared his belief "that the Scriptural evidence requires acceptance of a literal, six-day creation and a subsequent universal Deluge, regardless of ephemeral geological theories."[11]

Just two years later the ASA returned to Winona Lake, this time for a joint meeting with the Evangelical Theological Society (ETS). Ramm's recently published book was the topic of constant comment. Whitcomb was disappointed, but not surprised, to find the scientists in the ASA praising it, but he was delighted to discover that few of the theologians in the ETS were "willing to follow where Dr. Ramm leads." This knowledge steeled a growing determination to devote his Th.D.

dissertation to answering Ramm and defending "the position of George M. Price."[12]

To assess prevailing theological opinion, Whitcomb systematically queried professors of Old Testament, archaeology, and apologetics in evangelical schools about their beliefs on creation and the flood. His survey revealed that evangelical leaders, like Ramm himself, were "confused, very confused, on these basic matters." On the one extreme, only a few respondents confessed to a partiality for flood geology; on the other, some of the neo-evangelicals gave the junior scholar "a terrific beating" for having dared to ask such questions. Ramm's friend Edward J. Carnell (1919–1967), the recently installed president of the formerly fundamentalist Fuller Theological Seminary, "denounced the very idea of a questionnaire of this kind." In response to Whitcomb's charge that Ramm's book was becoming "a rallying-point for the 'New Deism,'" Carnell noted that he did not detect in Whitcomb's writings "a critical acquaintance . . . with the great empirical evidence which Kulp and others have in mind when they make their appeals." Such reactions convinced Whitcomb that it would "take more than wrath to undo the frightening effects" of *The Christian View of Science and Scripture.* "Even if I had no other reasons for wishing to write a dissertation on Creation and the Flood," he confided to Morris, "Dr. Ramm's book would be sufficient incentive for me."[13]

In 1957 Whitcomb completed a 450-page dissertation, "The Genesis Flood," and almost immediately set about condensing it for publication. Because the controversy over Ramm's volume seemed "to be subsiding a little," he wanted his own book to appear as quickly as possible; yet he did not want to rush into print with an imperfect product. "After all," he wrote to a friend, "if the book is to be published only for the purpose of answering Dr. Ramm, it won't be worth publishing." Two well-established evangelical publishers, Eerdmans and Moody, initially expressed interest in Whitcomb's manuscript, but the former withdrew after seeing it, perhaps as a result of pressure from ASA scientists. The editors at Moody expected criticism from the ASA but decided to take the work anyway, reasoning that Ramm's critics at least deserved a hearing.[14]

Both the Moody Press and Whitcomb agreed that his book "would be far more effective if the chapters dealing with the scientific aspects of the Flood were carefully checked or even written by a Ph.D. in science." The young scholar had no masochistic desire to become an object of ridicule like Price, and he harbored no illusions about his qualifications as a scientist. But he resented the contempt experts showed for his efforts to fit geological data into a rigid biblical framework. "The minute I open my mouth about the geological effects of the Flood, or put my thoughts into print," he complained to one correspondent, "I am told that I have no right to say anything, because I do not have a Ph.D. in geology!" Clearly, he needed the help of sympathetic specialists, but he despaired of discovering any "Ph.D.'s in geology today who take Genesis 6–9 seriously."[15]

The closest he could find was Douglas A. Block (b. 1921), a geologist at Wheaton College who had earned a master's degree in geology from the University of Iowa but had been discouraged from pursuing a Ph.D. there because of his Christian views. Block was reputedly the only scientist on the Wheaton faculty who still clung to the notion of a worldwide flood—perhaps because of his earlier seminary training—and even he denied that it had left any discernible geological effects. Nevertheless, Whitcomb decided to take a chance and let him read the manuscript of *The Genesis Flood*. Block found Price's recycled arguments almost more than he could stomach. "It would seem," wrote the upset geologist, "that somewhere along the line there would have been a genuinely well-trained geologist who would have seen the implications of flood-geology and, if tenable, would have worked them into a reasonable system that was positive rather than negative in character." He assured Whitcomb that he and his colleagues at Wheaton were not ignoring Price. In fact, they required every geology student to read at least one of his books, and they repeatedly tested his ideas in seminars and in the field. By the time Block finished Whitcomb's manuscript, he had grown so agitated he offered to drive down to Winona Lake to instruct Whitcomb on the basics of historical geology. But the visit only deepened the theologian's suspicions about earth scientists.[16]

Determined to get *some* scientific advice, even if not from trained geologists, Whitcomb sought out a number of known creationists, including Dudley Joseph Whitney, William J. Tinkle, Morris, Price, and the Lutheran biologist John W. Klotz (b. 1918). The elderly Price could offer little more than moral support. "As I am now almost 87," he wrote, "I can not take any active part in the work of present-day research. I can only sit on the sidelines and watch the procession go by. Yet I am glad to see younger men, like you and Henry M. Morris . . . coming forward to carry the torch of truth." Whitney at first tried to "beg off" from reading the manuscript on the grounds that such "routine pro-creation stuff" wasted his time and he could not imagine anything of value coming from the Moody Press. Later he reluctantly agreed to take a look at it, but was dismayed to find that Whitcomb had the entire universe being created recently. "Why, why, why," he asked, "should the saints be so prone to take positions which discredit the Bible?" To add to Whitcomb's frustration, Tinkle kept pushing him to allow for geological catastrophes before Adam.[17]

Morris, who had been corresponding with Whitcomb off and on since meeting him in 1953, proved to be the most helpful of the group. For years he had been working on his own flood-geology book, tentatively called "The Creation and Destruction of the World," and he shared Whitcomb's antipathy toward Ramm and his feeling that geologists were running a "closed-shop." He also respected Whitcomb's skill as an Old Testament scholar. "It would seem to me," he wrote after reading the first three chapters of his dissertation, "that, granted the presumption of Biblical inspiration, you have proved the case for a universal flood to such a high degree of certainty that only the wilfully blind could not see it." He strongly advised, however, that Whitcomb refrain from indulging in sarcasm, "especially when referring to specific individuals and their arguments." Having himself been the target of barbs from Ramm and various ASA writers, he did not want strict creationists to descend "to their level of disputation." Besides, he suspected that Price, Rimmer, and Whitney had injured themselves by resorting to sarcasm and ridicule.[18]

When Morris had a chance to read the rest of the dissertation, which treated geological issues, he winced at the extent to which Whitcomb had relied on Price and other antievolution writers from the 1920s as well as on the recently controversial Immanuel Velikovsky (1895–1979), who was rewriting earth history in terms of epochal cataclysms. "Price and Velikovsky," he reminded Whitcomb, "are both considered by scientists generally as crackpots, although no one ever takes the trouble to answer their arguments save by ridicule and summary dismissal." Short of developing an entirely new perspective on the subject, he thought it might be best "simply to point out Price's arguments as a matter of historical record, and then leave your main emphasis on the Scriptural framework and the geological implications thereof." In that way Whitcomb might avoid getting trapped in the quicksands of geology.[19]

Even before receiving this gentle reminder of his inadequacies as a geological critic, Whitcomb had conceived the idea of asking Morris to join him as a collaborator. His "greatest weakness," Whitcomb candidly admitted, was his inability to deal adequately with the technical objections to flood geology raised by Kulp. If Morris would help him, he would willingly "wait a year or so" to publish his book. Morris liked the idea of "a dual-authorship" volume, but he could not see his way clear to undertake such a venture in the near future. Whitcomb, unwilling to throw in the towel so quickly, responded by asking Morris to "prayerfully consider" writing one or more chapters. "If you could contribute a chapter on 'Radioactivity' and one on 'Stratification' and one on 'Uniformitarianism' and even one on 'Vulcanism,' " Whitcomb wrote, "it would put some 'teeth' into the last half of the book." Though "highly honored" by this proposal, Morris still hesitated. He recognized the vast amount of work it would take to present flood geology in a convincing manner, and he knew he could not get to the project right away. Nevertheless, he left open the possibility of joining forces—if Whitcomb were willing to wait for him.[20]

By this time Whitcomb was growing impatient with the prospect of a long delay; so before accepting Morris's offer, he explored coauthorship with John W. Klotz, who taught biology at Concordia Teachers College in River Forest, Illinois. An ordained Lutheran minister in the Missouri Synod and an unbending biblical literalist, Klotz had earned a Ph.D. in genetics from the University of Pittsburgh and a B.D. from Concordia Theological Seminary, where he had written an antievolution thesis on Genesis and genetics under the direction of Theodore Graebner. In 1955 he had published his first major book, *Genes, Genesis, and Evolution*, in which he defended a universal flood, a young earth, and a recent creation in twenty-four-hour days. When Whitcomb approached him about the possibility of contributing a couple of scientifically oriented chapters to *The Genesis Flood*, Klotz protested that geology was not his field. To take his place, he recommended Wilbert H. Rusch (b. 1913), a fellow Missouri Lutheran with some training in geology.[21]

Even before receiving Klotz's reply, Whitcomb impatiently reopened negotiations with Morris, promising that any collaboration would be "a strictly *equal* enterprise." He made it clear, however, that this latest invitation was contingent on a negative decision from Klotz. In the meantime Morris provisionally accepted Whitcomb's offer of joint authorship and drafted an outline for the three chapters he planned to write. When Whitcomb finally heard from Klotz in early December 1957, he quickly notified Morris that "the way is cleared for us to go ahead, subject to the approval of Moody Press." Thus was sealed the pact that would soon make "Whitcomb and Morris" a byword among evangelical Christians.[22]

HENRY M. MORRIS

Whitcomb's new partner, Henry M. Morris, was a Texas-bred engineer who had grown up in the Southern Baptist church. As a boy he lived in a number of Texas towns before his family finally settled in Houston, where his father worked as a realtor. During the Great Depression the Morrises slipped into poverty, making it impossible for young Henry to fulfill his ambition of studying journalism at the University of Texas in Austin. As a consolation, he enrolled in the local Rice Institute, where tuition was free and he could live at home. In 1939, when he graduated with a bachelor's degree in civil engineering, the yearbook showed a sober, earnest-looking young man with an outstanding academic record, including election to Phi Beta Kappa. His first job, as a junior engineer with the International Boundary and Water Commission, took him to El Paso. Having left his fiancée in Houston, the lonely bachelor spent his evenings reflecting on the meaning of life and studying his Bible. He had never given much thought to the question of origins, though as a youth he had drifted unthinkingly into evolutionism and religious indifference. A thorough study of the Bible convinced him of its absolute truth and prompted him to reevaluate his belief in evolution. After an intense period of soul-searching, he concluded that creation had taken place in six literal days, because the Bible clearly said so and "God doesn't lie." Following marriage in 1940 he and his new wife began teaching Sunday-school classes to young people, and he joined the Gideons, an evangelical fellowship famous for its distribution of Bibles. In early 1942 Irwin A. Moon brought his illustrated "Sermons from Science" to town and convinced Morris of the importance of harmonizing science and the Bible. From the itinerant Moody preacher he learned for the first time about the geological effects of Noah's flood and the vapor canopy that had enveloped the earth until the deluge.[23]

With America's entry into World War II, Morris expected to join the U.S. Navy; instead he returned to Rice as an instructor in civil engineering to teach young men headed for active duty. During his three years at Rice, known locally as a hotbed of infidelity, he continued working with the Gideons, became active in the Inter-Varsity Christian Fellowship, and served as the faculty advisor to the Baptist Student Union, where he met Walter R. Hearn. In those days Morris fairly bubbled over with religious enthusiasm. As one of his fellow instructors later recalled, it was impossible "*not* to talk with H. M. about religion." He routinely passed out Gideon Bibles and invited students to attend evening Bible classes in his home. (At least one student went because he thought "it wouldn't hurt my grade in surveying.") On occasion he would inquire if students had accepted Jesus Christ as their personal Savior. Such behavior struck some associates as a little peculiar, but it provoked no controversy. A mitigating factor was Morris's reputation as a mathematical "whiz." According to student lore, the brilliant H. M. had even constructed a mathematical proof for the existence of God.[24]

Increasingly Morris's thoughts turned to the biblical stories of the creation and flood. While sitting in his office at Rice, he would sometimes study the butterflies and wasps that flew in through the window. Being familiar with structural design, he calculated the improbability of such complex creatures developing by chance and concluded that nature as well as the Bible argued for special creation. When he turned for assistance in answering the claims of evolutionists, he found little creationist literature of value apart from the writings of Rimmer and Price. He especially liked Rimmer's *Theory of Evolution and the Facts of Science*, which more than any other work convinced him "once and for all that evolution was false." In 1943 he helped to bring Rimmer to Houston for "an unforgettable week" of meetings at the First Baptist Church, which left him hoping someday to have a "testimony in the university world somewhat like [Rimmer's] own." From a reference in one of Rimmer's books, Morris discovered Price's *New Geology*—"a life-changing experience for me"—and soon thereafter joined the Deluge Geology Society.[25]

By this time Morris was indulging his passions for witnessing and writing by composing an apologetical guide for college students called *That You Might Believe* (1946). It was, so far as he knew, the first book "published since the Scopes trial in which a scientist from a secular university advocated recent special creation and a worldwide flood." Emulating Rimmer, he discussed such chestnuts as the identity of Cain's wife, the "great fish" that swallowed Jonah, and the long day of Joshua. If the Bible is "God-given," he argued, "it is unthinkable that it should contain scientific mistakes; either it is scientifically accurate whenever it happens to touch on some phase of science, or it is purely the product of human beings and no better than any other book of ethics." In dealing with earth history, he again followed Rimmer in maintaining that the absence of sunlight before the fourth day of creation, "which would certainly be quite lethal to the vegetation created on the third day," definitively ruled out the day-age theory. Beyond that, he remained uncertain. He granted that "considerable Scriptural evidence" supported the ruin-and-restoration hypothesis; yet he saw "no objective reason . . . to assert dogmatically that all or most of the fossil forms could not have been living simultaneously instead of being spread out over millions of years." In writing about the deluge, he tended to follow Price, whom he grandly described as one of the world's leading authorities on historical geology. Morris concluded his book with a premillennialist testimony to his belief in the literal Second Coming of Christ.[26]

The twenty-eight-year-old Morris had hoped to place his first book with the well-known evangelical publisher Zondervan, but they turned him down. Eventually a fledgling outfit called Good Books agreed to take his manuscript, but only after Morris obtained a subvention from a well-to-do friend in the Gideons. While his book was in press, he read a critique of radiometric dating by Price's disciple Clifford L. Burdick that convinced him he "no longer had to dabble with the gap theory or some other means of allowing a great age for the earth." Thereafter he never strayed from strict young-earth creationism. And he no longer worried about how to reconcile the pre-Edenic destruction of fossilized animals with the biblical statement that death entered the world as a result of Adam's sin.[27]

When the Moody Press brought out a revised and expanded edition of his book, which had sold in excess of ten thousand copies, he deleted his earlier discussion of the gap theory in favor of a simple affirmation that "a literal reading of the biblical record will yield a date of about 4,000 B.C. for the creation." In the revised version, retitled *The Bible and Modern Science* (1951), he also added several pages on the second law of thermodynamics, the significance of which he had come to appreciate from reading R. E. D. Clark. "This law of energy degeneration," he wrote, "conveys us back inexorably to an affirmation of the necessary truth of the existence of a Creator, and a definite creation." He said nothing about the time the second law went into effect. In a still later abridged edition of his much revised and reprinted first book, Morris deleted his original paragraph praising Price and acknowledging his influence. Ironically, the more Morris identified himself with flood geology, the less he was willing to associate publicly with its disreputable father.[28]

While teaching at Rice, Morris fell unexpectedly in love with the academic life. Thus in 1946, with a wife and two young children in tow, he headed north to begin graduate training in hydraulic engineering at the University of Minnesota. The St. Anthony Falls Hydraulic Laboratory in Minneapolis was then the premier such facility in the country, and Morris hoped to combine his vocational interest in engineering with his avocational curiosity about the effects of water in earth history. To support his major in hydraulics, he chose minors in geology and mathematics.[29]

Soon after arriving in Minneapolis, Morris attended a series of sermons given by the celebrated creationist Arthur I. Brown, who shared Morris's fascination with biblical eschatology. In view of the imminent end of the world, Morris had been wondering if he should not leave engineering for the full-time ministry, as Brown had abandoned the practice of medicine. Morris took his problem to the fatherly physician, who counseled him to continue in graduate school. Brown pointed out that a doctoral degree would significantly enhance the young man's Christian witness and that one of the greatest needs was for "dedicated Bible-believing scientists." Morris heeded this advice and remained in school but devoted many hours a week to church-related activities.[30]

Another veteran of the antievolution wars, William Bell Riley, tried to lure Morris into academic administration. In 1947 the octogenarian pastor of the First Baptist Church in Minneapolis, sensing his end was near, began looking for a younger man to succeed him as president of the Northwestern Bible College and related institutions. Having been impressed by *That You Might Believe* and hearing that the author was living in town, he invited Morris to pay him a visit. Unbeknownst to Morris, Riley was scrutinizing him for the presidency. After Morris made clear his intention to remain at the university and obtain his Ph.D., Riley bestowed his mantle on the twenty-nine-year-old Billy Graham.[31]

While living in Minnesota, Morris decided to join the ASA—in the very year, 1949, that Kulp delivered his apparent *coup de grâce* to flood geology. For months Morris had been corresponding with F. Alton Everest in an attempt to get the ASA to reconsider its position on the subject, as reflected in *Modern Science and Christian Faith;* and on reading Kulp's paper, he immediately sent Everest an extensive rebuttal. Kulp's dismissive remarks about the Adventist origins of flood geology proved nothing beyond his own prejudice, wrote Morris. "The fact that Adventist theology teaches a recent Creation is not the basis for the Flood theory, but rather the fact that the Bible seems on the surface at least . . . to postulate a recent, six-day creation and subsequent worldwide, devastatingly destructive Deluge." Everest replied that he intended to recommend publication of Morris's response in the affiliation's *Journal* but wanted Kulp to read it first. Soon thereafter Everest traveled to Minneapolis, and he and Ramm, then teaching in nearby St. Paul, visited Morris in what he thought was an attempt to dissuade him from continuing his unrealistic fight against uniformitarian geology. However, the meeting went so well Morris offered to let Everest read the new material on the flood he had prepared for the revised edition of *That You Might Believe*. After seeing the manuscript, Everest complimented Morris on his reasonable approach to the subject and indicated that, "if the facts are straight," he would like to publish a condensed version in the *Journal*. Neither the rebuttal to Kulp nor the article on the flood ever appeared.[32]

In 1950 Morris received his Ph.D. from the University of Minnesota. Despite having made his creationist views widely known, he had earned excellent grades and had encountered no overt prejudice. After graduation he prepared to depart for Afghanistan, where he was to organize a civil-engineering curriculum for a new technological institute. When that opportunity fell through at the last minute, he accepted an appointment as head of the department of civil engineering at Southwestern Louisiana Institute (now the University of Southwestern Louisiana) in the heart of Cajun country. It was while living in Louisiana that he attended his first ASA meeting and met Whitcomb. The paper Morris presented on that occasion, like his earlier submissions to the *Journal of the American Scientific Association*, failed to meet the ASA's criteria for acceptance. When he read Ramm's book the next year, he angrily lined the margins with his private thoughts. Beside Ramm's claim that "the hyperorthodox have made a virtue of disagreeing with science," he jotted: "No one condemns 'science' or 'scientists'—only anti-Christian and evolutionary philosophies of science, biases toward which often cause misinterpretation of scientific 'facts.' " When he later caught the philosopher himself railing against "modern scientific thought" for ignoring final causes, he accused him of sounding "like a science-baiting Fundamentalist." Writing to Price soon afterward, he expressed the prophetic hope that Ramm's book would at least "stimulate some capable men, who take Genesis seriously, to begin some intensive and much-needed studies in flood geology and related issues."[33]

When Morris signed on with Whitcomb in late 1957, Morris had just moved to Blacksburg, Virginia, to head up one of the largest civil-engineering programs in the country, at Virginia Polytechnic Institute (VPI). Despite heavy teaching and administrative duties, he made steady progress on his promised chapters for *The Genesis Flood*, which expanded from a planned 100 pages to nearly 350 and eventually overshadowed Whitcomb's shrinking contribution by better than a two-to-one margin. Although Morris's name would appear second on the title page, he served in many respects as the senior author. From the beginning of their collaboration, Morris addressed his writing partner as John, while Whitcomb persisted in using the salutations "Brother Morris" and "Dr. Morris" until just weeks before their book rolled off the presses. After several years of wallowing in insecurity over his lack of scientific credentials, Whitcomb rejoiced finally to have an expert on hydraulics helping him to defend the flood. "Your manuscript convinced me of the importance of having a scientific background before writing on scientific subjects!" he wrote to Morris after reading a draft of Morris's chapters. "My own effort in dealing with uniformitarianism certainly reveals the folly of a Biblical scholar trying to come to grips with the geologists!"[34]

By early 1959 Morris had completed a draft of his chapters and sent them to Whitcomb for criticism. Whitcomb, still smarting from having his knuckles rapped by Morris for invoking Price and Velikovsky so much, caught Morris engaging in the same practice. "Even the references to Velikovsky should be thought through carefully, because his name, like that of G. M. Price, waves a red flag immediately before some people's eyes," advised the highly sensitized Whitcomb, who worried that his own material still contained too many allusions to Price and the Adventist tradition:

> I am becoming more and more persuaded that my chapter on "Flood Geology in the Twentieth Century" will hinder rather than help our book, at least in its present form. Here is what I mean. For many people, our position would be somewhat discredited by the fact that "Price and Seventh-Day Adventism" (the title of one of the sections in that chapter) play such a prominent role in its support. My suggestion would be to supply for the book a fairly complete *annotated bibliography* of twentieth-century works advocating Flood-geology, without so much as a mention of the denominational affiliation of the various authors. After all, what *real* difference does the denominational aspect make?

Morris, who fully shared Whitcomb's concerns, suggested going even further and discarding the Pricean tag: "The very term 'Flood Geology' seems to have unpleasant connotations to many people," he wrote, "and it might be better to use such terms as 'creationist geology,' 'Biblical geology,' and the like instead,—both in order to try to avoid the S.D.A. label and also to point up the fact that we don't try to account for *all* the geologic data by the Flood." But the old description proved too useful to abandon completely.[35]

Their concern about appearances prompted Whitcomb and Morris to sanitize their manuscript by deleting all but a few incidental references to Price and any mention of his Adventist connections. When Morris sent Price a copy of his chapters for comment, he self-consciously apologized for saying so little about Price's pioneering contributions to flood geology. "This of course is not because of lack of appreciation for them," he explained awkwardly, "because I still regard them as masterpieces, but rather in order to gain perhaps a better hearing by taking a somewhat fresh start on the problem." He expressed the hope that a new approach might finally stir up some interest in the subject. Price, grateful at his age for any belated recognition, gave no indication of feeling betrayed; indeed, he praised the book and its authors effusively. Some friends of Price's, however, found the slight offensive. The Old Testament scholar Oswald T. Allis (1880–1973), who as editor of the *Princeton Theological Review* in the 1920s had published three of Price's articles, chided the young men for inadequately acknowledging their intellectual debts. In response to his criticism, Whitcomb grudgingly agreed to mention Price in a footnote, so long as it did not draw attention to the old man's peculiar religious beliefs.[36]

In addition to Price and Allis, Whitcomb and Morris circulated their manuscript, or parts of it, to thirty-some colleagues interested in creationism. Many of the scientific readers were alumni of the Deluge Geology Society: Benjamin F. Allen, Molleurus Couperus, Burdick, Clark, Tinkle, Walter E. Lammerts, and Whitney. Others included the Lutherans Klotz and Rusch, the Wheaton geologist Block, the Calvin biologist Edwin Y. Monsma, the Canadian engineer Arthur C. Custance, and such newcomers on the creationist scene as Harold S. Slusher (b. 1934), a physicist at Texas Western College, and Wayne F. Frair (b. 1926), a biologist at The King's College. Except for Block, who rejected their conclusions, and Burdick, who lacked a graduate degree, Whitcomb and Morris had difficulty finding "a real geologist" to review their work. Whitcomb was reluctant to say much about flood geology in the twentieth century for fear that he would "unduly expose the weakness of our movement." The embarrassing absence of support from modern geologists in a book on historical geology led him at one point to suggest, perhaps jokingly, that he bring his chapters "to a great climax" by pointing out "that we now have one man who almost has earned a Ph.D. in geology (Burdick)!"[37]

By the time Whitcomb and Morris were ready to go to press, Moody was starting to have second thoughts about the book: its length, its insistence on literal creative days, and its likely reception. The longer Moody postponed publication, the more impatient the two authors grew. By early 1960 they were ready to jump ship. One of the readers of their manuscript, Rousas J. Rushdoony (b. 1916), recommended the Presbyterian and Reformed Publishing Company in Philadelphia, a small operation owned and operated by Charles H. Craig (1912–1983). Rushdoony, who later won notoriety as a leader of the Christian Reconstructionist movement (best known for favoring the death penalty for such Old Testament offenses as sodomy and Sabbath breaking), had published with Craig and recommended him highly. Craig, for his part, had long been an admirer of Price. Although he had majored in geology at Princeton, he had always preferred catastrophism to uniformitarianism and had hoped someday to acquire a manuscript "that would show the fallacy of both evolution and the geologic evidence in its behalf." *The Genesis Flood* fulfilled his dream.[38]

THE GENESIS FLOOD

The Genesis Flood, which finally appeared in February 1961, opened with an affirmation of belief in "the verbal inerrancy of Scripture." As Whitcomb and Morris elaborated in a preface to the second printing, "the basic argument of this volume is based upon the presupposition that the Scriptures are true." Both authors shared an unyielding commitment to the authority, inerrancy, and infallibility of the biblical record, "verbally inspired in the original autographs." But the two men differed somewhat on epistemology. Morris believed that God had revealed himself in two books, nature and the Bible, which could be studied independently, though with priority given to the Scriptures. Because God was the author of both works, Morris thought it inconceivable that God's Word would contradict his world. To a great extent Whitcomb shared this conviction, but as a strong presuppositionalist, he rejected reason and experience as sufficient means of discovering theological truth. For him, the character and purposes of God could be found only through the Bible, the reliability of which he accepted on the basis of its own claims rather than on the basis of external evidences.[39]

Given their respective specialties, it is not surprising that Whitcomb undertook to determine what the Bible taught concerning the flood, leaving Morris the task of trying to fit the scientific data into a biblical framework. Whitcomb began his section by marshaling scriptural arguments in favor of a universal flood: from the height of the floodwaters, which covered "all the high mountains that were under the whole heaven," to the need for an ark, which would have been "ridiculous if the Flood was confined to some section of the Near East." He singled out Custance, a local-flood partisan, for special criticism. Whitcomb assailed the Canadian creationist for underestimating the size of the ark, for assuming that only domesticated animals boarded the ark, and especially for suggesting that God had deceived Noah into believing that the flood would be worldwide as a means of motivating him to work on the ark. "I hit Arthur Custance pretty hard," he bragged to Morris, explaining that he did so in part to broaden his attack beyond Ramm.[40]

Having demonstrated the geographical extent of the flood to his satisfaction, Whitcomb next turned his attention to its anthropological scope. Here he zeroed in on his primary target, Ramm, who he thought was misleading evangelicals into supposing that the deluge had left North America and its inhabitants untouched. Whitcomb, sometimes assisted by Morris, also addressed such issues as the capacity of Noah's ark (equal to eight freight trains with sixty-five stockcars each), the source of water for a universal flood (an antediluvian canopy of water surrounding the earth and the "fountains of the great deep"), and the distribution of animals after the catastrophe (by diffusion from the mountains of Ararat rather than by special creation). He noted ironically that both flood geologists and evolutionists agreed on the likelihood of migration across ancient land bridges; they differed only on the length of time it took and the amount of evolution allowed. Never shy about invoking divine explanations, Whitcomb postulated that God had providentially guided the survivors of the deluge to their new homes. He cautiously discounted the persistent rumors about discovering the ark. "We fear that any hope of its preservation for the thousands of years of post-diluvian history is merely wishful thinking," he wrote. "Even if it *had* been preserved, through burial and freezing, it would be so hard to find that nothing less than divine direction could ever lead explorers to its true location."[41]

Morris introduced his section by highlighting the "serious dilemma" Whitcomb's conclusions posed for Bible-believing Christians. If the Scriptures clearly taught a universal flood, as Whitcomb maintained, then such believers had only two choices: reject either God's inspired Word or the testimony of "many thousands of trained geologists." Morris unhesitatingly recommended the latter course, "regardless of the 'deluge' of scholarly wrath and ridicule" that such a decision might entail. Citing the Harvard sociologist Barrington Moore, Jr. (b. 1913), he argued that the choice between alternate theories depended on "moral and emotional" factors as well as scientific data. "Few people today are likely to argue that the acceptance of scientific theories, even by scientists themselves, depends entirely upon the logical evidence adduced in support of these theories," he quoted Moore as writing in a recent issue of *Scientific Monthly*. "Extraneous factors related to the philosophical climate and society in which the scientist lives always plays [sic] at least some part." To help ease any psychological discomfort readers might experience in rejecting a widely held teaching of modern science, Morris reminded them that only interpretations, not facts, were at stake. And to replace what they were giving up, he offered "a new scheme of historical geology," true to God's revelation in both the Bible and nature. Actually, his scheme consisted of little more than Price's shelf-worn flood geology, neatly repackaged for the discerning evangelical of the 1960s. In the opening chapters of *The Genesis Flood* Whitcomb had answered Ramm; Morris now defended Price.[42]

Although Morris had deleted all but a few direct references to Price, his section read like an updated version of *The New Geology*. In arguing for a worldwide flood that deposited most of the fossil-bearing rocks, he followed Price in discarding the principle of uniformity, in questioning the notion of multiple ice ages, and in rejecting the so-called geological column. The apparent order of the column he attributed to such factors as the early death of marine creatures, buried by sediments deposited during the first stages of the flood; the hydrodynamic selectivity of moving water, which sorted out particles of similar sizes and shapes; and the superior mobility of vertebrates, which allowed them to escape early destruction. In dismissing the mechanism of thrust faulting, he, too, appealed to Chief Mountain in Glacier National Park, where "old" Precambrian limestone rested in apparent conformity on "young" Cretaceous strata. As authority, he cited the horticulturist Lammerts, who, after personally inspecting the mountain, came away convinced that Price was "even more right than he thought."[43]

The "real bombshell" in the book, as Rusch described it, was the evidence that humans and dinosaurs had once lived together, a possibility disallowed by conventional interpretations of earth history. From acquaintances who had belonged to the Deluge Geology Society, Whitcomb had learned in 1957 about Burdick's explorations in the Paluxy River bed, near Glen Rose, Texas, where he claimed to have found giant human footprints side by side with those of dinosaurs. Burdick eagerly supplied Whitcomb and Morris with information about the tracks, as well as photographs of the footprints, but insisted that his identity not be revealed. At the time he was working on his Ph.D. in geology at the University of Arizona, and he feared that public exposure as a flood geologist would jeopardize his obtaining his degree. He asked Whitcomb and Morris at one point to credit him under a pen name, but at the last minute he let them use his real name.[44]

On the basis of Burdick's testimony, Whitcomb and Morris claimed not only that human and dinosaur footprints were found close together in the Paluxy River but that "in some cases" they overlapped. They also quoted Burdick to the effect that scientists from the American Museum of Natural History had "even dug up a dinosaur track superimposed upon a human track." Soon after their book appeared, however, Rusch warned them that the overzealous Burdick might have misled them. Whitcomb was disappointed but hardly surprised. "As you know," he wrote Morris, "I have long been rather frustrated by our failure to get clear, definite, *final* confirmation of the Glen Rose 'human' footprints." He appreciated Burdick's willingness to help but wished that the Arizona geologist would pay more attention to "accuracy in names, dates, places, etc." In the months after the appearance of *The Genesis Flood,* Burdick continued to keep Whitcomb and Morris informed about the disputed tracks because, as he sheepishly wrote Morris, "you sort of stuck your neck out in publishing those Glen Rose tracks." Indeed they had, and for the third printing they silently revised the text.[45]

According to Morris, three events—"the Creation, the Fall, and the Flood"—dominated early world history. In six literal days, using methods unknown and unknowable, God had created the entire universe and populated the earth with "full-grown" plants, animals, and human beings. For an unspecified period Adam and Eve had lived in a perfect world, unmarred by sin and suffering. But the fall introduced a period of "decay and deterioration," implicitly associated with the second law of thermodynamics. Because there had been no death before this time, Morris felt "compelled to date all of the rock strata which contain fossils of once-living creatures as subsequent to Adam's fall." Most of these formations he attributed to the flood, which he and Whitcomb placed at between five thousand and seven thousand years before the present, a minimum of about seven hundred years before Ussher's date. This scenario, though strikingly similar to Price's, differed from his in two important respects. Because Ellen G. White had written of inhabited worlds that antedated the Edenic creation, Price and his fellow Adventists limited the six-day creation to the solar system, the earth, or even to life on earth. Thus they never insisted on a young universe, nor did they ever associate entropy with the fall.[46]

In sketching the history of life on earth between the creation and the flood, Morris repeatedly alluded to "a great vapor canopy" that enveloped the antediluvian earth. Formed on the second day of creation, when God divided the waters under the firmament from the waters above the firmament (Genesis 1:7), this canopy produced a "greenhouse effect" that gave the entire world a relatively mild and uniform climate, protected antediluvians from the harmful effects of cosmic radiation and thus greatly extended their lives, and provided a vast reservoir of water that deluged the earth during the flood of Noah. Morris also invoked the canopy to explain away the great ages given by carbon-14 dating methods. By shielding the earth from cosmic radiation, he argued, the canopy "inhibited the formation of radiocarbon in the high atmosphere" and "reduced the ratio of radiocarbon to ordinary carbon to a much smaller fraction than now obtains." Because carbon-14 dating assumed a constant ratio over time, it erroneously gave the appearance of great antiquity.[47]

In bringing *The Genesis Flood* to a climax, Morris stressed the central role of flood geology in adjudicating the conflict between creation and evolution. If it could be established that the fossil-bearing strata had been deposited during the brief period of Noah's flood, he argued passionately, then "the last refuge of the case for evolution immediately vanishes away, and the record of the rocks becomes a tremendous witness . . . to the holiness and justice and power of the living God of Creation!" Clearly, Christians could ignore the issue only at their peril.[48]

Despite its lack of conceptual novelty—one reader described it as "a reissue of G. M. Price's views brought up to date"—*The Genesis Flood* sold tens of thousands of copies during its first decade in print and provoked a furious debate among evangelicals. Strict creationists praised it for making biblical catastrophism intellectually respectable again, while progressive creationists and theistic evolutionists denounced it as a travesty on geology that threatened to return Christian science to the Dark Ages. The appeal of the book, suggested one cynic, lay primarily in the fact that, unlike previous creationist works, it "looked *legitimate* as a scientific contribution," accompanied as it was by footnotes and other scholarly appurtenances. No doubt many sales resulted from the feverish rhetoric of those who disliked it most. At first Whitcomb and Morris were dismayed to learn that the release of their book coincided with the publication of a widely used quarterly for Sunday-school teachers, prepared by the geologist Block, that blasted flood geology. But such outbursts only led curious Christians to seek out the cause of the commotion.[49]

Several dozen Christian magazines reviewed *The Genesis Flood*, and most of them praised the book for defending the scriptural account of the flood. Few reviewers, however, gave any indication of recognizing the radical implications of Whitcomb and Morris's thesis: that accepting flood geology meant rejecting the popular gap and day-age theories as commonly understood. For example, the Wesleyan Methodist physicist S. Hugh Paine (b. 1911), a onetime student of L. Allen Higley's, praised Whitcomb and Morris for their rejection of uniformitarianism. But he himself continued to promote a gap model of earth history that assigned most of the fossil-bearing rocks to a pre-Edenic flood rather than to Noah's deluge. Similarly, R. Laird Harris (b. 1911), a conservative Old Testament scholar who had given Whitcomb and Morris's manuscript a sympathetic reading, surprised them shortly after publication by confessing that he still adhered to the day-age theory. The failure of so many readers to recognize the incompatibility between the flood theory and competing old-earth schemes prompted Whitcomb to consider adding an appendix clearly stating that flood geology was a substitute for, not a supplement to, traditional views of Genesis. "As you know," he wrote Morris a few months after their book appeared, "I have been somewhat embarrassed about this situation, since practically everyone I know takes the gap theory or the day-age theory, even though they seem to be happy about our position on the Flood!" Embarrassed or not, he let the idea for an appendix pass without acting.[50]

More than almost anything else, Whitcomb and Morris would have loved the approval of the leaders associated with the ASA and *Christianity Today,* the voice of the neo-evangelical movement. But they expected rejection and were not surprised when they received it. Although *Christianity Today* selected *The Genesis Flood* as one of its "Choice Evangelical Books of 1961," its reviewer, Wheaton geologist Donald C. Boardman (1913–1988), had little good to say about Whitcomb and Morris's scholarship. Rather than debating the issues they had raised, he focused on their heavy use of secondary sources and their habit of lifting arguments out of context. His review, sniffed the two authors, "could just as well have been written by one who had never heard of Genesis." As a partial consolation, the editor of *Christianity Today,* Carl F. H. Henry (b. 1913), promised to publish paired essays by Morris and Kulp on the pros and cons of flood geology; but Kulp never produced his promised piece, and Henry refused to print Morris's alone. "Geologists are interesting people!" fumed an exasperated Morris, who eventually received a token $25.00 for his unpublished effort.[51]

Whitcomb and Morris encountered an even rougher welcome at the ASA. After waiting over two years for a review to appear in the ASA *Journal,* and watching his book pass through four printings, Morris "about decided the A.S.A. was officially ignoring it." However, the new book-review editor, Hearn, informed him that the geochemist Wayne U. Ault (b. 1923), a Kulp protégé, had promised a review, which all parties knew would be negative. "I wish I liked your book better, Henry," Hearn wrote in a letter to Morris. "I sure do like *you*, though!" Out of loyalty to his old Texas friend, and to make sure the book received the attention it deserved, Hearn was casting about for something positive to run with Ault's appraisal. In view of the delays at *Christianity Today*, Morris suggested that the essay he had submitted there might be appropriate. Hearn liked the idea and even offered "to write a partial rebuttal to the anti-evolutionary aspects," if the editor, David O. Moberg (b. 1922), decided to accept it. "I think we ought to publish it even if it makes the professional geologists irritated at us," Hearn advised Moberg. "In the light of the success of the book and its radical departure from other 'hyper-orthodox' approaches, I think we ought to deal at length with it to put it in perspective." Though neither Hearn nor Moberg liked *The Genesis Flood*, both were determined to take the high road in handling this "delicate" matter. Moberg even coached the delinquent Ault on the approach he might take. "I am sure that you feel as I do that we can reject the ideas of a person without rejecting the person," he wrote. "If somehow this note of kindness or Christian love can carry through in what you say even though your critique may be devastating, I think we will have saved the day."[52]

Three years after *The Genesis Flood* reached Christian bookstores, the ASA *Journal* finally noticed it. But instead of running positive and negative assessments, as planned, it featured two hostile reviews, introduced by Hearn and "edited extensively . . . to tone them down a bit." Given Whitcomb and Morris's position on a "full-grown" creation with the appearance of age at birth, Hearn professed puzzlement at why they even bothered challenging the accuracy of dating techniques. "Why not assume that the methods are essentially reliable for determining the *apparent* age which God structured into the rocks on the particular 24-hour day of their creation?" he asked. With tongue in cheek the jovial biochemist metaphorically summed up their accomplishment:

> Those who dwell inside the house of geological science have been in the process of remodeling it continuously ever since it was built. Now Henry Morris and John Whitcomb have come along insisting in the name of the Master Architect that the whole thing is on a shaky foundation and must be bulldozed to the ground. Detailed plans for the fine new edifice which should be built in its

place, they claim, were found by them in the pages of the family
Bible.

The embattled authors failed to appreciate Hearn's attempt at humor.[53]

In his long-awaited review, Ault questioned both the theological and
scientific assumptions of what he called "the Price-Morris-Whitcomb
catastrophic geology." Writing as a trained geochemist with a doctorate
from Columbia University, he pointed out that "at least six independent geochronometric methods" demonstrated the antiquity of the
earth. Although he was confident that *The Genesis Flood* would "pass
quite unnoticed" by secular scientists, he feared that it would have a
regrettable impact on the thinking of evangelical Christians, perhaps
even retarding "the development of true Christian scholarship in the
younger generation."[54]

In answering such critics, Whitcomb and Morris resolutely refused
to be drawn into a scientific debate. "The real issue," they insisted over
and over again, "is not the correctness of the interpretation of various
details of the geological data, but simply what God has revealed in His
Word concerning these matters." Whitcomb, in particular, could not
understand why the leaders of the ASA regarded "an appeal to clear
statements of Scripture" as irrelevant. Did they not think that "the
doctrine of a direct creation of Adam's body" was biblical? he asked
Moberg. In language that reflected Whitcomb's presuppositionalist
orientation, he and Morris maintained that the differences between
themselves and their detractors involved "assumptions and presuppositions," not science. "We take this revealed framework of history as
our basic datum, and then try to see how all the pertinent data can be
understood in this context," they explained. In like manner their opponents started "with the assumption of uniformity (and therefore,
implicitly, evolution) and then proceed[ed] to interpret all the data to
fit into *that* context." For both sides, they argued, faith alone determined the choice of initial assumptions: "It is not a scientific decision
at all, but a spiritual one."[55]

Whitcomb and Morris accurately attributed the impasse between
themselves and their critics to competing cosmologies. As they cogently observed, both sides viewed the world through distinctive "sets
of spectacles" that uniquely colored everything in sight. Mainstream
scientists had no quarrel with such a characterization—but cried foul
when the flood geologists went on to insist that the selection of lenses
lay outside the domain of science. Because Whitcomb and Morris
arbitrarily limited science to the study of "present and reproducible

phenomena," the two authors felt that geologists who wrote earth *history* were operating as nonscientists. And because the fathers of historical geology had included "such non-geologists as Charles Lyell (a lawyer), William Smith (a surveyor), James Hutton (an agriculturalist), John Playfair (a mathematician), Georges Cuvier (a comparative anatomist), Charles Darwin (an apostate divinity student turned naturalist), and various theologians (Buckland, Fleming, Pye Smith, and Sedgwick)," they could see no reason why their own specialties should disqualify them from evaluating what these amateurs had constructed.[56]

Discussions of *The Genesis Flood* continued to appear in the ASA *Journal* through the decade, culminating in 1969 with the publication of a devastating exposé by a Dutch Reformed geologist at the Free University of Amsterdam, J. R. van de Fliert (b. 1919). In the most critical evaluation of flood geology since Kulp's dissection of Price twenty years earlier, the evangelical Dutchman effectively dismissed Whitcomb and Morris as "pseudo-scientific" pretenders. To ensure that no readers missed his point, the ASA *Journal* ran boldfaced side-bars by evangelical geologists applauding van de Fliert's bare-knuckled approach. In the hardest-hitting blurb, Roger J. Cuffey (b. 1939), a geologist at Pennsylvania State University, praised van de Fliert for finally exposing the inanities of the flood geologists, who, "instead of being properly laughed out of court, were widely accepted in the intelligent Christian community." The inability of many college-educated persons to "recognize as big a blunder as this one" made Cuffey question the quality of science education in the country.[57]

On reading this latest attack, Morris dashed off a letter of protest to the editor of the ASA *Journal,* now Richard H. Bube (b. 1927), claiming gross distortion of the arguments presented in *The Genesis Flood* and requesting an opportunity to reply. In his more than twenty years as an ASA member, complained Morris, he had never seen an article in the ASA *Journal* supporting a universal flood, much less flood geology. Bube, invoking editorial policy, replied that any full-scale article would have to be reviewed, but he offered quick publication of a brief letter to the editor. Realistically assessing the chances of ever having a major essay accepted by the ASA *Journal,* Morris settled for the certainty of a letter. Like earlier critics, wrote Morris, van de Fliert had picked at "supposed flaws" in the geological argument while ignoring the unambiguous teachings of the Bible regarding the Noachian cataclysm. "The real crux of the matter," he repeated, "is 'What saith Scripture?' "[58]

The Fruits of Fame

The Genesis Flood forever changed the lives of Whitcomb and Morris. The continuing controversy surrounding their book, which went through twenty-nine printings and over 200,000 copies in a quarter-century, turned them into highly sought-after celebrities, famous among fundamentalists as the Davids who slew the Goliath of evolution. Since completing their flood manuscript, they had planned to write a sequel on the creation, but the unanticipated demands on their time after 1961 effectively killed that project. From his academic base at Grace Theological Seminary, where he remained through the 1980s, Whitcomb presided over the board of Spanish World Gospel Mission and produced a stream of books, tapes, and Bible-chronology charts distributed by Whitcomb Ministries. Although his interests ranged from Esther to eschatology, public demand and personal partiality kept him coming back to the topic of origins. In 1972 he brought out *The Early Earth*, in which he took another swipe at Custance and the gap theory; and the next year he followed with *The World That Perished*, a popular treatment of flood geology that served as the occasion to answer his critics. In 1978 he coauthored *The Moon: Its Creation, Form and Significance*, lauded by astronauts James B. Irwin and Charles M. Duke, Jr.[59]

Morris, who had shunned airplanes all his life, found himself in the 1960s surveying the country by air as he flew from one engagement to another. Most often he spoke at Baptist churches and schools, but his itinerary also included stops at conservative Presbyterian, Lutheran, Reformed, Episcopal, Wesleyan, Mennonite, and even Pentecostal institutions. Impressive lectures at Tennessee Temple College in Chattanooga, Biola College near Los Angeles, LeTourneau College in Texas, Bob Jones University in South Carolina, and Los Angeles Baptist College and Seminary all led to invitations to join those faculties. Tennessee Temple and Biola wanted him primarily to teach science, which he had little interest in doing. LeTourneau, one of the few evangelical colleges with an engineering program, asked him to become dean of its engineering school; but when the administration refused, for financial reasons, to set up a creation studies center, he decided against the position. Bob Jones, where two of Morris's sons were enrolled, offered to put him in charge of a new department of apologetics in the school of religion; but some friends discouraged him from affiliating with such an "extremist" institution. Los Angeles Baptist College, where the faculty, "from the president on down," had switched from the gap theory to flood geology as a result of reading *The Genesis Flood*, called him to chair its natural science department and promised him half time off for writing and speaking. But for various reasons, including the urging of some creationist colleagues to remain on a secular campus for the good of the cause, he decided to stay at VPI for the time being.[60]

In 1967 Morris gave his most rewarding performance, as the W. H. Griffith Thomas Lecturer at the Dallas Theological Seminary, where he spoke on "Biblical Cosmology and Modern Science." Dallas, "the largest nondenominational, conservative seminary" in the world and a fundamentalist powerhouse, had come under the influence of *The Genesis Flood* and had begun to contract earth history. "At the close of the final lecture," Morris reported excitedly to Whitcomb, "the entire student body and faculty responded with a lengthy and most embarrassing standing ovation!" With a student body of nearly four hundred, "most of whom could be considered of top calibre academically and spiritually," and a faculty that almost unanimously accepted a "recent six-day creation and the world Flood," Dallas seemed to offer "tremendous" potential for popularizing flood geology. Besides, it now published the once venerable *Bibliotheca Sacra*, formerly edited by George Frederick Wright, which for at least a decade averaged one Morris article per year.[61]

Outside conservative religious circles, *The Genesis Flood* and its authors created hardly a ripple of recognition. Morris did give one invited lecture on biblical catastrophism to the Houston Geological Society, but despite tying flood geology to anticommunism, he stirred up little interest. The president of what was reputed to be "the largest local geological society in America" had opposed Morris's coming and, in introducing his unwanted guest to the five-hundred-person audience, begged not to be held responsible for anything the visitor said. At the close of the talk the president quipped that "evidently Dr. Morris doesn't know that we know it takes 6000 years to make an inch of limestone." His call for questions met with total silence—because, as one friendly member afterward explained to Morris, the audience was "too stunned to speak!"[62]

Back in Blacksburg, Morris divided his time between home and church, writing and teaching. While Henry instructed engineers and battled evolutionists, his wife, Mary Louise, successfully fought off cancer and raised their three boys and three girls in a home filled with love, music, and piety. "The home life of the Morris family has, since before the children were born, centered around times of Bible reading and prayer at the morning and evening meals," reported a local newspaper in 1968, by which time all but two of the girls had gone off to college. "Scripture memorization and faithfulness in church responsibilities have always been stressed. As a result the family is a very close-knit group and all enjoy the occasional visits home for a family fellowship and song-fest."[63]

Throughout the 1960s Morris aggressively marketed his brand of creationism in a variety of packages: a series of Sunday-school lessons, a text for youth groups, and a collection of articles he had published in such popular evangelical journals as *His* (Inter-Varsity Christian Fellowship), *Collegiate Challenge* (Campus Crusade for Christ), and *Christian Life Magazine*, an interdenominational monthly. To arouse complacent readers, he devoted almost as much energy to discrediting the popular gap and day-age theories as to promoting flood geology. Occasionally he added a titillating twist to his now-standard argument or threw in a fresh bibliographic reference. In his Sunday-school lessons, which became "the all-time best seller" for its publisher, he suggested that the antediluvian "giants" mentioned in Genesis 6 —perhaps the very ones that had walked the Paluxy River bed—might have resulted from evil angels possessing men's bodies:

> The men whose bodies were possessed were evidently thereby made so attractive to women that they could take any they chose as wives. These "sons of God" thus controlled not only the men whose bodies they had acquired for their own usage, but also the wives they took to themselves, and then all the children they bore.

These children, he continued, "became the giants, the mighty men of old," mentioned in the Bible. In "The Death of Evolution," a chapter in *The Twilight of Evolution* (1963), Morris cited the recent "autopsy" by the British biologist Gerald A. Kerkut, whose iconoclastic *Implications of Evolution* (1960) made him the William Bateson of the 1960s. Though not a creationist himself, the University of Southampton scientist became a fundamentalist hero for questioning attempts "to explain all living forms in terms of an evolution *from a unique source*."[64]

Since boyhood Morris had affiliated with the theologically heterogeneous Southern Baptist Convention. On moving to Blacksburg, he and his family joined the local Baptist church as a matter of course. Although he found the milieu too modernistic for his tastes, he immediately joined in, first teaching a Sunday-night Bible class for "VPI faculty members and wives" and later taking over the college Sunday-school class. When his copies of *The Genesis Flood* arrived, he presented one to his liberal pastor, who dismissed it with the comment that "all the scientists couldn't be wrong." The preacher no doubt sensed trouble ahead. That same year the Sunday School Board of the Southern Baptist Convention had published Ralph H. Elliott's explosive book, *The Message of Genesis,* in which the seminary professor from Kansas City argued that the first chapter of Genesis dealt with "theological fact, not day-by-day physical history." The ensuing controversy divided the convention and created a supercharged atmosphere in which any statement on Genesis set off potentially dangerous sparks. The pastor, apparently hoping to prevent a schism in Blacksburg, relieved his contentious Sunday-school teacher of his duties and in effect ushered him to the church door. Along with another disgruntled family and some Inter-Varsity students, Morris started an independent College Baptist Church, which prospered so much that within a few years it had moved from Morris's living room to an impressive edifice on six acres of land. To help boost attendance with an evangelistic crusade, Morris at one point brought in a dynamic young acquaintance from nearby Lynchburg, Jerry Falwell (b. 1933), whom he had met when two of his daughters attended Falwell's summer camp and in whose huge Thomas Road Baptist Church he had spoken. Thus while losing the support of his local Southern Baptist minister, Morris gained the backing of someone who would soon emerge as the most influential fundamentalist in America.[65]

"Between the new church, and the GENESIS FLOOD," Morris wrote to Whitcomb in the fall of 1961, "I don't think I had better enter any popularity contest here in Blacksburg for a while." Fortunately for him, things were going well at work. His colleagues in the department of civil engineering, many of whom were professing Christians, appreciated his leadership and his policy of leaving them alone. Even though he rarely conducted original research, his professional activities as an officer of the American Society for Engineering Education, an editorial board member of the *Journal of Engineering Education,* and author of a widely adopted applied-hydraulics textbook brought prestige to both his department and the institute as a whole. His sometimes quaint behavior, such as inserting a long note on the theological implications of the laws of thermodynamics in his text, *Applied Hydraulics in Engineering,* elicited more smiles than sneers.[66]

The tolerant atmosphere at VPI began to evaporate early in 1963, when Morris accepted a student invitation to address the department of geology on the significance of Noah's flood. About 150 persons, including most of the geology and biology professors, showed up to hear an oral abstract of *The Genesis Flood*. The scientists "vehemently challenged" Morris's claim that the evolutionary and creationist models of earth history rested equally on nonscientific assumptions. The affair left him disillusioned about ever again discussing such matters with persons who did not share his beliefs and further eroded his good name in Blacksburg.[67]

A few months later a new dean of engineering, Willis G. Worcester (1918–1970), arrived on campus. Shortly after the *New York Times* quoted Morris as estimating the age of the earth at ten thousand to fifteen thousand years, the dean called Morris into his office to warn him about collegial concerns that his "peculiar and unscientific beliefs" were embarrassing the VPI community. Morris's weekly question-and-answer column in a county paper also brought criticism and led to a sharp exchange with David A. West (b. 1933), a biologist who suggested that Morris "stop writing 'pseudo-science' and return to the quiet world of Faith." But administrative practices more than ideological differences finally proved Morris's undoing. Early in 1968 Morris frustrated Worcester's attempt to abolish his department, but he quickly realized he had won a Pyrrhic victory. "My relations with the administration are more tenuous than ever," he confidentially informed Whitcomb, "and the eventual outcome will probably be that either I or the Dean will have to leave." A year later, in the spring of 1969, Morris voluntarily resigned as head of the civil engineering department in return for a year's sabbatical at full pay. "I'll then continue on the V.P.I. faculty unless the Lord leads otherwise," he informed Whitcomb. By the time he learned of Worcester's death in a plane crash the next year, he had already decided to change careers and serve the Lord in southern California.[68]

ELEVEN

The Creation Research Society

In the fall of 1957 John C. Whitcomb, Jr., solicited some expert scientific advice from the geneticist Walter E. Lammerts, a creationist who had been active in both the defunct Deluge Geology Society and the despised American Scientific Affiliation (ASA). Lammerts, eager for what he regarded as orthodox fellowship, proposed the creation of "an informal association" of persons interested in flood geology, who would exchange ideas by correspondence or at occasional meetings and who might engage in actual research on the effects of the Genesis flood. Whitcomb found the suggestion strongly appealing, though he feared that any new organization, like the old Deluge Geology Society, might fall under the control of Seventh-day Adventists. Henry M. Morris, who also liked the idea, expressed confidence that "some plan could be devised to prevent domination by the S.D.A.'s or by any other special interest group." He suspected that there were "quite a number" of like-minded men both "in and out of the A.S.A.," and he hoped that the forthcoming publication of *The Genesis Flood* "might awaken interest" in flood geology and swell the pool of possible recruits.[1]

At the time nothing came of Lammerts's proposal. Lammerts begged off taking the lead, claiming that he possessed neither the contacts nor the energy to do it himself. He encouraged Morris "to start the ball rolling," but the overcommitted engineer had just assumed his new position at the Virginia Polytechnic Institute and begun to write his chapters for *The Genesis Flood*. Several years passed with no progress. Then about 1961 Lammerts received a letter from his old friend William J. Tinkle lamenting the lack of activity in support of strict creationism and wondering what could be done to improve the situation. Lammerts replied: "Give me ten men who are active and we will do more than all others are doing at present." Within a year and a half Lammerts and Tinkle assembled the desired "Team of Ten," which became the nucleus for the Creation Research Society, the leading creationist organization of the late twentieth century.[2]

WALTER E. LAMMERTS

Walter E. Lammerts grew up on a small asparagus and strawberry farm in the state of Washington in a devout Missouri Lutheran family. As a child he would listen in the evenings as his grandmother read aloud from Martin Luther's German commentary on Genesis, and he never forgot what the great reformer said about a literal six-day creation. When he was ready for college, he enrolled in the University of California to study entomology. During his sophomore year he signed up for a course in geology, where talk of geological ages and uniformitarian processes unsettled his mind. If what his respected professor said were true, Lammerts reasoned, then it would be "foolish indeed to continue belief in the Bible." At the time, he was working his way through school shelving books in the university library, and one day he spied a copy of George McCready Price's recently published *The New Geology*, which he eagerly devoured. In Price's scheme Lammerts saw a way of reconciling his literal Lutheran theology with the testimony of the rocks. For the rest of his life he remained a grateful disciple.[3]

After graduating from the University of California as a Phi Beta Kappa, Lammerts stayed on at Berkeley as a research assistant while he pursued a doctorate in cytogenetics. Although the combative young creationist engaged in "many knock-down and drag-out arguments" with fellow graduate students over the issue of evolution, he never experienced any discrimination because of his peculiar views. Following receipt of his doctorate in 1930, making him one of the first strict creationists to earn a Ph.D. in biology, he won a two-year fellowship from the National Research Council for postdoctoral work at the California Institute of Technology. After a brief stint as a research associate back at Berkeley, by which time he had published articles in such journals as the *American Naturalist*, *Genetics*, and *Cytologia*, he turned to practical plant breeding. His success in developing a strain of rustproof snapdragon allowed him to pay off the mortgage on his parents' farm and to set them up in the nursery business. It also brought him to the attention of the owners of the Armstrong Nurseries in Ontario, California, who hired him to begin a research and breeding program. In five years with Armstrong he produced such notable developments as the Robin peach and the Charlotte Armstrong rose. From 1940 to 1945 he taught ornamental horticulture at the University of California, Los Angeles, then left academic life to help plan the Descanso Gardens in La Cañada.[4]

While teaching at UCLA in the early 1940s, Lammerts maintained an uncharacteristic silence in professional circles about his creationist beliefs. "Needless to say I keep my anti-evolution views to myself there," he confessed to one correspondent, explaining that "it will do no good for me to chirp until I've become a big enough 'bird' to at least make a loud squawk!" In private, however, he became actively involved in creationist causes for the first time. By 1941 he had joined the Deluge Geology Society and was encouraging fellow Lutherans to do likewise. He admitted that most of the several hundred members were Seventh-day Adventists, but he found them a congenial group that included "some fine students of nature." Later in life he would say that if he ever left Lutheranism, he would adopt Adventism. In urging Theodore Graebner to join the society, he expressed the hope that it would form "a united front of anti-evolution students" that would in the future become "a force potent enough to cause some of the evolutionists to at least consider seriously the many arguments against the theory." By the mid-1940s Lammerts had also signed on with the infant American Scientific Affiliation, having been assured by F. Alton Everest of that group's commitment to creationism.[5]

Although Lammerts coauthored the "Biology and Creation" chapter for the ASA's handbook *Modern Science and Christian Faith* (1948), his participation in creationist activities declined after leaving UCLA in 1945 and especially after moving to northern California in 1953. Settling in Livermore, just east of the Bay Area, he conducted horticultural research first for Descanso Distributors and later for Germain's. Nevertheless, he remained keenly interested in issues relating to flood geology. Reading Alfred M. Rehwinkel's *The Flood* (1951), a popularization of Price's catastrophism by a Missouri Lutheran theologian, got him to thinking again about flood geology, as did a vacation trip in 1956 to Glacier National Park, where he took photographs of what a park ranger identified as the famous Lewis overthrust. His observations strengthened his conviction that Price was right in doubting that overthrusting had occurred and conjured up fantasies of writing a book on the subject. "It may well be possible, if my good fortune, financially, continues for the next four or five years, for me to personally make a study of a large number of areas," he wrote Price on returning home. A detailed investigation of the contact lines in the areas of alleged overthrusting would, he thought, "give some very amazing evidence indicating quite clearly . . . that the stratified rocks do not occur in any definite order." Though an admitted novice in geology, he felt that his long experience in genetics equipped him with "the ability to see quite clearly natural evidence for what it is actually worth." Still, he hoped to return to the park "with some men having actual geological experience."[6]

In 1957 Lammerts began supplying Whitcomb with technical advice on events associated with the Genesis flood, especially the survival of plant life and the construction of the geological column. At first he expressed a reluctance to have Whitcomb cite him publicly because his rose-breeding did not leave him sufficient time to engage in debates over what he described as "sort of a hobby of mine." Within months, however, he changed his mind and began promoting the formation of a creationist society. When *The Genesis Flood* appeared in 1961, it carried several appeals to Lammerts's scientific expertise as well as three of his pictures from Glacier National Park, supposedly illustrating the contact line of the Lewis overthrust.[7]

Lammerts's wish to explore the overthrust area in the company of trained earth scientists came true in 1962, when he arranged to meet two young Seventh-day Adventists, Richard M. Ritland (b. 1925) and P. Edgar Hare (b. 1933), for a trek up Chief Mountain to inspect the Lewis overthrust. Ritland had majored in both biology and theology at Walla Walla College, where he learned the ecological zonation version of flood geology from one of Harold W. Clark's former students. He briefly studied biblical languages and archaeology at the Seventh-day Adventist Theological Seminary before switching to vertebrate zoology at Oregon State College, where he received a master's degree. He then entered Harvard University to work with the vertebrate paleontologist Alfred S. Romer (1894–1973), eventually writing a philosophically safe dissertation in comparative morphology and vertebrate paleontology, for which he received his Ph.D. degree in 1954. During the late 1950s he taught at the Adventist Loma Linda University, where, at a series of seminars on science and religion, he got acquainted with Price and occasionally saw Lammerts. By this time his explorations in the field had convinced him that both Price and Clark were wrong in attributing most of the sedimentary rock to Noah's flood. Though he continued to believe in the biblical deluge and passed as an unquestioned "Flood geologist" among most of his Adventist colleagues, he increasingly stripped the event of geological significance, attributing the lower part of the geological column to antediluvian times and the upper part to the postdiluvian era. To make room for these developments, he pushed back creation far beyond six thousand years.[8]

Hare, scion of a prominent Adventist family from New Zealand, had majored in chemistry and studied with Clark as an undergraduate at Pacific Union College before obtaining an M.S. degree in physical chemistry from the University of California. At Berkeley he acquired an interest in earth history and resolved to make the reconciliation of Genesis and geology his lifework. In the fall of 1955 he returned to the Napa Valley to teach chemistry at his alma mater but commuted to Berkeley from time to time to take additional courses in geology. Within a couple of years he decided to earn a doctorate in geochemistry, which he thought might allow him to answer some questions raised by radioactive methods of age dating. In 1958 the Seventh-day Adventist church, eager to set up a center for the study of earth history but lacking a single church member with a Ph.D. in geology, agreed to sponsor Hare for a doctorate in geochemistry at the California Institute of Technology. While there he began developing a dating method based on changes in the ratios of amino acids in ancient shells, which he at first hoped would yield relatively recent ages. However, as he informed a church official, his methods gave dates virtually identical to those obtained from radiocarbon data—and as high as thirty-five thousand to forty thousand years. "Is it just a coincidence that this came out this way?" he asked suggestively. Encouraged by some liberal Adventist theologians and influenced by his observations in the laboratory and field, he, like his friend Ritland, began limiting the effects of the flood and expanding the history of life on earth.[9]

On the morning of July 5, 1962, Lammerts met Ritland and Hare at Glacier National Park as planned. Together the men hiked up to the overthrust area at the south rim of the park, where the contact line between Precambrian and Cretaceous can be seen for miles. To Ritland and Hare, the evidence of overthrusting, especially signs of grooving and scouring, was "overwhelmingly clear." Lammerts, though appreciative of his young companions' scientific approach to the problem, found himself more confused than convinced. He thought it especially puzzling that Ritland and Hare seemed "so anxious to prove that Price was wrong and that this wrong order formation was really the result of overthrusting." As he descended the mountain, Lammerts appeared "badly shaken." Not only had he just gone on record in *The Genesis Flood* as discounting the evidence for overthrusting, but, as Ritland and Hare pointed out, the supporting photographs he had given Whitcomb and Morris were of rocks two hundred feet above the contact line. Besides, he had an article in press at *Christianity Today* in which he described the thrust faults in Glacier National Park as "purely imaginary."[10]

His initial reaction was to correct the piece for *Christianity Today* in light of what he had seen, but he eventually decided there was sufficient ambiguity to justify publishing what he had originally written. This decision "badly disillusioned" Ritland, who felt further chagrined when he read Lammerts's description of him in the article as a Harvard-trained Ph.D. who agreed with Price that "most" of the sedimentary rocks had resulted from Noah's flood. For months after his visit to the Lewis overthrust, Lammerts toyed with a novel flood model of overthrusting. "Sometimes I wonder if this sort of thing might not have occurred when the strata was [sic] still soft and relatively unconsolidated," he confessed in a letter to Ritland. "Water film as a sliding surface would greatly aid in this type of thing should the physical evidence really indicate that these rocks were not actually deposited in the order in which we now see them." In the end, however, he remained loyal to the Pricean view.[11]

To counter the arguments of Ritland and Hare and to acquire the credentials needed to debate geological questions, Lammerts the next year signed up for a course on structural geology at the University of California. Instead of finding his confidence in flood geology shaken by what he learned, he professed to be "more and more amazed" each week to discover how much evidence seemed to support the flood theory. He was especially struck by the degree to which presuppositions influenced the interpretation of data. According to his geology professor, " 'we bring about 80 to 90 per cent of our interpretation of what we see to every rock structure we look at,' " he informed Whitcomb. "If this is true, then the reports which are published of rock structures are 10 percent factual and 90 per cent in the mind of the man or woman who looks upon them and reports them." By the time Lammerts completed his reeducation, he had added courses in mineralogy, invertebrate paleontology, and paleobotany to his curriculum vitae.[12]

Lammerts's interest in creationism may have been revived by reading Rehwinkel and corresponding with Whitcomb, but domestic developments convinced him that evolution was a present danger. His wife's sudden death in 1954 had sent him into an emotional "tailspin" and left him with the responsibility of raising two teenage daughters. When they started coming home from the Livermore high school with textbooks claiming that "all scientists accept the fact of evolution," their shocked father, who had no idea such things were being taught in the schools, began to wonder if he were "the sole remaining intellectual dinosaur surviving in an otherwise completely enlightened age of mammals." Thus he sought out other creationist scientists and, to convince his skeptical younger daughter, twice addressed her class on the subject of evolution.[13]

Having successfully dealt with this family crisis, Lammerts promptly lost interest in what the school system taught. But he soon found a new object of concern: the virtual takeover of the ASA by persons who subscribed to the "ridiculous" notion of theistic evolution. In his opinion, the apostate affiliation was "sick, sick, sick!" His doubts about the health of the organization were confirmed when he presented a paper at the 1960 annual meeting attempting to disprove evolution by showing that no positive mutations resulted from irradiating rose buds. His talk provoked "the nearest thing to a violent outburst" that one member had ever seen and prompted Walter R. Hearn to suggest that in the future the association should provide two rooms for such divisive discussions: one for members who could talk about evolution calmly, the other for persons who refused to use the taboo word. Lammerts maintained his membership in the ASA until 1970, but he no longer respected an organization that would tolerate "the 'spouting off' of such rabid evolutionists as Dr. Walter Hearn."[14]

As Lammerts readily acknowledged, his views on evolution were extreme even by the standards of creationism. At a time when the strictest creationists were allowing for microevolution and the natural development of some species, he held out for "the absolute fixity of species." In his only departure from this principle, he invoked the miraculous manipulation of DNA molecules to explain the appearance of new "so called species" and races since the original Edenic creation. "As a part of His providential care God may at various times rearrange the DNA in order to adapt organisms including man to special conditions," he explained to one Christian biologist, giving the biblical incident at Babel as an example. "This however is not evolution but a designed change and would have to be effected rapidly and perfectly in order to result in a functional organism."[15]

Lammerts's stated reasons for opposing evolution varied with his audience, but he consistently downplayed biblical considerations in favor of scientific ones. While other creationists tried to expand Ussher's chronology by a few thousand years, he professed to have difficulty seeing how the earth could be even six thousand years old. As an eminently practical man, he hated to see students wasting so much time on a mere theory "when they could be learning something useful." Through personal observation he also concluded that the idea of evolution had "had a dulling effect on the minds and work of many younger and less imaginative plant and animal breeders." Because they mistakenly thought that organic change required millions of years, they approached their work timidly. He attributed his own undisputed success as a plant breeder in part to his recognition, as a creationist, "that within the limits of variability possible for a given species, change can be effected quite rapidly." Beyond such practical considerations, he had little desire to convert the mainstream scientific community to creationism. "If a man is such a stupid fool he can't see that evolution is wrong," he said, "I'm not going to try to convince him."[16]

Unlike many creationists, who typically spent their lives toiling in the academic backwaters of small church colleges and felt insecure whenever they ventured into the larger world of science, Lammerts oozed self-confidence. "He may be in error sometimes," noted one ambivalent colleague, "but he is never in doubt!" A large, imposing man, Lammerts possessed not only sterling scientific credentials and an international reputation as a horticulturalist but financial independence as well. Royalties from his prizewinning roses allowed him to go into the investment business and build up a nest egg of several hundred thousand dollars. By the early 1960s, growing disillusionment with the Missouri Synod, in part over creeping evolutionism, led him to decide to give more of his money to creationist causes and less to his church. Though a committed Christian, the self-described "cantankerous cuss" did not always conform to conventional forms of piety. In contrast to the godly Morris, who comported himself like a minister, Lammerts spoke his mind freely and punctuated his conversation with *hells* and *damns*.[17]

Politically and socially, Lammerts also broke with the conservative mold of the stereotypical creationist. As a La Follette Progressive turned Kennedy Democrat, he strongly supported civil rights, conservation, and progressive social legislation and abhorred John Birchers and other extreme right-wingers. During the early 1960s he served on an interfaith committee for "fair play to Negroes" and promoted conservation groups, such as the Sierra Club, that were trying to save the wildlife being destroyed by "our western type of mechanized civilization." Though a philosophical and practicing capitalist, he had little use for creationists who tried to hitch their cause to anticommunism. "I do not believe . . . that communism resulted from evolutionary concepts," he declared in one of the few political statements to appear in his creationist writings.

> Rather it or its spread, was the result of the horrible oppression of the common people by the Czars and ruling classes in Russia. Likewise in this country capitalism as late as 1890–1910 was very harsh, forcing workers to labor 12 hours per day for a mere pittance, often under the most unhealthy and dangerous conditions. Fortunately, progressive capitalists such as Henry Ford and legislators such as Robert Marion Lafollette [sic] gradually modified capitalism reducing the length of the working day, improving working conditions, and increasing the workers' share of each company's income.

Despite his distinctive attitudes, Lammerts would nevertheless emerge in the 1960s, along with Morris, as one of the two most influential scientific creationists in America.[18]

THE CREATION RESEARCH ADVISORY COMMITTEE

When Lammerts and Tinkle set out to recruit a team of ten creationists, the two men themselves agreed on little more than their common opposition to evolution. Although both of them had been active in the Deluge Geology Society and the American Scientific Affiliation, they scarcely knew each other because Tinkle had already left southern California for Taylor University in Indiana by the time Lammerts became involved in either organization. In the late 1940s they had collaborated by mail on an antievolution essay for the ASA's first book, but even afterward Lammerts suspected Tinkle of being a fifth columnist for continuing to believe in long geological ages. Despite Tinkle's antipathy to evolution and admiration for Price, he had never fully embraced flood geology.[19]

At the beginning of his venture with Lammerts, William J. Tinkle was already approaching his seventieth birthday. One of the few active creationists to have lived as an adult through the antievolution controversies of the 1920s, he had first contacted Price in 1932 at the suggestion of William Bell Riley. By that time he had already earned M.A. and Ph.D. degrees in zoology from Ohio State University and was serving on the faculty of Marshall College in West Virginia. Like Lammerts, he specialized in genetics, but his particular interests ran more toward breeding people than plants. He had devoted his master's thesis, popularized in the *Literary Digest*, to the inheritance of habitual wandering and his doctoral dissertation to a study of deafness as a problem in eugenics. As late as 1939, by which time most reputable geneticists had backed away from eugenics, he was still advocating selective human breeding in his creationist textbook, *Fundamentals of Zoology*. A seminary-trained minister in the Church of the Brethren, or Dunkers, he spent the latter part of his career teaching at schools associated with the Wesleyan Holiness tradition. Like many in that branch of American Protestantism, he cared little about the details of a person's creationist beliefs—as long as he or she did not include macroevolution.[20]

Tinkle's own views on earth history derived more from the old catastrophism of the French naturalist Georges Cuvier than from the new catastrophism of Price. Though far from consistent in his reading of the rocks, Tinkle tended, like Cuvier, to believe that the earth had undergone a series of geologically significant cataclysms rather than just one. "If the Bible stated that the major part of the sedimentary rocks were laid down during the flood, I would believe it; but I have not found such a statement," he informed Whitcomb. "Since the Bible does not claim to be a complete scientific record, there may have been other cataclysms besides the flood. It hardly seems reasonable that so much geological work was done in a single year." When Whitcomb pointed out the theological difficulty associated with having the fossilized animals die before sin entered the world, Tinkle awkwardly sidestepped the problem by suggesting that the fossil-bearing strata had been deposited *after* the Edenic creation. "Concerning catastrophes before Adam," he wrote, "I think there may have been much action of this sort, narrated in the first ten verses of Genesis 1. But of course this would not account for the fossiliferous rocks." Later he tried to harmonize Cuvier and Price by reversing the sequence of cataclysms: Rather than having Noah's flood represent the last of the catastrophes, as Cuvier had done, he placed it first, "followed by a series of major but relatively local events." But regardless of how he juggled cataclysms, he still needed far more time than flood geology allowed.[21]

In 1961 Tinkle and Lammerts possessed only the vaguest agenda for their proposed group. Tinkle, frustrated by the ASA's turn to theistic evolution but unwilling to jeopardize its future by leading a schismatic revolt, at first envisioned only an informal group within the affiliation. But he soon saw a grander future. After watching with the rest of the world as John Glenn orbited the earth in early 1962, he could not help thinking that he and Lammerts were embarked on an even more momentous mission. He wondered just how high they should set their sights. "Can we turn the majority of scientists against evolution?" he asked Lammerts. "Can we turn Christians toward creation and let non-Christians go their way?" Lammerts, ever the realist, replied that he had "no illusions" about turning most scientists against evolution. His immediate goals were to keep Christian young people from rejecting a literal creation and to let the world know about the existence of creationist scientists. Members of the group could also alert one another to recent articles of interest and critique one another's papers. He hoped that in the future they would be able to place pro-creation articles in such popular magazines as the *Saturday Evening Post* and the *Saturday Review of Literature*, and he suspected that even *Time* "might find it newsworthy to indicate in their science section that a number of actively working scientists have their various reasons for not believing in the Evolution theory." To get things started, he simply wanted a letterhead carrying the names of ten scientifically reputable creationists, which would indicate to recipients "that each one of us is not merely some individual 'oddball.' "[22]

In the fall of 1961 Tinkle wrote to eight known or suspected creationists inviting them to join Lammerts and himself in forming an antievolution caucus within the ASA. The eight invitees were Henry M. Morris, Frank Lewis Marsh, Molleurus Couperus, Edwin Y. Monsma, R. Laird Harris, Duane T. Gish (b. 1921), Philip V. Livdahl (b. 1923), and Edward L. Kessel (b. 1904). By late January he had heard from all but Livdahl, a physicist, and Kessel, an entomologist who turned out to be a theistic evolutionist. Of those responding, only Morris had expressed any hesitation. Because he believed that the ASA "was too permeated with evolutionism ever to be reclaimed," he urged Tinkle to start a new society. He feared that even "Tinkle's Cuvierianism" entailed too much compromise with unbiblical ideas, but he eventually set aside any reservations in hopes that Tinkle and Lammerts were at least moving in the right direction.[23]

By early March Lammerts was referring to the "committee of 8"— which soon shrank to seven when Couperus reconsidered his participation. Since his days with Tinkle and Lammerts in the Deluge Geology Society, he had moved far from conventional flood geology, and in May he asked Lammerts not to include his name on the proposed letterhead. "I am happy to be a member," he wrote diplomatically, "but I feel that there is no particular advantage in advertising the committee membership to those who are on the outside." Of the surviving seven, only Lammerts, Morris, Marsh, and Monsma publicly subscribed to a recent creation and a cataclysmic flood. The Seventh-day Adventist biologist Marsh had been defending flood geology since studying with Price as a young man, and the Christian Reformed botanist Monsma had long advocated strict creationism within the ASA. Tinkle openly questioned the wisdom of attributing most of the fossil-bearing rocks to a single flood, while Harris and Gish generally avoided the topic.[24]

Harris, a chemical engineer turned archaeologist, held a Ph.D. degree in Old Testament studies from the Dropsie College for Hebrew and Cognate Learning (now the Annenberg Research Institute) and taught at Covenant Theological Seminary, a Reformed Presbyterian institution in St. Louis. Along with most of the other members of the team he had read a prepublication draft of *The Genesis Flood* and given Whitcomb and Morris the impression that he generally agreed with their conclusions. However, within months of joining the Tinkle-Lammerts group, he informed Lammerts of a reluctance to endorse flood geology. Although he acknowledged that the Noachian deluge had probably been "universal in the sense that it destroyed life all over the world," he suspected that earlier floods had also contributed to the geological record. Lammerts assured him that the group had not yet reached a consensus on the subject and pointed out that even Tinkle accepted geological ages. Morris, surprised and disappointed by Harris's confession, speculated that he had come under the influence of his colleague J. Oliver Buswell, Jr., the former president of Wheaton College and, in Morris's opinion, a notorious defender of the day-age theory.[25]

Duane T. Gish, the only one of the founding seven not to have read *The Genesis Flood* in manuscript, was a relative newcomer to organized creationism. A Berkeley-trained biochemist (Ph.D., 1953) and onetime assistant professor at the medical college of Cornell University, he had since 1960 been conducting protein and polypeptide research for the Upjohn Company, a pharmaceutical giant in Kalamazoo, Michigan. Since becoming a confessing Christian in the Methodist church at age ten, he had taken the creation story as a factual historical account. And as an adult fundamentalist in the Regular Baptist church, he had affirmed belief in the Bible "as the inerrant inspired Word of God." In the late 1950s, as a result of reading John R. Howitt's anonymously published booklet *Evolution: Science Falsely So-Called*, he resolved to become an active antievolutionist and joined the ASA, mistakenly assuming it to be a creationist society. Through the ASA he met Tinkle, who welcomed him to the band of dissenters protesting evolution.[26]

Because Howitt had developed his case against evolution without invoking deluge geology, Gish saw little reason to make an issue of the flood theory. As far as Lammerts could tell, Gish believed that the idea of a cataclysmic deluge was "very interesting but rather a matter of indifference as regards the main argument." This attitude led Gish occasionally to consort with old-earth creationists, who were anathema to such strict constructionists as Whitcomb and Morris. "When I am asked about the apparent age of the earth," he explained to Whitcomb in the mid-1960s,

> I mention the flood geology theory, the gap theory, the day-age theory and finally emphasize the fact that very important assumptions are made in all dating methods and if these assumptions are wrong the whole method is wrong. I insist on no one particular position, and state that regardless of the geological dates assumed, the fossil evidence is incompatible with the theory of evolution. In doing so I am adopting a time-honored strategy—concentrating your effort at the enemy's weakest point.

Gish did not question the biblical legitimacy of flood geology; he simply left its advocacy to others.[27]

By late September the team of ten, which Lammerts had rechristened the Creation Research Advisory Committee after a similarly named camellia group of which he was a member, still had three unfilled slots. Lammerts, who had predictably taken charge of the group, especially wanted to sign up a physicist and a geologist. Livdahl had declined to join the group, as had Thomas Edward Gilmer (b. 1925), a colleague of Morris's in the VPI department of physics who had given advice on *The Genesis Flood*. Thus Morris, on behalf of the committee, invited John J. Grebe (1900–1984), director of nuclear and basic research for the Dow Chemical Company in Midland, Michigan, to become the eighth member. Grebe, who had emigrated from Germany as a youth, was a highly respected physical chemist and inventor with an honorary Sc.D. degree from the Case School of Applied Science (now part of Case Western Reserve University) in Cleveland. An active Missouri Lutheran, he had impressed Morris by volunteering to write a testimonial to be used in advertising *The Genesis Flood*. Only after Grebe had agreed to join the committee did Lammerts discover that their new recruit, though fond of flood geology, allowed the first three days of creation to span an immense period of time, a concession Lammerts regarded as being "practically the same as the evolutionary concepts."[28]

Landing a geologist who believed in flood geology proved to be Lammerts's most difficult task. He toyed for a while with asking one of the three Adventists in the field—Ritland, Hare, and Clifford L. Burdick—but after hearing about James H. McGuirt (1910–1974), the consulting geologist who had hosted Morris's visit to the Houston Geological Society, he decided that a man with ties to industry "would command much more respect." McGuirt readily accepted the invitation but, like Couperus, had second thoughts when he realized that his name would appear on a letterhead. Eventually he begged off, claiming that membership might obligate him to endorse ideas he still questioned and that his own Episcopal church did not subscribe to the strict creationism that the committee represented. Even Lammerts's assurance that the group required only a commitment to direct creation could not persuade him to rescind his resignation.[29]

To fill the ninth position on the committee, Lammerts next turned to Ritland, his guide to the Lewis overthrust a few months earlier. Although Ritland was more of a paleontologist than a geologist and less of a flood geologist than Lammerts ideally wanted, his Harvard credentials made him particularly attractive. As soon as McGuirt backed off the committee, Lammerts telephoned Ritland, then living in southwestern Michigan, and spent a half hour or so trying to lure him on board. Ritland saw the advantages of pooling resources and providing mutual criticism, but he disliked the idea of flaunting the names of the committee members on a letterhead. "In scientific endeavor," he explained to Lammerts in a follow-up letter, "I believe people will be more impressed by the contents of the package rather than its wrapping and I further wonder if some perceptive individuals might not feel that we are trying to make up by authority what they suppose we lack in the actual 'bill of goods.' " In truth, Ritland used the stationery issue to camouflage his primary reason for declining to serve: his unwillingness to associate professionally with someone who interpreted nature as uncritically as Lammerts had done during the recent excursion to Glacier National Park. But the unsuspecting Lammerts graciously accepted Ritland's excuse at face value and urged the reluctant paleontologist to consider himself "a member in a consultant sort of capacity."[30]

Spurned by Ritland and impatient to complete the committee after over a year of trying, Lammerts turned to his own religious backyard and enlisted two fellow members of the Missouri Synod: John W. Klotz and Wilbert H. Rusch. These additions unbalanced the committee denominationally (with four Lutherans) and professionally (with three geneticists—Klotz, Tinkle, and Lammerts himself), but by this time

Lammerts hardly seemed to care. He also seemed undisturbed by his inability to recruit a single qualified geologist to a committee ostensibly dedicated to promoting flood geology. Rusch, a sometime graduate student in geology and the only one of the ten without some kind of doctorate, came the closest of any committee member to representing that discipline. The son of a Lutheran science teacher, Rusch had earned a bachelor's degree in physical science from the Illinois Institute of Technology and a master's degree in biology from the University of Michigan. Then while teaching at the Lutheran Concordia Teachers College in Seward, Nebraska, in 1957, he had begun to work on a second M.S., in geology, at the University of Nebraska. Three years later he assumed the leadership of a Missouri Lutheran project, called the Rock Symposium, dedicated to reevaluating the relationship between Genesis and geology. When he left Nebraska at the end of 1962 to teach biology and geology at Concordia Lutheran Junior College in Ann Arbor, Michigan, he forfeited his chance for a graduate degree in geology. But because of his training and experience, Lammerts asked him to serve as the committee's designated geologist.[31]

On February 9, 1963, Lammerts proudly announced to Tinkle that the additions of Klotz and Rusch finally brought the committee up to full strength. Of the ten founding members, five (Klotz, Lammerts, Marsh, Monsma, and Tinkle) possessed doctorates in biology; a sixth, Gish, had earned a Ph.D. degree in biochemistry; and a seventh, Rusch, held a master's degree in biology. Thus whatever their reason for rejecting evolution, it was not lack of intelligence or training. "I think one of the first objectives," Lammerts went on to advise Tinkle, "is simply to let the public know that we exist and thus dispell [sic] the myth that 'all scientists accept the fact of evolution.' " As a means to this end, he immediately ordered the long-desired stationery announcing the existence of the ten-member Creation Research Committee. In part because he had access to the services of a secretary at Germain's, he assumed the role of chairman and assigned the retired Tinkle the position of secretary. At the top of the letterhead appeared the motto:

HAEC CREDIMUS

"For in six days the Lord made heaven and earth, the sea, and all that in them is, and rested on the seventh day."—Exodus 20:11

Each committee member may have interpreted the verse differently, but at least no one disputed its essential truth.[32]

In his letter to Tinkle announcing the completion of the team of ten, Lammerts proposed that the Creation Research Advisory Committee expand into a full-fledged society "by no later than 1964." Although it had taken him over a year to round up just ten strict creationists, he now hoped optimistically to find ninety more who were "definitely committed to the proposition that 'God created the universe, solar system, the earth and all that is therein within a six-day period.'" By the end of 1963 he not only presided over a thriving society, created months ahead of schedule, but was already halfway toward his goal of a hundred members.[33]

The Creation Research Society formally came into existence in mid-June 1963 in back-to-back meetings in Kentucky and Michigan. The ASA had scheduled its annual conference that year for June 19–21 on the campus of Asbury College and Theological Seminary in Wilmore, Kentucky, and Tinkle was helping to organize the program. Four members of the Creation Research Advisory Committee—Harris, Klotz, Lammerts, and Morris—were slated to present papers, and three others—Grebe, Marsh, and Rusch—planned to attend. The occasion seemed the perfect time to convene the committee and lay plans for a broader society. Besides, Grebe had invited the group to retire after the meeting to his home in Midland, Michigan, where Gish and Monsma might be able to join them. In April Morris notified Lammerts that he would be unable to participate because of a conflicting four-week conference in southern New Mexico, near El Paso. Lammerts urged him to fly back for the weekend to help neutralize Grebe's old-earth views. Lammerts admired the industrial scientist's devotion to creationism, and looked forward to tapping his bank account, but he feared that in the absence of any other mathematically trained member Grebe might "feel that we are simply 'completely nuts' in discarding the usual time concepts or radioactive disintegration studies and similar phenomena." But Morris could not change his plans.[34]

Despite Morris's absence, the June meetings went well. The committee officially reconstituted itself as the Creation Research Society (CRS) and elected Lammerts as president, Tinkle as secretary, and Rusch as treasurer. Lammerts, whose management style tended to the dictatorial, quickly asserted his authority. Everyone went along with Lammerts at the first meeting, because, suspected Morris, he "is a large, forthright, outspoken man, with firm opinions on just about everything, and can be quite intimidating." Following Lammerts's lead, the group dedicated itself to realigning "science based on theistic creation concepts" and to publishing creationist textbooks. It limited voting membership to "scientists having an M.S. (or equivalent in experience), Ph.D., Sc.D., Ed.D., or M.D. Degrees," but allowed nonscientists, such as Whitcomb, to join as nonvoting members.[35]

Lammerts's biggest headache during the early days of the society was drawing up a statement of belief acceptable to all parties. Even before the June meetings the committee had been wrangling over wording. Everyone agreed on the necessity of keeping out evolutionists, but beyond that, there was little consensus. Morris wanted to commit the society unambiguously to the plenary inspiration of Scripture and to flood geology, arguing that "unless these two items are clearly spelled out, we are in effect compromising with the evolutionary position, as it may later crop up." Lammerts, with good reason, worried that "such thorough-going creationists" as Tinkle, Gish, and Harris would have difficulty attributing most of the geological column to the flood. Besides, Lammerts himself, though willing to assign all the geological strata to the deluge, believed that this plank in the creationist platform was "not too critical." Klotz, ambivalent about the flood theory, feared that the society would "be criticized for trying to solve all geographical, geological and paleontological problems by the flood." He favored simply saying that "the great flood of Genesis is an historical event which probably brought about great geographical and geological changes." Lammerts hoped to include the creation of the universe, solar system, and earth within "a six-day period," but Marsh refused to include the universe, and Grebe rejected the very notion of literal days, at least for the first three periods of creation. Rusch proposed opening the society to non-Christian creationists, such as Jews and Muslims, but others wanted membership restricted to persons who accepted Jesus Christ as their personal Savior.[36]

The final statement of belief, binding upon all new members, failed to require the acceptance of flood geology or the recent creation of the universe, but it did exclude non-Christians:

1.   The Bible is the written Word of God, and because it is inspired throughout, all its assertions are historically and scientifically true in all the original autographs. To the student of nature this means that the account of origins in Genesis is a factual presentation of simple historical truths.

2.   All basic types of living things, including man, were made by direct creative acts of God during the Creation Week described

in Genesis. Whatever biological changes have occurred since Creation Week have accomplished only changes within the original created kinds.

3. The great Flood described in Genesis, commonly referred to as the Noachian Flood, was an historic event worldwide in its extent and effect.

4. We are an organization of Christian men of science who accept Jesus Christ as our Lord and Saviour. The account of the special creation of Adam and Eve as one man and woman and their subsequent fall into sin is the basis for our belief in the necessity of a Saviour for all mankind. Therefore, salvation can come only through accepting Jesus Christ as our Saviour.

For years Morris agitated to revise the statement to make belief in flood geology and the recent creation "of all things" a condition of membership, but the most he won was a board decision to the effect that "no publication of the Society . . . would ever advocate the 'old-earth,' geological-ages position."[37]

At the organizational meeting in Midland three new members joined what came to be known as "the inner-core steering committee" of the CRS: Karl W. Linsenmann (1909–1990), a Lutheran friend of Grebe's who practiced medicine in Midland; and John N. Moore (b. 1920) and David A. Warriner (b. 1922), who both taught in the natural-science program at Michigan State University. After earning a master's degree in botany from Michigan State, Moore had stayed on to obtain a doctorate in education, specializing in philosophical and pedagogical issues related to science. Nominally a Methodist, he had in 1962 experienced "a spiritual re-birth," which led him later in the decade to join an independent Bible church based on New Testament principles. As a creationist, he had devoted himself to "searching out writings by true, bonified [sic] scientists regarding limitations, qualifications, short comings and inadequacies of the theory of organic evolution," a theory that he linked to left-wing politics. As Rusch discovered during a visit to East Lansing, Moore was an archconservative and a militant anti-communist. Warriner, if anything, stood politically to the right of Moore. According to neighbors, he was not only right-wing, but racist, a trait some acquaintances attributed to his Southern roots. Although he had earned a doctorate from Cornell University in educational psychology and had studied at the fundamentalist Dallas Theological Seminary, he fancied himself an expert on scientific methodology and a qualified judge of the merits of evolution.[38]

Meanwhile, out West, Morris was rounding up several additional committee members. In El Paso he corralled three Southern Baptists: Harold S. Slusher, Thomas G. Barnes, and Willis L. Webb. Slusher and Barnes both held appointments in the physics department at Texas Western College (now the University of Texas at El Paso). Slusher, whose highest degree was an M.S. from Oklahoma State University, taught astronomy and geophysics. To the annoyance of some of his colleagues, he required students in his geophysics course to read *The Genesis Flood* and recommended that they also look at Price's *Evolutionary Geology and the New Catastrophism*. His colleague Barnes (b. 1911) headed the Schellenger Research Laboratories at Texas Western, where he worked primarily in applied physics. In addition to an M.S. degree in physics from Brown University, he possessed an honorary Sc.D. degree from Hardin-Simmons University, his undergraduate alma mater in Abilene, Texas. At the time of joining the CRS, he was completing a textbook on electricity and magnetism, published by D. C. Heath. Webb (1923–1983), who at the time possessed only a bachelor's degree, was a well-known meteorologist with the U.S. Army, commuting daily to the White Sands missile facility in New Mexico, just north of El Paso.[39]

Against Lammerts's initial objections, Morris also convinced the group to invite the maverick Seventh-day Adventist geologist Burdick to serve on the committee. On a visit to Tucson, Morris had been struck by Burdick's apparent humility and sincerity, and he hoped that the CRS would be able to assist the impecunious creationist financially while at the same time benefiting from his dogged pursuit of geological evidence against evolution. Finally, the chemist-theologian Paul A. Zimmerman (b. 1918), president of the Missouri Lutheran junior college where Rusch taught, joined the committee, bringing the total to eighteen.[40]

The composition of the original steering committee reflected, albeit imperfectly, the denominational, regional, and professional bases of the creationist revival in the early 1960s. Along with six Missouri Lutherans, six Baptists (four Southern, one Regular, and one independent), and two Seventh-day Adventists, there were single representatives from the Reformed Presbyterian church, the Christian Reformed church, the Methodist church, and the Church of the Brethren. Twelve of the eighteen lived in the Midwest (nine in Michigan alone), four in the Southwest, one in California, and one in Virginia. The committee included five men with earned Ph.D. degrees in biology, plus a sixth with a doctorate in biochemistry and an additional two with master's degrees in biology. Burdick, who falsely claimed to have a master's degree in geology, was the only earth scientist. Seven members were affiliated with church-related colleges and five with state institutions; the rest worked for industry, the government, or themselves. All, it seems, became or remained antievolutionists primarily for biblical reasons. Although creationists increasingly stressed the scientific evidence for their position, one estimated that "only about five percent of evolutionists-turned-creationists did so on the basis of the overwhelming evidence for creation in the world of nature."[41]

GROWTH AND CONTROVERSY

From the beginning the CRS stressed education and research rather than evangelistic and political activities. Instead of sponsoring public meetings, it concentrated its energies on publishing texts and journals. Capitalizing at first on conservative disapproval of the ASA's flirtation with evolution and later on reaction to the growing emphasis on evolution in the public schools, the CRS expanded rapidly. After only three years it boasted a subscription list of 680 names, including about 200 voting members having the requisite scientific credentials. Only 60 of the 200, however, possessed doctoral degrees in science. By the time of its tenth anniversary, in 1973, total membership stood at 1,999, with 412 eligible to vote.[42]

The rapid growth of the society emboldened Lammerts to take a hard line against inactive and deviant board members. Monsma and Warriner were the first to go. The ailing Monsma had contributed nothing; and Warriner, who had lost his job at Michigan State, had fallen out of favor with Lammerts for opportunistically suggesting that the society hire him as a paid promoter. Lammerts next engineered the departure of Webb (for inactivity) and Harris (for heresy). Lammerts found the latter unacceptable on two counts: Harris not only opposed the majority view of a literal six-day creation but, worse yet, earned his living mainly as a theologian, a professional type Lammerts held in low esteem.[43]

The four replacements included three more biologists: George F. Howe (b. 1931), a Baptist; Bolton Davidheiser (b. 1912), a peripatetic churchgoer who briefly attended Plymouth Brethren, Quaker, and Baptist congregations; and H. Douglas Dean (b. 1908), a member of the Church of Christ. Howe, a Wheaton alumnus with a Ph.D. degree in botany from Ohio State University, taught at Westmont College in Santa Barbara. At the time of joining the CRS in 1963, he still doubted the validity of flood geology and the necessity of compressing creation into six twenty-four-hour days, but he soon embraced the new creationist orthodoxy. In 1968 he succeeded Lammerts as editor of the *Creation Research Society Quarterly.* Dean, a Pepperdine College professor who never won Lammerts's confidence, survived on the board for only two years. Though a strict creationist, he ranked lower than a theistic evolutionist in Lammerts's hierarchy because he allegedly believed "some garbage about God creating 7 or 8 basic types such as a basic bird type and all birds evolving from this!!" Davidheiser, a Johns Hopkins-trained zoologist who preferred working in the secular environment of Disneyland to teaching at such liberal evangelical colleges as Westmont and Biola, lasted no longer. A man of prickly principles, he refused to continue on the CRS board so long as it tolerated such dangerous anti-Christian cultists as the Seventh-day Adventists, whose teachings he abhorred but never quite understood.[44]

Ironically, the very year Davidheiser stepped down from the board, the Adventist Marsh resigned because of unfounded fears that the society was turning against Saturday worshippers who refused to honor "the Pope's sabbath." Because the board had started holding its annual meetings over the weekend, he suspected a sinister attempt "to push out the conservative Saturday believers from participation in all business of the Society," and he worried that fellow board members were on the verge of "taking an open stand in their publications against Saturday as the true Sabbath."[45]

No issue divided the society more during its early years than time. The original board itself included several members (Grebe, Harris, and Tinkle) who believed in an ancient earth and at least two more (Gish and Rusch) who assigned low priority to the question of age. Even the advocates of a young earth split over the antiquity of the universe. Morris, for example, insisted on the recent creation of the entire cosmos, while Marsh limited the creation story of Genesis 1 to the solar system. Though Lammerts had little sympathy for creationists who claimed to need more than six thousand years—and, indeed, pressured Grebe into silence on the subject of age—he at first tended to agree with Rusch that "just as evolutionists all differ among themselves as to the details of evolution and how they think it occurred, so creationists can be expected to differ also." By the end of 1964, however, the society appeared to be robust enough to survive the pruning of any members unwilling to abandon gap and day-age thinking. "I am determined," Lammerts declared, "to get our organization so clearly committed to not only creation but flood geology and the young earth concept that it will later be difficult to deviate from these commitments."[46]

To ensure that the society remained true to flood geology, Lammerts arranged in 1967 for Morris to succeed him as chairman of the board. On major issues the two titans of strict creationism generally stood side by side, but occasionally even they differed over secondary matters. Morris, for instance, attached great apologetical significance to the second law of thermodynamics, which he associated with the Fall, while Lammerts dismissed such "confounded thermodynamics junk" as worthless prattle. Their most public quarrel erupted in the mid-1970s when Morris wrote a letter to the editor of the *Creation Research Society Quarterly* contrasting his and Lammerts's views on the effects of the flood and condemning the geneticist's "imaginative reconstruction of the variegated activity of the one creation and cataclysm clearly described in the Bible." Whereas Lammerts stuck to "an Ussher-type chronology of only about 7,000 years" and assigned all of the fossil-bearing rocks to the deluge, Morris allowed for an additional two or three thousand years of earth history and placed "the deposits of the Pleistocene and possibly the Pliocene" after the flood. Morris permitted considerable organic development after the deluge, while Lammerts, to avoid any evolution at all, attributed post-flood diversity to divine genetic engineering. Morris viewed such miraculous intervention—amounting, in his opinion, to a second creation—as theologically suspect because it violated the "economy of the miraculous in God's orderly world."[47]

This stinging criticism pushed Lammerts into a renewed defense of pluralism as "a healthy state of affairs." In responding to Morris, he denied that God's genetic tinkering after the flood could be considered "creations," though it had produced new forms of plant, animal, and human life. "I will say that if Morris can explain how Noah could have been heterozygous for all the distinctive characteristics of human beings, and then how the races of mankind could have originated by *natural means* I will be most happy to accept such an explanation," he replied. "Evolutionists are hard put to account for this complexity in hundreds of thousands of years, yet Morris would have this occur in the 5300 or so years since the Flood, by natural selection of the variation potential which existed in the survivors of the Flood." As for not going beyond the biblical record, he wondered how Morris could accept the occurrence of post-flood glaciation, even of limited duration, when the Bible said nothing of ice ages.[48]

Despite such disagreements, even at the highest echelons, the CRS acquired a well-deserved reputation for welcoming only committed flood geologists. Creationists who remained unpersuaded by the arguments for a young earth increasingly found themselves without an institutional home: too conservative for the ASA but too liberal for the CRS. In 1965 the Canadian gap theorist and self-described fundamentalist Howitt complained that while many ASA members were embracing theistic evolution "or some such rubbish, the true fundamentalists are all going over to flood geology. Oh me, oh my!" At first he tried to convince CRS leaders not to make flood geology a virtual condition of membership. "Let it be truly a Creation Research Society," he urged, "seeking for the truth, no matter where it ends up,—as flood geology, the gap theory, theistic evolution or what not." On one occasion he suggested "a Pugwash Conference on the Gap vs the Flood" similar to the famous gatherings in Nova Scotia that brought Cold War scientists together. But by the early 1970s he had abandoned hope of any reconciliation. "The flooders are getting pretty dogmatic these days," he noted sadly.[49]

One victim of this diluvial dogmatism was Daniel E. Wonderly (b. 1922), a biology teacher at Grace College, Whitcomb's academic home. A graduate of Wheaton College, Wonderly had obtained advanced training in both theology and science. Though a theologically conservative Baptist, he accepted the evidence for an old earth, believing that it could be harmonized with revelation by means of the day-age theory. The administrators at Grace knew of this and of Wonderly's opposition to flood geology, but they decided to hire him anyway—on condition that he would not undermine the school's commitment to young-earth creationism. When his true views leaked out to students, the president formally forbade him from discussing them, either orally or in writing. Under the circumstances, Wonderly felt compelled to resign. Not surprisingly, his plight elicited little sympathy from the flood geologists in the CRS. "Those Christians who believe in the geological age system, with all the wresting of Scripture that system necessarily entails, need not complain (as Wonderly does) that they don't get a fair hearing or that we 'recent' creationists are ignorant of the evidence," argued Morris in defense of exclusionist policies. "Their side is heard almost constantly, everywhere and all of us have been thoroughly exposed to it. It is, rather, the recent creationist who doesn't get a fair hearing, and this is the purpose of our own Quarterly."[50]

During the early years of the CRS, the officers kept a watchful eye out to avoid political and theological entanglements that might deflect the society from its appointed mission. They requested, for example, not to be listed in a national directory of right-wing groups, and they refused to get involved with efforts to outlaw evolution in the public schools or mandate the teaching of creation. "Laws against the teaching of evolution are not effective and could be enforced fully only in a police state," wrote Tinkle, the secretary, in a statement of policy: "Instructors teach what they believe. It is the prerogative of the Creation Research Society to point out to them in detail that planning and forming by an intelligent Creator is a more reasonable interpretation than the chance events which occur in nature." As a matter of principle, the board voted in 1973 not to request federal funds to support its activities. Years later what had once been a source of pride came to be seen as a sign of persecution. "Creationists do not have access to Federal money taken from the taxpayer's pockets that is available to evolutionists," complained the editor of the CRS *Quarterly* in the late 1980s.[51]

At first Morris feared that the Seventh-day Adventists might "take over the CRS like they had in the old Deluge Society," but the cliquish Saturday worshippers generally kept to themselves. The Missouri Lutherans posed more of a problem. In view of their prominence in the society—members occupied one-third of the eighteen seats on the steering committee and served as chairman, editor, and treasurer—even Lammerts worried that outsiders might identify the group as Lutheran. Particularly troublesome in this regard was an ambitious Lutheran pastor from Idaho named Walter Lang (b. 1913), who aspired to lead the creationist revival. Unhampered, as he put it, "by the notion that science is 'objective,' " he prized spiritual insight above scientific expertise. Upon hearing of the birth of the CRS in 1963, he wrote Lammerts offering his services as a publicity agent for the new society and announcing that he had recently decided to specialize "in the Bible-Science field" and to launch a newsletter promoting young-earth creationism. Lammerts, determined to keep the CRS both non-clerical and nondenominational, politely rebuffed Lang's attempt to insinuate himself into the society, explaining that the CRS admitted only scientists.[52]

The early issues of Lang's *Bible-Science Newsletter* seemed to justify Lammerts's caution. Their typographical and intellectual sloppiness, to say nothing of Lang's bumptious manner, offended some image-conscious leaders of the CRS. And they were scandalized when in the mid-1960s he opened his magazine to hard-core biblical literalists who placed the earth at the center of a finite universe. Citing such verses as Joshua 10:12 (where the Israelite warrior commanded the sun to stand still) and Psalms 93:1 ("the world also is stablished, that it cannot be moved"), these latter-day disciples of the sixteenth-century astronomer Tycho Brahe insisted that the sun revolved around the earth, not vice versa. "I fear," wrote a despairing Lammerts, who refused space to geocentrism in the CRS *Quarterly*, "that our friend Lang will never learn the difference between scientific theory . . . and purely scientific drivel."[53]

Such concerns notwithstanding, the CRS did not want unnecessarily to alienate an energetic ally who each month distributed thousands of copies of creationist propaganda. Consequently, in 1966 Lammerts and representatives of Lang's Bible-Science Association (BSA) negotiated an arrangement whereby the CRS would advise the BSA and allow Lang to reprint and summarize CRS research in return for the BSA's help with publicity and fund-raising. Before long, however, Lammerts came to believe that the CRS should divorce itself *"completely"* from the BSA, before its implied endorsement of geocentrism tarnished all creationists. Morris refused to take such drastic action, and thus the two societies continued their codependent relationship, with the CRS catering to the needs of scientists while the BSA carried creationism to the masses. After Lammerts stepped down as editor of the CRS *Quarterly*, it, too, carried occasional pieces in support of a stationary earth "at the center of Creation just where the Bible puts it." The author of these contributions was James N. Hanson (b. 1933), a Missouri Lutheran turned Baptist who taught computer and information science at Cleveland State University and served as vice president of the BSA.[54]

THE TEXTBOOK PROJECT

During its first decade the CRS focused its limited resources on two projects: publishing the CRS *Quarterly* and preparing a high-school biology text. For over a quarter-century after the Scopes trial in 1925, American textbook publishers tried to avoid antagonizing conservative Christians by saying as little as possible about evolution. This policy of "neutrality based on silence" began to crumble in the late 1950s, after the Soviet Union in 1957 successfully launched Sputnik, the first artificial satellite to circle the earth. An embarrassed United States sought to regain world leadership in science and technology by pouring millions of dollars into improving science education. Backed by generous funding from the National Science Foundation, a group of biologists in the American Institute of Biological Sciences established a center at the University of Colorado, the Biological Sciences Curriculum Study (BSCS), to produce state-of-the-art biology texts. Responding in part to complaints from leading biologists that "one hundred years without Darwinism are enough," the BSCS authors wove evolution into their material as "the warp and woof of modern biology." After extensive testing in over a thousand schools, the BSCS in 1963 issued three versions of its tenth-grade text, each identified by the dominant color of its cover: blue, yellow, or green. Before long nearly half of the high schools in America were using these books or other curriculum materials developed by the BSCS—and introducing hundreds of thousands of high-school students to their apelike ancestors. Like Bernard Ramm's attack on flood geology, which had provoked Whitcomb and Morris into defending Pricean catastrophism, these controversial texts created a furious backlash against the very theory they were designed to promote.[55]

Not surprisingly, concerned creationists viewed this latest offensive as an "attempt to ram evolution down the throats of our children." "It seems clear," wrote Rita Rhodes Ward (b. 1910), a creationist biology teacher in El Paso, "that all three of these books are dedicated to the promulgation of total organic evolution to the exclusion of objectivity in biology, if need be, in order to eliminate any belief in fiat creation." In making God unnecessary, she argued, the texts stopped just short of espousing atheism. An indignant Lammerts scatologically dismissed the BSCS texts as "mostly BS."[56]

When the education establishment in Texas moved to adopt the BSCS books, local creationists, prodded by Melvin and Norma Gabler, a couple of self-appointed textbook censors from Longview, sprang into action. On learning that the state textbook selection committee would hold a public hearing in Austin in October 1964, an aroused Ward not only talked Barnes into testifying but coached the physicist on what to say. She also arranged for her Church of Christ to fly in fellow believer Dean from California. During the hearing, when committee members asked for the titles of acceptable alternatives to the BSCS texts, the creationists could name none. In fact, only two candidates came to mind: Tinkle's out-of-print *Fundamentals of Zoology* (1939) and Ernest S. Booth's *Biology: The Story of Life* (1950), a Seventh-day Adventist text that contained unacceptable references to the creationist origins of the seventh-day Sabbath and to the testimonies of the prophet Ellen G. White.[57]

To meet the apparent demands for a creationist text, Lammerts appointed a CRS textbook committee initially comprising Barnes, Ward, Dean, Tinkle, and himself. In anticipation that Texas might lead the way in adopting a creationist work to supplement the BSCS books, Lammerts asked Barnes to head up the project. He himself took on the job of trying to raise $10,000, the minimum amount he thought would be needed to compete with books "prepared under government grants totaling approximately $8,000,000." Ward, the only high-school teacher on the committee, outlined the desired volume and assisted Barnes in lining up over a dozen contributing authors, including three biologists from Church of Christ colleges. Moore, serving as senior editor, melded the disparate parts into a relatively unified whole.[58]

The CRS at first anticipated little difficulty finding a major publisher for the textbook. Encouraging reports circulated that a well-known publishing house had queried Dean in Austin about the possibility of writing a creationist text. "They will publish anything they think will make money," Barnes assured Lammerts. "Surely there would be a sizeable market for a top flight Biology text [written] from our point of view." But when Barnes contacted the fifteen leading publishers of high-school texts, not one expressed interest. According to Morris, editors at these houses, remembering the collective action taken against the publisher of Immanuel Velikovsky's cataclysmic *Worlds in Collision* (1950), feared "that all of their textbooks would be boycotted if they would dare to publish a creationist book." To salvage the project, the CRS finally turned to the Christian publisher Zondervan, which in 1970 issued an attractively produced volume titled *Biology: A Search for Order in Complexity.*[59]

The book met with mixed success. Although Zondervan sold out the first printing of ten thousand copies in a year and immediately ran off twenty-five thousand more, few public schools adopted the text. Several state textbook committees, including Indiana's, approved its use, but when one district in southern Indiana decided to rely on just the CRS text, a state court banned further use in public schools. "The question is whether a text obviously designed to present *only* the view of Biblical Creationism in a favorable light is constitutionally acceptable in the public schools of Indiana," declared one of the judges involved. "Two hundred years of constitutional government demand the answer be *no.*" When other states followed Indiana's lead, a CRS committee recommended revising the text to give fairer coverage to evolution, expressing the hope that this could be done without alienating the Christian school market. The committee also recommended deleting biblical references and playing down flood geology while paying more attention to strictly biological issues. But for over a decade nothing came of these suggestions. In the meantime royalties from sales of the text—amounting to $6,500 during the first two years alone—provided the society with the means to fulfill its dream of supporting research on what increasingly came to be called "creation science."[60]

TWELVE

Creation Science and Scientific Creationism

In his preface to the first printing of *Biology: A Search for Order in Complexity* (1970) Creation Research Society (CRS) President Henry M. Morris warned that the text would at times transgress the limits of science—"the organized body of observation and experiment on present processes"—to take up the philosophical question of origins, which dealt with unobservable events in the past. Philosophies of origins, he claimed, came in two guises: "the doctrine of evolution and the doctrine of special creation." The former postulated gradual development over time; the latter assumed "the essentially instantaneous origin of life and of the major kinds of living organisms by special creative processes utilized directly by the Creator Himself." This approach to origins placed creation on an equal footing with evolution by depriving the latter of its scientific standing. It did not assign scientific status to the biblical doctrine of creation.[1]

When the third printing of *Biology* appeared just four years later, a new preface, by John N. Moore, called attention to a novel way of conceptualizing the differences between creation and evolution, the so-called two-model approach. "Scientists," Moore explained,

work in terms of "models," and each proposed model is evaluated in terms of the effectiveness with which available data may be correlated into the model as a frame of reference. Following this example, the two basic viewpoints of origins may be called the "evolution model" and the "creation model." A choice between these two models may be made in terms of the effectiveness with which each may be used to correlate available data.

Although Moore stopped short of applying the recently coined term "creation science" to the creation model and continued, using Morris's own words, to distinguish between the science of biology and the philosophy of origins, his description of model-building as a scientific activity nudged creationism in the direction of science.[2]

The appearance of Moore's preface coincided with the publication of an innovative handbook for high-school teachers, prepared by Morris, called *Scientific Creationism* (1974). In it Morris sought to peel off the biblical wrappings of creationism and repackage it as science. Using the two-model approach, he argued that creationism could "be taught without reference to the book of Genesis or to other religious literature or to religious doctrines." In stressing that public schools should teach only "the basic scientific creation model," stripped of all allusions to its biblical origins, he hoped simultaneously to preempt the field for flood geology and to disqualify competing creation myths. If overt biblical creationism were taught, he feared that it "would open the door to a wide variety of interpretations of Genesis" and produce unwanted demands for the inclusion of non-Christian cosmogonies.[3]

By the mid-1970s the advocates of flood geology, such as Morris and Moore, had securely attached the synonymous tags "creation science" and "scientific creationism" to the Bible-based views of George Mc-Cready Price. This relabeling reflected more than euphemistic preference; it signified a major tactical shift among strict six-day creationists. Instead of denying evolution its scientific credentials, as biblical creationists had done for a century, the scientific creationists granted creation and evolution equal scientific standing. Instead of trying to bar evolution from the classroom, as their predecessors had done in the 1920s, they fought to bring creation into the schoolhouse and repudiated the epithet "antievolutionist." Instead of appealing to the authority of the Bible, as John C. Whitcomb, Jr., and Morris had done in launching the creationist revival, they downplayed the Genesis story in favor of emphasizing the scientific aspects of creationism. Perhaps, as one sociologist has suggested, creationists began stressing the scientific legitimacy of their enterprise because in an increasingly secularized culture "their theological legitimation of reality was no longer sufficient for maintaining their world and passing on their world view to their children."[4] But, as we shall see, the appeal to science arose primarily in response to specific educational and legal developments.

SELLING SCIENCE

In 1963, in a decision prompted in part by the protests of the atheist Madalyn Murray (b. 1919), the United States Supreme Court ruled that mandatory Bible reading and prayers in public schools breached the constitutional wall separating the government from religion. The court did not, however, grant secularists an advantage over religionists. It explicitly endorsed a policy of religious neutrality, which prohibited states from establishing "a 'religion of secularism' in the sense of affirmatively opposing or showing hostility to religion." Thus in a landmark victory for atheists and others opposed to public expressions of religion, the court ironically opened the door for aggrieved Christians to seek legal protection whenever they felt the public schools were teaching views, such as evolution, hostile to religion.[5]

Among the first to seize this opportunity was a Baptist mother from southern California, Nell J. Segraves (b. 1922), troubled by some of the things her children were learning in school. Murray's success in shielding her son from unwelcome religious exposure suggested to Segraves that creationist parents such as herself could also use the law to shield their offspring. Together with Jean E. Sumrall (b. 1927), a Missouri Lutheran friend and former Sunday-school student of Walter E. Lammerts with whom she was associated in both the Bible-Science Association and the CRS, Segraves petitioned the California State Board of Education to require that evolution be designated a theory in all state-approved biology texts. Their efforts elicited a positive response from the U.S. attorney general's office and from the California Superintendent of Public Instruction, Max L. Rafferty, (1917–1982), who in 1966 encouraged the two women to demand equal time for creation. His reading of the 1964 Civil Rights Act, which included a provision allowing teachers to mention religion as long as they did not promote specific doctrines, prompted his suggestion. Besides, the creationists suspected that the Supreme Court would momentarily declare restrictions on the teaching of evolution to be unconstitutional. Late in 1965 Susan Epperson (b. 1941), a young biology teacher in Little Rock, had challenged the 1928 Arkansas law banning instruction in evolution. Given the judicial climate, creationists expected that the Supreme Court would strike down the old statute, which it did in 1968.[6]

Segraves and Sumrall failed in their first effort to persuade the Board of Education to incorporate creation into the curriculum, but a second opportunity arose in 1969. In the fall of that year an advisory committee on science education submitted to the board a document titled *The Science Framework for California Schools*. The dogmatic tone of the section on teaching evolution offended several board members, including John R. Ford (b. 1923), a Seventh-day Adventist physician from San Diego and one of the few blacks prominent in the creationist movement. The ensuing controversy caught the attention of Vernon L. Grose (b. 1928), a Pentecostal systems expert in the aerospace industry and a member of the American Scientific Affiliation. Though not a young-earth creationist, he offered substitute wording for the *Framework* that satisfied, though hardly pleased, Segraves and Sumrall as well as the board. "While the Bible and other philosophical treatises also mention creation, science has independently postulated the various theories of creation," read the revised *Framework*, released in 1970. "Therefore, creation in scientific terms is not a religious or philosophical belief." Such language inflamed evolutionists and kept the California textbook controversy raging throughout the early 1970s. The dispute eventually ended in a draw: Evolutionists kept creation out of public-school biology texts, but creationists succeeded in demoting evolution to the level of a mere speculative theory.[7]

The transmogrification of creationism from religion to science took place in direct response to the events in California, which encouraged creationists to believe that they could squeeze into science classrooms simply by shedding superfluous biblical weight. "Creationism is on the way back," announced Morris, "this time not primarily as a religious belief, but as an alternative scientific explanation of the world in which we live." The new labels for this alternative science first appeared about 1969. In anticipation of a favorable ruling by the California State Board of Education, Segraves, Sumrall, and other associates of the Bible-Science Association in southern California set up Creation Science, Inc., to prepare creationist textbooks. In 1970 this organization merged with the planned creation studies center at Christian Heritage College in San Diego to form the Creation-Science Research Center. Morris, who had agreed to move to San Diego to become academic vice president of the college if he could also organize a creation center there, served as director. In the fall Morris offered a course at Christian Heritage titled "Scientific Creationism," apparently his first public use of the term. In the September 1971 issue of the *Creation Research Society Quarterly*, he introduced the two-model approach to his colleagues in the CRS, arguing that evolution and creation were equally "scientific" and equally "religious." Shortly thereafter he described evolution and creation as "competing scientific hypotheses." At the spring 1972 meeting of the CRS board, members were instructed to begin using "scientific creationism," a phrase creationists came to use interchangeably with "creation science." As Morris explained, some creationists preferred the former because they believed that neither evolution nor creation was a demonstrable "science." Others, who felt that "creationism" sounded too religious, preferred the latter. Morris himself deemed neither term ideal, "for it is not possible to use any one simple term to identify such a complex and comprehensive subject."[8]

One of the most precise explications of creation science appeared in a 1981 Arkansas law mandating "balanced treatment" in teaching creation and evolution. The Arkansas statute specified six cardinal tenets of creation science:

(1) Sudden creation of the universe, energy, and life from nothing; (2) The insufficiency of mutation and natural selection in bringing about development of all living kinds from a single organism; (3) Changes only within fixed limits of originally created kinds of plants and animals; (4) Separate ancestry for man and apes; (5) Explanation of the earth's geology by catastrophism, including the occurrence of a worldwide flood; and (6) A relatively recent inception of the earth and living kinds.

At first glance this looks suspiciously like old-fashioned biblical creationism. But there are significant and subtle differences. Few biblical creationists before the 1960s would have included an appeal to geological catastrophism. Scientific creationists, in contrast, identified the Genesis flood as "the real crux of the conflict between the evolutionist and creationist cosmologies." According to Morris, the scientific creationist also excluded such biblical data as "the six days of creation, the names of the first man and woman, the record of God's curse on the earth because of human sin, the story of Noah's ark, and other such events which could never be determined scientifically." At the same time they included some evidence, such as the fossil record, not explicitly mentioned in the Bible.[9]

Although scientific creationists insisted that creationism could be taught without reference to the Bible, they did not deny that the Bible had molded their thinking. Morris, for example, readily conceded that "*only* in the Bible can one find this concept of special creation." Practically, scientific and biblical creationism differed so little that the book

Scientific Creationism came in two nearly identical editions: one for public schools, containing no references to the Bible, and another for Christian schools, to which was added a chapter, "Creation according to Scripture."[10]

Not all creationists, even flood geologists, agreed on the essential elements of scientific creationism or on the desirability of soft-pedaling its religious roots. The Arkansas law, for example, required a young earth, but Morris, for one, assigned that concept to biblical, not scientific, creationism. His sometime collaborator, Whitcomb, expressed reservations about the entire repackaging effort. In his opinion, so-called scientific creationism sacrificed both certainty and Christianity. "One might just as well be a Jewish or even a Muslim creation scientist as far as this model is concerned," he observed disgustedly. Simply for public acceptance, Morris had sold the creationist birthright for a mess of institutional pottage. "By avoiding any mention of the Bible, or of Christ as the Creator, we may be able to gain equal time in some public school classrooms," wrote Whitcomb. "But the cost would seem to be exceedingly high, for absolute certainty is lost and the spiritual impact that only the living and powerful Word of God can give is blunted." He, too, wanted to reach "the millions of students who are being systematically brainwashed in evolutionary humanism in public schools and universities," but he refused to deny or downplay the biblical basis of creationism to do so. "Far from being a hindrance and an embarrassment to scientific creationism," biblical theology "is actually its only source of final authority, power, and victory," he argued.[11]

Even when promoting their beliefs as science, creationists sometimes denied similar legitimacy to evolution. They did so by limiting science to Baconian fact-gathering or, more typically, by appealing to dictionary definitions of science as factual knowledge. As late as the 1980s, two creationists assured readers of the *Creation Research Society Quarterly* that evolution could "be beaten to a pulp with the dictionary." But with the shift to a scientific apologetic, creationists increasingly took more interest in establishing scientific equality with evolutionists than in evolution-bashing. Thus references to Francis Bacon gave way to citations to two new philosopher-heroes: Karl R. Popper (b. 1902) and Thomas S. Kuhn (b. 1922). In the process creationists sacrificed the possibility of absolute truth for "the paraphernalia of conceptual relativism," as one worried creationist phrased it, but most fellow believers seemed neither to notice nor to care.[12]

To solve the age-old problem of distinguishing science from metaphysics or pseudoscience, Popper invoked the criterion of falsifiability as a substitute for the less rigorous test of verifiability. To qualify as science in his terms, a theory had to be refutable by experience. Because Darwinism failed to satisfy this requirement, he denied it scientific status. Instead, he classified it as a *"metaphysical research programme,"* which, he explained, might nevertheless serve science well by providing "a possible framework for testable scientific theories." Popper's formulation handed creationists a two-edged weapon, which they could employ either to disqualify evolution as science or to promote creation as an alternative metaphysical research program. When Popper got wind of how creationists were using him, he hastily affirmed that theories regarding the history of life on earth were scientific because "their hypotheses can in many cases be *tested*." But his clarification, which stopped short of granting scientific standing to Darwinism, had no noticeable effect on his popularity among creationists, who continued to engage in what one sociologist derisively called "Popper-chopping."[13]

The historian and philosopher of science Kuhn, in his immensely influential book *The Structure of Scientific Revolutions* (1962), described scientific development partially in terms of competing models or paradigms rather than the accumulation of objective knowledge. In his scheme, scientific revolutionaries, even while constituting only a small embattled minority, occupied scientific ground. Thus creationists saw no reason why their model of origins, though supported by a minuscule fraction of practicing scientists, should not be accorded space within the scientific curriculum along with the majority view of evolution. They also liked Kuhn's description of the various factors that influenced scientists in selecting one paradigm over another:

> Individual scientists embrace a new paradigm for all sorts of reasons and usually for several at once. Some of these reasons—for example, the sun worship that helped make Kepler a Copernican—lie outside the apparent sphere of science entirely. Others must depend upon idiosyncrasies of autobiography and personality. Even the nationality or the prior reputation of the innovator and his teachers can sometimes play a significant role.[14]

Not surprisingly, the American scientific establishment refused to share its hard-won position of privilege and power with what it regarded as a disreputable band of scientific pretenders and religious zealots. Ignoring centuries of history that found the present-day categories of "science" and "religion" thoroughly entangled, the council of the National Academy of Sciences in 1981 declared prescriptively that "religion and science are separate and mutually exclusive realms of human thought whose presentation in the same context leads to misunderstanding of both scientific theory and religious belief." In a mass-circulated booklet, *Science and Creationism* (1984), the academy contrasted the provisional and testable claims of scientists with the unchanging and unsubstantiated conclusions of creationists. "Examples of events changing scientific thought are legion," claimed the authors, apparently unmindful that the same could be said of creationist thought. In fact, creationist opinion probably changed more radically in the half-century from 1930 to 1980 than views of evolution.[15]

Critics of creationism tried to head off the move toward science by erecting various definitional barriers that excluded creationists from scientific territory. For example, one entomologist argued that scientific creationism was not scientific because it demanded "acceptance of a premise that lies outside of science," namely, that "a particular interpretation of particular passages of scripture be accepted as constituting an absolute truth." A geologist insisted not only that creation science was not science but also that creation scientists were not scientists, regardless of their degrees. They were not scientists, he explained, "precisely because they have abandoned the scientific method and the scientific attitude, criteria far more crucial to the definition of scientist than the location or duration of one's training or the identity of one's employer." Such "absurd and arbitrary" pronouncements made Morris long for the good old days when science meant knowledge, not naturalism or secularism.[16]

Some evolutionists faulted the creationists for simultaneously claiming "to have given evidence that the theory of evolution *is false* and also that it *is not falsifiable*." But creationists were not the only ones guilty of such tactics. The very author who accused the creationists of logical inconsistency, a philosopher, implied that creationism was not scientific because it failed to satisfy Popper's criterion of falsifiability. Yet his comments appeared in a collection of essays in which scientific colleagues argued that creationism had been falsified by the findings of anthropology, biology, geology, and physics. "An examination of both the fossil record and living organisms falsifies the hypotheses of creationists," concluded a contributor from botany. Similarly, in *Science and Creationism*, the authors asserted that "the hypothesis of special creation has, over nearly two centuries, been repeatedly and sympathetically considered and rejected on evidential grounds by qualified observers and experimentalists." But just four pages later the same writers claimed that special creation was not "a testable hypothesis for the origin of the universe, the earth, or of life thereon." The Harvard paleontologist Stephen Jay Gould (b. 1941), who at times also described scientific creationism as both false and unfalsifiable, attempted to harmonize such apparently contradictory opinions by pointing out that "creationism reveals its nonscientific character in two ways: its central tenets cannot be tested and its peripheral claims, which can be tested, have been proven false."[17]

In selling the two-model approach to school boards and state legislatures, creationists repeatedly appealed to the scientific status of creationism and the scientific standing of its leading proponents. "Stress that creationists are not proposing to teach the 'creation story of Genesis' in the schools," advised Morris, "but only to show that the facts of science can be explained in terms of the scientific model of creation." An enthusiastic member of the Bible-Science Association board vigorously drove the message home:

> Sell more SCIENCE. . . . Who can object to teaching more science? What is controversial about that? . . . do not use the word "creationism." Speak only of science. Explain that withholding scientific information contradicting evolution amounts to "censorship" and smacks of getting into the province of religious dogma. Use the "censorship" label as one who is against censoring science. YOU are for science; anyone else who wants to censor scientific data is an old fogey and too doctrinaire to consider.[18]

This approach proved extremely effective, at least initially. The state legislatures of Arkansas and Louisiana and various school boards adopted the two-model approach. An informal poll of school-board members in 1980 showed that only 25 percent favored teaching nothing but evolution. In 1982, however, a federal judge declared the Arkansas law, requiring the "balanced treatment" of creation and evolution, to be unconstitutional. Three years later a court in Louisiana reached a similar judgment. The United States Supreme Court upheld these decisions in 1987, while allowing, in the words of one justice, that "teaching a variety of scientific theories about the origins of humankind to schoolchildren might be validly done with the clear secular intent of enhancing the effectiveness of science instruction."[19]

The creation trials of the 1980s hinged on the question of whether creation science was really science or religion masquerading as science. Because the United States Constitution, as interpreted by the Supreme Court, banned the teaching of religion—but not *bad* science—in public schools, opponents of the Arkansas and Louisiana laws, to win in court, needed to show that creation science was inher-

ently religious. During the Arkansas trial philosopher of science
Michael Ruse (b. 1940) instructed the judge on the Popperian method
of separating science from nonscience. Under Ruse's tutelage Judge
William R. Overton (1939–1987) decided that "the essential characteristics of science" included naturalness, tentativeness, testability, and
falsifiability. Because creation science failed to meet these criteria, the
judge concluded that it was not science and that teaching it was unconstitutional because it served only to advance religion.[20]

Most anticreationists saw only cause to celebrate in the Little Rock
victory, but the arbitrary way in which Ruse and Overton drew the
contested boundaries of science worried some scholars. In highly negative reviews of Ruse's courtroom performance, Larry Laudan (b.
1941), a fellow philosopher of science, charged his colleague with "unconscionable behavior" for failing to disclose the vehement disagreements among experts regarding scientific boundaries in general and
Popper's lines in particular. By emphasizing the nonfalsifiability of
creationism in order to deny its scientific credentials, argued Laudan,
Ruse and Overton had neglected the "strongest argument against Creationism," namely, that its claims had already been falsified. "The core
issue is not whether Creationism satisfies some undemanding and
highly controversial definitions of what is scientific; the real question is
whether the existing evidence provides stronger arguments for evolutionary theory than Creationism." Laudan, who dismissed the demarcation question as a "pseudo-problem" and a "red herring," also
pointed out that dogmatism often plays a constructive role in science
and that critics of creationism helped themselves little "by pretending
that science is characterized by an uncompromising openmindedness."
Ruse, in rebuttal, dismissed Laudan's strategy as "simply not strong
enough for legal purposes."[21]

In the opinion of some sociologists of science, including Thomas F.
Gieryn, the Arkansas trial and related activities provided a revealing
glimpse of scientists vigilantly guarding their boundaries. Just as creationists pushed to expand the limits of science to accommodate their
religiously inspired agenda, so their opponents invoked a narrow definition of science to maintain their "monopoly over the market for
'scientific' knowledge in Arkansas schoolrooms." By discrediting the
creationists as "pseudo-scientists" unworthy of public patronage, argued Gieryn and his collaborators, the established scientific community hoped to eliminate a politically powerful competitor for scarce
resources. Whatever the respective merits of the two sides, their struggle illustrated the historically contingent nature of "science" and the
futility of assigning the term an invariant meaning.[22]

The Arkansas trial also shattered myths about the so-called warfare between science and religion. As Judge Overton pointed out in his opinion, the plaintiffs (who opposed the teaching of creation science) included "the resident Arkansas Bishops of the United Methodist, Episcopal, Roman Catholic and African Methodist Episcopal Churches, the principal official of the Presbyterian Churches in Arkansas, other United Methodist, Southern Baptist and Presbyterian clergy," and various Jewish organizations. Joining them were one high-school biology teacher and one scientific society, the National Association of Biology Teachers. No religious groups appeared on the list of defendants. Testifying at the trial against creationism, in addition to four scientists and various other experts, were a Methodist bishop, a Catholic priest, a Protestant theologian, and an evangelical church historian. In contrast, most of the witnesses for the defense (supporting creation science) were well-credentialed, if not all well-known, scientists. Given the surprising composition of the two sides, the theologian-participant Langdon Gilkey (b. 1919) characterized the controversy as involving "two bizarre, unaccustomed and visibly uneasy *partnerships*: on the one side a union of what we might call elite religion and elite science, and on the other side a union of 'popular' (fundamentalist) religion with 'popular science.' " There may have been conflict in Little Rock, but it scarcely conformed to any simplistic science-versus-religion formula.[23]

CREATION RESEARCH

The phrase "creation research," as critics tauntingly charged and creationists occasionally conceded, bordered on the oxymoronic. Even Lammerts, who coined the term, admitted the impossibility of researching creation because "we were not there to watch God do it!" As he explained to a quizzical ASA officer, the CRS investigated extant evidence of God's creative activity, not the act of creation itself. For years creationists tended to expend their limited resources on library research (to find fallacies or inconsistencies in the writings of evolutionists) or on low-cost field studies (to find evidence that might support flood geology). In 1967 the CRS established a research committee, which raised $700 in its first year, but larger amounts did not become available until the early 1970s, when royalties from the biology textbook began flowing into the treasury. In the early 1980s, frustrated by repeated failures to encourage coordinated research efforts, the CRS voted to set up its own research facilities: a Grand Canyon Experiment

Station in Arizona, to investigate canyon formation and plant succession, and a Grasslands Experiment Station in Oklahoma, to study the survival of organisms in simulated floodlike conditions.[24]

Although creation research more frequently utilized libraries than laboratories, some notable exceptions appeared in the literature. For example, the early volumes of the *Creation Research Society Quarterly* featured experimental reports from Lammerts intending to cast doubt on the likelihood of beneficial mutations required by evolutionary theory. Using the facilities of the Lawrence Radiation Laboratory in Livermore, California, he induced mutations in roses by irradiating buds with high-energy neutrons. He found that the resulting changes stayed within fixed limits and rarely, if ever, produced commercially desirable features. "The fact that they change only within the limits of their variability potential and that most mutations are harmful should make evolution minded scientists reconsider their basic asumptions," he concluded.[25]

Of greater significance, both scientifically and symbolically, were the experimental findings of Robert V. Gentry (b. 1933), who identified radioactive halos in primitive granite as "God's fingerprints," left at the time of creation. A hyperintense, one-track man, Gentry earned a master's degree in physics from the University of Florida before taking up employment in the defense industry. While conducting nuclear-weapons research in 1959, he fell under the spell of a television evangelist, who converted him to Seventh-day Adventism and strict creationism by pointing out the linkage between a six-day creation and the seventh-day Sabbath. Gentry subsequently entered the doctoral program in physics at the Georgia Institute of Technology but left in disappointment after two years when the department chairman refused to let him work on the age of the earth for his dissertation. By this time Gentry had become convinced that halos—microscopic multicolored rings in rocks produced by radioactive decay—might be "the key" to unlock the truth about the age of the earth and might, as he informed Morris, vindicate flood geology. Working with a small microscope in a back-room "laboratory" at home, he focused his attention on halos formed by the decay of polonium isotopes, which have brief half-lives ranging from microseconds to days. Before long he recognized that he was staring at a "true enigma": If "the Precambrian granites containing these special halos had crystallized gradually as hot magma slowly cooled over long ages," as evolutionists maintained, then "the radioactivity which produced these special radiohalos had such a fleeting existence that it would have disappeared long before the hot magma had time to cool sufficiently to form a solid rock."[26]

His epiphany came on a spring afternoon in 1965. As he puzzled over the origin of the halos, the idea struck him that perhaps the three-minute half-life of polonium 218 represented the interval between God's creation of the chemical elements and the formation of the granites. *"In my search for the truth about the age of the earth, had I discovered evidence for its instantaneous creation?"* he wondered. *"Were the tiny polonium halos God's fingerprints in Earth's primordial rocks? Could it be that the Precambrian granites were the Genesis rocks of our planet?"* Stunned by such grandiose thoughts, he reflected on how they might revolutionize the way in which scientists viewed evolution. Just as Price over a half-century earlier had used his insights into deceptive conformities and thrust faults to compress the millions of years spanned by the fossil-bearing rock into the brief span of Noah's flood, so Gentry hoped that his hypothesis about the instantaneous creation of the granites would collapse "several billion years of earth history to almost nothing."[27] (See chart, last page of insert.)

In 1969 the Oak Ridge National Laboratory, intrigued by the possible implications of Gentry's research on halos for discovering superheavy elements, invited him to use its facilities as a guest scientist while retaining his affiliation with an Adventist college in Maryland that had employed him for the past three years. Since beginning his work on halos, he had resolved to submit his findings only to established scientific journals as a means of guaranteeing that religious bias would not contaminate his results. At first he naïvely hoped that he could sneak his creationist message into his articles, but he quickly learned otherwise. In a manuscript submitted to *Applied Physics Letters*, he noted the difficulty of reconciling his results with conventional views of the earth's crustal formation and suggested that the polonium halos were "more nearly in accord with a cosmological model which would envision an instantaneous fiat creation of the earth." An alert referee spotted this "wild speculation" but nevertheless recommended that a revised version of the article be submitted to *Nature*, which eventually published the piece. By the time Gentry moved to Oak Ridge, he had already reported his empirical findings—minus their creationist implications—in some of the most prestigious science journals in the world, including *Science*, and he continued to do so into the early 1980s, when the notoriety associated with his testimony at the Arkansas trial closed such doors and terminated his relationship with the Oak Ridge laboratory. Unable thereafter to obtain public support for his research, he accepted the patronage of the Adventist baker and philanthropist R. Ellsworth McKee (b. 1932), maker of Little Debbie cookies and cakes.[28]

In the late 1970s Gentry dramatically challenged the scientific community to a test of his claims. If anyone could synthesize even "a hand-sized specimen of a typical biotite-bearing granite," like the halo-bearing rock he associated with the divine creation of the earth, he promised to accept the evidence as falsification of his thesis. To his dismay, skeptical scientists simply dismissed his proposal as "a worthless experiment" that would only demonstrate the ability to produce a large piece of granite synthetically. "As far as I am concerned, Gentry's challenge is silly," sneered the geologist G. Brent Dalrymple (b. 1937). "He has proposed an absurd and inconclusive experiment to test a perfectly ridiculous and unscientific hypothesis that ignores virtually the entire body of geological knowledge." In Dalrymple's expert opinion, Gentry had done no more than pose "a very tiny mystery."[29]

Despite Gentry's starring role at the Arkansas trial and his heroic status in some fundamentalist circles, a number of creationists, including fellow Seventh-day Adventists, disparaged his work. Several Adventist scientists accused him of willfully ignoring pertinent evidence and of inconsistently and arbitrarily assuming nonuniform decay rates for all radioactive isotopes except polonium. "If, because of theological reasons, he claims that at the Creation, the Fall, and the Flood, the decay rate for uranium changed, shortening the time required to form uranium halos, then he has left the domain of science," they argued. Other creationists, annoyed by his frequent whining about discrimination, suggested that his scientific snubs resulted more from his own abrasive style than from his peculiar beliefs.[30]

Most creation researchers eschewed the laboratory in favor of back-porch theorizing. A favorite topic was the cause of the flood. In the early 1970s the editor of the *Creation Research Society Quarterly* complained about being swamped by zany models of the deluge. The most noteworthy of such ideas came from Donald W. Patten (b. 1929), an M.A.-level geographer in Seattle. In *The Biblical Flood and the Ice Epoch* (1966) Patten appealed to "astral catastrophism"—in which gravitational and magnetic forces suddenly and simultaneously disturbed the air, oceans, and magma—to explain the deluge and its aftermath. His speculations provoked considerable debate, but they failed to convince many leaders of the CRS.[31]

Far more influential was Thomas G. Barnes's revolutionary hypothesis concerning the decay of the earth's magnetic field, hailed as "one of the most elegant ideas to come out of modern Creationism." Assuming an exponential rate of decay in the earth's main magnetic field and ignoring evidence of fluctuations over time, the physicist estimated that "the life of the earth's magnetic field should be reckoned in thousands, not millions or billions, of years." And because a stronger magnetic field in the past would have shielded the earth from cosmic radiation and produced lower amounts of carbon 14, he argued that the long ages given by radiocarbon dating were invalid. Despite some harsh criticism from both evolutionists and creationists—one of the latter described Barnes's theory as "riddled with problems and major inconsistencies"—many flood geologists came to regard it as proof that the earth was no older than ten thousand years. One authoritative work on creation science ranked it number one in a list of sixty-eight scientific arguments for a young earth. Barnes, who served as president of the CRS during the mid-1970s, enhanced his reputation as a bold thinker by also repudiating relativity and quantum theory, with their associated notions of relativism and indeterminancy. "If he were not a creationist," wrote one admirer hyperbolically, "Dr. Barnes would undoubtedly receive a Nobel Prize for his excellent work."[32]

During the late 1960s and early 1970s the CRS research effort was headed by Larry G. Butler (b. 1933), a self-effacing "country boy" from Oklahoma who earned a Ph.D. in biochemistry from UCLA in 1964. Two years later he joined the department of biochemistry in the Purdue University School of Agriculture, which gave him a secular platform envied (and exploited) by fellow creationists. Reared in a family of devout Southern Baptists, he subsequently affiliated with the more fundamentalist Regular Baptists, the denominational home of Duane T. Gish. While teaching briefly at Los Angeles Baptist College in the mid-1960s, he joined the CRS. In 1969, shortly after his election to the board of directors, he accepted the vice presidency of the society and the chairmanship of the Committee on Research. Although he found the rationale for flood geology convincing, he attached little weight to arguments for a young earth; he embraced strict creationism primarily because of his desire to read the Bible as "straightforwardly" as possible and because of the absence of convincing evidence for a mechanistic origin of life. He became actively involved in the movement more because of others' expectations than because of his own "overwhelming conviction." Nevertheless, for the better part of a decade he conscientiously occupied one of the most visible and influential positions in organized creationism.[33]

One of the first research proposals to cross Butler's desk came from Arthur C. Custance, who remained skeptical of the evidence for flood geology. Apparently confused by all the talk of "upside-down" strata, a phrase creation scientists used to designate the out-of-order rocks that geologists attributed to overthrusting, the Canadian suggested that the CRS undertake a study of "the supposed overturned beds (of large area) to see if there were particular kinds of fossils that had been fossilized in an upside-down posture," for example, with their legs sticking up. Even after being told by CRS leaders that truly upside-down fossils would be found only "in strata that have actually been folded to the point where they have literally overturned," a phenomenon rarely associated with overthrusting and not generally disputed by flood geologists, Custance remained convinced that his proposed research "would settle the issue once for all."[34]

As head of research for the CRS, Butler sought to "present an image of scientific respectability as much as possible without Biblical compromise." Above all, he hoped "to exclude authentic psychopaths, cranks, & kooks" looking for a forum to present their farfetched ideas. As Butler quickly discovered, too many creationists suffered from a weakness for the sensational: "We make astonishing observations (human footprints contemporary with dinosaurs); we postulate dramatic upheavals (sudden deposits of masses of ice from a planetary visitor); we propose sweeping scientific generalizations (negation of the entire system of 14C dating)." Even among fellow board members he detected a fondness for "fanciful" ideas. Shortly after the board met in 1971, he complained to Morris of hearing "everything from a statement that the earth does not appear to be 2000 years old, to an enthusiastic but unintelligible explanation of why the first days of creation were indefinite periods of time."[35]

Despite Butler's pointed criticisms of creation research, he strongly supported a recent special creation in six natural days. Thus Morris in the early 1970s anointed him as his successor to the CRS presidency. Morris could think of only three suitable candidates who held "strong professorial positions in secular universities"—Moore at Michigan State, Barnes at Texas Western, and Butler at Purdue—and of the three only Butler possessed an earned Ph.D. But Butler demurred. Though honored by Morris's confidence in him, he had no desire for higher office and no interest in being used for symbolic purposes. Besides, he questioned the direction in which organized creationism seemed to be moving. "I consider the efforts to force, by law, the teaching of creationism in public schools to be a disaster," he confided to an acquaintance in 1972. "It seems to me that creationism taught by a hostile teacher is worse than it not being taught at all."[36]

Horrified by the continuing influence of what he called "the lunatic fringe" of creationism, Butler in January 1972 circulated a stunning "Critique of Creationist Research" to CRS board members. In it he damned creationists for their "negative, critical perspective" toward evolution, for their delusional search "for some dramatic discovery which will completely overthrow the theory of evolution once and for all at a single blow," and for their emphasis on historical and descriptive evidence rather than experimental and manipulative studies. "Investigations relying on evidence involving affidavits, sworn statements, etc., are not scientific," he observed. Given the difficulties associated with investigating processes no longer in operation, he found it remarkable that "creationist research" existed at all. It reminded him of Samuel Johnson's observation in the eighteenth century that female preachers were "like a dog walking on his hind legs: One is not surprised that it is done badly; the surprising thing is that it is done at all." Continuing to draw on historical illustrations, Butler suggested that creationists might emulate Louis Pasteur's approach to the problem of spontaneous generation—using failure to achieve an effect as evidence against the phenomenon—"and try to prove that evolution is true!"[37]

Although a few CRS leaders congratulated Butler on his jeremiad, it produced little tangible change, and the crusading biochemist increasingly felt like "a thorn in flesh" of his brethren. Discouraged by the failure of his efforts to raise the scientific standards of creationist research, he resigned from the board of directors in 1975 and later allowed his membership in the society to lapse. During the ensuing years he came more and more to suspect that God had used gradual evolutionary development in creating the world but had kept the exact mechanism a mystery. His own research on enzymes revealed "amazing similarities . . . which suggest strong inherited relationships between organisms as different and distinct as bacteria and snakes and higher animals." Along with discarding belief in the recent beginning of the world, he jettisoned any expectation of an imminent end. Convinced that humans would inhabit the earth for a very long time to come, he turned his reforming impulses toward the "morally wrong" agricultural policy of the United States. He remained a Christian but, for reasons unrelated to creationism, quit attending church.[38]

THIRTEEN

Deception and Discrimination

Critics of creation science have often accused its devotees of lacking "a self-critical, self-policing ethos like that of mainstream scientists" or, worse, of systematically twisting and distorting evidence and sometimes even lying. One sociologist has recently argued that in contrast to the larger scientific community, which has agencies to "ferret out deception" and punish offenders, creation scientists are "unwilling to punish systematic deception in their very midst." Historically, there is some truth to these charges, though the abuses are less prevalent than the above comments imply. And rather than condoning sloppy or deceptive work, leaders such as Larry G. Butler (and even Henry M. Morris and Walter E. Lammerts in their own ways) have privately agonized over what one embarrassed creationist called the "low grade or pseudo-science [that] has been published by individuals who call themselves Creationists." As we shall see, some of the most telling criticisms of creation science have come from creationists themselves and have appeared in their own journals.[1]

CLIFFORD L. BURDICK AND
THE INTEGRITY OF CREATION RESEARCH

No one epitomized Butler's concerns about creation science more than Clifford L. Burdick, the most energetic researcher in the Creation Research Society (CRS) and the most frequent recipient of its funding. And, not surprisingly, no one took greater offense at Butler's critique of creationist research. "Larry I dont feel that we voluntarily seek sensationalism," protested the aging geologist, whose fame rested in large part on his extravagant claims about human and dinosaur tracks found in the same ancient rocks. "If this seems sensational I feel its only because its the kind of evidence that God in his providence has dumped in our laps." Sensationalist or not, Burdick, according to Morris, for a decade did "more research for the Society, at greater personal sacrifice, than anyone else, all of it in areas of critical importance." During the 1960s and 1970s he not only continued his search for out-of-place fossil footprints, but repeatedly scoured Mount Ararat for Noah's ark, investigated reported thrust faults, and announced an evolution-shattering discovery of pollen from modern plants in Precambrian rock. To the consternation of creationists seeking scientific respectability, controversy dogged his every step.[2]

Since his association with the Deluge Geology Society in the 1940s, Burdick had eked out a living as a handyman and consulting geologist in Arizona, hustling friends and acquaintances to invest in his various mining ventures or to underwrite his creationist schemes. In the mid-1950s, partly to hone his skills in defending his mentor George Mc-Cready Price against the criticisms of Harold W. Clark, the sexagenarian returned to the classroom to study geology and paleontology at the University of Arizona. In the fall of 1956 he reported to a friend that he was carrying four courses, each with a laboratory section: "I am continually gathering material from these courses that could be the basis of a number of papers against evolution as it concerns geology, stratigraphy, etc." Although never formally admitted to a degree-granting program, by 1960 he had completed enough advanced work to sit for the comprehensive examinations required for a Ph.D. in geology. Three days before the scheduled orals a professor on his examining committee discovered an article on flood geology that Burdick had written for an Adventist magazine, *Signs of the Times*, and reportedly announced that he could never vote to award a doctorate to the author

of such a scientifically heretical work. The news of this discovery pan-
icked Burdick. For years he had carefully concealed his creationist
learnings—and his earlier failure at the University of Wisconsin—and
he reckoned the odds at a thousand to one that his professors might
discover his true thoughts. "I might go thru a dozen more institutions
without they [sic] ever getting next to my inner feelings," he later said
wistfully.[3]

"I saw the handwriting on the wall; I knew my goose was cooked,"
he wrote of the traumatic ordeal:

> The emotional shock induced a severe case of acute indigestion,
> and I was unable to eat hardly anything for the three days prior to
> the test. The graduate school granted me a postpon[e]ment until
> I got back on my feet, but the geology dept. would not O.K. it,
> even though I was sick. I think they really wanted to take advan-
> tage of that opportunity to "scrub" me. I sensed the air of hostility
> the moment I entered the examination room. I should have stayed
> in bed.
>
> I "browned out" several times during the exam, and could not
> answer even the most simple questions, that I knew as well as my
> own name. Even at that I was told I passed as far as knowledge of
> geology went, but I just ran out of gas and could not answer the
> reasoning questions, and being sick did not make too good an
> impression.

The committee, no doubt grateful that Burdick's poor performance
spared them the embarrassment of passing a student who repudiated
the very foundations of historical geology, refused to grant him a second chance, despite repeated appeals and a lawsuit claiming religious
discrimination.[4]

Burdick's case soon became a *cause célèbre* among creationists, incontrovertible proof of the academic prejudice awaiting anyone who
dared challenge the dogma of evolution. "Although he had completed
his doctoral thesis and his oral examination, he was denied his doctorate," went the much-told story. "Because he was a creationist he was
denied his degree and he was no longer welcome at the university."
Privately, Burdick conceded that the *Signs* article was not "the main
reason" for his academic troubles, and he confessed to having failed not
only the Ph.D. oral but a defense of his M.S. thesis as well. He
attributed his second failure to a developing case of hepatitis, brought
on by exposure during a winter field trip to the mountains of Idaho. He
had returned to Tucson to find a letter from the department of geology

asking him to sit for an examination right away. "The medicine I was taking seemed to paralyze my thinking apparatus," he explained. "Neither exam was difficult, I would have passed either one 98 times out of 100. I happened to hit the 2 unlucky times, that's all."[5]

By the time the CRS organized in 1963, Burdick needed the society almost as much as the society needed a geologist. Distressed by his devastating experience at the University of Arizona and by the ostracism of leading scientists in his own church, he turned to the CRS as a "City of Refuge." He had gone back to graduate school in hopes of securing employment at a Seventh-day Adventist institution, only to find himself regarded as "Persona Non-Grata" because of his outrageous and occasionally deceptive claims. He attributed his shunning largely to the machinations of Richard M. Ritland, whom he correctly suspected of campaigning "in the Adventist Denomination to downgrade me so my writings are not as welcome in that group as formerly." But even his old friend Frank Lewis Marsh, with whom he had shared Price's classroom in the 1920s, warned Lammerts about Burdick's "tendency to lean into the fantastic in geology." Marsh feared that Burdick's celebrated fossil footprints would "turn out to be those of the large ground sloth."[6]

When the CRS selected Burdick as its primary geologist, it acquired not only an indefatigable and high-profile researcher but all the liabilities that went with him. First, there was the matter of his credentials. Society leaders seem never to have suspected that the master's degree he claimed from the University of Wisconsin was nonexistent, but they did worry from the beginning about his lack of a doctorate. Lammerts could hardly believe Burdick's "fantastic" tale of the prejudicial treatment he had received at the University of Arizona, but he naïvely offered to intervene on Burdick's behalf. Only then did he learn of the two failures in Tucson. Nevertheless, he agreed to write various professors and administrators at the university. To allay faculty concerns about Burdick's views on the age of the earth and the origin of species, Lammerts asked him if he could not assure the university that he did "not advocate that the earth is only 6,000 years old" and did not insist on the special creation of every species. With Burdick's concurrence and assistance in drafting a letter, Lammerts casuistically informed the relevant parties that "since Clifford Burdick has had more training in geology at Tucson, his attitudes have liberalized considerably. . . . For example, he does not now hold to a 6,000 year age for the earth as some have assumed. Neither does he deny that there has been some evolution." When his efforts failed to produced the desired results and he better understood the university's side of the story, Lammerts urged

Burdick simply to forget the degree and get on with his work on fossil footprints.[7]

Burdick, however, would not give up so easily. In a second effort to obtain a Ph.D. degree from Arizona, he signed on with a paleobotanist working on fossil spores and pollens in the Petrified Forest. But just when his prospects looked brightest, "at the psychological moment Satan interposed, and made it clear again that he has title to all the geological departments of the land." It seems one of the geologists at Arizona received a copy of John C. Whitcomb, Jr., and Morris's *Genesis Flood* as a Christmas present in 1964—and noticed the credits to Burdick for providing pictures of giant human tracks and for reviewing the manuscript. The ensuing furor convinced Burdick that his days at the university were finally over, but he somehow managed to stay on unofficially by telling his professors, apparently at Lammerts's suggestion, that he "was just the free-lance photographer who chanced to sell the pictures to Morris, before I ever knew him." Thus the affair, he happily reported, turned out to be "more or less of a tempest in a teapot."[8]

Into the late 1960s Burdick clung to the hope of obtaining a doctorate from the University of Arizona, or at least of salvaging an M.S. degree. But the university refused to budge. In the meantime he solved the title problem by acquiring, in 1966, a doctor of philosophy in geology degree from the University of Physical Science in Phoenix. Burdick acknowledged the hand of God behind this good fortune, but he also felt that he had done "enough work at Wisconsin and Arizona for one and a half such degrees." Although his diploma looked to Lammerts like a driver's license, and Lammerts suspected that the Phoenix institution was some "sort of 'degree mill,' " he at first agreed that no harm would come from allowing Burdick to affix the Ph.D. to his name when publishing in CRS journals. But when Burdick began claiming that the state of Arizona stood behind his degree, Lammerts resolved to investigate the matter. His inquiries revealed the University of Physical Science to be nothing more than a registered trademark. As described in its own mimeographed bulletin, "The University is not an educational institution, but a society of individuals of common interest for the advancement of physical science. There are no campus, professors, or tuition fee." Indeed, the university had no permanent address or telephone number.[9]

Thoroughly disgusted by his findings, Lammerts instructed Burdick to quit using the fake degree or risk being dropped from the rolls of the CRS. "We simply cannot afford to have any of our members going under false pretenses," he stated. As a compromise, he suggested that

Burdick put "honorary Ph.D." after his name. Unfortunately, the *Creation Research Society Annual* for 1969 had just gone to press, carrying an article coauthored by "Clifford L. Burdick, Ph.D." To atone for this lapse, in the next issue of the *Creation Research Society Quarterly* the editors ran an erratum calling "attention to the fact that the degree identified with Clifford Burdick on p. 49 of the *Creation Research Society Annual*, 1969, is an honorary degree. Therefore the designation 'Hon. Ph.D.' should have been used, as will be found in the footnote identification of Mr. Burdick in this issue." Ironically, the inside back cover of that very issue listed Thomas G. Barnes, D.Sc., and John J. Grebe, D.Sc., as directors, without calling attention to the honorary nature of their degrees.[10]

Of graver concern than Burdick's credentials was his trustworthiness. For years some acquaintances had worried that he let his enthusiasm cloud his judgment when it came to evaluating fossil footprints, but the integrity issue did not come to the fore until the late 1960s and then in association with his research on ancient pollen grains and spores. As mentioned earlier, about 1964 he had joined a team of paleobotanists at the University of Arizona, headed by Gerhard O. W. Kremp (b. 1913), studying microfossils in the Petrified Forest. While working on the project, Burdick became skilled in separating these tiny objects from the rocks in which they were embedded. This prompted Kremp to suggest that Burdick try his hand at analyzing rock samples collected from the Grand Canyon. To the professor's chagrin, Burdick reported finding pollen grains of conifers in rock as old as the Precambrian Hakatai shale. This evidence contradicted evolutionary theory, which did not allow for the appearance of modern pinelike trees until hundreds of millions of years later.[11]

Kremp, suspecting that samples contaminated by later pollen had been used, told Burdick not to publish his conclusions, but the impetuous creationist could not resist an opportunity to disprove the evolutionary time scale. Evolutionists like Kremp might be bullied by theory, but not he. "With me," he bragged to Lammerts, "the facts are more sacrosant [sic] than the theory that houses them." Confident that his discovery would "be enough to bury evolutionary geology forever," he surreptitiously submitted an announcement to Lammerts for publication by the CRS. The report appeared in 1966, accompanied by an editorial blurb touting it as "truly revolutionary." "For several years," wrote Lammerts disingenuously, "I have been following closely Burdick's painstaking work, and more than anyone else realize how difficult it has been for him to consent to the publication of these data so totally at variance with the usual ideas of plant evolution." In recognition of Burdick's sacrificial devotion to "this original and fundamental research discovery," Lammerts urged the CRS to assign highest priority to supporting Burdick's future studies. [12]

Shortly after endorsing Burdick's work, Lammerts began to have second thoughts about his creationist colleague. As editor of CRS publications, he had rewritten enough of Burdick's prose to know that the Arizonan could not write proper English, but the extent of his dullness and ignorance did not sink in until the two men spent several days together at a creation seminar in Arkansas early in 1967. "I was appalled," Lammerts reported to Morris,

> at the slowness of Burdick mentally when at the Creation seminar and hope he is not misleading us on some [of] his opinions. He for instance had evidently never heard of the series of horse-like animals found and was at a complete loss to explain them. Evidently he has not kept up with his reading very much. . . . We have some very good men—only Burdick is weak but maybe his mind just works slowly.

Having risked his own reputation in defense of Burdick's research, Lammerts hoped that Burdick was "really academically honest" and that he did "not have delusions of some easy road to fame." But the "doctors degree stuff" made him wonder. [13]

In 1969, in response to growing doubts about the validity of what Burdick was calling his "science-shaking original-pioneer work in palynology in Grand Canyon," the CRS Committee on Research approved funds to have two independent scientists, neither a creationist, collaborate with Burdick on replicating his pollen studies. By that time Lammerts, convinced that he had caught Burdick in a lie about society matters, was worrying less about contaminated samples than about the possibility of an outright "hoax." Eventually, he concluded that Burdick was just incredibly sloppy, not basically dishonest. Burdick not only botched the follow-up study, but, in reporting his progress, falsely claimed that two scientists from Loma Linda University had corroborated his original claims. Burdick's alleged discovery of Precambrian pollen, regarded by supporters as a "landmark" in the history of creationism, continued to divide creationists into the 1980s, as did the question of Burdick's integrity. In 1981 Arthur V. Chadwick (b. 1943), a creationist biologist who had once worked with Burdick in the Grand Canyon, published a searching critique of Burdick's methods and reported finding not a "single example of an authentic pollen grain" in any of fifty samples taken from the same strata Burdick had studied.

But just a few years later an official CRS investigation, headed by George F. Howe, claimed to have vindicated Burdick's original discovery.[14]

As mentioned earlier, Burdick's credibility also figured prominently in another ballyhooed area of creationist research: giant human footprints found with dinosaur tracks in Cretaceous strata. Because evolutionists insisted that the last dinosaur had perished tens of millions of years before the appearance of the first human, such evidence threatened the very structure of the geological column. Many of the best specimens came from the Paluxy River, near Glen Rose, Texas, where Burdick had been visiting since the 1940s. His inaccurate description of these anomalous impressions had already embarrassed Whitcomb and Morris, who relied on Burdick's testimony in writing *The Genesis Flood* (see Chapter 10), but many creationists, including Whitcomb and Morris, still gave credence to his purported evidence that humans and dinosaurs had once trod the earth together. In response to reports of carved imitation tracks, which further undermined Burdick's credibility, he returned to Texas in the summer of 1968 to confirm his earlier observations. "I believe all reasonable doubts have now been removed, and the evolutionary geologists will be backed into a corner," he assured Lammerts after reviewing the evidence. "One hundred million years collapsed from geologic column, since man and dinosaurs were evidently contemporary."[15]

Despite the importance many creationists were attaching to the giant human tracks, several leading creationists remained skeptical. Wilbert H. Rusch visited the Paluxy River in 1970, "hoping and praying" that he would find some "indisputable evidence" in support of Burdick. But, as he informed Whitcomb, he could only "return the old Scotch verdict 'not proven.'" Like Lammerts and Butler, to name only two, Rusch detested the "apocryphal tales" that so often circulated as creation science. "We need these episodes for our cause like we need a hole in our heads," he wrote. "Premature statements, too strong statements on insufficient evidence do us as much harm." Lammerts, too, harbored serious doubts, but for another reason. "The whole footprint business raises more problems for our side than for evolutionists," he insisted in a letter to Morris. "On the basis of a world wide flood what were people doing *walking* around yet after so much sediment deposited? Burdick has never answered this question nor has any of the footprint enthusiasts."[16]

In the early 1970s Burdick learned that men engaged in surface mining near Moab, Utah, had discovered two human skeletons in Cretaceous rocks, deposited long before the estimated appearance of humans. Burdick visited the site and confirmed to his satisfaction that "the bodies were buried at the time of the emplacement of the sandstone rock," thus providing additional evidence that "the geologic column has been 'built' on too meager and perhaps even a 'flimsy' foundation." Rusch and Duane T. Gish, who followed Burdick on the scene, could not rule out the possibility that the rocks surrounding the skeletons had been disturbed and, thus, that the bones were much younger than the rocks. Butler, as head of the CRS Committee on Research, argued for publishing a report by Gish and Rusch, even if it did not " 'disprove evolution.' " "Perhaps the strongest reason is that if we don't we are liable to be charged with suppressing evidence which does not favor our position," he advised the editor of the *Creation Research Society Quarterly*. "Scientifically, publication of neutral or even slightly negative results could be a very healthy thing to do." No formal Gish-Rusch report ever appeared, but the *Creation Research Society Quarterly* did carry a sensationalist piece by Burdick boldly announcing the "Discovery of Human Skeletons in Cretaceous Formation." Lammerts, however, inserted the following cautionary note:

Admittedly this discovery offers as much of a problem for Flood geologists as for those of the orthodox point of view. For it is difficult to explain how two men could still be alive after such a depth of strata had been deposited. And if already drowned, why were they not buried later in the Mesa Verde formation? A more detailed and clear cut concept of just how the Flood accomplished its work is badly needed in order to be able to see how such finds as these fit into theoretical expectations, or creationists will be guilty of the same ad hoc explanations as evolutionary minded colleagues.

Clearly Lammerts was tiring of constantly having to clean up after one of the biggest stars in the creationist show.[17]

The most telling attack on the alleged hominoid tracks in Texas came from flood geologist and geneticist Berney Neufeld (b. 1941), who in 1975 reevaluated the evidence Burdick and others had cited in favor of human footprints. Extensive field and laboratory studies led him to conclude that the Paluxy River yielded no "good evidence for the past existence of giant men." Nor did it "provide evidence for the coexistence of such men (or other large mammals) and the giant dinosaurs." Neufeld, a loyal Seventh-day Adventist, went out of his way to avoid casting doubt on the flood story in general; he simply disputed Burdick's interpretation of the evidence.[18]

A decade later, after intensive investigation by skeptics and believers alike, Henry M. Morris and his son John, himself a longtime student of the tracks, formally announced what many persons had suspected for years: The ambiguous Paluxy data provided no certain evidence against evolution. In a special supplement to the widely circulated *Acts & Facts*, published by the Institute for Creation Research, John D. Morris (b. 1946) downgraded the puzzling depressions in the Texas river bed to the status of a "mystery" and urged fellow creationists to face up to the unpleasant truth. In a letter accompanying his son's forthright concession, the elder Morris tried to soften the blow to creationist apologetics by assuring his followers that "this question in no way affects the basic creation/evolution issue. These tracks have always been only illustrative, not definitive, and the over-all scientific case against evolution, which is overwhelmingly strong, is not affected in any way."[19]

In many ways Burdick's activities typified creationist research during the renaissance of flood geology following the publication of *The Genesis Flood*. His unrelenting attacks on the geological column, including repeated CRS-funded assaults on thrust faulting, represented the Price tradition at its purest, and satisfied the criterion of concentrating effort "wherever the potential impact in terms of Christian witness would be greatest in relation to expenditures." As Henry Morris once explained, this meant focusing

on geological field studies, especially on so-called overthrust formations and on anomalous fossils. If it can be shown, by studies at the thrust planes, that enough of these "thrust faults" are not really overthrusts but only normal sedimentary sequences, or if it can be shown that several supposed out-of-place fossils have not really been displaced at all but have been normally deposited with the encasing sediments, then the standard geologic age system would eventually have to be abandoned in favor of the creation/flood model of earth history.

For a decade and a half after the founding of the CRS no one contributed more than the eccentric Burdick to furthering this agenda. Unfortunately for the reputation of creationism, no one among the leadership was so lacking in credentials, critical thinking, and caution. Even to creationist colleagues he remained an enigma. "In a sense he represents virtues which we desire: patience, a sacrificial spirit, and a gracious rather than bitter attitude toward his critics," observed a puzzled Butler. "In another sense, however, he epitomizes the very image we wish to avoid: a non-rigorous approach to research, unscholarly, and more important, he seems to seek out and thrive on sensationalism."[20]

PERSECUTION OR FAILURE?

Burdick's name rarely appears in creationist literature without mention of the indignities he supposedly suffered at the hands of evolutionists at the University of Arizona. Because of his tale and similar reports, including the revelations surrounding the notorious Velikovsky affair, the CRS from the beginning kept its membership list confidential. Creationist leaders warned graduate students to keep their views to themselves, "because if you don't, in almost 99 percent of the cases you will be asked to leave." To avoid harassment or worse, several graduate students took to using pseudonyms when writing for creationist publications.[21]

Creationists had good reason to be wary. One public opinion poll showed that well-educated Americans tolerated religious fundamentalists less than any other group. Fifteen percent of the respondents said they would dislike having a fundamentalist move into the neighborhood. Otherwise sensitive college and university teachers ridiculed creationist beliefs, and one professor at Iowa State University went so far as to recommend failing students in geology and biology courses who denied the truth of evolution or the great antiquity of the earth. If such "gross misunderstandings" were discovered after completion of the course or after graduation, he proposed "retracting grades and possibly even degrees." Around such persons, creationists could not feel safe.[22]

Creationist literature abounds with stories of believers who allegedly ran afoul of the scientific establishment. Sinister motives seem to explain virtually every nonadmission to graduate school, every unsuccessful quest for a job, and every rejection of a manuscript. In the mythology of creationism, failure, deserved or otherwise, leads to martyrdom. Typical in this respect are two creationist books that appeared in 1984. In his comprehensive *History of Modern Creationism*, Henry Morris covers not only the obligatory Burdick episode (without any mention of Burdick's poor performance) but the experiences of creationists at Northeastern Oklahoma State University and the University of Michigan who suffered for their faith. In describing his own departure from Virginia Tech, Morris suggests that evolutionists in the geology and biology departments conspired to drive him out of Blacksburg: "I am convinced (though, of course, with no proof) that the subsequent difficulties I began having with the administration were essentially part of an attempt on their part to get me to leave or else to gradually break down the department." Other sympathetic witnesses, however, attribute his problems almost entirely to nonideological, administrative matters.[23]

Gerald R. (Jerry) Bergman (b. 1946), a creationist educator who was denied tenure at Bowling Green State University, devoted an entire book, *The Criterion: Religious Discrimination in America* (1984), to chronicling cases similar to his own. Of more than one hundred creationists interviewed, he writes, "All, without exception, reported that they had experienced some discrimination—including open derision from their colleagues, being made the brunt of jokes, and, in some instances, outright firings." But Bergman's compendium of alleged persecutions often lacks the documentation needed to support his claims. And in a number of celebrated cases additional testimony either falsifies or substantially modifies his account.[24]

As far back as the 1920s, writes Bergman, Byron C. Nelson, later an officer of the Religion and Science Association, "was denied his Master of Science in genetics at Rutgers University, despite an almost straight A grade point average," and all because of his creationist principles. But shortly before his death Nelson told a far different story. While pastoring in New Jersey, he discovered that an old friend from Wisconsin was head of the zoology department at nearby Rutgers University. At his friend's invitation Nelson quietly enrolled in a genetics course taught by a notoriously anticlerical professor, Alan A. Boyden (1897–1986):

> From time to time Boyden would say nasty things about ministers. At the end of the semester I chanced to sit beside him at a lecture. He asked me, "By the way, what is your business, Mr. Nelson?" When I told him I was a Lutheran minister, his face got as red as a beet. Then he urged me to join a seminar class in which I breed [sic] fruit flies by the thousands, attempting to bring about mutations by feeding them arsenic. While in this advanced class I was invited to join the whole faculty once a week and take my turn lecturing to the bunch. One day I was absence [sic] and the whole faculty divided on evolution—encouraged by my taking a stand. The Nelsons and Boydens became very friendly—fellowshipping at picnics etc. He helped me to make some improvements in

AFTER ITS KIND. Boyden had several copies put into the university library.

Though recounting his experience to a fellow creationist who would have been sympathetic to charges of discrimination, Nelson said not a word about being denied a degree.[25]

According to Bergman, two of the founding fathers of the CRS, Burdick and David A. Warriner, paid dearly for espousing creationism. Soon after the founding of the CRS, Warriner lost his untenured position on the faculty of Michigan State University. He ascribed the nonrenewal of his contract directly to the publication of a pamphlet he had written on the origin of life, but even colleagues on the CRS board suspected that other factors had played a more significant role. Rusch, for example, surmised that Warriner's conservative social and political views might have gotten him into trouble, while Lammerts chalked Warriner's and Burdick's predicaments up to their rigid personalities. "Though perhaps it is unfair to say so, I believe that most of the difficulties which have been related such as those of Warriner and Burdick are largely due to other personal problems," he wrote to Rusch. "I believe that many of these men perhaps lack a sense of humor on this whole question."[26]

Because neither Lammerts nor Rusch had personally experienced academic discrimination, they tended to think that many of the complainers were creating their own difficulties. On entering the graduate program in geology at the University of Nebraska, Rusch had feared reprisals from professors who he knew had read one of his essays on creationism. But, to his relief, he encountered no hostility. "I would ask pertinent and sometimes embarrassing questions, they would answer, and that was it," he informed a concerned young creationist. "I did not ever believe that I had the right to make a platform of another man's classroom to deliver a harangue, and they probably appreciated that. At any rate, they remember me with kindness, and I had good relations."[27]

Another frequently mentioned victim of anticreationist prejudice is Ervil D. Clark (1927–1981), the son of Harold W. Clark. Although young Clark had excelled as a graduate student in ecology at Stanford University, writes Bergman, "several members of his Ph.D. committee" conspired to ensure his failure on the oral defense of his dissertation—"because he was a creationist." In recounting the bitter incident, Clark himself expressed far more uncertainty about how much weight to assign to his creationist views. Early in his program he had taken a course on evolution with Paul R. Ehrlich (b. 1932), from whom he received a *B*. Later, when Clark attempted to defend his dissertation on the ecology of a single county in northern California, Ehrlich zeroed in, quizzing him on ecology around the world, about which the young biologist admittedly knew relatively little. From what he later learned, four of the five committee members voted to pass him, but the department required a unanimous decision. The next year he retook the examination with the same results, which led to automatic termination. Clark's supportive major professor attributed his student's weak performance in part to inadequate immersion in the program. While studying at Stanford, Clark had continued to live at home in an Adventist community north of San Francisco, thus missing out on the informal, though invaluable, education that comes from interaction with fellow graduate students. Determined to earn the doctorate his father never possessed, Clark later enrolled in a general science program at Oregon State University, from which he received a Ph.D. in 1971 for a dissertation on radiation biology.[28]

Some creationists have included the Canadian gap theorist Arthur C. Custance in the pantheon of antievolutionist martyrs, claiming that the University of Toronto refused to award him a Ph.D. degree because of his religious beliefs. A mechanical engineer by vocation, Custance earned B.A. and M.A. degrees in biblical languages from the University of Toronto, receiving the latter in 1940. His interest in science and religion led him in 1951 to reenter the University of Toronto to begin working on a doctorate in anthropology. Because of his age (early forties) and his long-standing interest in ancient cultures, the department of anthropology apparently agreed to let him compress a seven-year program into three. Despite some problems over unexcused absences from classes, he thought he was doing well until the oral part of his comprehensive examinations, when a professor asked, "Do you believe Adam and Eve were real people?" When Custance answered in the affirmative, one of the committee members reportedly threatened to resign if they "put through a man for a Ph.D. who believes this kind of crap." Concluding that Custance's lapse in judgment reflected "insufficient scientific training"—and unwilling to have him represent the department as its first Ph.D.—the faculty strongly encouraged him to withdraw, which he eventually did.[29]

Shortly thereafter Custance took a job in Ottawa as head of the Human Engineering Laboratory of the Defence Research Board, where he conducted research on heat stress and sweating for the Canadian military. Still eager to acquire a doctorate, he arranged with a professor in the University of Ottawa School of Psychology and Education to take some evening and weekend classes and submit a previously written thesis titled "Does Science Transcend Culture?" In 1959 he received his Ph.D. for this work, though later he liked to say that he had been "invited to complete his Ph.D. as a guest of the Physiology Dept. in the Medical School at Ottawa University."[30]

Two of the most celebrated cases of creationist persecution in Great Britain concerned a schoolteacher, David C. C. Watson (b. 1920), and a biologist, Arthur J. Jones (b. 1946). In neither instance were the circumstances as simple as sometimes described. In 1976 Watson, a member of the Evolution Protest Movement who taught religious education at a state-run school, lost his job for refusing, as one creationist magazine reported, "to teach that Genesis is myth." Although fellow creationists rallied to his support and some even testified on his behalf when he appealed his dismissal to an industrial tribunal, in private they whispered about his inability to control classroom behavior and his failure to stick to the prescribed syllabus. One newspaper reported that students interrupted his classes with shouts of "rubbish."[31]

Jones supposedly fell victim to the antipathy of British universities toward creationism. While earning two degrees in biology from the University of Birmingham, including a Ph.D. in zoology and comparative physiology in 1972, he occasionally found himself the object of warnings and innuendoes. Though repeatedly pressured by his professors to abandon his creationist views, he regarded their efforts as "all quite legitimate." At the completion of his studies he ignored well-meaning advice not to risk his degree by revealing his colors in his thesis, but he suffered no ill effects for doing so. The external examiner from the British Museum did not even seem upset. Nevertheless, Jones remained convinced that his positive experience was "almost unique" in the annals of British creationism.[32]

Ever since Price's dispute with James McKeen Cattell over access to *Science* in the early 1920s (see Chapter 5), creationists have complained about the barriers preventing them from publishing in mainstream scientific journals. Indeed, as Robert V. Gentry and other professionally active creationists learned, about the only way to smuggle creation-related research into the scientific literature was to conceal carefully its creationist implications. At the Arkansas creation-evolution trial in 1981, creation scientists defended their virtual invisibility in standard scientific publications by accusing the editors of being "close-minded." Yet, as the presiding judge noted, "no witness produced a scientific article for which publication had been refused." In fact, creation scientists rarely tried to publish the results of their research through conventional outlets. A survey of sixty-eight leading scientific journals revealed that only 18 of approximately 135,000 manuscripts submitted during a three-year period openly advocated scientific creationism. Thirteen of the 18 articles had been sent to science-education journals, 4 to anthropology journals, and only 1 to a biology journal. None had been accepted, but an examination of the referee reports turned up "no evidence for 'out of hand' rejections. The articles appeared to have been rejected on their lack of merit." If barriers to publication existed, creationists themselves helped to erect them.[33]

FLOOD GEOLOGY WITHOUT FLOOD GEOLOGISTS

One of the greatest anomalies in the history of scientific creationism, or flood geology, is the nonpresence of geologists. "Where are the Ph.D.'s in geology today who take Genesis 6-9 seriously?" a concerned young Whitcomb asked Morris in 1955. "Where are the geologists on the side of the angels?" echoed Dudley Joseph Whitney in a letter to Price that same year. Besides the Adventists Clark, Burdick, and Price himself, none of whom possessed even an undergraduate degree in geology, Whitney could think of no one. "In Protestantism out of Adventism is there a single geologist who has not surrendered, lock, stock and barrel, to the ages theory?" he asked rhetorically.[34]

The striking absence of geologists in an enterprise based largely on geological arguments and evidence goes a long way toward explaining the troublesome Burdick's prominence and longevity in the movement. Because of his unique expertise—and his claim to a master's degree in geology from the University of Wisconsin—creation revivalists such as Whitcomb and Morris cultivated his cooperation. If only Burdick could earn his doctorate, wrote Morris wistfully in 1958, "he would be the only Ph.D. geologist I know of who believes in deluge geology and as such ought to be able to command attention."[35]

In putting together the original CRS inner-core steering committee in 1963, Lammerts spent more time trying to fill the geology slot than any other position. The Adventists Ritland and P. Edgar Hare, though not geologists in the strictest sense, ranked high on his recruiting list, but neither wanted anything to do with Lammerts (see Chapter 11). In the end Burdick saved the day. He might have been the weakest of the group, as Lammerts ruefully discovered, but at least he called himself a geologist. And, as Morris noted in justifying Burdick's selection, he was "perhaps better informed on geological matters than any of the rest of us."[36]

In view of the high percentage of Ph.D.-carrying biologists among

the founders of the CRS, the scarcity of qualified flood geologists requires an explanation. Two reasons for this discrepancy seem especially plausible. Because of the demand for Christian physicians, a
number of theologically conservative colleges had long offered premedical training and hired biologists to do the teaching; these same
schools had little incentive to add potentially controversial courses in
the earth sciences to their curricula. Moreover, creationist biologists
could with relative ease select an area of specialization far removed
from the contaminating influence of evolution, while geologists could
scarcely escape the ubiquitous evidence for biological and geological
development. Indeed, the very language for describing rocks reflected
the historical nature of the discipline. Before about 1970 the few disciples of Price brave enough to enter graduate programs in geology
usually found evolution so pervasive they either abandoned geology or
discarded flood geology.

The experience of the Missouri Synod Lutheran Alfred H. Meyer
(1893–1988) poignantly illustrates the first type of response. Having
majored in geology as an undergraduate at the University of Illinois, he
aspired to a career in the field. He concluded that he could never teach
the subject "because of the incompatibility of the current theories with
revealed scriptures," but he thought it might be safe to go into oil
geology. However, early in 1923, as he was completing the requirements for a master's degree in geology from Illinois, he plunged into a
heartrending crisis of confidence. Having recently read antievolution
books by Price and fellow Lutheran Theodore Graebner, he despaired
of being able "in good faith even [to] take up the practise as I had
planned to do, that is, to seek employment as an oil geologist to do
field, laboratory, or office work." He greatly admired Price's "high
Christian ideal" and hoped that the crusader's forthcoming *New Geology* would "reform science as Luther reformed religion." Sinking
slowly into a quagmire of scrupulosity, he came to fear that even using
the language of historical geology was sinful. After submitting a copy of
his thesis to the university library, he began obsessing about a chapter
that mentioned the phrase "Biological Development." So great was his
"spiritual anguish," he considered petitioning to have his thesis with-
drawn and destroyed. Despite assurances from Graebner that his faith
would not be compromised by using "the vernacular of modern geology," Meyer vowed to leave the evil field of geology entirely. "I want
to get as far away from heretical ideas as I can and endeavor to be true
to 100% Missouri Lutheranism," he informed Graebner. "If I could
but only recall my past mistakes."[37]

True to his word, Meyer left geology for geography and transferred to the University of Michigan, where he wrote a theologically safe dissertation on land utilization in northern Indiana and Illinois. But he did not entirely escape unwanted problems. As head of the department of geography and geology at the Lutheran Valparaiso University in the 1930s, he found himself still wrestling on occasion with such issues as how to reconcile the evidence for an ice age with Ussher's chronology. "Much as I have been trying to support the cause of orthodoxy within our church by fighting evolution and the like," he wrote Graebner, "I must confess this age business has put me on the spot." The battle-hardened apologist urged Meyer to compartmentalize his scientific and biblical worlds and postpone any attempt to harmonize the two. Meyer found this solution uncongenial, but he ultimately came to embrace it. Writing to the president of his university in the early 1960s, he described his change of thought:

> It was my earlier conviction that "all things in Science and the Bible must be brought into a unitary focus." It seems to me the sooner all of us once are reconciled to the fact that Bible teachings and Science teachings fall into two quite different categories—the one based on Revelation, the other on Rationalization, the better off we shall be.

As he knew all too well, this approach would "save a lot of headaches and heartaches, if not entirely eliminate all our misunderstandings and apprehensions."[38]

As late as the 1940s, when the American Scientific Affiliation (ASA) was getting off the ground, Meyer remained the only known strict creationist in the United States with a graduate degree in geology. "Meyer of Valparaiso U. is the only professional geologist . . . I know who professes full belief in the Scriptures," wrote the knowledgeable Whitney in 1944. Though Meyer had apparently made some concessions to geological ages, Whitney thought he admitted "lots of flood action." When F. Alton Everest ran into difficulty lining up a qualified person to contribute a chapter on Genesis and geology to the ASA's first book, *Modern Science and the Christian Faith* (1948), the well-connected Russell L. Mixter could think of only three possible candidates: S. James Bole, a horticulturalist, L. Allen Higley, a chemist, and Paul M. Wright (b. 1904), another chemist, who sometimes taught a course on geology at Wheaton College. To the consternation of flood geologists, Everest eventually chose Edwin K. Gedney, a day-age creationist on the faculty of Gordon College of Theology and Missions who had studied geology in the 1920s (see Chapter 9).[39]

From time to time promising young evangelicals with an interest in geology ventured to undertake advanced study in the field, only, like the intrepid J. Laurence Kulp, to emerge from the exhilarating ordeal with their faith in strict creationism badly shaken. In the mid-1960s, however, flood geologists felt certain they had finally discovered a young man who would break the pattern: the impeccably pedigreed Davis A. Young (b. 1941). The son of the eminent Old Testament scholar E. J. Young (1907–1968), Davis had entered Princeton University in the late 1950s to study geological engineering. While there he briefly came "under the sway of the uniformity idea," but reading the work of theologians such as his father's colleague Cornelius Van Til (1895–1987) convinced him that "science must first subject itself to the authority of the Bible if it is to make any real progress." This conviction grew even stronger after graduating with honors from Princeton in 1962 and moving to Pennsylvania State University to work on a master's degree in mineralogy. As a consequence he immersed himself in *The Genesis Flood* and, after two intense readings, emerged a true believer. "As I studied your book," he wrote Whitcomb, whom he had heard speak a couple of times at Princeton, "I began to realize that your basic thesis was indeed correct, and the old objections I harbored against it in college (the evidence just doesn't support the flood, I thought!—smugly) began to disappear." It was about time, Young informed Morris, for Christian geologists to stop "capitulating to modern science and begin looking for the evidence of the Flood."[40]

Young's enthusiasm for flood geology soon began to wane, however. In 1965 he enrolled in the Ph.D. program in the geological sciences at Brown University, where he worked with the chairman of the department, F. Donald Eckelmann (b. 1929), a Wheaton alumnus who had become a Christian evolutionist. Though Young during his years at Brown sometimes spoke to church groups on the significance of the Genesis flood, he grew increasingly suspicious that the evidence from the rocks did not fit the deluge theory. When Morris in 1969 invited him to join the CRS, Young replied that he was unsure whether he still qualified for membership. "Although a few years ago I was in substantial agreement with the position of *The Genesis Flood*," he explained, "I find that on the basis of Scriptural considerations I can no longer accept it." Because certain New Testament passages implied that the seventh day of creation was still in effect, and thus spanned far more than twenty-four hours, Young felt free to interpret the first six days of creation as long periods of time. But unlike his major professor and the liberals in the ASA, he remained, as an Orthodox Presbyterian, "totally committed to the infallibility and inerrancy of Scripture and to the historicity of the Genesis accounts of creation and the Flood." He simply no longer attached much geological significance to the latter event.[41]

The news of this defection greatly saddened Morris. Just when Young, with doctorate in hand and a good job at New York University, could safely risk speaking out in favor of strict creationism, he turned his back on the cause. "I have been hoping," wrote the disappointed Morris, "that you were one whom the Lord was raising up to take the lead in this return to true Biblical science."[42]

With the passage of time Morris's disappointment in Young turned into frustration and anger. In 1972 Young published a letter in a Presbyterian magazine warning readers that the geologically uninformed CRS might be dangerous to orthodox Christianity. He especially worried that young people would suffer religious trauma when they discovered that they had placed their faith in "faulty" geology. Five years later he fired a broadside at flood geology. In a book called *Creation and the Flood: An Alternative to Flood Geology and Theistic Evolution* (1977) he charged flood geologists with teaching "bad geological science." Hoping to lead evangelical Christians to the middle ground between the bad geology of the CRS and the bad theology of the ASA (or at least of its more liberal members), he proposed accepting "the long-day theory of Genesis 1" and correlating the sequence of creative acts with the fossil record. Morris was not impressed. To his way of thinking, this "strategy of compromise" served as little more than a "half-way house" on the road to theistic evolution.[43]

In 1978 Young moved to the Christian Reformed Calvin College, a hotbed of anti-Morris sentiment since the early 1960s, when the flood geologist accused the faculty of heresy. From his vantage point in Grand Rapids, Young kept up a steady attack on flood geology, while his own thinking drifted from progressive creationism to virtual agnosticism about the scientific evidence of God's creative activity. Influenced by Old Testament scholarship as much as by science, he concluded that the first chapter of Genesis yielded no geological information, not even sequence, and that the seven days of creation were nothing more than a literary convention adopted by Moses. When he proposed in 1987 that evangelicals "stop treating Genesis 1 and the flood story as scientific and historic reports," Morris could not resist drawing parallels between Young's intellectual journey and that of Darwin over a century earlier. "Starting out as a strict creationist, he converted to progressive creationism in graduate school, then to theistic evolution," wrote Morris with scarcely concealed contempt. "Now, finding that no such compromise really works or is acceptable to the secular evolutionists from whom he longs for approval, Young proposes to give up Genesis altogether, so far as any actual scientific relevance is concerned."[44]

Within months of hearing the good news about Young's conversion to deluge geology in the mid-1960s, creationist leaders learned of a second geologist, a young Dutchman named Nicolaas A. Rupke (b. 1944), likewise eager to cast his lot with the flood theorists. "You may rest assured," Rupke wrote Lammerts, "that as soon as possible I'll join in our common task to rewrite science on the more realistic base . . . of God's revelation in His Word and in Nature." When Rupke submitted a sophisticated manuscript on cataclysmal sedimentation for publication by the CRS, Lammerts assumed he was dealing with an experienced scientist. Only later did he learn that the author was still an undergraduate at the University of Groningen.[45]

The son of a Dutch minister in the strict Free Reformed Church, Rupke grew up believing the Bible to be God's infallible Word and accepting the creation and flood stories as factual. As a teenager he discovered Immanuel Velikovsky's *Earth in Upheaval* (1955), which led him first to Price's writings and from there to Whitcomb and Morris's *Genesis Flood*. On entering Groningen, Rupke attached himself to Philip Henry Kuenen (1902–1976), a renowned authority on turbidity currents, who kindled his interest in marine geology and sedimentology. When Kuenen saw Rupke's piece in the 1966 *Creation Research Society Annual* identifying the author with the University of Groningen, he accused his student of dishonesty, even though Rupke had absolved Kuenen of any responsibility for the views presented. The professor's objections to Rupke's "unscientific" Bible-based approach to geology and his warnings that Rupke would face only misery if he pursued a career in the earth sciences prompted his insecure pupil to contact creationists in the United States about the possibility of continuing his education there.[46]

The Americans instinctively assumed that Rupke had fallen victim to academic bigotry. Believing that Kuenen had forced Rupke out of *graduate* school because of the article he had written for the CRS, Lammerts urged Rupke to transfer to an American university. "You would have to keep your views strictly to yourself . . . until *after* your degree is granted," he warned. When Rupke realized that the American creationists were blowing his case out of proportion, he informed them that Kuenen was not withholding his bachelor's degree, and he requested them not to publicize his case. Still, Lammerts inferred the worst. "Later no doubt when he is safely on the road to a doctorate perhaps this can be brought up as one of many instances of the lack of academic freedom," he suggested to Morris, adding "we could stand a sharp young geologist like him on our side."[47]

With the support of Kuenen, who not only passed Rupke with distinction but arranged for him to publish a "catastrophist" article in the respected *Journal of Geology*, Rupke gained admission to the doctoral program in geology at Princeton University. Before enrolling in Princeton, he spent the summer as a guest of the Adventist biologist Harold G. Coffin (b. 1926), sharing a fourteen-foot travel trailer and spare vegetarian meals while the two of them studied the famous stacks of fossilized trees found near Joggins, Nova Scotia, and in Yellowstone National Park. Most scientists took for granted that the trees had grown in place over long periods of time and that each level represented a new forest. Coffin and Rupke hoped to show that the trees had been "washed in place," presumably during the brief period of Noah's flood. Indeed, at Joggins they found considerable evidence of roots having been transported by currents to their present resting place. To protect his young colleague from possible unpleasantness at Princeton, Coffin generously suggested that Rupke publish these findings in the *Bulletin of the Geological Society of America* without any acknowledgment of their collaboration. Coffin himself reported their study, without any mention of Rupke, in the *Creation Research Society Annual*. Their work scarcely revolutionized geology, but it did demonstrate that publishable scientific research could be conducted within the flood-geology model.[48]

In the autumn of 1968 Rupke entered Princeton, still intending "to demonstrate that the sedimentary record had accumulated rapidly and catastrophically, and that the earth was very much younger than was generally assumed." If he achieved his goal, he would strike a blow against organic evolution by greatly reducing the time available. But rather than undermining evolution, he found his own religious convictions, including his belief in creationism, slowly ebbing away. Freed from the cloistered fundamentalism of his youth, he began exploring life in "the outside world" and discovered it was not as malevolent as he had been led to believe—especially when he was in the company of his first girlfriend, a worldly French woman. By the time he received his doctorate in 1972, having worked under the supervision of the mild-mannered and tolerant Franklyn B. Van Houten (b. 1914) and the more temperamental but equally tolerant Alfred G. Fischer (b. 1920), he had come to accept organic evolution and had forsaken the faith of his family. Later, while at Oxford, he left geology for a distinguished career in the history of science.[49]

Flood geologists celebrated in 1979, when a well-known young-earth creationist, Stephen A. Austin (b. 1948), finally acquired a legitimate Ph.D. degree in geology—with faith intact—from Pennsylvania State University. Austin had come to the attention of the CRS in 1970, when he submitted a manuscript to the *Creation Research Society Quarterly* under the pen name Stuart E. Nevins, hoping thereby to protect himself from a fate like Burdick's. Leaders of the CRS found Austin's work so impressive they commissioned him to study the origin of fossil reefs, which presented flood geologists with a problem because of the length of time they required for growth. After investigating the Capitan Limestone in southeastern New Mexico and western Texas, Austin delivered the hoped-for verdict: "The available data certainly do not require many thousands of years for the Capitan to accumulate, and, therefore, seem to present little problem for Biblical chronology." Though other creationists questioned his judgment, Austin emerged from the dispute as one of the fair-haired boys of creation science.[50]

At the time of his first contact with the CRS, Austin possessed only a bachelor's degree in geology from the University of Washington, but he soon thereafter acquired an M.S. from San Jose State University, where he wrote a thesis critical of uniformitarianism. By concealing his biblical views, he confided to Young, he had passed through the program unscathed. In 1975 he enrolled in the doctoral program at Penn State, hoping to demonstrate a connection between the flood and coal deposits. Reading such creationist classics as Price's *New Geology* and Whitcomb and Morris's *Genesis Flood* had piqued his interest in the topic, as had Kulp's skeptical assertion in 1950 that "the origin of coal, when the evidence is understood, also precludes flood geology from being correct." Because Austin wanted to do research related to creationism, and not the projects for which he might obtain departmental funding, he asked the CRS Committee on Research to pay for his tuition and living expenses. He begged them, however, to be discreet in announcing his award, suggesting that they say only that Stuart Nevins had received support for work on coal. Although he felt the Lord had led him to a department where the faculty were sympathetic to his catastrophic point of view, he did not want his professors to discover his biblical beliefs regarding earth history. He was sure that if they learned the truth, they would deny him his degree.[51]

Morris hailed Austin's dissertation on the formation of a coal bed in western Kentucky—written without CRS support—as "a real scientific breakthrough" for flood geology. Austin had not only found evidence "that coal was formed from plant debris deposited under mats of vegetation floating in sea water," as posited by the flood model, but had, in Morris's opinion, taken a giant step toward the development of the first-ever theory capable of predicting the location and quality of coal. But most important of all, Austin's graduation from Penn State signaled the beginning of a new era in the history of scientific creationism, when at least some flood geologists could claim to speak as real geologists.[52]

During the 1980s several other qualified geologists—including Douglas A. Block, who in the 1950s had scorned Whitcomb's efforts to defend flood geology—joined the creationist movement. None acquired greater celebrity than the Harvard-trained Kurt P. Wise (b. 1959), protégé of the archcritic of creation science, Stephen Jay Gould. Reared in a fundamentalist Baptist family in rural Illinois, Wise embraced flood geology as a teenager after being exposed to the concept by creation scientists from Bob Jones University at a midwestern conclave for Christian youth. Even before finishing high school, Wise taught a fifteen-week evening course on creationism at his local church. In 1981 he graduated from the University of Chicago with honors in the geophysical sciences and immediately headed east to work with Gould at Harvard. Fellow graduate students sometimes taunted their creationist colleague, but the bemused Gould always treated him with respect. Proud creationists speculated that God had planted Wise "right in the middle of S. J. Gould's paleontology program as a testimony to a man who otherwise might not have been reached."[53]

Although Wise thought that most pre-Cenozoic fossils had been buried during Noah's flood and believed that the entire universe was no older than six thousand years, give or take a couple hundred, he insisted on honestly looking at the scientific evidence. Like Ritland, who had preceded him by a generation at Harvard, Wise accepted the order of the geological column and the existence of overthrusts, including that favorite exhibit of flood geologists, the Lewis overthrust in northern Montana and southern Alberta. "The existence of an inverted section in a thrust belt region with slickensides, dragfolds, and sheared rubble along the unconformity leaves no reasonable doubt that the Lewis Overthurst is in fact a result of overthrusting," he declared in one of his earliest contributions to creationism. "It cannot be considered a contradiction to the geological column." He also cast doubt on Gentry's polonium halo research and other popular weapons in the creationist arsenal. Not all old-timers appreciated Wise's youthful iconoclasm. After he curtly dismissed the arguments of a senior creationist regarding rapid cave formation, the offended man retaliated by accusing his "arrogant" Ivy League critic of "warmed-over uniformitarianism."[54]

In the fall of 1989 Wise left cosmopolitan Cambridge with a Ph.D. in hand to take a position as assistant professor of science at tiny Bryan

College in Dayton, Tennessee, just a stone's throw from the site of the notorious Scopes trial. As director of a new Origins Research and Resource Center at the college, he hoped "to formulate a model of earth history which is consistent with both the Scriptures and the physical data—constructed according to a code of excellence and integrity in ethics and practice." He forswore the evolution-bashing characteristic of early scientific creationism and pledged to shrink from no evidence, however distasteful. With fellow geologists Austin and Andrew Snelling (b. 1952), an Australian, he dreamed of bringing out a revised and updated edition of *The Genesis Flood* that would present a "new creationism" for the twenty-first century. Critics of creation science found his candor both amazing and refreshing. He is, wrote one, "the closest thing to an ally that science has in the creationist movement."[55]

FOURTEEN

Creation Research Institutes

In the late twentieth century no one did more to popularize scientific creationism than Henry M. Morris and his colleagues at the Institute for Creation Research in greater San Diego. In a rare study of creationism at the grass roots, one anthropologist discovered that in North Carolina "for information, the local creationists refer to Henry Morris's organization to lead them through *Genesis* and geochronology; for inspiration, they turn to Morris himself to steer them past doubt and difficulty. No other authority or influence matters nearly as much."[1] Less well known to the public, but of great importance to scientific creationists, was the Seventh-day Adventists' Geoscience Research Institute, which for years set the standard for creationist research. The histories of these two institutions not only shed light on the structure of the creationist movement but illuminate the place of creationism in the Baptist and Adventist traditions.

INSTITUTE FOR CREATION RESEARCH

In 1970 Morris turned down an offer of an endowed chair in civil engineering at Auburn University and moved west to help Tim F. LaHaye (b. 1926), pastor of the Scott Memorial Baptist Church in San

Diego, set up a Bible college and center for creation research. Creation-ist colleagues such as Walter E. Lammerts worried that the move would result in a loss of prestige to the "creationist and deluge geology posi-tion," but Morris felt that the time had come to devote all of his energies to the creationist cause. When Nell Segraves and her son Kelly (b. 1942), a Southern Baptist minister who headed the Bible-Science Radio program in Anaheim, heard of Morris's impending move, Morris be-lieved they "manipulated and maneuvered themselves" into becoming partners. Together he and the Segraveses set up the Creation Science Research Center (CSRC) as an adjunct to LaHaye's Christian Heritage College. Morris served as director, Kelly L. Segraves as assistant di-rector, and Nell Segraves as research librarian. On the twelve-person board of directors, Morris and LaHaye controlled only four position; the Segraves faction, eight. To assist with the intellectual work of the center and enhance its reputation, Morris recruited a dozen technical advisors, including Larry G. Butler, Thomas G. Barnes, and Duane T. Gish. "All of these men have doctorates and are strong creationists, premillenni-alists, etc.," Morris assured John C. Whitcomb, Jr., another advisor. Though not all the staff and advisors were Baptists, the new center had a distinctive Baptist flavor.[2]

The brochure announcing the birth of the center promised that activities would focus on three major areas: First, the staff would im-mediately begin preparing creationist curriculum materials for grades one through six. Because of the recently published Creation Research Society (CRS) textbook for high-school students, the center would cater to the elementary grades. Second, the center would undertake research on the physical aspects of the flood. And third, it would support various "extension ministries," such as radio programs, semi-nars, and literature. The first issue of the *Creation-Science Report*, the center's bimonthly newsletter, declared that the goal for 1972 was to establish "a grass roots movement across the United States to demon-strate how creation can be taught in the public schools." The *Report* explained that creation research would not focus on the long-ago act of creation but on "the post-effects of creation, the fall, and the Flood." Appropriate topics included "radiological dates and dating methods, geological features and theories related to the Flood, anomalous fossils, the range of intra-specific variations, the search for the Ark, and an-thropological research on a baboon living today that possessed 'humanoid-like' teeth."[3]

Within fifteen months of its opening, the center staff had mush-roomed to the equivalent of sixteen full-time employees. But the Morris-Segraves alliance proved unstable. Morris quickly came to distrust the ambitious young Segraves, who aspired "to be a leader in the creationist movement, despite his complete lack of any background in science or teaching." The Segraveses tended to regard Morris as a "figurehead" director and refused to accept his cautious leadership. They favored political and promotional activities; he preferred educational and scientific efforts. They wanted to rush into print with supplemental creationist texts; he urged delay, if necessary, to get things right. The increasingly tense relationship finally came to a head when, during Morris's absence from the center, the Segraveses rushed into print some books he considered unacceptably sloppy and unconscionably expensive. On the pretext that any connection with religious organizations such as Christian Heritage College and Scott Memorial Baptist Church might jeopardize the sale of books to public schools, the Segraveses moved to take control of the center and sever all relationship with the church and college. On a predictable eight-to-four vote in the spring of 1972, the board of directors backed the Segraveses, who took the name Creation Science Research Center and moved their operation across town.[4]

Soon thereafter the Segraveses announced the "grand opening" of their new headquarters, a facility necessitated, they explained, by the success of their textbooks. But within a few years the center was struggling to survive. With their mailing list a fraction of what it had been and debts of $200,000, the Segraveses canceled their radio ministry, suspended publication of their magazine, reduced staff, and moved to smaller quarters. On a shoestring budget the center continued to fight evolution (along with sex education, abortion, women's rights, and gay rights) and to investigate such topics as the relationship between the teaching of evolution and the incidence of venereal disease. Nell Segraves reported in 1977 that research conducted by the CSRC had "demonstrated that the results of evolutionary interpretations of science data result in a widespread breakdown in law and order," leading to "divorce, abortion, and rampant venereal disease." Despite the hiring of Robert E. Kofahl (b. 1924), a chemist with a doctorate from the California Institute of Technology, as a part-time "science coordinator," the CSRC rapidly slipped to the margins of the creation-science movement. The publication of such works as Kelly Segraves's sensationalist *Sons of God Return* (1975), an attempt to link UFOlogy, flood geology, and demonology, and Kofahl's juvenile *Handy Dandy Evolution Refuter* (1977)—"Be the first on your block to really know the truth about *life* on *other planets*"—did little to enhance the center's reputation for sober science.[5]

After the breakup, Morris reorganized the remaining staff, including the recently hired Gish and Harold S. Slusher, into the Institute for Creation Research (ICR). With obvious relief he promised loyalists that the new institute would be "controlled and operated by scientists" and would engage in research and education, not political action. Not all supporters were reassured by the presence of Gish, who had left a research position with the Upjohn Pharmaceutical Company in Michigan to join Morris in southern California. Lammerts, for one, had never quite trusted the biochemist's commitment to flood geology. "How can he work with you at the center?" he asked Morris incredulously. Morris, who had formerly had his own doubts about Gish, explained that his new colleague had experienced a conversion. "I think he has changed some of his former views on the geological ages, mostly because he was not really informed on the significance of this problem before," he wrote. "His approach previously was to say that even if we assume geological ages as they are, there was still no evidence of evolution in the fossil record." In his best-selling *Evolution: The Fossils Say No!* (1972) Gish indeed committed himself to the view that "a sound Biblical exegesis" required the acceptance of flood geology.[6]

The work of the ICR, funded largely by small gifts, staff honoraria, and royalties from institute publications, encompassed three different "ministries": research, writing, and speaking. Despite its name, the institute for years conducted little research outside the confines of its modest library. During its first decade the ICR did sponsor a couple of trips to Mount Ararat in search of Noah's ark and "at least 15 field expeditions for geological or archaeological research," but, as Gish explained apologetically in 1978, the staff devoted much of its research effort to scouring the scientific literature for references favorable to creationism. "It is our dream and fervent hope that some day we will have the facilities, the personnel and the funds to carry out bench-type research right here in the Institute," he added. For a long time, however, the geologist Stephen A. Austin, a visiting scientist off and on during the 1970s who joined the staff full time in 1979, single-handedly conducted most of the institute's nonliterary research. "The main trouble," complained the ever-critical Lammerts, "is that Henry looks at this whole thing as a sort of 'missionary' effort rather than a scientific one." Gish, too, tended to take an evangelistic approach, rationalizing that his writing and speaking contributed more to creationism than any research he might do in the laboratory or field. "If I can win 100 good Ph.D. research scientists to the creationist viewpoint," he argued, "they will be able to do 100 times more research than I could ever have accomplished."[7]

Despite its meager research record, the ICR turned out stacks of books—fifty-five in little over a decade—in addition to its widely circulated newsletter, *Acts & Facts*. In 1981 Morris estimated that over a million copies of books by ICR authors, including the authoritative *Scientific Creationism* (1974), were in circulation. By that time his own books, which continued to stress the importance of choosing young-earth creationism over the compromising gap and day-age theories, were available in Chinese, Czech, Dutch, French, German, Japanese, Korean, Portuguese, Russian, and Spanish editions. Frequently, his books appeared with glowing endorsements by such famous fundamentalists as Charles C. Ryrie, Josh McDowell, Jerry Falwell, and Tim LaHaye.[8]

With the exception of the legal battles to get scientific creationism into public schools, nothing brought more attention to the creationists than their debates with prominent evolutionists, usually held on university campuses. During the 1970s the ICR staff alone participated in more than a hundred of these contests and, according to their own reckoning, never lost one. (Later, Morris grudgingly admitted that he met his match in 1981, when he debated the Catholic biologist Kenneth R. Miller [b. 1948] at Brown University. In addition to being "handsome in appearance, charismatic in manner, and very glib of speech," Miller had come "exceedingly well-prepared.") Although Morris preferred delivering straight lectures—and likened debates to the bloody confrontations between Christians and lions in ancient Rome—he recognized the value of forensics for carrying the creationist message to "more non-Christians and non-creationists than almost any other method." Fortunately for Morris, Gish relished such confrontations. If the mild-mannered, professorial Morris was the Darwin of the creationist movement, then the bumptious Gish was its T. H. Huxley. He "hits the floor running" just like a bulldog, observed an admiring colleague; and "I go for the jugular vein," added Gish himself. Such enthusiasm helped draw crowds of up to five thousand. Together with lectures to various college and church groups—to say nothing of radio and television programs—the debates helped the ICR directly reach more than 600,000 persons during its first ten years.[9]

Early in 1981 the ICR announced the fulfillment of a recurring dream among creationists: a program offering graduate degrees in various creation-oriented sciences. Besides hoping to meet an anticipated demand for teachers trained in scientific creationism, the ICR wished to provide an academic setting where creationist students would be free from discrimination. The first catalog for the new graduate school advertised four M.S. degrees: in biology, geology, astro/geophysics, and science education. Austin headed the program in geology; Slusher, in astro/geophysics. Gary E. Parker (b. 1940), an ex-evolutionist who had earned an Ed. D. degree from Ball State University, chaired the biology department; while Richard B. Bliss (b. 1923), a former science consultant for the Racine, Wisconsin, unified school district, directed the science-education program. Bliss had recently obtained an Ed. D. degree from an experimental university in Florida for a dissertation on the two-model approach to teaching origins. Barnes served as dean of the ICR graduate school.[10]

Slusher, whose highest degree was an M.S. from Oklahoma State University, decided it was time to attach a Ph. D. to his name. In 1982 the ICR announced that he had finally acquired a Ph. D. in geophysics, from the Graduate School of Columbia Pacific University in Marin County, California. His dissertation, devoted to proving a young earth, was titled "A New Cosmic Time Scale: A Reexamination of the Evidence Relating to the Time Scales of Cosmology." According to the ICR announcement, the graduate school of this "nontraditional university" operated in a manner similar "to that of such European universities as the University of Vienna, University of Bologna, and the Sorbonne (Paris)." In fact, Columbia Pacific was an unaccredited correspondence school that recruited students with the lure of a degree "in less than a year." Slusher's dissertation consisted of a manila folder containing copies of five mimeographed ICR "technical monographs" and a copy of the ICR graduate school catalog, all held together with a rubber band. The supervising professor was his creationist colleague from El Paso and the ICR, Barnes, who himself possessed only an honorary doctorate.[11]

Owing to both internal instability and external vulnerability, the graduate school barely survived its first decade of life. After only a few unhappy years as dean, Barnes returned to El Paso to devote his remaining years to rescuing classical physics from uncertainty and relativity. His friend Slusher took his place but soon resigned under a cloud and followed Barnes back to Texas. A short time later Parker, restive with his limited role and heavy duties at the ICR, left for Australia. By the time of his departure the graduate school was coming under the close scrutiny of the California State Department of Education, particularly that of the superintendent of public instruction, Bill Honig (b. 1937). California required all degree-granting institutions to obtain state approval, which the ICR graduate school had initially done with little trouble. But as its notoriety spread, critics inundated the state with complaints, creating what one politician described as "a two-foot-thick stack of outrage" against the institution.

When the ICR sought reapproval for its school in 1988, a five-man visiting team, including two partisans of flood geology, discovered that the graduate-school faculty comprised only five full-time professors, that students sometimes relied on videotaped courses, that laboratory instruction and facilities were inadequate, and that academic freedom existed only within the limits of the school's doctrinal tenets. Nevertheless, the committee voted three to two in favor of reapproval—only to reverse itself when one member, apparently at Honig's urging, changed his vote. In 1990 the state took away the school's license to teach, but two years later a federal judge restored it.[12]

Morris and his colleagues at the ICR attributed their ordeal to religious intolerance rather than academic incompetence. They found it especially galling to discover that while Honig was trying to shut them down, he had "invited leading homosexual educators into his San Francisco home to discuss ways of better including the teaching of the homosexual agenda in California public school classrooms." To their way of thinking, homosexuality itself represented one of the worst results of evolution. Nearly all homosexual leaders, argued one of Morris's sons, justified their life-style in Darwinian terms: " 'We are descended from animals,' they say; 'We have certain animal desires that must be satisfied. Few animals mate for life, and many times animals of the same sex interact sexually. And didn't all *bisexual* animals evolve from *asexual* animals?' " If the state wanted to help an oppressed minority, it could start with the creationists.[13]

With its graduate program in jeopardy and creation science under a constitutional cloud, the ICR began recultivating old-fashioned biblical creationism. In 1988 the institute launched a series of Back to Genesis seminars that attracted huge crowds and raised thousands of dollars. On successive weekends in March 1989, for example, church-sponsored seminars in Seattle and Grand Rapids brought together a total of ten thousand persons and generated more money from fees and book sales than the entire ICR staff could earn in honoraria in a year. The churches hosting these events (and smaller single-speaker engagements) ranged across the spectrum of evangelical denominations, but Baptists predominated, as they did at the ICR itself. Although Morris had earlier left the Southern Baptist Convention to become an independent, he lent apologetic support to the advocates of inerrancy in the "holy war" between fundamentalists and moderates that broke out in the 1960s. As an independent, he also had ready access to the loose coalition of churches and chapels that formed the backbone of the New Religious Right. According to one scholar, Morris's contact with the network of independent Baptist organizations "helps account for the ICR's pre-eminence among creationist groups."[14]

Institutionally, Morris scored some of his biggest victories in such fundamentalist Baptist colleges as Cedarville, Baptist Bible, Los Angeles Baptist (later The Master's College), and Liberty Baptist (renamed Liberty University in 1985). From the beginning, Liberty Baptist College, founded by Jerry Falwell, committed itself to Morris's brand of scientific creationism. Professors were required to believe it; students, to learn it. In the early 1980s, however, the American Civil Liberties Union filed a complaint with the Virginia State Board of Education challenging the qualifications to teach science of Liberty graduates trained in creationism. In response the college moved creationism out of the biology department and placed it in a new extradepartmental unit called the Center for Creation Studies, which boasted having the "World's Largest Creation Museum." The center offered a creation-science course on the "History of Life" taken by all Liberty students. Directing the operation was Lane P. Lester (b. 1938), a Purdue-trained geneticist and alumnus of the BSCS textbook project, who had been rescued from evolution by Gish.[15]

As Henry Morris entered his eighth decade, he rejoiced that the ICR had grown to become "the world's largest creationist organization." Yet at the same time he felt the force of "Satanic oppression and opposition" manifesting itself at the institute, and he worried about finding a worthy successor. Beset by declining energy and failing eyesight, he turned increasingly to his son John D. Morris for counsel and administrative assistance. Before joining the ICR staff in 1984 as associate dean of the graduate school and chairman of the general science department, young Morris had earned a doctorate in geological engineering from the University of Oklahoma and had taught in that institution for several years. In 1989, having established himself as a creationist authority on Noah's ark and fossil footprints, he became administrative vice president of the ICR and heir apparent to his father's realm.[16]

GEOSCIENCE RESEARCH INSTITUTE

Early in 1978 the staff of the ICR drove north from San Diego to Loma Linda, California, to observe the western branch of the Geoscience Research Institute (GRI) run by the Seventh-day Adventist church. "We were greatly impressed by the tremendous facilities available to their outstanding staff of scientists and by the excellent scientific research that is being done by the 7th Day Adventist scientists," Gish enthusiastically reported to Lammerts. "They are probably doing more in the way of active research than any group of creationists in the world." With generous financial support from their church, which had given birth to modern flood geology, the Adventist scientists had developed an unparalleled creationist institute, complete with laboratories and affiliated graduate programs.[17]

The roots of the GRI went back to 1957. Responding to the concerns of Adventist science teachers about the absence of qualified earth scientists in the denomination, church leaders set up a Committee on the Teaching of Geology and Paleontology and allocated $13,500 "to send two mature, experienced men of proved loyalty, to take special studies in the above-mentioned fields in qualified institutions for advanced study." Within a year the committee had selected two suitable candidates: the fifty-eight-year-old biologist Frank Lewis Marsh and the young chemist P. Edgar Hare, who aspired to a career in geochemistry. George McCready Price recommended Clifford L. Burdick, but the committee ruled him out in part because of his already advanced age. With the committee's backing, Marsh commuted to Michigan State University to audit geology courses incognito, while Hare matriculated in the California Institute of Technology. In 1960 a third member, the Harvard-trained comparative anatomist and paleontologist Richard M. Ritland, joined the group.[18]

In the euphoria of creating a new institute, the three men, despite considerable differences in age and outlook, worked together in reasonable harmony. In the spring of 1958 Marsh and Hare spent three days at Marsh's house combing through Ellen G. White's writings, collectively called the Spirit of Prophecy, for clues about how to proceed with research into origins. They also pored over J. Laurence Kulp's critique of flood geology, which led to a five-page agenda for future research. While acknowledging the validity of Kulp's charge that flood geology "assumes in its major premise that which is to be proved," they defended such an approach on the grounds that "if the Bible is inspired then we *must* accept its assertions regarding the early history of our earth as major premises." Hare agreed to see if he could use his research on the amino-acid content of marine animals to "show that fossils found in the different strata are essentially the same age"; Marsh resolved to put Price's cherished claims about out-of-order strata to an empirical test. He confessed to having made a mistake in previously writing "positively on the subject" solely on the basis of Price's armchair theorizing. Until personally examining some alleged overthrusts, he refused to certify that there was "even one clear case of 'reversed order' where, in conformable strata, complex fossils are found below more simple animals of the same phylum." During the summer of 1959 he and Ritland, who had not yet formally joined the group, traversed the continent from New Mexico to Canada inspecting geological formations and mulling over the meaning of controvertible statements in the Bible and the books of White. All the while Marsh felt he was in the company of a kindred spirit.[19]

The goodwill began to evaporate the next year during a three-and-a-half-week geological tour of the West with a group of Adventist college teachers and administrators. While on the road the creationist trio spent hours together in the cab of a pickup truck discussing philosophies and phenomena. In contrast to Marsh, who insisted on using the historic Adventist interpretations of the Bible and the inspired writings of White as the starting point for any investigation into origins, Hare and Ritland expressed a willingness to reinterpret the biblical and prophetic accounts of creation and the flood in the light of scientific evidence. Specifically, they argued for greatly extending the age of the earth from thousands to billions of years and for drastically reducing the number of fossil-bearing strata deposited by the flood. Having privately concluded that flood geology was a parody of earth history, they hoped to use the new institute as a vehicle for moving their church to a more scientifically defensible position. The discussions left Marsh confused and angry. He felt as though his two younger colleagues were trying to "brainwash" him into accepting a "satanic" approach to the study of nature. He could not understand, for instance, why both men seriously considered evidence from radioactive time clocks that placed "Creation Week hundreds of millions of years ago" in direct contradiction to the Bible and White.[20]

Other tour members, unaware of the tempest brewing in the lead truck, reveled in the opportunity to see for themselves what they had hitherto only read about. One of the group hailed the event as a milestone in the history of Adventist science: "For the first time our teachers, vitally interested in Flood geology, could have the opportunity for firsthand investigation of many fossil areas, with time for lectures and discussion under the direction of men who have been trained in paleontology and related fields." In Yellowstone National Park they examined the famous series of fossil forests and, guided by Ritland, discovered that the trees had not been washed into place during the flood, as many had previously believed, but had grown *in situ*. As the veteran creationist Harold W. Clark explained to Morris, the trees were "practically all standing, and in many cases the roots can be seen, rooted into what appears to be good soil." He hastened to add, however, that he was sure the entire sequence could be accounted for in "not more than 3000 years." On the basis of the close resemblance between Tertiary fossils and modern animals, Clark concluded that all rocks above the Cretaceous were postdiluvial in origin. The reformation of Adventist science seemed to be under way.[21]

By 1962 the ever-prudent Ritland felt sufficiently confident to spell out his views in an unpublished paper on "Problems and Methods in Earth History," discreetly circulated to a small circle of influential Adventists. A careful study of the sedimentary strata, he pointed out, clearly indicated that multiple catastrophes, not just the Genesis flood, had sculpted the crust of the earth. Thus to continue defending flood geology would "only bring embarrassment and discredit to the cause of God." He had come to this position only after years of intense study and soul-searching—and only because his previous exposure to biblical studies and his continuing association with a handful of liberal Adventist theologians convinced him he could do so without "giving up the faith."[22]

Marsh needed no additional evidence to convince him of the impossibility of forming a consensus at the GRI. In view of the intellectual impasse, he suggested that the governing committee divide the work of the institute into separate spheres of authority, with Ritland and Hare splitting the field of geology and himself focusing on "evolution versus special creation." In the absence of a director, a position still not formally filled, each man would report directly to the committee as the "single view and undivided front" in his respective area. When Marsh saw his colleagues' growing influence in the church, however, he quickly reneged on his promise not to discuss the age of the earth. Conscience would not allow him to follow his associates down "the road leading to uniformitarianism and organic evolution," even if it meant rejecting the apparent facts of science. "In my opinion we cannot use our senses in the manner of uniformitarians in interpreting what we see in the earth," he exclaimed. "This is an extremely important point. *Special* Revelation takes precedence over *natural* revelation because natural science can be correct only when in harmony with special Revelation."[23]

By early 1963 relations at GRI headquarters had deteriorated to the point that Ritland feared he might inadvertently be setting the stage "for Marsh to become a martyr to the cause of defending Bible and Spirit of Prophecy." During the past few years Ritland and Hare had come to question not only their senior associate's methods but his competence and character as well. Tensions developed over everything from office management to Marsh's right to litter his colleagues' desks with unwanted advice and quotations from White. Marsh, who had long thought Morris sounded like an Adventist, enthusiastically promoted *The Genesis Flood* in denominational circles, while Ritland denounced it as a pseudo-erudite work full of "flagrant errors which the uninitiated person is scarcely prepared to detect." When Lammerts learned of the troubles among the Adventists, his heart sank. He had been following the progress of the institute with interest, hoping to anchor the fledgling Creation Research Society to the GRI "as a base of operations." But with Ritland's and Hare's leftward turn, his bubble burst.[24]

By this time Hare's research on amino-acid ratios in marine shells was clearly indicating that life had been on earth far longer than traditionally allowed by Adventists. He felt compelled to notify church leaders of the potential problem. "Frankly," he confessed,

> I am beginning to wonder if our whole approach to this problem is in error. We have been taught for years that almost everything in the geologic record is the result of the flood. I've seen enough in the field to realize that quite substantial portions of the geologic record are not the direct result of the flood. We also have been led to believe by men like Marsh and Burdick that the evidence for the extreme age of the earth is extremely tenuous and really not worthy of any credence at all. I have tried to make a rather careful study of this evidence over the past several years, and I feel the evidence is not ambiguous but that it is just as clear as is the evidence that the earth is round.

Within a year of Hare's writing this letter the president of the church, Reuben R. Figuhr (1896–1983), notified him that he was free to remain with the Carnegie Institution in Washington, where he had gone to conduct his laboratory studies. The primary purpose of the GRI, explained the president, was to read, write, and study—"looking for inconsistencies in the evolutionary writings that appear"—rather than do original research. Hare took the hint and stayed at the Carnegie Institution, where he built an international reputation as a geochemist. Though his thinking moved relentlessly in the direction of theistic evolution, he chose to retain his membership in the Seventh-day Adventist church.[25]

Throughout the early 1960s Marsh and Ritland jockeyed for recognition as the Adventist authority on origins. Despite claiming Price's mantle, Marsh increasingly lost ground to his younger rival, who outmaneuvered him in getting close to the pious but pragmatic Figuhr.

Finally, in 1964, on the occasion of Marsh's sixty-fifth birthday, church leaders shunted the querulous crusader off to the biology department of Andrews University, which impressed Marsh as being little better than "banishment into the farthest corner of Siberia." Many factors, behavioral as well as ideological, contributed to his fall, but Marsh chose to view it as the result of "a no-holds-barred process of indoctrination" carried on by the ambitious Ritland.[26]

As the newly named head of the GRI, Ritland moved immediately to create an atmosphere congenial to his revolutionary plans. Even before officially taking over, he arranged to bring in Harold G. Coffin, a biologist he had known since college days, to replace the departed Hare. Though theologically conservative and cautious by nature, Coffin struck Ritland as an honest man who could be "emancipated" by the evidence. Ritland then recruited two protégés, Harold E. James, Jr. (b. 1936), and Edward N. Lugenbeal (b. 1940), who, like himself, had attended the Seventh-day Adventist Theological Seminary. James went off to Princeton to obtain a doctorate in geology; Lugenbeal, to the University of Wisconsin to study prehistoric archaeology. Ritland gave a part-time appointment to Ariel A. Roth (b. 1927), a former student of Clark's who had earned a doctorate in parasitology at the University of Michigan and who used his time with the GRI to retool as a geologist at the University of California at Riverside. Unfortunately for Ritland, he proved to be only an average judge of character. Although James and Lugenbeal lined up solidly behind his revisionist program, Coffin and Roth ultimately found it psychologically and theologically impossible to break substantively with the flood-geology model. When Roth "backed off from the brink" and retreated to the shelter of White's cosmogony, the badly disappointed Ritland knew the tide had turned.[27]

The nearly simultaneous publication of two GRI books at the end of the decade—one by Coffin, the other by Ritland—publicly exposed the widening fissure between the two institute factions. In 1969 Coffin, assisted by Roth, Clark, and a few other traditionalists, brought out a comprehensive work entitled *Creation—Accident or Design?* As Clark had been urging for decades, Coffin modified Price's scheme by accepting post-flood glaciation and an orderly geological column; and, following Ritland, he allowed for substantial pre- and post-flood fossilization. But he refused to deviate one jot from a strict reading of the Bible and the writings of White. Only the imprecision of White's statements on the age of the earth gave him the latitude to add a few years to the six thousand she frequently mentioned.[28]

The following year Ritland issued his long-gestating book, *A Search for Meaning in Nature: A New Look at Creation and Evolution* (1970). Instead of poking holes in evolutionary theory or trying to fit the fossils into a flood model of earth history, he highlighted the purported evidences of design in the natural world. For him, the "fundamental question" was not creation versus evolution but "whether the earth, with its continents, rivers, and seas, its atmosphere and climate, all so well-fitted for life, is a product of accident, or of plan and design." Although he accepted divine creation as "a reasonable explanation of the facts," he conceded that the fossil record did not rule out the possibility of some kind of evolution. Privately, he had come to accept the antiquity of life on earth and to see Noah's flood as a local event of limited geological significance.[29]

The exiled Marsh wondered with amazement how "one man could carry such a wallop" in denominational circles, but Ritland's star was already descending. In 1966 a new church president, the doctrinally rigid Robert H. Pierson (1911–1989), replaced the more flexible Figuhr. Two years later Pierson accompanied Ritland and the institute staff on the now-standard tour of geological landmarks in the Rockies, and he did not like what he saw and heard. Upon returning home, he laid down new guidelines for studying origins. "In our controversy with proponents of the evolution theory," he announced in the official church paper, "we must keep in clear perspective—*the Bible and the Spirit of Prophecy are not on trial.*" It did not take a Harvard Ph.D. to see that Pierson wanted apologetics, not research. In 1971 Ritland, finding it impossible to function within the new constraints imposed by the church hierarchy, resigned as director of the GRI and joined Marsh in what was fast becoming an Adventist gulag, the biology department of Andrews University. The Adventists' brief experiment with "openmindedness" thus came to an end.[30]

Under the direction of Robert H. Brown (b. 1915), a physicist fascinated by radioactive time clocks, the GRI quickly swung into line behind the Pierson administration. Brown eagerly committed the institute to the apologetical mission of showing that life on earth "originated within six consecutive rotations of the planet," that the earth "experienced a universal destruction as portrayed in Genesis 6-8," and that life on earth was no older "than 10,000 years." Because the dates yielded by radiocarbon dating contradicted the "testimony given by Moses and Ellen G. White" regarding the age of life on earth, he dismissed them as invalid. Yet at the same time he took advantage of revelatory silence to embrace the evidence of radioactive time clocks indicating that the raw materials of the earth had been around for billions of years. This schizoid attitude toward dating pleased neither conservatives such as Marsh and Robert V. Gentry, who insisted on a young earth, nor liberals such as Ritland and Hare, who argued for the antiquity of life on earth. The "snag" with Brown's position, as Marsh saw it, was that the fossils had to be as old as the rocks that embedded them.[31]

For a man who consistently elevated religious knowledge above scientific belief, Brown invested a surprising amount of energy in polishing the tarnished scientific image of creationism. He and his colleagues at the GRI especially liked to contrast their careful studies with the sometimes slipshod presentations of Morris and his staff at the ICR. Indeed, the GRI's journal *Origins*, founded by Brown and edited by Roth, set a new standard for critical creationist scholarship. From Berney Neufeld's early exposé of the Paluxy River tracks to Arthur V. Chadwick's debunking of Burdick's Precambrian pollen and Brown's own assault on Gentry's radioactive halos, it carried some of the most trenchant analyses of creationist claims to appear in print. By the 1980s it had become a publication of choice even for non-Adventist creationists.[32]

Under Brown and his successor, Roth, who moved the institute's main offices to Loma Linda, California, the GRI devoted itself to salvaging what it could of flood geology. Staff scientists who resisted the deluge soon found themselves without a platform and eventually without a job. The geologist James was the first to go. Shortly after earning his doctorate from Princeton at institute expense, he learned that the GRI no longer desired his services. Lugenbeal went next. Having concluded that flood geology was "desperately weak and improbable," he wondered how he could "in good conscience continue to absorb the Church's resources in what seems to me a futile and self-deceptive effort to disprove the obvious in science and an emotionally and ethically debilitating attempt to bolster our people's faith by telling them a series of partial truths about science." In the fall of 1979 he submitted his resignation, thereby ending an "emotionally, spiritually, and intellectually" ravaging period of his life.[33]

In 1953, shortly before the birth of the GRI, the Seventh-day Adventist church had published its first Bible commentary. In introductory essays on creation and the flood, Marsh and Price had routinely questioned the integrity of evolutionists and systematically rejected such "fantastic" ideas as the order of the fossils, continental glaciation, and thrust faults. A quarter-century later the church brought out an updated edition with new essays by scientists from the GRI. Gone were the snide comments about evolutionists and their fanciful geological column. But for the most part the changes involved style rather than substance. The revisionists, like Marsh and Price before them, continued to compress the history of life on earth into a few thousand years and to invoke Noah's flood in explaining the origin of many fossils. The Seventh-day Adventist church as a whole also stayed resolutely in the deluge-geology camp, probably more so than any other denomination in the world. A survey conducted in 1980 showed that the overwhelming majority of members, about 94 percent, believed in the recent creation of life; an equal number interpreted the days of Genesis to be periods of twenty-four hours; and nearly as many subscribed to a universal flood, though only 85 percent attributed *most* of the sedimentary rocks to the deluge. Adventists may have lost their innocence during the intellectual upheavals of the 1960s and 1970s, as one observer claimed, but most of them remained faithful to the basic vision of Ellen G. White and George McCready Price.[34]

FIFTEEN

Creationism in the Churches

T he creationist revival sparked by the publication of *The Genesis Flood* elevated the deluge geology of George McCready Price to a position of fundamentalist orthodoxy and endowed special creationism with a measure of respectability unknown since the deaths of Arnold Guyot and John William Dawson in the late nineteenth century. Although most Christians remained untouched by the revival and even millions of evangelicals, from Billy Graham to Jimmy Lee Swaggart (b. 1935), continued to subscribe to old-earth creationism, the flood geologists exerted enormous influence. By the last decades of the twentieth century they had virtually co-opted the creationist label. Writing as a day-age believer in the early 1980s, Davis A. Young regretfully conceded the name to the flood geologists, who, in the public mind at least, had "come to be known generally as creationists."[1]

It is impossible to say how much the creationists' success stemmed from converting evolutionists as opposed to mobilizing the already converted—or how much it owed to such factors as the widespread disillusionment with elite science in the 1960s and 1970s. We do know that the scientific creationists launched their crusade with a large reservoir of potential support. A survey of church members in northern California in 1963, at the very beginning of the creationist revival, revealed that over a fourth of those polled—30 percent of Protestants and 28 percent of Catholics—already opposed evolution. Broken down by denomination, it indicated the relative strength of creationist sentiment in the churches:

Liberal Protestants (Congregationalists, Methodists, Episcopalians, Disciples)	11%
Moderate Protestants (Presbyterians, American Lutherans, American Baptists)	29%
Church of God	57%
Missouri-Synod Lutherans	64%
Southern Baptists	72%
Church of Christ	78%
Nazarenes	80%
Assemblies of God	91%
Seventh-day Adventists	94%

The Adventists, Assemblies of God, and Nazarenes may not have ranked among the largest denominations in the United States, but they were some of the fastest growing; and the eight-million-member Southern Baptist Convention stood on the verge of becoming the largest Protestant body in America.[2]

The scanty evidence available suggests that belief in creationism may have increased by as much as 50 percent during the next couple of decades. A nationwide Gallup poll in 1991 showed that 47 percent of Americans professed belief in a recent special creation, with another 40 percent preferring theistic evolution. Blacks, women, and the poor seemed most receptive to creationism. Support for teaching creationism in public schools extended well beyond the circle of strict creationists. When running on the Republican ticket for the presidency of the United States in 1980, Ronald Reagan insisted that "if evolution is taught in public schools, creation also should be taught." That same year the *American School Board Journal* asked readers, presumably including many school-board members, "How should public schools handle the teaching of the origin of man?" Two-thirds of the respondents favored including the creation story; 19 percent wanted no evolution at all; 48 percent thought that both Darwin and the Bible should be taught. A poll of lawyers conducted for the American Bar Association in 1986 found that nearly two-thirds saw "no First Amendment obstacle to the teaching of creationism in public schools."[3]

Unfortunately, none of the surveys distinguished between old-earth and young-earth creationists, and there is reason to suspect that not all persons polled knew what they were saying. While surveying the opin-

ions of San Francisco–area residents in the early 1970s, two sociologists
found that 94 percent of conservative Protestants agreed with the view
that "God created the first man and woman" and that 10 percent
believed that "man evolved from lower animals." Presumably, the
researchers noted wryly, "the 4 percent of conservatives who accepted
both theories are especially agreeable people."[4] The recent surveys
also did not break down responses by denominational affiliation. However, anecdotal evidence suggests that Baptists contributed the greatest number of strict creationists, while Seventh-day Adventists
possessed the greatest purity. We have already seen some of the ways
in which Baptists and Adventists reacted to the creationist revival; here
we shall explore the influence of scientific creationism on other religious groups, from Lutherans and Mormons to Jehovah's Witnesses
and Jews.

THE LUTHERAN CHURCH—MISSOURI SYNOD

In the autumn of 1962 Price happened across a Lutheran defense of
deluge geology that appeared in the evangelical weekly *Christianity
Today.* "Are the Lutherans to get ahead of us Adventists in teaching
Flood Geology?" he anxiously asked his friend Frank Lewis Marsh.
The answer was no, but Price had ample cause for concern. Since the
days of Theodore Graebner, the Missouri Lutherans, who sometimes
defended Bible science to the point of rejecting a heliocentric solar
system, had been among the most united American denominations in
opposing evolution. In 1932 the synod had adopted a doctrinal statement binding members to the proposition "that God has created
heaven and earth, and that in the manner and in the space of time
recorded in the Holy Scriptures, especially Genesis 1 and 2, namely,
by His almighty creative Word, and in six days." During the 1950s
Alfred M. Rehwinkel's *The Flood* (1951) had introduced Price's deluge
geology to a new generation of Lutherans, who became a prime market
for the sale of Adventist books on creationism. When the Creation
Research Society was founded in 1963, one-third of its eighteen charter
members, including the president, Walter E. Lammerts, the treasurer, Wilbert H. Rusch, and steering committee members John W.
Klotz and Paul A. Zimmerman, came from the Lutheran church—
Missouri Synod.[5]

Until after World War II the synod remained relatively untouched
by evolution. But in the years around mid-century, as students from
secular high schools enrolled in synod colleges and prospective faculty

members pursued graduate degrees outside the church, evolution began to trouble some members. Before long a few intellectuals, theologians and scientists alike, were pushing the synod to "interpret the first chapters of Genesis in the light of the evolutionary theory." The hitherto confined debate broke into the open in 1955 with the publication of *Genes, Genesis, and Evolution* by the conservative biologist-theologian Klotz. Klotz knew enough science to wish that he did not have to defend special creation and a universal flood, but he felt he had no choice because "Scripture speaks and that settles it for me." Deferring to Genesis over genetics, he argued in his book for a recent creation in six literal days and for organic development bounded by the originally created "kinds" of plants and animals. Although he assumed that Noah's flood had destroyed "all air-breathing animals that were not on the ark," he avoided addressing the geological effects of the deluge, a silence some flood geologists attributed to a desire to avoid "the tremendous reaction against Rehwinkel's book in scholarly Lutheran circles."[6]

Klotz no doubt hoped to ward off Rehwinkel's critics, but he did not succeed. Leading the anticreationist charge against Klotz's book was Carl H. Krekeler (b. 1920), who, like Klotz, had combined advanced training in biology and theology. As a young man studying at Concordia Seminary in St. Louis, he had learned of the "evils of evolution" directly from Graebner and Rehwinkel. But studying biology in graduate school, first at Washington University and later at the University of Chicago, he had begun to rethink "the party line on evolution." A course on vertebrate paleontology especially affected his thinking. As he later recalled, "The documentation, not only of changes within a lineage such as the horses, but of transitions between the classes of vertebrates—particularly the details of the transition between reptiles and mammals—forced me to abandon thinking of evolution as occurring only within 'kinds.' " By the time he reviewed *Genes, Genesis, and Evolution* for the Lutheran *Cresset*, he was teaching biology at the church-related Valparaiso University in Indiana.[7]

Krekeler deplored Klotz's work on several counts. Theologically, he charged Klotz with creating a false problem by judging science against a fallible and arbitrary reading of Scripture. Scientifically, he faulted Klotz for "fighting a losing battle." While granting that science had not yet delivered a "knockout punch" in the sense of absolutely proving evolution, he noted that "a steady stream of blows has the anti-evolutionist reeling." Time and again even Klotz was forced by the pummeling to step back and say, "These are facts which cannot easily be reconciled with what Scripture tells us." Ethically, Kreckler accused Klotz of "reprehensible" behavior for unfairly treating the arguments and evidence in favor of evolution. He feared that such acts might produce "immeasurable harm by binding the consciences of those struggling with problems in this area."[8]

Four years later, on the occasion of the centennial of Darwin's *Origin of Species*, the Missouri Synod published a collection of creationist essays titled *Darwin, Evolution, and Creation* (1959). Included in the volume were contributions by Klotz, Zimmerman, then serving as president of a Lutheran teachers college in Seward, Nebraska, and Rusch, who taught science at Zimmerman's school. Again Krekeler rose to object to this synodical condemnation of evolution, arguing that even Lutheran theologians disagreed "as to whether Scripture reveals the 'how' and 'when' of creation." He especially condemned the perversion of scholarship in defense of the faith: the "dozens of places where half-truths are spoken, where quotations supporting the authors' views are taken from the context of books presenting contrary views, and where there is misrepresentation." Krekeler's sharp words provoked both Klotz and Zimmerman into responding publicly. Klotz, making no effort to conceal his hurt, especially resented the accusations of "dishonesty, deceit, and insincerity." Zimmerman, hoping to inflict as much pain as he had suffered, branded the Valparaiso professor a heretic for implicitly rejecting the synod's 1932 statement on creation. "Has the Missouri Synod been in error all these years?" he asked ominously. Such pyrotechnics ensured a wide readership for the book, which in Lutheran circles played a role in promoting creationism analogous to that of *The Genesis Flood* among more conventional fundamentalists.[9]

In 1963 Krekeler virtually confirmed his heretical standing by coauthoring a controversial biology text that embraced evolution in everything but name. Written with William W. Bloom (b. 1910), a colleague at Valparaiso who had also lost his faith in creation while studying biology at the University of Chicago, *General Biology* advocated what the authors self-protectively called "continuous creation." In a chapter called "The Theory of Evolution," drafted by Krekeler, they maintained that the Genesis story, though true in a sense, was "not intended to describe events and give proximate explanations." Thus Christian scientists had the liberty to search for this information in nature. According to the theory of "continuous creation," which the authors contrasted with the more familiar notion of "discontinuous creation," God's creative activity blended into his "work of preservation." Except linguistically, continuous creation seemed to differ little from theistic evolution. In fact, Bloom and Krekeler conceded that there was "no essential conflict" between the two theories.[10]

The appearance of Bloom and Krekeler's book ignited what Rusch called a "brush fire" in the Missouri Synod. Thrilled at last to have the enemy out in the open, Rusch hastily mailed the synod president a copy of the book, along with his own damning comments. Krekeler's chapter on evolution provided "plenty of ammunition" for persons interested in disproving "that Valpo is still kosher in its treatment of Creation," he informed Lammerts, who by this time had joined Klotz, Zimmerman, and Rusch in the battle to save the synod for creationism. Krekeler tried to contain the damage to Valparaiso by writing a reassuring statement on the evolution controversy for distribution to Lutheran clergy. While admitting that biologists at the university taught evolution "as a working hypothesis," he insisted that there was "no doubt whatsoever that Scripture teaches that God was and is the Creator." Scripture simply did not reveal his method of creation.[11]

After years of sniping at a distance, the protagonists finally met face to face in October 1965 at a meeting of pastors and teachers held at Valparaiso. Shedding the camouflage of "continuous creation," Bloom and Krekeler forthrightly supported "the process of evolution" while succinctly presenting the evidence in its favor. Klotz, Rusch, and Zimmerman not only questioned the data allegedly supporting evolution but also claimed that an analysis of the confessions showed that "it is not Lutheran to accept theistic evolution." Klotz, who had once bemoaned the synod's anti-intellectualism, now believed that the pendulum had swung in the other direction. "We are making a god of scholarship and are forgetting the limitations of the human mind," he complained to Lammerts. If Lutherans abandoned a literal creation, he feared they might next reconsider "such central matters as the physical resurrection of Christ."[12]

Although the Lutheran conservatives refused to compromise on the issue of a recent special creation, they tended not to insist on the flood model of earth history. As Klotz explained to John C. Whitcomb, Jr., most communicants accepted the idea of a universal flood but split over the issue of "whether all of the rock strata were laid down by the flood." For a decade the synod supported a seven-man committee to investigate the geological and paleontological record. Despite the presence of Klotz, Rusch, and Zimmerman in the group, flood geology received little explicit support, except for one member's suggestion that the theory deserved "a serious reappraisal." In 1977 the synod in convention voted against establishing a permanent creationist organization like the Adventist Geoscience Research Institute.[13]

Throughout the 1960s and early 1970s scarcely a regular convention passed without some group introducing a resolution relating to evolution. In 1971 the convention received a report from the Commission on Theology and Church Relations ruling out any change in the organic world affecting "the essential nature of the creatures in question." This opinion and a 1973 "Statement of Scriptural and Confessional Principles" came the closest since 1932 to an official synod position on creation. By this time, however, the synod was in the early throes of a schism that a few years later resulted in liberal faculty and students walking out of Concordia Seminary and the synod—and taking 4 percent of the synod's membership with them. The Lutheran conservatives, like their Adventist brothers and sisters, had repulsed liberalism and evolutionism, but at great price.[14]

CREATION, EVOLUTION, AND HOLY GHOST RELIGION

The bulk of twentieth-century converts to flood geology came from the cluster of churches associated with the evangelical Lutheran and Calvinist traditions, which put a premium on correct belief. But many committed creationists could also be found among the more experientially oriented Wesleyan evangelicals, who ranged from the relatively moderate and prosperous Nazarenes in the Holiness movement to the socially marginal Churches of God and Assemblies of God in the Pentecostal party. But with rare exceptions, such as the loosely affiliated Dudley Joseph Whitney and John N. Moore, they remained aloof from the catastrophist crusade of Price and Morris. Even the most ardent antievolutionists in the conservative Wesleyan camp relegated creation to a position far behind salvation. During the 1920s, for example, no one in the Holiness community agitated more vigorously against "the biological baboon boosters" than the Kentucky evangelist Andrew Johnson (1875–1959), but he felt obliged to promise that his efforts to "shake the monkey out of the cocoanut tree" would not eclipse the gospel. "There is nothing like an old-fashioned, soul-saving revival of Holy Ghost religion," declared the bombastic preacher in the *Pentecostal Herald*. "So, let it be distinctly understood that the lectures on Evolution are absolutely secondary to the main line work of intense, soul-saving evangelism to which we have been called and in which we expect to remain."[15]

Holiness people paid little attention to *The Genesis Flood* when it came out in 1961, but those who did notice it wished it well. S. Hugh Paine, a physics professor at the Wesleyan Methodist Houghton College in New York, praised Whitcomb and Morris for their "objective and scholarly manner" and particularly for their arguments against "the uniformitarian hypothesis—an unprovable thesis which by its very nature forbids any literal acceptance of the Genesis record." He hoped that their catastrophist critique would "carry in the ranks of Evangelical scholarship," but he feared that "deep-seated prejudices" would prevent most readers from accepting its radical thesis. Like his old Wheaton College teacher L. Allen Higley, Paine himself favored what he called the "Gap-Flood" model of earth history, which assigned most of the fossil-bearing rocks to a pre-Edenic flood rather than to Noah's deluge. This scheme possessed the advantage of allowing creationists to accept the evidence for "the apparent great age of the earth, the apparent tenure of ancient populations of pre-Adamic animals, and the apparent pre-Adamic hominid paleontology."[16]

By the 1960s virtually all of the major Holiness denominations had set aside their early reservations about higher education and had invested heavily in liberal arts colleges. In part to provide adequate training for premedical students, these schools had gradually developed programs in the biological sciences. For example, Asbury College, where the dissidents in the American Scientific Affiliation had first met to form the Creation Research Society, began teaching biology in the 1920s; but it was not until after World War II, in 1946, that the school hired its first adequately trained biologist, Cecil B. Hamann (1913–1984), who had earned a doctorate at Purdue University. A self-styled progressive creationist, he allowed the period of creation to extend over thousands, even millions, of years, occasionally punctuated by special creative acts. Until his retirement in the 1980s, he effectively defended this middle-of-the-road position against theistic evolutionists, on the one flank, and flood geologists, on the other. Other Holiness colleges sometimes hired theistic evolutionists to teach biology, but these professors generally maintained a low profile to avoid becoming the target of a more conservative constituency. Few true flood geologists could be found in these schools.[17]

Theologians in the Wesleyan-Holiness tradition, like their scientific colleagues, tended to stick to the familiar gap and day-age readings of Genesis 1 rather than switch to the novel flood theory. *The Wesleyan Bible Commentary* (1967), the first such work produced by evangelical Wesleyans after the beginning of the creationist revival, ignored flood geology altogether. The author of the section on Genesis in the *Beacon*

Bible Commentary (1969), published by the Nazarenes, twice cited *The Genesis Flood* as a factual authority but seemed oblivious to its hermeneutic significance. In one jarring passage on Noah's flood, he artlessly juxtaposed references to the Whitcomb-Morris book and Bernard Ramm's *Christian View of Science and Scripture* without even hinting that the two works represented opposite schools of thought on the topic. The comprehensive *Contemporary Wesleyan Theology* (1983) simply stressed "the *moral* and religious dimensions of creation." Clearly, flood geology was winning few converts among the scientists and theologians in the Holiness movement.[18]

Pentecostals also kept their distance from the fundamentalist flood geologists, at least at first. Despite superficial similarities, Bible-thumping fundamentalists with their insistence on scriptural inerrancy, and tongues-speaking Pentecostals, with their talk of Spirit baptism, had distrusted each other since their nearly simultaneous births in the early twentieth century. The Pentecostals tended to see the fundamentalists as cold and doctrinaire, while the fundamentalists often dismissed glossolalia and healings as satanic delusions. Morris, for instance, looked upon charismatic behavior as akin to modernist thinking, both, in his opinion, stemming from personal beliefs and experience rather than the Bible. In view of this prejudice, it is small wonder that his message made little headway among Pentecostals.[19]

In 1972 the president of the Assemblies of God Evangel College invited the staff of Morris's Institute for Creation Research to use the facilities in Springfield, Missouri, for the first-ever Summer Institute on Scientific Creationism. On arriving, the flood geologists discovered no supporters among the science faculty. Later, when Turner Collins (b. 1939), a plant pathologist, co-offered a seminar on science and religion, he used texts and guests that were anathema to flood geologists. About the same time, L. Duane Thurman (b. 1933), a botanist from Berkeley on the faculty of Oral Roberts University, wrote *How to Think about Evolution*, in which he irenically evaluated a number of acceptable interpretations of Genesis. Myrtle M. Fleming (b. 1913), who earned a Ph.D. in biology at the University of Georgia and who built up the science programs at both Lee College, operated by the Church of God, Cleveland, Tennessee, and Emmanuel College in Georgia, run by the Pentecostal Holiness Church, warned Pentecostal theologians not to be taken in by claims for a young earth. In writing and teaching about origins, she said, "great care should be taken to distinguish between facts and theory, original works and philosophers' thinking."[20]

On the surface, Pentecostals seemed undisturbed by the growing popularity of flood geology. Many of the most vocal and visible preachers saw no reason to abandon the ruin-and-restoration gap theory found in the beloved Scofield Reference Bible for the creation science of the flood geologists. The endorsement of the gap theory in *Dake's Annotated Reference Bible* (1963), an immensely influential work that began displacing *Scofield* among conservative Pentecostals in the early 1960s, no doubt contributed significantly to its continuing popularity. Prepared by Finis Jennings Dake (1902–1987), a Pentecostal evangelist with a checkered legal and denominational history, *Dake's*, as it was commonly called, gave eighteen "proofs" of the existence of a pre-Adamic world destroyed by Lucifer's flood. Such Pentecostal stalwarts as Jimmy Lee Swaggart, Lester F. Sumrall (b. 1913), Kenneth E. Hagin (b. 1917), and Gordon Lindsay (1906–1973) also remained loyal to a pre-Edenic gap. Some commentators, such as Dake, restricted the days of creation to twenty-four hours each, but a few, including Lindsay, followed the turn-of-the-century Pentecostal pioneer Charles F. Parham (1873–1929) in expanding the "re-creative days" to six periods of one thousand years.[21]

Below the surface, however, creation science was winning considerable support among rank-and-file Pentecostals. The Assemblies of God, though taking no official position on the issue, provided Whitcomb with an early forum in the *Pentecostal Evangel* and continued into the 1970s and 1980s to print positive notices about the Creation Research Society and Institute for Creation Research. Flood geology also appeared in works by influential Church of God (Cleveland) and Church of God of Prophecy writers, as well as by various independent charismatic groups. William A. (Winkie) Pratney (b. 1944) of Last Days Ministries, who "gave up a promising career in research chemistry" to become an evangelist, increasingly promoted flood geology in the successive editions of his slick mass-circulated tract *Creation or Evolution?* By the 1980s Gordon Lindsay's son, Dennis (b. 1946), was teaching a course on "Scientific Creationism" at his late father's Christ for the Nations Institute in Dallas.[22]

LATTER-DAY SAINTS

In 1935 only 36 percent of the students at the Mormons' Brigham Young University denied that humans had been "created in a process of evolution from lower life forms." By 1973 the figure had risen sharply to 81 percent. No doubt many factors pushed young Mormons toward fundamentalism and antievolutionism. But the most significant scientifically was the far-reaching influence of Price, whose flood geology seeped through the virtually impermeable wall separating the followers of the rival American prophets Ellen G. White and Joseph Smith (1805–1844).[23]

Before the twentieth century, evolution rarely troubled the lives of the Saints living around the Great Salt Lake in Utah. With the exception of allowing each creative day to represent a thousand years, early leaders of the Church of Jesus Christ of Latter-day Saints tended to take the Genesis story at face value. They attached little importance to the niceties of prehuman earth history. As Smith's successor, Brigham Young (1801–1877), said, "It matters not whether it took six days, six months, six years, or six thousand years." Shortly after the turn of the century, when educated Mormons began trickling back from secular universities in the East infected with evolution, the First Presidency became sufficiently alarmed to issue its first formal statement on the subject, "The Origin of Man." This noncreedal document, published in 1909, both affirmed that "all men were created in the *beginning* after the image of God" and denied that "the original human being was a development from lower orders of the animal kingdom." A couple of years later a heresy hunt drove several purported evolutionists from the faculty of Brigham Young University, but still the church adopted no official position and made no attempt to excommunicate the accused teachers. Even after the University of Utah geologist Frederick J. Pack (1875–1938) wrote a book in the 1920s approving of theistic evolution, including "the anthropoid origin of man," he remained a highly respected member of the Mormon community.[24]

One of the first signs of coming controversy appeared in the late 1920s, when the contentious Mormon scholar Brigham H. Roberts (1857–1933) drafted a monumental manuscript in which he argued for the existence of a race of "pre-adamite" human beings destroyed by a cleansing cataclysm just prior to the Edenic creation. This suggestion of death before the fall of Adam distressed Joseph Fielding Smith (1876–1972), grandnephew of the founding prophet and since 1910 a member of the Quorum of the Twelve Apostles, a body second only to the three-man First Presidency in the church hierarchy. For over a decade Smith had been denouncing evolution as an "idiotic hypothesis," incompatible with orthodox Mormonism. "If you accept the teachings of the Prophet Joseph Smith," he had written in 1920, "then you must believe that man is the offspring of God, and if so he was not evolved from a tadpole, or from the scum of the sea." His discovery of Price's critique of evolution—especially the dictum "No Adam, no fall; no fall, no atonement; no atonement, no savior"—emboldened him to speak out more forcefully than ever. The flood geologist's diatribes against orthodox geology confirmed Smith's impression that there was "no department of modern science so creed-bound, so much in need of a logical house-cleaning." In letters to Price, Smith urged the embattled creationist to continue attacking the "false theory" of evolution until its defenders surrendered.[25]

Early in 1931, as part of a campaign to discredit Roberts, Smith began preaching to his fellow apostles about the merits of flood geology. At least one member of the Quorum of the Twelve, mining geologist and engineer James E. Talmage (1862–1933), thought he smelled a charlatan. As a young man Talmage had dedicated himself to enlightening Mormons about the harmony between Genesis and geology, "a subject upon which so many of our people have mistaken ideas." To equip himself for a scientific career, he enrolled in Lehigh University and Johns Hopkins University. While studying in Baltimore, he noted in his journal that he could "see no reason why the evolution of animal bodies cannot be true—as indeed the facts of observation make it difficult to deny—and still the soul of man is of divine origin." He later retreated to safer theological ground, but apparently he never abandoned his belief in the existence of pre-Adamite races. Among Mormon intellectuals, he had few peers. Before becoming an apostle in 1911, he served as president of both the Latter-day Saints College and the University of Utah, where he had also occupied the chair of geology. In 1896 he picked up one of the first Mormon doctorates, a degree granted by Illinois Wesleyan University "for nonresident work."[26]

After his election to the Twelve, Talmage had little time to keep up with the geological literature. Thus when Smith tried to sneak Price's "extravaganzas" into the inner sanctum of Mormonism, Talmage turned to his son Sterling for assistance. Fresh from graduate school at Harvard and an assistant professorship at Northwestern University, Sterling B. Talmage (1889–1956) had recently joined the geology faculty at the New Mexico School of Mines. Early in February 1931 he received an urgent letter from his father cryptically asking him to check out the reputation of one George McCready Price. "Please get a copy of *The New Geology*, study it with attention, consult with your colleagues, and tell me how Price is held in the opinion of geologists in general," he instructed his son. More specifically, he requested an opinion on the contested Lewis overthrust in Montana and Alberta, the centerpiece of Price's antievolutionism. Though bound by the rule of confidentiality from disclosing the exact reason for his sudden interest in flood geology, Talmage did indicate that some Mormons had recently expressed "considerable confidence" in Price's views. He worried that Price's new catastrophism, though primarily associated with Noah's flood, might "appeal to some of our ardent readers of the Book of Mormon, as having direct bearing on the seismic disturbances on this continent incident to the Crucifixion."[27]

Within days Sterling replied that he had known about *The New Geology* for years but had never heard the "book discussed at Northwestern or elsewhere, without the element of comedy being dragged in." Having mentioned comedy, he could not resist passing on a joke of his own about *The New Geology*, which contained nothing "new" nor any real "geology": "*With these two corrections*, the title remains the best part of the book." On a more serious note, young Talmage assured his father that "neither the book nor its author has any standing whatever among American geologists." He knew of only one school, the fundamentalist Wheaton College, that used Price's text; and at Northwestern there had even been talk about "the advisability of removing it from the open shelves." He himself regarded Price's work "as the most amazing piece of conceited bombast" he had ever seen in print, with one possible exception. As for the Lewis overthrust, he thought that lithologic evidence alone, to say nothing of fossils, demonstrated that it had not been formed in place. Meanwhile, Pack, who had succeeded James Talmage as head of the geology department at Utah, was checking with the Canadian Geological Survey to see what they knew about the overthrust.[28]

Sterling Talmage implored his father to tell him what lay behind the inquiry, but the apostle would say only that "some of our zealous members have presented Price's views in support of the scriptural account of creation, and have rashly condemned conclusions of geologists, especially regarding the order of stratigraphical sequence." Not until Sterling revealed that he, too, knew of Smith's enthusiasm for Price did his father confirm that the investigation related to the Roberts-Smith debate over pre-Adamites. Like Roberts, James Talmage believed that the 5,931-year-old Adamic race had been preceded on earth by pre-Adamite life. The usually cautious father boasted to his son that in discussions among the Twelve he had been "bold enough to point out" some conclusive evidence against Smith's position. He had personally inspected a pile of stones at Spring Hill, Missouri, declared by Joseph Smith to be part of "the altar on which Adam offered sacrifices," and had seen that it contained fossilized animals. "If those stones be part of the first altar," he reasoned, "Adam built it of stones containing corpses, and therefore death must have prevailed in the earth before Adam's time."[29]

Just as James Talmage enlisted his son for behind-the-scenes advice on how to handle Smith, so Smith turned to Price for help in responding to the Talmages' allegations about the Lewis overthrust and Price's lack of scientific credentials. The elder Talmage, reported Smith, not only questioned Price's scientific training but ridiculed his "foolish" interpretation of the overthrusting in Montana and Alberta. Smith tried to defend Price and his heterodox views, "but not being a geologist, my word does not mean very much when I speak of this science." Before Price could be of much assistance to Smith, the fearful First Presidency suspended the apostolic debate and declared a draw: Neither the existence nor the nonexistence of pre-Adamites would constitute church doctrine.[30]

Aroused by his father's reports of "the pre-adamite near-controversy" in Salt Lake City, Sterling Talmage waited impatiently for an opportunity to enter the fray himself. The more he thought about the issue, the more convinced he became that the church should take a stand on evolution. "To my mind, the *principle* of evolution, considered separate and apart from some of the controversial and hypothetical phases related to it, is tremendously faith-promoting," he confessed to longtime confidant John A. Widtsoe (1872–1952), an apostle who had graduated with honors in chemistry from Harvard and then earned a Ph.D. in biochemistry from the University of Göttingen. Talmage's chance came in 1934, when the *Deseret News*, a church-owned newspaper, carried an article questioning the antiquity of the earth. Unfettered by the restraining hand of his recently deceased father, he dashed off a reply "written at white heat" and sent it to Widtsoe. Talmage's published polemic reignited the old debate about the age of the earth and prompted Smith to call in the Gentile firebrand Whitney for support. The *Deseret News* ran several spirited pieces from Whitney before the board of directors halted the series. As the general manager explained, Whitney's articles had raised "a lot of _____ ," so "we just decided to stop the deal and wait until we could draw a better hand." As Whitney saw it, his "critics got the brakes on just in time." Had seven or eight installments appeared, he was sure "the teaching of evolution would have been pretty badly demoralized in the Inter Mountain states."[31]

Without Whitney's dampening influence evolution did indeed prosper in the Inter-Mountain West, at least among Mormon academics. But Smith and his creationist cohorts did not give up. In 1954 Smith broke years of relative silence on the subject with a screed against evolution called *Man: His Origin and Destiny*. Enlisting authorities ranging from his granduncle the prophet to such non-Mormons as

Price and Byron C. Nelson, Smith tried to prove that evolution was both unscientific and un-Mormon. "For the first time in Mormon history," noted one scholar, "Mormonism had a book that was openly antagonistic to much of science." By opting "for schism rather than synthesis," Smith broke with the long tradition of recognizing the validity of scientific knowledge and "sparked a wave of religious fundamentalism" that reverberated through Mormonism for decades. Although the president of the church pointedly denied that the book carried any imprimatur, many Mormons thought otherwise. Smith's elevation to the First Presidency in 1965 and to the presidency itself five years later gave added credence to this impression, as did its endorsement by son-in-law Bruce R. McConkie (1915–1985) in his authoritative, but also unofficial, *Mormon Doctrine* (1958). In one of the all-time best-sellers in Mormon history, McConkie followed his father-in-law in limiting the history of life on earth to thousands of years and in appealing to flood geology to reconcile science and revelation. Those who sought to harmonize revealed religion with organic evolution he damned as "devilish."[32]

The protests of Mormon scientists who feared that Smith's book would force the church "into a position that cannot and should not be defended" failed to drown out the ringing endorsement of Melvin A. Cook (b. 1911), professor of metallurgy at the University of Utah and an internationally acclaimed expert on high explosives. In an introduction to *Man*, Cook vouched for the scientific sophistication of the author and the validity of his conclusions. This testimonial infuriated many Mormon scientists and catapulted Cook into the center of the creation-evolution controversy. Despite having earned a Ph.D. in chemistry from Yale University, Cook at first felt "unprepared to meet attacks by philosophers and earth scientists." But not wishing to "appear defeated," he plunged "deeply into a self-study of the earth and life sciences." This led him to question the validity of radioactive time clocks and to conclude that "the Noachian Flood was responsible for most of the fossil deposition and strata differences as well." Following the Mormon custom of interpreting each day of creation as a thousand years, he calculated that the earth was about thirteen thousand years old.[33]

By the early 1960s Cook felt sufficiently informed to announce publicly that all radiocarbon dates should be "telescoped" down to thirteen thousand years. His scientific colleagues, including Henry Eyring (1901–1981), dean of Mormon scientists as well as of the graduate school at Utah, distanced themselves from the maverick metallurgist. A contingent from the Utah geology department denounced his article as "pseudoscience at its worst" and dismissed his geology as "so full of error as to be almost laughable." But Cook would not be silenced. A few years later he brought out two creationist books in rapid succession: *Prehistory and Earth Models* (1966), a technical critique of evolution, and *Science and Mormonism* (1967), a defense of flood geology from "the viewpoint of LDS literalists." The latter was sprinkled with puffs to the Mormon theologians Smith and McConkie and citations to such creation scientists as Morris and Burdick.[34]

When Lammerts discovered these writings, he invited Cook to join the Creation Research Society, a "strictly nondenominational" organization that included "some fine Mormon members." Before long Cook was contributing to the *Creation Research Society Quarterly* as the society's most prominent Latter-day Saint. Although some CRS leaders disliked his theology, they valued his scientific expertise and, especially, his growing fame. The very year the *Quarterly* first carried Cook's byline, he won the Nitro Nobel Gold Medal, awarded for outstanding contributions to the field of explosives. The citation described his invention of slurry or water-gel explosives as "perhaps the most epoch-making innovation in the field of commercial explosives since Alfred Nobel's invention of the dynamite." The creationists finally had their own (Nitro) Nobel laureate.[35]

OTHER PROTESTANTS, CATHOLICS, AND JEWS

In the late 1980s one frustrated fundamentalist noted that "not a single Christian denomination or association" had yet adopted recent special creationism as a cardinal doctrine. On the basis of that somewhat hyperbolic observation he concluded, no doubt correctly, that "the vast majority of Christians still do not believe that the doctrine of the six-day creation is relevant for Christian spiritual life."[36] As we have seen, even many creationists, especially in the Holiness and Pentecostal traditions, resisted the flood geologists' pleas to discard the gap and day-age interpretations of Genesis. But the flood geologists did make substantial gains, especially among Baptists, Adventists, and Lutherans. Pockets of strength could also be found in Protestant groups ranging from the Churches of Christ to some of the more conservative Presbyterian bodies, especially in the South.

The Churches of Christ, dedicated to the restoration of New Testament Christianity, had always favored a nonallegoric reading of Genesis but not necessarily one that limited earth history to a few thousand years. James D. Bales, the Church of Christ preacher-professor who at one time represented the Evolution Protest Movement in the United States, never endorsed flood geology. But in the 1960s a number of Church of Christ scientists joined the Creation Research Society. The CRS textbook committee that prepared *Biology: A Search for Order in Complexity* (1970) included not only Rita Rhodes Ward, a Church of Christ high-school biology teacher, but three biologists from various Church of Christ colleges. In 1972 these professors joined Bales and other fellow believers in producing *A Critical Look at Evolution* (1972), edited by a Church of Christ minister enamored of Price's arguments against evolution. Later the most outspoken Church of Christ proponent of scientific creationism was Charles Albert (Bert) Thompson (b. 1949), a food microbiologist who briefly taught at Texas A & M University and who on occasion placed classified ads in *Science Digest* to publicize creationism. In the early 1980s he and a minister friend, Wayne Jackson (b. 1937), started a new magazine, *Reason & Revelation*, that spread young-earth creationism throughout the Churches of Christ. Members of the brotherhood who resisted scientific creationism, such as some of the science faculty at Abilene Christian University, found themselves branded as evolutionists and worse. Thompson and Jackson followed the Whitcomb-Morris line, with one exception: They refused to link "pure, biblical creationism" to "any kind of 'premillennial assumption.' "[37]

A group of fundamentalist extremists called the Christian Reconstructionists also broke with Whitcomb and Morris on the issue of eschatology. Founded in the 1960s by Rousas J. Rushdoony, an Orthodox Presbyterian minister, this postmillennial movement worked for the complete Christianization of society. Eventually, the Reconstructionists expected to set up a totalitarian government based on Old Testament law, under which certain categories of sinners—including homosexuals, Sabbath breakers, and incorrigible children—would be put to death. Their presuppositionalist epistemology led them, like Whitcomb, to regard scientific creationism as a corruption of biblical creationism. For years the Reconstructionists applauded the efforts of Whitcomb and Morris. In the early 1960s Rushdoony himself had brokered the publication of *The Genesis Flood*, and his economist son-in-law, Gary North (b. 1942), a major voice in the movement, credited Whitcomb and Morris with having written "the most important book in the revival of the six-day creation view of Genesis." By the late 1980s, however, North was sharply criticizing the scientific creationists for building their arguments against evolution on the "shaky pillar" of the second law of thermodynamics. In emphasizing the fall of Adam, he argued, they had neglected the resurrection of Christ, which falsified the principle of entropy. Though North granted science little role in constructing a creationist apologetic, he expected creationism to yield tangible technological benefits. "If six-day creationism could be used to locate oil and mineral deposits less expensively than the methodology of evolutionism does, we would begin to see the abandonment of evolutionism," he wrote wistfully. "What we need is for evolutionism to start drilling more dry holes than we do."[38]

During the second half of the twentieth century, few denominations promoted strict creationism more vigorously than the apocalyptic Jehovah's Witnesses and the Worldwide Church of God, members of the same millenarian family that included the Seventh-day Adventists. The Witnesses sprang indirectly from the Advent Christian Church, which, like the Seventh-day Adventists, had grown out of the ashes of the Millerite movement of the 1840s. The Worldwide Church of God (formerly the Radio Church of God) was a third-generation offshoot of the Seventh-day Adventists by way of the Church of God (Seventh Day). Both sects, though often indebted to the scientific creationists for arguments and evidence, remained ambivalent toward flood geology and antagonistic toward its promoters, a feeling amply reciprocated. Duane T. Gish summed up the common attitude: "Of course, they don't want anything to do with us, and we don't want anything to do with them."[39]

Jehovah's Witnesses, like their Seventh-day Adventist cousins, readily granted the antiquity of the universe while limiting the age of life on earth to tens of thousands of years. On the basis of some exegetical gymnastics that led them to interpret the creative "days" of Genesis as periods of seven thousand years each, they estimated the history of the earth to have spanned about forty-eight thousand years (six seven-thousand-year days plus six thousand years since Adam and Eve). The founders of the movement borrowed their geology from a Quaker schoolteacher in Pennsylvania named Isaac Newton Vail (1840–1912), who extrapolated from the nebular hypothesis to formulate what he called the annular system of earth history. According to Vail, the earth was originally surrounded by Saturn-like rings or canopies of aqueous vapor, which one by one collapsed on the earth, burying fossils in "a succession of stupendous cataclysms, separated by unknown periods of time." The deluge of Noah resulted from "the last remnant" of this vapor canopy. Though geologically significant, the flood in Vail's system accounted for far less of the fossil record than Price assigned to it.[40]

The fear of evolution and other heresies prompted leaders of the Watchtower Bible and Tract Society, as the Witnesses are known legally, to discourage their youth from attending school longer than required by law. Thus, aside from the occasional adult convert, the society produced few scientists or other educated persons capable of addressing technical issues in a sophisticated manner. Most of the Watchtower literature on evolution appeared anonymously. During the second half of the twentieth century the Witnesses distributed gargantuan numbers of antievolution tracts. Eighteen million copies of *Did Man Get Here by Evolution or by Creation?* (1967) rolled off the presses in Brooklyn before it was replaced in 1985 by an updated and expanded work that had a first run of two million. In these profusely illustrated booklets the Witnesses recycled many arguments from scientific creationism, but they departed from flood geology in allowing for nonliteral "days," the creation of animals "thousands of years before man," and, following Vail, a flood that was only one of many "catastrophes that caused great climatic and terrestrial changes."[41]

The Worldwide Church of God evolved from the efforts of one man: Herbert W. Armstrong (1892–1986), an enterprising Church of God (Seventh Day) preacher, who in the 1930s began an independent ministry that grew into one of the largest Saturday-keeping Christian bodies in the world. By the time of his death in the mid-1980s, his *Plain Truth* magazine had a circulation of over 7,500,000 and his radio and television program, "The World Tomorrow," reached millions more. An early encounter with an evolutionist (as well as his wife's discovery of Seventh-day Adventism) had driven him to an intense study of the Bible, which resulted in his drawing up a list of seven fundamental doctrines. Number two, right after an affirmation of God's existence, held that "evolution stands disproved—an error—a false theory." Like some gap theorists, Armstrong and his followers allowed for the existence of an ancient earth, populated by Satan and his fallen angels. Although they at times relied heavily on the writings of creation scientists, the Armstrong people defended creationism primarily on biblical grounds, refusing to dilute "Biblical truth to satisfy scientists by equating it with scientific theory."[42]

Flood geology only rarely spread beyond the confines of conservative Protestant churches and Mormon temples. Roman Catholicism contributed a few camp followers, such as Paul Ellwanger, a respiratory therapist from Anderson, South Carolina, who drafted the creationist bill adopted by the Arkansas legislature in 1981, and Paula Haigh (b. 1926), a reference librarian at a Catholic college in Louisville, Kentucky, who in the mid-1970s founded the ephemeral Catholic Center for Creation Research and briefly edited *The Catholic Creationist*. A self-described "disciple of Dr. Henry Morris, Dr. Duane Gish, and the other Creationist scientists," Haigh single-handedly tried

to arouse fellow Catholics from their lethargy toward creationism. After a year of tireless effort, she found herself "still waiting for some
Catholic theologian or philosopher or scientist or team of such to appear publicly and in print *for* Creation and *against* evolution." Apparently, she waited in vain, because after a few more years her
organization died for lack of support.[43]

The Creation Research Society's restrictive policy of admitting only
persons who accepted Jesus Christ as their Savior kept non-Christian
creationists on, or beyond, the periphery of the creation-science movement. Although many orthodox Jews took the Torah to be literally
true, they rarely deferred to the scientific creationists' reading of the
text. One Jewish writer not unsympathetic to the Christian creationists
declared categorically that "no one in the Jewish world agrees with the
version of creationism held by Christian fundamentalists." But on occasion even Jewish liberals expressed admiration for the fundamentalists' refusal to kowtow to scientific authority. "However absurd the
strict content of their views may be, and however evil may be the
association of these views with right-wing militarism and anticommunism and with a servile dependency on fundamentalist preachers who purport to speak for an authoritarian God," wrote one
contributor to the liberal Jewish journal *Tikkun*, "there is something
correct and admirable in their refusal to accept the hegemony of science as a privileged source of truth." From time to time Jewish creationists shared evidence against evolution with their Christian
counterparts. For example, in 1983 an Israeli from Ben Gurion University of the Negev informed the CRS about a recent meeting of
orthodox Jewish scientists in Jerusalem at which evidence was presented indicating that the transitional Archaeopteryx, linking reptiles
and birds, was based on fraudulent fossil specimens. But in general
Jewish creationists remained untouched by the floodwaters of scientific
creationism.[44]

SIXTEEN

Creation Science Floods the World

T he creationist revival that erupted in the early 1960s at first attracted little attention beyond the pews of the most conservative churches. Most scientists and educators remained oblivious to the growing movement until the California textbook controversies in the late 1960s and early 1970s alerted them to what one biologist called "the necessity for preserving the integrity of the classroom from such onslaughts." In 1972 the National Academy of Sciences (NAS) and the American Association for the Advancement of Science (AAAS) both spoke out against the intermingling of science and religion as found in creationism. But most scientists remained quietly on the sidelines. Some feared that they might inadvertently legitimize creationism by responding to its advocates; others disliked the idea of suppressing dissent; while still others jokingly dismissed creationists as a bunch of pseudoscientists who got "their doctorates in a box of Cracker Jacks." Meanwhile, as the *Wall Street Journal* reported in a front-page story in 1979, the creationists were building up "a full head of steam" in their push to enter the public-school classroom.[1]

The creationist movement entered a new political phase in the late 1970s, when Wendell R. Bird (b. 1954), a clever young law student at Yale, formulated a strategy for introducing creationism into public-school curricula. In 1978 he published a prizewinning essay on the subject in the *Yale Law Journal*. The "cornerstone" of Bird's argument, in the opinion of one legal scholar, was the contention that "scientific creationism was science, not religion, and teaching it did not violate the constitutional restrictions against religious instruction, while not teaching it violated the free-exercise rights of creationist students." Upon graduating from law school, Bird went to work for the Institute for Creation Research, where he put his legal skills to use improving an equal-time resolution that Henry M. Morris had drafted for adoption by local school boards. The ICR hoped to use Bird's work merely to persuade board members to adopt the two-model approach, but Paul Ellwanger, the Catholic creationist who headed Citizens for Fairness in Education, adapted it for introduction into state legislatures. His model bill called for the balanced treatment of what he called "creation-science" and "evolution-science." Within two years lawmakers in Arkansas and Louisiana had approved versions of Ellwanger's bill, and more than twenty other states were considering it. "Creationism," reported one science journal incredulously, "is breaking out from coast to coast."[2]

This alarming turn of events finally aroused evolutionists en masse. The summer of 1980 saw the appearance of *Creation/Evolution*, the first journal devoted exclusively to answering "all the significant arguments creationists usually put forth in their publications and debates." Early the next year, at a meeting of the AAAS, William V. Mayer (b. 1920), director of the Biological Sciences Curriculum Study, alerted members to the "Madison Avenue approach" creationists were using to further their religious aims at taxpayers' expense. The scientific community, he warned, could no longer afford to dismiss creationism as "nonsense on a par with the concerns of the flat earth society." The potential for damaging science education was too great to remain silent. In October 1981 the National Academy of Sciences and the National Association of Biology Teachers both called emergency meetings in Washington to plot a policy of containment. The AAAS journal *Science* reported a palpable "crisis atmosphere" within the scientific establishment: "In addition to the pending legislation in almost half the states in the Union, the scientific community can witness Duane Gish presenting creationists in the mold of Galileo, facing the 'dogma of the scientific establishment.' " One scientist even spoke of creationist "terrorist tactics."[3]

At the grass-roots level Stanley L. Weinberg (b. 1911), a veteran high-school biology teacher in Iowa, began in the fall of 1980 to organize statewide Committees of Correspondence "committed to the defense of education in evolutionary theory." Modeled after the citizens organizations of the same name in pre-Revolutionary America, Weinberg's committees aimed to keep interested parties abreast of creationist plans and share effective countermeasures. More clearly than most, Weinberg saw the struggle against scientific creationism as a political contest that would be won or lost at the local level. Within a few years he had networks of volunteers operating in almost every state in the union, all linked by a *Creation/Evolution Newsletter*. By the mid-1980s the Committees of Correspondence had incorporated as the National Center for Science Education (NCSE) and installed Weinberg as the first president.[4]

In 1984 the National Academy of Sciences blitzed American school superintendents and science teachers with over forty thousand copies of a glossy booklet titled *Science and Creationism: A View from the National Academy of Sciences*. Written by a special committee of distinguished scientists and lawyers, the report sought to justify the teaching of evolution and to expose special creation as a threat to both science and society. "In a nation whose people depend on scientific progress for their health, economic gains, and national security," explained the president of the academy in the preface, "it is of utmost importance that our students understand science as a system of study, so that by building on past achievements they can maintain the pace of scientific progress and ensure the continued emergence of results that can benefit mankind."[5]

The academy's unrestrained endorsement of evolution not only angered the fundamentalists in the Creation Research Society but stirred up the previously quiescent evangelicals in the American Scientific Affiliation (ASA). Two years after receiving *Science and Creationism*, science educators found a copycat work, *Teaching Science in a Climate of Controversy: A View from the American Scientific Affiliation*, in their mailboxes. Prepared by a group of evangelical scientists that included Walter R. Hearn, James O. Buswell III, and Daniel E. Wonderly—three of the names most despised by flood geologists—the ASA booklet urged teachers to adopt "a cautious and humble attitude" (in contrast to the cocksure position of the NAS scientists) when dealing with such sensitive topics as human evolution. In offering "Some Classroom Guidelines," the evangelicals encouraged teachers to explore the "broad middle ground" between godless evolutionism and scientific creationism, "where real science can coexist with real faith in God."[6]

This irenic attempt to defuse the creation-evolution controversy outraged many evolutionists, who denounced the ASA for promoting creationism in disguise. A cluster of big-name critics writing for *The*

Science Teacher accused the ASA scientists of hiding their intentions under "the veneer of sweet reasonableness," of "telling lies to naïve and trusting young persons," and of obfuscating, distorting, and waffling "to the point of pure nonsense." William J. Bennetta (b. 1938), a passionate skeptic from San Francisco who organized the critics' response, dismissed the ASA effort as "an ordinary exercise in creationist pseudoscience," made all the more dangerous because it presented "that pseudoscience in a package so slick that it can seem respectable to people who would dismiss an I.C.R. tract as ridiculous." In his opinion, the Creation Research Society, the Bible-Science Association, and the Institute for Creation Research had all "been eclipsed by one that now is much more virulent: the A.S.A." When the National Center for Science Education, which included a number of theistic evolutionists among its members, refused to join Bennetta in his witch-hunt, the anticreationists fell to squabbling among themselves over the relative merits of pragmatism and ideological purity. To the executive director of the NCSE, sitting in Berkeley, the scene was all too reminiscent of the Trotskyites versus the Maoists in the 1960s. Indeed, it was a strange world that could seem to turn Hearn, perhaps the most notorious and uncompromising evolutionist in the evangelical world, into a dangerous crypto-creationist.[7]

That both the NAS and the ASA directed their advice to teachers rather than lawmakers reflected a change in venue for creation-evolution contests. Faced with a series of losses in state assemblies and a string of negative decisions in federal courts, creationists in the mid-1980s shifted from headline-grabbing legislative battles to quiet persuasion among teachers and school-board members. The Supreme Court decision of 1987 left open the possibility of *voluntarily* teaching creation science, and many teachers opted to do just that. Polls of public-school teachers in various states showed that substantial percentages—ranging from 30 in Illinois to 69 in Kentucky—favored including creationism in the curriculum. Right-wing creationists in public education banded together in the National Association of Christian Educators to drive not only evolutionism but secular humanism, forced unionism, and socialism from the schools.[8]

Support sometimes came from Christian parents who wanted their children exposed to creationism, a growing number of whom turned to home schooling. In an era when pedagogical policy dictated that students learn morals and values by critically choosing among alternatives, it seemed inconsistent to some parents that science allowed for no choice. One urban science professor, Martin Eger (b. 1936), observed such skepticism in a rural community in upstate New York. "If it is a good thing for children to consider all sorts of alternatives in moral decision-making, no matter how repugnant—stealing, cheating, betraying one's friends—all for the sake of developing critical reason and autonomy, then why, suddenly, when we come to evolution, is it far more important to learn *right answers* than to think critically?" asked puzzled parents. "Why *just here* are certain alternatives taboo, even for the sake of discussion—despite the fact that being wrong about a scientific theory of remote origins can never have consequences as grievous as being wrong in one's moral outlook?" If there were compelling reasons for teaching morality and science differently, then educators had a duty to disclose them, argued Eger.[9]

Well into the 1980s some evolutionists consoled themselves with the notion that "creationism is an American institution, and it is not only American but specifically southern and southwestern."[10] Creationism may have grown most luxuriantly in the Sun Belt of the United States, but it also thrived in communities throughout the nation and, increasingly, around the world. Unlike the antievolution crusade of the 1920s, which remained confined mainly to North America, the creationist revival of the last third of the century rapidly spread overseas as American creationists and their books circled the globe. By the 1990s scientific creationism, though made in America, had become a small-scale international phenomenon.

CREATIONISM·IN GREAT BRITAIN

The early 1960s found British creationism in a state of lethargy. The Evolution Protest Movement (EPM) had fallen into the hands of Albert G. Tilney (1891–1976), a theologically dogmatic and scientifically illiterate schoolmaster and pastor, who from the mid-1950s through the mid-1970s ran the organization "as a one-man band." He personally composed more than a hundred EPM pamphlets, dominated thematically by an unwavering devotion to the gap theory. Although he reported impressive membership gains in the late 1950s, from about two hundred in 1955 to over five hundred in 1960, the rolls included the names of many inactive members. For years the EPM did little besides publish its four-page pamphlets, pass out copies of Vice President John R. Howitt's booklet *Evolution*, and hold an annual public meeting, where leaders often found themselves "preaching to the converted." Several of Britain's most prominent creationists, including Robert E. D. Clark, declined to serve as officers, as did William R. Thompson (1887–1972), a well-known Canadian entomologist who in 1956 scan-

dalized evolutionists by attaching a negative introduction to an edition
of Darwin's *Origin of Species*. "The Movement has been largely ignored or ridiculed by both the scientific and ecclesiastical establishments," mourned a dejected officer. "Compared with 'flat earthists' by
the one and labelled 'fundamentalists' by the other, it has had either a
bad press or none at all."[11]

Rumors of revivals, both foreign and domestic, soon raised the flag-
ging spirits of British creationists. By the early 1960s conservative
evangelicals in Britain were massing to retake some of the ground lost
to liberals in recent decades. In journals such as the *Banner of Truth*
and in pulpits filled by such men as Martyn Lloyd-Jones (1899–1981),
a Welsh physician-turned-minister who "awed packed congregations
with his blend of logic, fire and close attention to the text of scripture,"
the conservatives mounted a spirited counterattack. Though they remained a small minority among British churchgoers and failed to agree
among themselves on the acceptability of evolution, their emphasis on
the plain meaning of Scripture prepared the way for a second revival:
young-earth creationism.[12]

Within a year of the publication of *The Genesis Flood* in 1961, Tilney
was predicting that Whitcomb and Morris's "revolutionary re-inter-
pretation" of earth history would usher in "a new era." Already the
EPM had distributed scores of copies, and short supplies had forced
the secretary to ration the remaining stock. But Tilney tempered his
enthusiasm with caution. He worried that the American creationists
might be "cranky" and "credulous." And he feared that their fondness
for flood geology would lead them to overlook that Genesis 1 described
a "Restoration-Week," not a "Creation-Week." Over the next few
years, however, while still remaining partial to the gap theory, he
"steadily and gladly accepted the fossils as being (mostly) the story of
the sudden departure of life from this planet."[13]

By the late 1960s, when *The Genesis Flood* appeared in an English
edition, Whitcomb and Morris had become "household names among
British evangelicals." As one British creationist observed, "More than
any other single factor, this scholarly but highly controversial volume
lifted creationism from the Gospel Hall tract-rack to the College seminar room." Not all British creationists welcomed this importation of
American fundamentalism. Clark, for example, thought Whitcomb and
Morris were as "silly and dishonest" as the evolutionists they opposed.
He especially resented the way Morris had twisted his argument
against evolution from the second law of thermodynamics into "base-
less rubbish" by associating it with the Fall of Adam and Eve.[14]

In the fall of 1973 Morris himself visited London to address an international meeting of Baptist ministers at the Metropolitan Tabernacle, where A. C. Dixon, editor of *The Fundamentals*, had once pastored. While there Morris met with about twenty creationist scientists, including Edgar C. Powell (b. 1947), a Reformed Baptist teacher of secondary-school geography and geology. Alienated by Tilney's tendency to wrap creationism in his own peculiar interpretations of Scripture, Powell took the lead in organizing a new society, the Newton Scientific Association (NSA). The Newtonians limited membership to professing Christians who affirmed the 1846 Evangelical Alliance Statement of Faith. To avoid the EPM's volatile mix of science and theology, the NSA restricted discussions to scientific issues only. Even when it reprinted technical monographs from the Institute for Creation Research, the association deleted all biblical references. The NSA took no official position on the age of the earth, but most members favored flood geology. In part because the NSA never expanded its base very far beyond the Metropolitan Tabernacle, tightly controlled by an American-style fundamentalist preacher, the NSA died within a decade.[15]

Far more successful was the Biblical Creation Society (BCS), founded in 1977 by the Scottish minister Nigel M. de S. Cameron and a group of evangelical students concerned about the popularity of theistic evolution among conservative Christians. Repelled by the "wholly negative" attitude of the EPM and encouraged by Martyn Lloyd-Jones and Francis A. Schaeffer (1912–1984), the Swiss-based evangelical guru, the group began building a new creationist society from the bottom up. Although inspired by Whitcomb and Morris's method of examining scientific data in the light of the biblical narrative, BCS leaders refused to limit membership to "young earthers" and, in choosing a name, pointedly rejected the American effort to separate scientific creationism from its biblical roots. In this respect they followed Whitcomb more than Morris; indeed, on at least one occasion they were accused of promoting "Whitcombism." In the British context, where there were no constitutional barriers to teaching creationism in state-supported schools, the biblical orientation had considerable appeal. Within a decade the society had signed up approximately 750 members, and its journal, *Biblical Creation* (renamed *Origins* in 1987), was reaching thousands. The vast majority of members came from conservative evangelical traditions. Of the seventeen officers in the early 1980s, eleven were Reformed Baptists, three belonged to the Church of Scotland, and one each came from Presbyterian, Brethren, and Pentecostal churches.[16]

For its president, the BCS chose the nation's most respected creationist scientist of the late twentieth century, E. H. Andrews (b. 1932), longtime head of the department of materials at Queen Mary College, University of London. A Reformed Baptist, Andrews had come under the influence of Whitcomb and Morris in the 1960s and converted to flood geology. But he did not want to be known as a "camp follower" of the Americans or to have anything to do with the "nonsense element" that plagued creationism abroad. He insisted on dating the operation of the laws of thermodynamics from the creation, rather than the Fall, and he refused to make a dogma of a young earth. In fact, he broke with conventional six-day creationism in suggesting that the first day of creation "might have been of indefinite length."[17]

Another of the BCS's leading lights, David C. Watts (b. 1945), also expressed reservations about Whitcomb and Morris's style of flood geology, which he had embraced as a student. A senior lecturer in biomaterials science in the faculty of medicine at the University of Manchester, Watts went even further than Andrews in dissenting from the Americans, who he thought had made a "great mistake" in putting the age of the earth at the top of the creationist agenda. Admitting the possibility of life on earth before the Edenic creation, he claimed that "the '100%' Flood-Geology theory (as classically formulated by Morris and Whitcomb) is not the only view to legitimately claim the attention of Christian geologists." As he moved slowly from strict to progressive creationism, he came to regret that "many evangelical scientists who are not 'evolutionists' are having to say that they are not 'creationists' either by this narrow definition, despite their firm commitment to creation, the Creator and his word."[18]

Because of transatlantic influences as well as a commitment to the historicity of the Genesis record, the BCS could not escape dealing with the geological significance of the flood. After years of debate, society leaders at a strategy session in 1989 drafted a carefully worded statement aimed at accommodating both hard-core flood geologists and doubters like Watts:

> The Noahic flood and its after effects were responsible for the formation of substantial portions of the fossiliferous strata. This includes the possibility that some such formations, containing the remains of organisms that died in the years before the flood, were disrupted, reworked and recemented. Equally, regional but large scale catastrophism after the ark grounded on Mount Ararat, may also have produced some fossilisation, as the reverberations of the initial global event subsided over some hundreds of years.

Members might differ on the exact mix of miracle and natural law responsible for the deluge, but everyone would agree that, with the exception of the refugees on Noah's ark, the Genesis flood destroyed all living creatures.[19]

By the early 1970s even the Evolution Protest Movement was grudgingly recognizing deluge geology, as symbolized by the largely honorific vice presidencies awarded to the Americans Morris and Frank Lewis Marsh. In mid-decade several young-earthers joined the council. Their addition at one stroke dramatically dropped the average age of that superannuated body and shifted the balance of power away from Tilney and his gap-theory allies, who could only hope that the creationist cause would not be permanently damaged "by the reckless and sometimes abusive, ill-supported claims over the Young Earth theory." Led by insurgent council member David T. Rosevear (b. 1934), a chemist at Portsmouth Polytechnic who shunned "such apparent contrivances" as the gap and day-age theories, the young-earth contingent grew in strength and numbers. In 1980 the Evolution Protest Movement took the name Creation Science Movement (CSM). Five years later Rosevear's ascension to the chairmanship of the society left the advocates of geological ages without a platform. After a straw poll of members in the mid-1980s revealed that "young earthers outnumbered old earthers by more than four to one," the CSM officially incorporated flood geology into its "Deed of Trust," at the same time condemning the gap and day-age interpretations of Genesis as unscriptural. All officers were required to sign the document. Despite shutting down its foreign branches, the movement nearly doubled its membership in the late 1980s and early 1990s.[20]

Although all three of the British antievolution societies warmly welcomed flood geology, none made its acceptance a condition of membership. This unwillingness officially to endorse the theory distressed such Whitcomb-and-Morris loyalists as the combative Welsh chemist A. J. Monty White (b. 1944), whose writings, lectures, and BBC appearances gave him the highest profile of any British creationist in the 1980s. The son of a left-wing atheist, White grew up with "communist leanings" but converted to Pentecostal Christianity while studying chemistry at the University College of Wales, Aberystwyth. Shortly after his conversion he began taking courses in geology and immediately confronted the problem of evolution. In 1969 he married a strict creationist, who convinced him that organic evolution was incompatible with the biblical account of origins. For two years, however, he continued to wrestle with the problem of reconciling science and Scripture. Finally, in 1971, his friend and former roommate Watts gave him a copy of *The Genesis Flood*, and White "saw for the first time" how other Christians harmonized Genesis and geology. "At last," he wrote, "I was a Creationist with a reasoned faith."[21]

White promptly joined the Creation Research Society (CRS) and began contributing pieces to the *Creation Research Society Quarterly*. A couple of years later he agreed to speak to a small church troubled by the effects of evolution on its young people. The positive response to his talk convinced him that the Lord was calling him to a special ministry. He quickly ran off a mimeographed *Creation News Sheet* and mailed it to about forty persons in the United Kingdom; soon he was sending out thousands of copies worldwide. By this time he had joined an independent Reformed church, earned a Ph.D. degree in gas kinetics, and taken an administrative position at the University of Wales Institute of Science and Technology in Cardiff. As a specialist in kinetics, he had one quarrel with Whitcomb and Morris: the way they linked the second law of thermodynamics with the Fall. Because thermodynamic principles controlled all chemical processes, it seemed unlikely to him that, even in the perfect pre-Fall world, animals could have lived and the sun given light in the absence of thermodynamic laws.[22]

Convinced that all interpretations of Genesis that allowed for long periods of time were nothing but "sops to the evolutionists," White schemed to start something like a British CRS, committed exclusively to young-earth creationism. In December 1973, after meeting with Watts and some other creationist friends, he announced that "a number of Christians (including myself) in this country have started a British Creation Research Society." However, when he learned shortly thereafter of the birth of the Newton Scientific Association, he aborted the project and joined the NSA—only to withdraw after discovering that its leadership would allow no mention of "God, the Bible, Noah, or the Flood." A few years later he explored joining forces with the new Biblical Creation Society, but that group's refusal to take "a firm stand for a six day recent creation by Almighty God" convinced him that it, too, would never be more than just another antievolution society. Besides, he was personally offended by the behavior of its leaders, who, like Tilney and the EPM, seemed more interested in his mailing list than his ideas.[23]

By the late 1970s White had washed his hands of all three British societies. "These organisations are all anti-evolution and not one of them (including the recently formed, and in my opinion incorrectly named, Biblical Creation Society) is prepared to take a stand on a recent creation by Almighty God in six literal days," he declared judg-mentally in his *Creation News Sheet.* "Unlike the American-based Creation Research Society, these societies are not prepared to pro-claim what the Bible teaches about the creation and early history of the earth, but insist that to demolish evolutionary teachings by the use of scientific facts, data and arguments is sufficient." Scientific evidence might destroy evolution or demonstrate a youthful earth, but only the Bible taught that "in six days the Lord God created all things."[24]

A major impediment to the propagation of flood geology in the United Kingdom was the "Made in America" label. Critics of the Brit-ish disciples of Whitcomb and Morris accused them of "injecting an alien element into British evangelicalism" and importing "the shibbo-leths of American fundamentalism." Some opponents even alleged ties to "right-wing North American politics." Because of the prejudice against things American, British creationists went out of their way to distance themselves from the American movement, stressing the in-digenous roots of creationism in the local evangelical reformation and, where possible, quoting British rather than American authorities. The leaders of the BCS tried to deflect the anti-American argument by pointing to their advocacy of *biblical* creationism and their "cautious approach to Flood Geology and Scientific Creationism." In a public address, reported in *Nature,* Andrews warned against following the Americans in divorcing creationism from the Creator. But, as at least one young earther acknowledged, British flood geologists who denied their intellectual debt to America were reacting culturally, not histor-ically. Making the best of the embarrassing historical relationship, David C. C. Watson invoked the memory of Dwight L. Moody's re-vivals in the 1860s and Billy Graham's campaigns in the 1940s to an-swer the question "Can any good thing come out of America?" with a resounding "Yes!"[25]

Just as American creationists set their sights on equal time in state-run classrooms, so British creationists fought for air time on the state-controlled BBC. Radio and television exposure, they reasoned, would not only popularize their cause but legitimize it. Since almost the beginning of the EPM its leaders had been railing against the BBC for abusing its monopoly of the airwaves to suppress creationism. If British creationists envied one aspect of life in America it was "freedom of the air." In the United States any preacher with a few thousand dollars in his pocket could start a broadcast ministry; in Great Britain no amount of money could buy time on the BBC, which repeatedly refused "to grant creationists a fair chance to present their case." Such arbitrari-ness led the American visitor Gish, who had become something of a media star back home, to judge the BBC "the worst when it comes to excluding an alternative point of view." In 1981 White finally broke the broadcast ban, but even then creationists complained about the way his comments had been edited.[26]

INTO ALL THE WORLD

The British experience tended to repeat itself wherever the Evolution Protest Movement had established a foothold: in Canada, Australia, New Zealand, and South Africa. For years the EPM had enjoyed a virtual monopoly over the minuscule antievolution movements in these countries, but with the coming of flood geology in the 1960s the EPM branches struggled to stay afloat.[27]

Despite repeated efforts at colonization, the EPM had never done well in North America. James D. Bales's attempt in the late 1940s to establish an EPM beachhead in Searcy, Arkansas, had met with little success; and the Ontario psychiatrist Howitt, though a longtime vice president of the parent society, had never done much organizationally. Tilney, however, kept looking for the right person to get things going, and in the late 1960s he finally spotted his man: W. Dennis Burrowes (b. 1918), an Oxford-educated specialist in agricultural statistics and economics from Victoria, British Columbia, and a cousin-in-law of Robert E. D. Clark's. The Jamaican-born, Brethren-reared Burrowes had joined the EPM in 1946 while studying in England on a Rhodes scholarship, and during a return visit in 1969 he had been tapped by Tilney to be "a secretary and distributor for Canada." Burrowes questioned his qualifications for the job but eventually agreed to take it. "I feel quite free (and increasingly responsible) to make my protest to teachers and librarians against the monopoly & the falsity of evolutionary teaching in schools & libraries, and to draw attention to viable scientific alternatives," he explained, adding that in so doing he would stay away from "Scripture-oriented treatments."[28]

In the fall of 1969 Burrowes decided to launch his branch society by contacting EPM members living in Canada. But Tilney's records were in such disarray the secretary could not provide the names and addresses of the twenty or so members supposed to be in the country. In 1971 Burrowes invited a few other creationists, including Howitt and Arthur C. Custance, to form a North American EPM Council. Like Burrowes himself, these men felt uncomfortable with the young-earth orthodoxy of the Creation Research Society. As a result of reading *The Genesis Flood*, Burrowes had for a time moved toward flood geology, but contrary evidences, including Custance's defense of the gap theory in *Without Form and Void: A Study of the Meaning of Genesis* 1.2 (1970), had pulled him up short.[29]

Burrowes hoped to steer a middle course between the dogmatism of the CRS and the diffidence of the EPM. Writing to Custance, he described the contrasting styles of creationism on the two sides of the Atlantic:

> On this side we have all the zeal & assurance that all forms of macro-evolutionary doctrine are false, and [that] truth will prevail; on that side they are still much affected by the diffident, apologetic, compromising tendency that has made the [English] church's public testimony so feeble during this century, and on the other hand they seem to have all the (desired) critical faculty to see the "howlers" and limitations in the creationist literature from this side!

The ideal mix, he thought, would result from "shaking up together" both approaches.[30]

Custance, peeved by the meager attention the EPM had given to his frequent publications on science and religion, declined to get involved with any affiliated society. But even without Custance's collaboration, Burrowes succeeded in founding an EPM branch that came to be called the North American Creation Movement. True to Burrowes's intentions, the Victoria-based creationists escaped polarization into old-earth and young-earth camps and, in doing so, attracted about 120 subscribers in five years. Such unprecedented activity in North America gratified EMP leaders back in Great Britain, but left Burrowes, who witnessed the far more spectacular growth of flood-geology groups, a little cold. Already by the late 1960s, Earl G. Hallonquist (1907–1985), an industrial chemist across the straits in Vancouver, was planting the young-earth flag of the Bible-Science Association in Canada. Burrowes watched with mixed feelings as the aggressive flood geologists prospered while he settled for their leavings. Boosted by frequent visits from prominent creationists to the south, Hallonquist's efforts produced an independent Creation Science Association of Canada, with provincial divisions from British Columbia to Ontario. By the mid-1980s Morris could write that "creationists," meaning flood geologists, "have been more active and productive in Canada than in probably any country except the United States." At the close of 1988 Burrowes, unable to support the new young-earth orthodoxy within the CSM but unwilling to fight fellow creationists, suspended his operation.[31]

If the United States and Canada led the world in promoting strict creationism, then surely Australia and New Zealand could not have been far behind. Despite an inspirational visit by Tilney in the late 1960s and the dedicated efforts of John Byrt, a schoolmaster, the Australian branch of the EPM had, according to one account, dwindled "almost to the vanishing point" by 1970. Then came the flood geologists, beginning with Morris in 1973 and Gish two years later. Their lectures and books at first led to the revitalization of the EPM, which swelled to 350 members by 1976. But the presence of so many young earthers in a society ambivalent about creation science proved to be a mixed blessing. As Byrt explained to Burrowes, he was reluctant to make an issue of flood geology because so many EPM members in Australia "would be quite happy to endorse the CRS creed." He feared that emphasizing the differences between the traditional goals of the EPM and the CRS would only "invite their disappearance."[32]

The very next year followers of Whitcomb and Morris in South Australia, led by Carl Wieland, a physician, organized their own society, the Creation Science Association, which adopted a statement of belief explicitly modeled after that of the CRS. The founders vowed to follow the Creation Research Society and the Institute for Creation Research "in presenting the very substantial scientific case for creation." In 1978 they began publishing a magazine, *Ex Nihilo*, "to explain and promote special creation as a valid SCIENTIFIC explanation of origins." In 1980 the South Australians merged with a like-minded group from Queensland to form the Creation Science Foundation. By the mid-1980s the foundation was supporting a staff of fourteen, including a geologist, Andrew Snelling, and a master propagandist, Kenneth A. Ham (b. 1951). A consulting geologist with a doctorate from the University of Sydney, Snelling edited a technical journal (a companion to the lay-oriented *Creation Ex Nihilo*) and headed up the project to bring out an updated edition of Whitcomb and Morris's *Genesis Flood*. Ham, a charismatic public speaker and former biology teacher, later joined the staff of the ICR in southern California.[33]

Because Australia, like Great Britain, permitted the teaching of religion in public schools, the early creation scientists felt confident that attractive curriculum materials would almost certainly get them "a fair hearing in the Public School system." In what one critic called the "major Australian triumph" of creationism, the minister for education in the conservative state of Queensland came out in favor of teaching creation science in state schools. Creation science also prospered in private Christian schools, including some run by the Catholic church. In fact, the popularity of creationism among fundamentalist Catholics prompted the Catholic Education Office in Sydney to issue a biting anticreation booklet, *The Bumbling, Stumbling, Crumbling Theory of Creation Science* (1987), by Barry Price (b. 1935). The author, a teacher of science and religious studies, minced no words in damning creation science as "pseudo-science and utter nonsense" that relied on "fraud and deceit to achieve its ends." When the outcries of creationists led the Catholic church to withdraw his work from circulation, Price on his own expanded his sensationalist critique into *The Creation Science Controversy* (1990). Characterizing creation science as a sleazy "American import onto the Australian scene," he sought to expose it as "*a unique and brilliant marketing approach* for a particular brand of religious fundamentalism."[34]

At first Australian scientists tended to dismiss creationism as a "bit of a joke," but by the mid-1980s they were discussing it at length in learned journals and occasionally taking the time to engage in highly publicized debates. In one of the most celebrated encounters in the history of creation-evolution forensics, Ian R. Plimer (b. 1946), the rough-and-ready head of the geology department at the University of Newcastle, introduced Gish to a new style of debating: "You don't put up with bullshit, and you take no prisoners." A reporter for the *Sydney Morning Herald*, who described Plimer's street-fighting style as going in "boots and all, aiming for his opponents' kneecaps," gave the following colorful account:

> For more than 20 blistering minutes, Professor Plimer mocked, ridiculed, and challenged every tenet the movement holds dear, and made a string of blunt personal allegations about some of its more prominent members. At one point he even donned insulating gloves, took a live electric wire and offered Dr Gish the opportunity to electrocute himself.

Because Plimer viewed his debate with Gish as a political rather than scientific exercise, he refused to argue genteelly "about spots on butterflies or the speed of light" or to exchange common courtesies. An appalled Gish described the ordeal as "the most disgusting performance I've ever witnessed in my life."[35]

From nation to nation the spread of flood geology conformed to an almost predictable pattern. A six-week campaign by Morris in New Zealand in 1973 concluded with the formation of the Creation Literature Society of New Zealand. A visit by Morris and Gish to Korea in

1980 resulted in the organization of the Korea Association of Creation Research, which within a decade had recruited nearly one thousand members (including about three hundred with doctorates) and had established several branches in the United States. A South African tour by Gish in 1983 produced a national creation-science association where only a small outpost of the EPM had previously existed. By the 1990s the roster of creation-science societies and centers, large and small, included groups in Japan, Taiwan, Hong Kong, Singapore, the Philippines, India, Sri Lanka, Nigeria, Kenya, Brazil, Argentina, Mexico, El Salvador, the Dominican Republic, and Puerto Rico, to say nothing of European societies.[36]

In continental Europe the Dutch took the lead in propagating flood geology. Already by the mid-1960s the young Nicolaas A. Rupke was serving as the CRS "agent in Holland for new members" and contributing to the *Creation Research Society Quarterly*. The creationist cause suffered a serious setback in 1968 when J. R. van de Fliert, an evangelical geologist on the faculty of the Free University of Amsterdam, published his caustic commentary on *The Genesis Flood*. But with the translation of several creation-science books into Dutch, relatively liberal access to television time, and the leadership of such men as William J. Ouweneel, a sometime research biologist on the staff of the Royal Netherlands Academy of Sciences, flood geology rebounded. In 1974 Dutch creationists formed the Foundation for the Advancement of Studies Faithful to the Bible and the next year began publishing a creationist magazine edited by Ouweneel.[37]

No single person did more to popularize creationism in Europe than the English-born pharmacologist Arthur E. Wilder-Smith (b. 1915), often described as "Europe's leading creationist scientist." Reared in England as an Anglican, Wilder-Smith as a young man joined the fundamentalist Plymouth Brethren. His three earned doctorates—in physical organic chemistry from the University of Reading, in chemotherapy from the University of Geneva, and in pharmacology from the Eidgenossische Technische Hochschule in Zurich—opened many doors and brought him widespread recognition as a leading creationist theoretician. Fluent in German, Wilder-Smith began his creationist ministry lecturing to evangelicals in Germany after World War II; later he worked in Switzerland, the United States, Norway, and Turkey. While living in the United States, in 1965, he accompanied Clifford L. Burdick to the Paluxy River for a personal inspection of the alleged human tracks. Of his many books in several languages, one reportedly sold a million copies. As a result of Wilder-Smith's influence, as well as visiting lectures by Ouweneel and the translation of *The Genesis Flood*,

German creationists in the late 1970s formed a society and began publishing a monthly journal, *Factum*. Organized creationism on a smaller scale also appeared in such countries as Italy, France, Spain, Portugal, and Sweden. In 1984 European creationists began holding biennial congresses that brought together flood geologists from across the continent.[38]

From time to time creation science sprang up in the most unlikely places. In 1987 the Institute for Creation Research reported that for several years "the Turkish Ministry of Education and a private Islamic foundation have been translating ICR books [without biblical references] into Turkish," and that every science teacher in the public schools of Turkey received a copy of *Scientific Creationism* as a basic reference tool. In 1990 the ICR announced that "fourteen top Soviet scientists" had formed a Russian Creation Science Association. The most prominent member of the group, the biomedical scientist Dmitri A. Kouznetsov (b. 1955), contributed a monthly column on "Science without Atheism" to the *Protestant*, one of the most widely read Christian magazines in the Soviet Union. Taking advantage of *glasnost*, the ICR Graduate School appointed Kouznetsov an adjunct professor of biology.[39]

WHY FLOOD GEOLOGY?

In reviewing the history of special creationism during the past century or so, we have observed a movement characterized by remarkable growth, change, and heterogeneity. At the beginning of the twentieth century, the handful of scientifically trained creationists worked in obscurity and isolation. Less than a hundred years later, well-credentialed creationists numbered in the hundreds, if not thousands, and organized creationism could be found around the world. Instead of being ignored or dismissed as a bad joke, creationists found themselves the objects of intense criticism and concern.

During the early decades of the twentieth century, few creationists, even among hard-shell fundamentalists, insisted on a young earth or a fossil-producing flood. Some naïve readers of the Bible no doubt assumed that the date 4004 B.C. found in the margins of the first chapter referred to the original creation of the earth, but except for the Adventist disciples of Ellen G. White they almost never committed such beliefs to writing. By applying the unquestionably orthodox day-age and gap theories to Genesis 1, even the staunchest defenders of biblical inerrancy could accommodate the claims of historical geology. But by the end of the century, through the efforts of men such as George McCready Price, John C. Whitcomb, Jr., and Henry M. Morris, the very word *creationism* had come to signify the recent appearance of life on earth and a geologically significant deluge. Although believers and skeptics alike often referred to this view as young-earth creationism, the core belief was not a young earth (which even Price at times rejected) but young *life*. On almost every other issue—from the age of the universe and the origin of the laws of thermodynamics to the limits of organic variation and the number of fossils attributable to Noah's flood—the scientific creationists indulged in open and spirited debate.

Twentieth-century creationism cannot be reduced to a mere expression of the "anti-intellectual tradition" in America, as some critics have done, without considerable distortion.[40] Strict creationists may have opposed elite science, but they developed an alternative tradition that in some ways was just as "intellectual" as the one they rejected. What most distinguished the leading creationists from their evolutionary counterparts was not intellect or integrity but cosmology and epistemology. Both the geneticist Theodosius Dobzhansky, who exchanged lengthy letters with the flood geologist Frank Lewis Marsh, and the British evolutionist H. S. Shelton (1876–?), who frequently corresponded with the EMP's Douglas Dewar, conceded as much. "The difference between us is a little difficult to describe," observed Shelton, who found himself in surprising agreement with Dewar regarding factual matters but not interpretations. "You *see* things differently."[41] And creationists saw things differently because they viewed nature through Mosaic lenses.

In making their case, creationists occasionally appealed to scientific and philosophical authorities, but for the most part they worked outside the context of established science and philosophy. Turn-of-the-century antievolutionists such as Price might have invoked the memory of Francis Bacon, just as his late-century successors sometimes dropped the names of Karl R. Popper and Thomas S. Kuhn. But substantively the creationists cobbled their populist epistemology independently of philosophical experts and acquired their definition of science from the obvious place: the dictionary. Although they adorned their literature with the names of scientists who questioned evolutionary orthodoxy—William Bateson in the 1920s, Austin H. Clark in the 1930s, Richard B. Goldschmidt (1898–1958) in the 1940s, William R. Thompson in the 1950s, Gerald A. Kerkut in the 1960s, and Robert H. Dott, Jr. (b. 1929), and Stephen Jay Gould in the 1970s and 1980s—these citations were little more than literary ornaments. Both Gould and Dott, for example, vigorously opposed creationism of any kind, but scientific creationists nevertheless appropriated their "neo-catastrophism"—Dott's use of nonuniformitarian "episodic sedimentation" to explain the formation of geological strata and Gould's employment of "punctuated equilibria" to develop a theory of evolution by spurts—in defending their deluge model of earth history.[42] To understand twentieth-century creationism, little knowledge of formal science and philosophy is necessary; familiarity with the Byzantine world of popular religion is essential.

Like evolution, creation could readily be turned to social and political ends. Some creationists, for example, tried to link the practice of abortion with belief in evolution. "Why worry about disposing of 'unwanted' babies in the womb," asked one rhetorically, "since they are simply baby primates?" Others attempted to connect feminism and evolution. "Feminism is anticreation and antiChristian," declared a contributor to the *Creation Research Society Quarterly*, an ostensibly scientific journal:

> There can be no such thing as Christian feminism. They are conceptual contradictions like capitalistic communism and evangelical lesbianism. The divine order of the sexes is a corollary doctrine of the order of creation. It establishes the family as the natural unit within society and the church. . . . The husband/father is given the headship in the family and the church with the wife/mother as an assistant and chief educator of the children.

In a non-Christian society governed by "survival of the fittest," men and women would depart from this ideal to the extent of competing equally for jobs.[43] Not surprisingly, no women's names appeared on the roster of CRS leaders. Creationists may have liked their women domestic and their politics conservative (though not all did), but in general, politics lagged far behind religion as a motivating force.

Any explanation of the creationist revival must distinguish between the general question of why so many conservative Christians spoke out against evolution after about 1960 and the narrower issue of why increasing numbers of creationists shifted from accommodation to repudiation of historical geology. In the United States, the outcry against evolution undoubtedly owed its timing in large part to the abrupt reappearance of evolution in many high-school classrooms in the 1960s as federally funded biology texts featuring evolution replaced books that treated evolution in muted tones. The creationists also capitalized on the romantic rebellion against the scientific establishment that catapulted Immanuel Velikovsky's *Worlds In Collision* (1950) to the top of

the *New York Times* best-seller list and made the Jewish catastrophist
a cult hero on the fringes of many college campuses in the 1960s.
Later, antievolution agitation rode the political coattails of the New
Religious Right, which arose to national prominence in the 1970s. And
all along it benefited from the lingering suspicion, expressed by the
country singer Merle Haggard, that "evolution is a laughing matter for
anybody that's got a rational mind."[44] But such factors cannot explain
why so many creationists turned from the gap and day-age theories to
flood geology, which necessitated a total rejection of conventional earth
history.

There is, of course, no simple answer to the question "Why flood
geology?" The reasons for adopting that particular interpretation of
Genesis varied from one religious tradition to another and even from
one person to another. But the testimony of countless converts suggests that the lion's share of credit—or blame—for the popularity of
flood geology must go to John C. Whitcomb, Jr., and Henry M. Morris, who in *The Genesis Flood* gave George McCready Price's Adventist flood theory a proper fundamentalist baptism and then skillfully
promoted it as biblical orthodoxy. By starting with the text of Genesis
and then fitting scientific data into that framework, they appealed to
Christians "fed up with articles and books which tried to make scripture conform to the latest theory." They succeeded where others had
failed, explained one admirer, partly because they provided "a total
synthesis of biblical data and hard scientific facts . . . while arguing
against other syntheses. For the first time in years, creationists were
not poking away at isolated problems and scoring points in narrow
areas of conflict."[45]

For believers in the verbal inerrancy of the Bible, flood geology
required no assumptions of days that really meant ages or of temporal
gaps that went unmentioned. By showing how the deluge of Noah
compressed earth history into no more than ten thousand years, Whitcomb and Morris at one stroke eliminated the need for such "biblical
gymnastics" and deprived evolutionists of the time required for the
natural origin of species. David C. C. Watson succinctly summed up
the appeal of such maneuvers: "Tens of thousands of Christians have
been convinced by Morris & Whitcomb's books because *they make
sense of the Bible.*"[46]

Except for the relatively small band of postmillennial Reconstructionists, most flood geologists (in America at least) came from churches
awaiting Christ's soon return to earth. And for Christians expecting the
imminent end of the present age—whether premillennial Baptists and
Adventists or amillennial Lutherans and Church of Christ members—

Whitcomb and Morris offered a compelling view of earth history framed by symmetrical catastrophic events and connected by a common hermeneutics. "If you take Genesis literally," reasoned Morris, "you're more inclined to take Revelation literally." Or as an Australian Adventist put it, "The cataclysmic beginning of our world revealed in the book of Genesis *guarantees* the supernatural end of our planet when Eden lost will become Eden restored." In the opinion of the British creationist E. H. Andrews, logical consistency demanded that anyone looking "forward to a miraculous end to this present age" not "rigorously exclude miracle from the process of creation."[47]

Finally, for fundamentalists seeking what the anthropologist Christopher P. Toumey calls "scientific sanctification," flood geology came with the endorsement of real scientists, who assured them that nature, like Scripture, argued for a nonevolutionary history of life.[48] The confluence of these largely religious factors in a culture increasingly tolerant of challenges to the authority of elites raised the folk science of flood geology to heights of popularity only dreamed of by George McCready Price. Its shocking success, limited though it may have been, shattered facile beliefs about the inevitability of secularization and scientific progress and called into question long-cherished convictions about the relationship between science and religion.

ACKNOWLEDGMENTS

After years of benefiting from the kindness of others, I am happy to be able to thank publicly the many institutions and individuals who contributed to the making of this book. In the summer of 1970 I began collecting the materials on which this study is based. The department of history at Andrews University, where I was then completing my first year of teaching, provided me with a part-time research assistant, Tom Gammon, to help ferret out documents for an article on the history of Seventh-day Adventist responses to evolution. Ten years later the Graduate School Research Committee of the University of Wisconsin-Madison, to which I had moved in 1974, awarded a small grant that enabled me to employ one of my graduate students, Rennie B. Schoepflin, to assist in preparing an article on creationism in the twentieth century. That exercise and the promptings of my colleague David C. Lindberg convinced me to undertake a book-length study, for which, from 1983 to 1985, I received substantial support from the John Simon Guggenheim Foundation, the National Science Foundation (Grant No. SES–8308523), and the Graduate School Research Committee of the University of Wisconsin-Madison. During that time the late Paul K. Pruyser and the Menninger Foundation in Topeka, Kansas, generously provided me with a base of operations away from home. A. Hunter Dupree, David C. Lindberg, George M. Marsden, and Martin E. Marty all aided me in my quest to secure funding.

Judith Walzer Leavitt and my other colleagues in the department of the history of medicine at the University of Wisconsin-Madison good-naturedly indulged my interest in "pathological science" for longer than they (or I) expected; and Carolyn Hackler, the departmental secretary, repeatedly processed my words with care. At various times I took advantage of the computing expertise of Bernie Rozmovits and the research skills of Dawn Corley, Don DeLand, Tim Kruse, Lisa MacPherson, and Sarah Pfatteicher.

Many friends and a few strangers pointed out pertinent sources or provided otherwise useful information. Marian Finger, whom I have never met, sent me countless packets of documents clipped from newspapers or copied from magazines. William Kostlevy and others associated with the Wesleyan/Holiness Studies Project at Asbury Theological Seminary informed me about creationist sources in the Holiness and Pentecostal traditions. Eric Anderson passed on correspondence relating to the Price-Clark affair that he rescued from the soon-to-be-destroyed files of a former president of Pacific Union College. Among the many others who lent a hand are Michele Aldrich, Eileen Barker, Paul Boyer, John Burnham, Lester Bush, Vern Carner, Joel Carpenter, Robert Dott, Marguerite Dupree, William Edmondson, Richard Eyde, Lawrence Friedman, Willard Gatewood, Neal Gillespie, Alan Graebner, Jack Haas, Mark Kalthoff, Donald McAdams, Benjamin McArthur, Donald McNally, Robert Miller, Gregg Mitman, James Moore, Albert Moyer, Paul Nelson, Lynn Nyhart, Jon Roberts, the late Ernest Sandeen, Todd Savitt, Robert Schadewald, Roger Schultz, Robert Siegfried, Joseph G. Smoot, Michael Sokal, Lester Stephens, Rodney Stiling, Christopher Toumey, Paul Waggoner, David Watts, George Webb, and James Harvey Young.

My history of creationism rests in large part on evidence gleaned from dozens of face-to-face interviews and from manuscripts and letters in over eighty collections, many still in private hands. For allowing me to pry orally into their backgrounds and beliefs, I wish to thank the following persons, several of whom are now deceased:

Richard Acworth	Clifford L. Burdick
E. H. Andrews	W. Dennis Burrowes
Stephen A. Austin	Larry G. Butler
Alonzo L. Baker	J. Frank Cassel
James D. Bales	Alan J. Chapman
Cordelia Erdman Barber	Ervil D. Clark
Douglas A. Block	Harold W. Clark

Robert E. D. Clark
Molleurus Couperus
F. Alton Everest
Duane T. Gish
P. Edgar Hare
Walter R. Hearn
Robert E. Kofahl
J. Laurence Kulp
Walter E. Lammerts
Walter Lang
Frank Lewis Marsh
Russell L. Mixter
Henry M. Morris
Edwin A. Olson
Edgar C. Powell
Bernard Ramm
C. Brandon Rimmer

Richard M. Ritland
Donald H. Rohrer
David T. Rosevear
Ariel A. Roth
Nicolaas A. Rupke
Wilbert H. Rusch
Nell J. Segraves
James R. Sims
Harold S. Slusher
Edward J. Specht
C. E. A. Turner
R. D. Walker
David C. Watts
A. J. Monty White
John C. Whitcomb, Jr.
James M. Wiggert
Kurt P. Wise
Davis A. Young

A number of creationists—including Thomas G. Barnes, Jerry Bergman, Bolton Davidheiser, Vernon L. Grose, James N. Hanson, John N. Moore, Jean E. Sumrall, William J. Tinkle, Rita Rhodes Ward, and Randy L. Wysong—answered my questions over the telephone. Other participants in the creation-evolution debates—including Ernest S. Booth, Arthur J. Jones, Carl H. Krekeler, Lane P. Lester, Andrew Snelling, Arthur E. Wilder-Smith, and Daniel E. Wonderly—responded to my queries by letter.

The following manuscript collections yielded information of value:

Bernard Acworth Papers, courtesy of Richard Acworth
American Scientific Affiliation Collection, courtesy of Robert L. Herrmann; now deposited in Special Collections, Buswell Memorial Library, Wheaton College
James D. Bales Papers, courtesy of James D. Bales
E. S. Ballenger Papers, courtesy of Donald F. Mote
Charles A. Blanchard Papers, Wheaton College Archives
S. James Bole Papers, Nebraska State Historical Society
Samuel Palmer Brooks Papers, The Texas Collection, Baylor University
William Jennings Bryan Papers, Library of Congress
Clifford L. Burdick Papers, courtesy of Clifford L. Burdick

Enoch Fitch Burr Papers, Yale University Library
Larry G. Butler Papers, courtesy of Larry G. Butler
J. Frank Cassel Papers, courtesy of J. Frank Cassel; now deposited in
    Special Collections, Buswell Memorial Library, Wheaton College
James McKeen Cattell Papers, Library of Congress
Robert E. D. Clark Papers, courtesy of the late Robert E. D. Clark
Howard H. Claassen Papers, courtesy of Howard H. Claassen
Molleurus Couperus Papers, courtesy of Molleurus Couperus
J. B. Cranfill Papers, Texas History Center, University of Texas at
    Austin
Creation/Evolution Archive, Special Collections, Iowa State Univer-
    sity Library
Creation Research Society Papers, Concordia Historical Institute
Arthur C. Custance Papers, Special Collections, Redeemer College
James Dwight Dana Correspondence, Yale University Library
Clarence S. Darrow Papers, Library of Congress
Charles B. Davenport Papers, Library of the American Philosophi-
    cal Society
Dawson Collection, McLennan Library, McGill University
A. C. Dixon Collection, Dargan-Carver Library of the Historical
    Commission of the Southern Baptist Convention
Robert H. Dott, Jr., Papers, courtesy of Robert H. Dott, Jr.
V. Raymond Edman Collection, Wheaton College Archives
Evolution Protest Movement/Creation Science Movement Council
    Meeting Minutes, courtesy of David T. Rosevear
Alfred Fairhurst Papers, Transylvania University Library
Orval Eugene Faubus Papers, Special Collections, University of
    Arkansas Library
Sir Ambrose Fleming Correspondence and Papers, University Col-
    lege London Library
Martin Gardner Papers, courtesy of Martin Gardner
Geoscience Research Institute Papers, Archives of the General Con-
    ference of Seventh-day Adventists
Theodore Graebner Papers, Concordia Historical Institute
Asa Gray Correspondence, Gray Herbarium, Harvard University
Arnold Guyot Papers and Correspondence, Princeton University
    Library
P. Edgar Hare Papers, courtesy of P. Edgar Hare
Walter R. Hearn Papers, courtesy of Walter R. Hearn
Robert M. Hutchins Papers, University of Chicago Library
David Starr Jordan Papers, Stanford University Libraries

Howard A. Kelly Papers, Chesney Medical Archives, Johns Hopkins University

R. T. Ketcham Papers, Files of the National Representative, General Association of Regular Baptist Churches

Carl H. Krekeler Papers, courtesy of Carl H. Krekeler

Walter E. Lammerts Papers, Bancroft Library, University of California, Berkeley

McCrady Collection, Jessie Ball du Pont Library, University of the South

Edward McCrady, Sr., Collection, South Carolina Historical Society

Frank Lewis Marsh Papers, courtesy of Frank Lewis Marsh; now deposited in the Adventist Heritage Center, Andrews University

Kirtley F. Mather Papers, Harvard University Archives

William V. Mayer Papers, courtesy of William V. Mayer

Russell L. Mixter Papers, courtesy of Russell L. Mixter; now deposited in Special Collections, Buswell Library, Wheaton College

Byron C. Nelson Papers, Institute for Creation Research

Byron C. Nelson Papers, courtesy of Paul Nelson

J. Frank Norris Collection, Dargan-Carver Library of the Historical Commission of the Southern Baptist Convention

Edwin A. Olson Papers, courtesy of Edwin A. Olson

William Louis Poteat Papers, Baptist Historical Collection, Wake Forest University Library

Alfred M. Rehwinkel Papers, Concordia Historical Institute

William Bell Riley Collection, Northwestern College Library

Richard M. Ritland Papers, courtesy of Richard M. Ritland

Ariel A. Roth Papers, courtesy of Ariel A. Roth

Nicolaas A. Rupke Papers, courtesy of Nicolaas A. Rupke

Wilbert H. Rusch Papers, courtesy of Wilbert H. Rusch

Rennie B. Schoepflin Papers, courtesy of Rennie B. Schoepflin

Charles Schuchert Papers, Yale University Library

Edward J. Specht Papers, courtesy of Edward J. Specht

Lyman Stewart Correspondence, Biola University Library

William Lee Stokes Papers, courtesy of William Lee Stokes

John Roach Straton Papers, American Baptist Historical Society

Sterling B. Talmage Papers, courtesy of William Lee Stokes

David C. Watts Papers, courtesy of David C. Watts

David A. West Papers, courtesy of David A. West

John C. Whitcomb, Jr., Papers, courtesy of John C. Whitcomb, Jr.

A. J. Monty White Papers, courtesy of A. J. Monty White

James M. Wiggert Papers, courtesy of James M. Wiggert

Gerald B. Winrod Collection, Wichita State University
Davis A. Young Papers, courtesy of Davis A. Young
Henry P. Zuidema Papers, courtesy of Henry P. Zuidema

Additional data came from the Moody Bible Institute Archives
(James M. Gray and Alexander Patterson), the Department of Ar-
chives and Manuscripts of the Catholic University of America (George
Barry O'Toole), the Provincial Archives of New Brunswick (George
McCready Price), the correspondence files of the Publishing Department of the General Conference of Seventh-day Adventists (George
McCready Price and Harold W. Clark), the Archives of the General
Conference of Seventh-day Adventists (George McCready Price), the
secretary's records in the Archives of the American Association for the
Advancement of Science (Harry Rimmer), the Special Collections De-
partment of the Pittsburg State University Library (Harry Rimmer),
the personnel files of Wheaton College (S. J. Bole and L. Allen Higley), the Hal Bernard Dixon, Jr., Pentecostal Research Center at Lee
College (Myrtle M. Fleming), and Union University Archives (C. W.
Davis).

I am indebted to more archivists and librarians than I can remem-
ber, but several deserve special mention: Louise Dederen of the Ad-
ventist Heritage Center, Andrews University; Bert Haloviak of the
Archives of the General Conference of Seventh-day Adventists; Gene
DeGruson of the Special Collections Department, Pittsburg State Uni-
versity Library; Robin E. Rider and Sheila K. O'Neill of the Bancroft
Library, University of California, Berkeley; and Clark A. Elliott of the
Harvard College Archives.

In view of the sometimes sensitive nature of my research, I was
repeatedly surprised and gladdened by the cooperation I received
from persons about whom I was writing. Even Henry Morris, who
declined to let me see his correspondence, graciously allowed me the
use of his office and personal library and offered to make specific documents available on an item-by-item basis. John Whitcomb not only
granted me the use of his office and the services of his secretary but
permitted me to read and copy his invaluable letter files. At my suggestion Walter Lammerts donated two decades of correspondence relating to the Creation Research Society to the Bancroft Library of the
University of California, Berkeley, his alma mater. Wilbert H. Rusch,
who had saved few papers of his own, arranged access to the sealed
records of the Creation Research Society, housed in the Concordia
Historical Institute in St. Louis.

I was occasionally less pleased by the responses to my requests for permission to quote from the documents I found. I attempted to contact all of the living authors of manuscripts and letters that I used. Most readily consented to my request, but some refused outright to be quoted and others tried to bargain for more favorable treatment by withholding permission. I have indicated in the notes where permission was withheld.

Balancing the sometimes frustrating experiences with permissions are indelible memories of hospitality and trust: the days spent as a guest in the northern California mountain home of Dos and Molleurus Couperus; the visits with Dick and Juanita Ritland surrounded by the woods of southwestern Michigan; discovering the records of the American Scientific Affiliation in Bob Herrmann's basement and then being given the run of the house and the use of a copier while Bob and his wife went out for the evening; the elderly Robert Clark in his small apartment outside of Cambridge insisting on cooking me supper, just weeks before his death in 1984; sharing family dinner with the Richard Acworths at Potlack Cottage in Church Broughton; David Rosevear entrusting me with the minute books of the Evolution Protest Movement while I traveled around Great Britain.

Because so much of the documentation in this book comes from interviews and documents not readily available to other scholars, I have arranged to deposit most of the tapes and papers I collected in the Adventist Heritage Center, James White Library, Andrews University, Berrien Springs, Michigan. In a very few cases this proved impossible.

Being reluctant to waste anything, even words, I occasionally recycled prose that appeared in previously published articles of mine: " 'Sciences of Satanic Origin': Adventist Attitudes toward Evolutionary Biology and Geology," *Spectrum* 9 (January 1979): 17–30; "Creationism in 20th-Century America," *Science* 218 (1982): 538–44, copyright 1982 by the AAAS; "The Dilemma of Evangelical Scientists," in *Evangelicalism and Modern America*, ed. George Marsden (Grand Rapids, MI: William B. Eerdmans, 1984), pp. 150–60; "The Creationists," in *God and Nature: Historical Essays on the Encounter between Christianity and Science*, ed. David C. Lindberg and Ronald L. Numbers (Berkeley and Los Angeles: University of California Press, 1986), pp. 391–423; and "George Frederick Wright: From Christian Darwinist to Fundamentalist," *Isis* 79 (1988): 624–45. I am grateful to the respective publishers and editors for permission to do so.

In the interest of accuracy I circulated a penultimate draft of this book to a number of readers, both participants and observers. I was both horrified and gratified by what they found. The following persons commented on the entire manuscript: Edward Davis, Martin Gardner, Duane Gish, Walter Hearn, Edward Larson, Benjamin McArthur, Thomas McIver, Henry Morris, Richard Ritland, Jon Roberts, Nicolaas Rupke, Rodney Stiling, Terry Trivett, William Trollinger, Christopher Toumey, George Webb, John Whitcomb, and Davis Young. Others read selected chapters or sections: Thomas G. Alexander, Douglas Block, Larry Butler, Frank Cassel, Molleurus Couperus, Robert Dott, Alton Everest, Richard Hammill, Mark Kalthoff, John Klotz, Laurence Kulp, David Lindberg, Edward Lugenbeal, Russell Mixter, Paul Nelson, Bernard Ramm, David Rosevear, Roger Schultz, Michael Shank, C. E. A. Turner, Monty White, Stephen Wolfgang, and Seung-Hun Yang. It is customary at this point in the acknowledgments to grant such readers a dispensation from the blame for any problems that remain. In this instance, such a disclaimer is no mere formality. A number of the readers disagreed vehemently with my interpretations and even some of my "facts." Remaining errors and distortions are entirely my own fault.

At Knopf, I wish to thank Ashbel Green for first taking an interest in my manuscript, Denise Quirk and Karen Leh for meticulously copy-editing it, and especially Jane Garrett, my editor, for resolutely standing by her author.

Josiah Wedgwood's words in the dedication (found in *The Correspondence of Charles Darwin*, vol. 2: *1837–1843*, ed. Frederick Burkhardt and Sydney Smith [Cambridge: Cambridge University Press, 1986], p. 119) capture my own sentiments exactly. Without my family I might have finished this book sooner, but my life would have been far less meaningful and joyful.

NOTES

1. Henry M. Morris, *A History of Modern Creationism* (San Diego: Master Book Publishers, 1984), pp. 310–11; "Poll Finds Americans Split on Creation Idea," *New York Times*, August 29, 1982; "The Creation," *U.S. News & World Report*, December 23, 1991, p. 59 (Gallup poll). An informal survey of school-board members showed that 67 percent favored the inclusion of creationism in the curriculum; "Finding: Let Kids Decide How We Got Here," *American School Board Journal* 167 (March 1980): 52.

2. The Arkansas Balanced Treatment Act, *Creationism, Science, and the Law: The Arkansas Case*, ed. Marcel Chotkowski La Follette (Cambridge, MA: MIT Press, 1983), pp. 15–19. On the stated differences between scientific creationism and biblical creationism, see Henry M. Morris and Gary E. Parker, *What Is Creation Science?* (San Diego: Creation-Life Publishers, 1982), p. 264.

3. Henry M. Morris, ed., *Scientific Creationism*, general ed. (San Diego: Creation-Life Publishers, 1974), p. 252.

4. See, e.g., Davis A. Young's complaint of this phenomenon in *Christianity and the Age of the Earth* (Grand Rapids, MI: Zondervan Publishing House, 1982), p. 10.

5. See, e.g., Philip Kitcher, *Abusing Science: The Case against Creationism* (Cambridge, MA: MIT Press, 1982); Norman D. Newell, *Creation and Evolution: Myth or Reality?* (New York: Columbia University Press, 1982); Michael Ruse, *Darwinism Defended: A Guide to the Evolution Controversies* (Reading, MA: Addison-Wesley Publishing Co., 1982); Roland Mushat Frye, ed., *Is God a Creationist? The Religious Case against Creation-Science* (New York: Charles Scribner's Sons, 1983); Douglas J. Futuyma, *Science on Trial: The Case for Evolution* (New York: Pantheon Books,

1983); Laurie R. Godfrey, ed., *Scientists Confront Creationism* (New York: W. W. Norton, 1983); David B. Wilson, ed., *Did the Devil Make Darwin Do It? Modern Perspectives on the Creation-Evolution Controversy* (Ames: Iowa State University Press, 1983); J. Peter Zetterberg, ed., *Evolution versus Creationism: The Public Education Controversy* (Phoenix, AZ: Oryx Press, 1983), which includes selections from creationists; Chris McGowan, *In the Beginning . . . : A Scientist Shows Why the Creationists Are Wrong* (Buffalo, NY: Prometheus Books, 1984); Ashley Montagu, ed., *Science and Creationism* (New York: Oxford University Press, 1984); Arthur N. Strahler, *Science and Earth History—The Evolution/Creation Controversy* (Buffalo, NY: Prometheus Books, 1987); Michael Ruse, ed., *But Is It Science? The Philosophical Question in the Creation/Evolution Controversy* (Buffalo, NY: Prometheus Books, 1988); Howard J. Van Till, Davis A. Young, and Clarence Menninga, *Science Held Hostage: What's Wrong with Creation Science AND Evolutionism* (Downers Grove, IL: InterVarsity Press, 1988); and Raymond A. Eve and Francis B. Harrold, *The Creationist Movement in Modern America* (Boston: Twayne Publishers, 1991).

The best historical study is Edward J. Larson, *Trial and Error: The American Controversy over Creation and Evolution* (New York: Oxford University Press, 1985). Dorothy Nelkin, *The Creation Controversy: Science or Scripture in the Schools* (New York: W. W. Norton, 1982), is described on the dust jacket as "a history of the struggle between creationists and scientists from the nineteenth century to the Arkansas trial," but it focuses almost exclusively on events between 1957 and 1982. Morris, *History of Modern Creationism*, provides the most comprehensive history of recent creationism—but from a decidedly creationist perspective.

6. See, e.g., Laurie R. Godfrey and John R. Cole, "Picking a Bone with Philosophers of Science," *Creation/Evolution* 25 (Fall 1989): 53, who argue that "there are precious few bones of contention among creationists. . . ."

7. For critical evaluations of the warfare thesis, see James R. Moore, *The Post-Darwinian Controversies: A Study of the Protestant Struggle to Come to Terms with Darwin in Great Britain and America, 1870–1900* (Cambridge: Cambridge University Press, 1979), pp. 17–122; Introduction to *God and Nature: Historical Essays on the Encounter between Christianity and Science*, ed. David C. Lindberg and Ronald L. Numbers (Berkeley and Los Angeles: University of California Press, 1986), pp. 1–18; David C. Lindberg and Ronald L. Numbers, "Beyond War and Peace: A Reappraisal of the Encounter between Christianity and Science," *Church History* 55 (1986): 338–54; and Ronald L. Numbers, "Science and Religion," *Osiris*, 2nd ser., 1 (1985): 59–80.

8. Langdon Gilkey, *Creationism on Trial: Evolution and God at Little Rock* (Minneapolis: Winston Press, 1985), quotation on p. 169; see also pp. 21–2, 269.

9. Larry Laudan, "The Demise of the Demarcation Problem," in *The Demarcation between Science and Pseudo-Science*, ed. Rachel Laudan (Virginia Tech Center for the Study of Science in Society, Working Papers, vol. 2, no. 1, April 1983), pp. 7–35, quotation on p. 29; Thomas F. Gieryn, "Boundary-Work and the Demarcation of Science from Non-Science: Strains and Interests in Professional Ideologies of Scientists," *American Sociological Review* 48 (1983): 781–95; Thomas F. Gieryn, George M. Bevins, and Stephen C. Zehr, "Professionalization of American Scientists: Public Science in the Creation/Evolution Trials," *American Sociological Review* 50 (1985): 392–409. Other recent works on "pseudoscience" include Marsha P. Hanen, Margaret J. Osler, and Robert G. Weyant, eds., *Science, Pseudo-Science and Society* (Waterloo, Ontario: Wilfrid Laurier University Press for the Calgary Institute for the Humanities, 1980); and Roy Wallis, ed., *On the Margins of Science: The Social Construction of Rejected Knowledge*, Sociological Review Monograph 27 (Keele, England: University of Keele, 1979).

10. Judge William R. Overton's decision in *McLean v. Arkansas* is reprinted in La Follette, *Creationism, Science, and the Law*, pp. 45–73. See also the exchange between Larry Laudan and Michael Ruse, ibid., pp. 150–73; and Philip L. Quinn, "The Philosopher of Science as Expert Witness," in *Science and Reality: Recent Work in the Philosophy of Science*, ed. James T. Cushing, C. F. Delaney, and Gary M. Gutting (Notre Dame, IN: University of Notre Dame Press, 1984), pp. 32–53.

11. W. R. Bird, *The Origin of Species Revisited: The Theories of Evolution and of Abrupt Appearance*, 2 vols. (New York: Philosophical Library, 1987–1989), 2:211.

12. See, e.g., Martin Rudwick, "Senses of the Natural World and Senses of God: Another Look at the Historical Relation of Science and Religion," in *The Sciences and Theology in the Twentieth Century*, ed. A. R. Peacocke (Notre Dame, IN: University of Notre Dame Press, 1981), pp. 241–61.

One. CREATIONISM IN THE AGE OF DARWIN

1. [William Hayes Ward], "Whether It Is Right to Study the Bible," *Independent* 32 (February 26, 1880): 4. Michael Ruse, *The Darwinian Revolution: Science Red in Tooth and Claw* (Chicago: University of Chicago Press, 1979), p. 229, argues that most British biologists were evolutionists by the mid-1860s, while David L. Hull, Peter D. Tessner, and Arthur M. Diamond, "Planck's Principle," *Science* 202 (1978): 721, point out that more than a quarter of British scientists continued to reject the evolution of species as late as 1869. On the acceptance of evolution among religious leaders, see, e.g., Jon H. Roberts, *Darwinism and the Divine in America: Protestant Intellectuals and Organic Evolution, 1859–1900* (Madison: University of Wisconsin Press, 1988); and Owen Chadwick, *The Victorian Church*, pt. 2, 2nd ed. (London: Adam & Charles Black, 1972), pp. 23–4. On the American Catholic response, about which I say little, see John L. Morrison, "A History of American Catholic Opinion on the Theory of Evolution, 1859–1950" (Ph.D. dissertation, University of Missouri, 1951); and John Rickards Betts, "Darwinism, Evolution, and American Catholic Thought, 1860–1900," *Catholic Historical Review* 45 (1959–60): 161–85.

2. Charles Darwin, *On the Origin of Species*, a facsimile of the first edition, with an introduction by Ernst Mayr (Cambridge, MA: Harvard University Press, 1966), p. 484; A. Hunter Dupree, *Asa Gray, 1810–1888* (Cambridge, MA: Harvard University Press, 1959), pp. 300–1, 339. On Darwin's goals in writing the *Origin*, see Charles Darwin, *The Descent of Man, and Selection in Relation to Sex*, 2 vols. (London: John Murray, 1871), 1:152. On Pentateuchal language, see C. Darwin to J. D. Hooker, March 29, 1863, in *The Life and Letters of Charles Darwin*, ed. Francis Darwin, 2 vols. (New York: D. Appleton, 1889), 2:202–3.

3. Darwin, *The Descent of Man*, 2:389; P. R. Russel, "Darwinism Examined," *Advent Review and Sabbath Herald* 47 (1876): 153.

4. See Peter J. Bowler, *The Eclipse of Darwinism: Anti-Darwinian Evolution Theories in the Decades around 1900* (Baltimore: Johns Hopkins University Press, 1983).

5. Asa Gray, *Darwiniana: Essays and Reviews Pertaining to Darwinism*, ed. A. Hunter Dupree (Cambridge, MA: Harvard University Press, 1963), pp. 78–9, from an essay first published in 1860.

6. Simon Newcomb, Address of the Retiring President, American Association for the Advancement of Science, *Proceedings*, 27 (1878): 21.

7. E. D. Cope, *The Origin of the Fittest: Essays on Evolution* (New York: D. Appleton, 1887), p. 2, from an article, "Evolution and Its Consequences," first published in 1872; Louis Agassiz, "Evolution and the Permanence of Type," *Atlantic*

Monthly 33 (1874): 95; William F. Sanford, Jr., "Dana and Darwinism," *Journal of the History of Ideas* 26 (1965): 537; [William Hayes Ward], "Do Our Colleges Teach Evolution?" *Independent* 31 (December 18, 1879): 14–15. In 1873 Asa Gray could think of only two eminent naturalists, Dawson and presumably Agassiz, who still held to the fixity of species; *Darwiniana*, pp. 202–3. For a summary of the *Independent-Observer* debate, see "Scientific Teaching in the Colleges," *Popular Science Monthly* 16 (1880): 556–9. On the scientific reception of Darwinism, see Edward J. Pfeifer, "United States," in *The Comparative Reception of Darwinism*, ed. Thomas F. Glick (Austin: University of Texas Press, 1974), pp. 168–206; and Peter J. Bowler, "Scientific Attitudes to Darwinism in Britain and America," in *The Darwinian Heritage*, ed. David Kohn (Princeton, NJ: Princeton University Press, 1985), pp. 641–81.

8. Louis Agassiz, *Essay on Classification*, ed. Edward Lurie (1857; reprint, Cambridge, MA: Harvard University Press, 1962), pp. 95–6, 173–5; Edward Lurie, *Louis Agassiz: A Life in Science* (Chicago: University of Chicago Press, 1960), pp. 252–302. See also Mary Pickard Winsor, "Louis Agassiz and the Species Question," *Studies in History of Biology* 3 (1979): 89–117.

9. John McCrady to Louis Agassiz, April 7 and April 17, 1873, and Louis Agassiz to John McCrady, April 13, 1873, all in the John McCrady Papers, du Pont Library, University of the South. I am indebted to Lester Stephens for first telling me about McCrady and for pointing out the existence of this and other collections of McCrady papers. On McCrady, see "Prof. John McCrady," in *Cyclopedia of Eminent and Representative Men of the Carolinas of the Nineteenth Century*, 2 vols. (Madison, WI: Brant & Fuller, 1892), 1:158–60.

10. John McCrady to Edward McCrady, March 26, April 2, April 11, April [?], and May 4, 1877, all in the Edward McCrady, Sr., Collection, South Carolina Historical Society. McCrady's views are expressed in his letter to Louis Agassiz, May 12, 1869, his letter to the editor of Appleton's *Popular Science Monthly*, April 14, 1873, and his "Memorial Address Delivered in Agassiz' Lecture Room the Day after His Burial," all in the private possession of Professor Edward McCrady of Greensboro, North Carolina. For the letter to Agassiz and the Memorial Address, I have relied on Lester Stephens's notes, which he kindly shared with me.

11. Arnold Guyot, *Creation; or, The Biblical Cosmogony in the Light of Modern Science* (New York: Charles Scribner's Sons, 1884), pp. 116–28; James D. Dana, "Memoir of Arnold Guyot," National Academy of Sciences, *Biographical Memoirs* 2 (1886); 334. On Guyot's exegesis of Genesis 1, see Ronald L. Numbers, *Creation by Natural Law: Laplace's Nebular Hypothesis in American Thought* (Seattle: University of Washington Press, 1977), pp. 91–100. On the special creation of humans, see Arnold Guyot to J. D. Dana, February 16, 1889, in the James Dwight Dana Correspondence, Yale University Library.

12. Arnold Guyot to J. W. Dawson, April 3, 1878, and James McCosh to J. W. Dawson, April 4, 1878, both in the Dawson Collection, McLennan Library, McGill University. For biographical information, see Charles F. O'Brien, *Sir William Dawson: A Life in Science and Religion* (Philadelphia: American Philosophical Society, 1971); and William R. Shea, Introduction to *Modern Ideas of Evolution*, by J. William Dawson (New York: Prodist, 1977).

13. J. W. Dawson, *The Meeting-Place of Geology and History* (London: Religious Tract Society, 1874), p. 147; J. W. Dawson, *The Story of Earth and Man* (New York: Harper & Brothers, 1873), pp. 340–1; O'Brien, *Sir William Dawson*, pp. 60, 123. On Dawson's shifting views, see John F. Cornell, "From Creation to Evolution: Sir William Dawson and the Idea of Design in the Nineteenth Century," *Journal of the History of Biology* 16 (1983): 137–70.

14. J. Lawrence Smith, Address, American Association for the Advancement of Science, *Proceedings* (1873): 14–16; Thomas Coulson, *Joseph Henry: His Life and Work* (Princeton, NJ: Princeton University Press, 1950), pp. 294–5; Nathan Reingold, "Developing Science in the United States: Insights from the Papers of Joseph Henry on the Reception of Darwin," in *Proceedings of the 16th International Congress of the History of Science* (Bucharest, 1981), C–D, pp. 301–6. On Henry as an antievolutionist, see, e.g., D. S. Gregory, "Is Evolution Science?" *Independent* 32 (May 27, 1880): 2.

15. David N. Livingstone, *Darwin's Forgotten Defenders: The Encounter between Evangelical Theology and Evolutionary Thought* (Grand Rapids, MI: William B. Eerdmans, 1987), p. 85.

16. Edward Hitchcock, "The Law of Nature's Constancy Subordinate to the Higher Law of Change," *Bibliotheca Sacra* 20 (1863): 520–5; Edward Hitchcock, *Elementary Geology*, new ed. (New York: Ivison, Phinney, 1862), pp. 373–4, 377–93. See also Stanley M. Guralnick, "Geology and Religion before Darwin: The Case of Edward Hitchcock, Theologian and Geologist (1793–1864)," *Isis* 63 (1972): 529–43; Philip J. Lawrence, "Edward Hitchcock: The Christian Geologist," American Philosophical Society, *Proceedings* 116 (1972): 21–34; and Rodney Lee Stiling, "The Genesis Flood in Nineteenth-Century American Thought" (Ph.D. dissertation, University of Wisconsin-Madison, 1991).

17. E. F. Burr, *Pater Mundi; or, Doctrine of Evolution* (Boston: Noyes, Holmes, 1873), esp. p. 12; Bert James Loewenberg, "The Controversy over Evolution in New England, 1859–1873," *New England Quarterly* 8 (1935): 244–6, which includes the quotation from the *Religious Magazine and Monthly Review* 49 (1873): 492. See also "Enoch Fitch Burr," *Dictionary of American Biography*, 3:321–2; and James R. Moore, *The Post-Darwinian Controversies: A Study of the Protestant Struggle to Come to Terms with Darwin in Great Britain and America, 1870–1900* (Cambridge: Cambridge University Press, 1979), pp. 197–8.

18. George D. Armstrong, *The Two Books of Nature and Revelation Collated* (New York: Funk & Wagnalls, 1886), pp. 86, 96–7, 136. See also "George Dod Armstrong," in *Dictionary of American Biography*, 1:352–3; and Ernest Trice Thompson, *Presbyterians in the South, 1861–1890* (Richmond, VA: John Knox Press, 1973), p. 477. The northern Presbyterian clergyman John T. Duffield, who taught mathematics and mechanics at Princeton, believed that evolution was "irreconcilable with the direct teaching of the Scriptures as to the origin of man," but he focused his critique of evolution on this single issue. See Duffield, "Evolution Respecting Man, and the Bible," *Princeton Review*, 4th ser., 1 (1878): 173; and "John Thomas Duffield," *Appletons' Cyclopaedia of American Biography*, 2:249.

19. Herbert W. Morris, *Science and the Bible; or, The Mosaic Creation and Modern Discoveries* (Philadelphia: Ziegler & McCurdy, 1871), pp. 51, 80; Herbert W. Morris, *Harmonies of the Universe, as Displayed in the Laws of Nature, the Dominion of Providence, and the Dispensations of Grace* (Philadelphia: P. W. Ziegler, 1875), pp. 169, 512–28. In a British edition of *Science and the Bible*, published under the title *Work-Days of God; or, Science and the Bible* (London: Pickering & Inglis, n.d.), p. 6, Morris claims to have sold fifty thousand copies in a little more than three years.

20. Roberts, *Darwinism and the Divine in America*, pp. 91, 209–10; T. De Witt Talmage, *Sermons: 2nd Series* (New York: 1875), pp. 58–60, quoted in Roberts, *Darwinism and the Divine in America*, p. 211. See also T. De Witt Talmage, "Evolution: Anti-Bible, Anti-Science, Anti-Commonsense," *Advent Review and Sabbath Herald* 60 (1883): 261–2; and "Thomas De Witt Talmage," in *Dictionary of American Biography*, 18:287–8. For a succinct account of the Protestant response to evolution, see

Frederick Gregory, "The Impact of Darwinian Evolution on Protestant Theology in the Nineteenth Century," in *God and Nature: Historical Essays on the Encounter between Christianity and Science,* ed. David C. Lindberg and Ronald L. Numbers (Berkeley and Los Angeles: University of California Press, 1986), pp. 369–90.

21. Charles Hodge, *What Is Darwinism?* (New York: Scribner, Armstrong, 1874), p. 174; Robert L. Dabney to James Woodrow, ca. 1873, quoted in Thomas Cary Johnson, *The Life and Letters of Robert Lewis Dabney* (Richmond, VA: Presbyterian Committee of Publications, 1903), p. 346. On Hodge, see Livingstone, *Darwin's Forgotten Defenders,* pp. 100–5; and Mark A. Noll, ed., *The Princeton Theology, 1812–1921* (Grand Rapids, MI: Baker Book House, 1983), pp. 142–52. See also Robert L. Dabney, *The Sensualistic Philosophy of the Nineteenth Century* (New York: Anson D. F. Randolph, 1876), esp. chap. 9, "Evolution Theory Materialistic, and Therefore False."

22. Robert D. Whalen, "Millenarianism and Millennialism in America, 1790–1880" (Ph.D. dissertation, State University of New York at Stony Brook, 1972), p. 228; Dwight L. Moody, quoted in William G. McLoughlin, Jr., *Modern Revivalism: Charles Grandison Finney to Billy Graham* (New York: Ronald Press, 1959), p. 213. On Moody, see James F. Findlay, Jr., *Dwight L. Moody: American Evangelist, 1837–1899* (Chicago: University of Chicago Press, 1969). On premillennialism, see Ernest R. Sandeen, *The Roots of Fundamentalism: British and American Millenarianism, 1800–1930* (Chicago: University of Chicago Press, 1970).

23. On Hastings, see Clyde E. Hewitt, *Midnight and Morning: An Account of the Adventist Awakening and the Founding of the Advent Christian Denomination, 1831–1860* (Charlotte, NC: Venture Books, 1983), pp. 233, 271; and the promotional material at the end of H. L. Hastings, *Was Moses Mistaken? or, Creation and Evolution,* Anti-Infidel Library No. 36 (Boston: H. L. Hastings, 1896). On Millerism, see Ronald L. Numbers and Jonathan M. Butler, eds., *The Disappointed: Millerism and Millenarianism in the Nineteenth Century* (Bloomington: Indiana University Press, 1987). On Moody's association with Hastings and Townsend, see, e.g., C. K. Ober, "Professor Drummond in the American Colleges," *Our Day* 1 (1888): 306.

24. Robert Patterson, *The Errors of Evolution: An Examination of the Nebular Theory, Geological Evolution, the Origin of Life, and Darwinism,* 3rd ed. (Boston: Scriptural Tract Repository, 1893), pp. ix–x; Hastings, *Was Moses Mistaken?,* pp. 25–6, 35.

25. "Abstract of the Annual Report of the Education Committee of the Bible League of North America," *Bible Student and Teacher* 10 (1909): 372–3; Luther Tracy Townsend, *Evolution or Creation: A Critical Review of the Scientific and Scriptural Theories of Creation and Certain Related Subjects* (New York: Fleming H. Revell, 1896), pp. 13, 133–4, 154; L. T. Townsend, *Adam and Eve: History or Myth?* (Boston: Chapple, 1904), p. 83; L. T. Townsend, *Collapse of Evolution* (New York: American Bible League, 1905). See also "Luther Tracy Townsend," *Dictionary of American Biography,* 18:618–19.

26. A. C. Dixon to G. F. Wright, May 5, 1910, G. F. Wright Papers, Oberlin College Archives; Alexander Patterson, *The Other Side of Evolution: An Examination of Its Evidences,* with an Introduction by George Frederick Wright, 3rd ed. (Chicago: Bible Institute Colportage Association, 1912). On the association, see Findlay, *Dwight L. Moody,* pp. 398–9. Although Patterson's career remains obscure, his connection with Moody and the Moody Bible Institute is documented in a letter to A. P. Fitt, including "Recollections of D. L. Moody," November 14, [?], Moody Bible Institute Library, copy courtesy of Walter Osborn. The *Minutes of the General Assembly of the Presbyterian Church in the United States of America* first show Patterson as an evan-

gelist in Chicago in 1884; in 1904 he is identified as the pastor of a small church in Morgan Park. His name does not appear in E. C. Scott, comp., *Ministerial Directory of the Presbyterian Church, U.S., 1861–1941* (Austin, TX: Von Boeckmann-Jones, 1942), which lists all ordained ministers.

27. Alexander Patterson, *The Bible as It Is: A Simple Method of Mastering and Understanding the Bible* (Chicago: Winona Publishing Co., 1906), pp. 55–77, 103; Patterson, *The Other Side of Evolution*, pp. ix–xii, 11, 60; Roberts, *Darwinism and the Divine in America*, p. 96 (populist).

28. "David Nevins Lord," *Dictionary of American Biography*, 11:405; "Eleazar Lord," ibid., pp. 405–6; Sandeen, *The Roots of Fundamentalism*, pp. 90–1; Whalen, "Millernarianism and Millennialism in America," pp. 40–1.

29. Eleazar Lord, *The Epoch of Creation: The Scripture Doctrine Contrasted with the Geological Theory*, with an introduction by Richard W. Dickinson (New York: Charles Scribner, 1851), pp. 157, 209, 230. For an earlier statement of these views, see [Eleazar Lord], *Geological Cosmogony* (New York: Robert Carter, 1843), esp. pp. 151–6. In 1840 Edward Hitchcock apologized to geological readers of his new textbook for even mentioning that "the fossiliferous rocks were deposited by the deluge of Noah"; see his *Elementary Geology* (Amherst, MA: J. S. and C. Adams, 1840), p. 276.

30. David N. Lord, "The Bearing of the Geological Theory of the Age of the World on the Inspiration of the Bible," *Theological and Literary Journal* 9 (1856): 259; David N. Lord, *Geognosy; or, The Facts and Principles of Geology against Theories*, 2nd ed. (New York: Franklin Knight, 1857), pp. 309, 316–18, 393–5; David N. Lord, "Darwin on the Origin of Species," *Theological and Literary Journal* 13 (1860): 101–48. Battle Creek College adopted *Geognosy* as a text; see Maurice Hodgen, ed., *School Bells & Gospel Trumpets: A Documentary History of Seventh-day Adventist Education in North America* (Loma Linda, CA: Adventist Heritage Publications, 1978), p. 18.

31. Thomas Mitchell, a Methodist minister and crusader against infidelity, insisted on a recent creation of six literal days; see Mitchell, *Cosmogony: The Geological Antiquity of the World, Evolution, Atheism, Pantheism, Deism and Infidelity Refuted, by Science, Philosophy and Scripture*, 2 vols. (New York: American News Co., 1881), 1:159; and Mitchell, *Conflict of the Nineteenth Century—The Bible and Free Thought* (New York: Universal Book Co., 1893), pp. 348–9, 446. Martyn Paine (1794–1877), a prominent medical educator in New York City, also defended a recent six-day creation and universal flood, which formed the coal-bearing rocks; see Paine, *Physiology of the Soul and Instinct, As Distinguished from Materialism* (New York: Harper & Brothers, 1872).

Two. GEORGE FREDERICK WRIGHT: FROM CHRISTIAN DARWINIST TO FUNDAMENTALIST

1. A. C. Dixon to G. F. Wright, May 16 and 24, 1910, G. F. Wright Papers, Oberlin College Archives. Orr eventually contributed an essay: "Science and Christian Faith," *The Fundamentals*, 12 vols. (Chicago: Testimony Publishing Company, n.d.), 4:91–104. See also James Orr, *God's Image in Man: And Its Defacement in the Light of Modern Denials*, 2nd ed. (New York: Armstrong and Son, 1906), pp. 87–8, 105; and David N. Livingstone, *Darwin's Forgotten Defenders: The Encounter between Evangelical Theology and Evolutionary Thought* (Grand Rapids, MI: William B. Eerdmans, 1987), pp. 140–4. For a negative appraisal of Townsend by Wright, see Wright's review of Townsend's *Evolution or Creation?* in *Bibliotheca Sacra* 54 (1897): 202–3. For a somewhat fuller version of this chapter, see Ronald L. Numbers, "George

Frederick Wright: From Christian Darwinist to Fundamentalist," *Isis* 79 (1988): 624–45.

2. G. Frederick Wright, *Story of My Life and Work* (Oberlin, OH: Bibliotheca Sacra Co., 1916). For additional biographical data, see Michael McGiffert, "Christian Darwinism: The Partnership of Asa Gray and George Frederick Wright, 1874–1881" (Ph.D. dissertation, Yale University, 1958); and William James Morison, "George Frederick Wright: In Defense of Darwinism and Fundamentalism, 1838–1921" (Ph.D. dissertation, Vanderbilt University, 1971).

3. Wright, *Story of My Life*, pp. 116, 123, 132; G. F. Wright to Asa Gray, June 26, 1875 (influence of Gray), Archives, Gray Herbarium, Harvard University.

4. Asa Gray, *Darwiniana: Essays and Reviews Pertaining to Darwinism*, ed. A. Hunter Dupree (Cambridge, MA: Harvard University Press, 1963), p. 5; [G. F. Wright], Review of *Letters of Asa Gray*, ed. Jane Loring Gray, *Bibliotheca Sacra* 51 (1894): 182.

5. Gray, *Darwiniana*, p. 130; G. Frederick Wright, "The Debt of the Church to Asa Gray," *Bibliotheca Sacra* 45 (1888): 527.

6. Wright, *Story of My Life*, p. 132; G. F. Wright, "Recent Works on Prehistoric Archaeology," *Bibliotheca Sacra* 30 (1873): 381–4.

7. Wright, *Story of My Life*, pp. 116, 133–40; G. F. Wright to Asa Gray, June 26, 1875 (infidel class, Hodge, and Dawson), Gray Herbarium; Asa Gray to G. F. Wright, August 14, 1875 (evolutionary teleology), and June 1, 1876 (Hodge), Wright Papers.

8. George F. Wright, "Recent Books Bearing upon the Relation of Science to Religion: No. 1—The Nature and Degree of Scientific Proof," *Bibliotheca Sacra* 32 (1875): 548 (quoting Stanley Jevons regarding bookkeeping); George F. Wright, "Recent Works Bearing on the Relation of Science to Religion: No. II—The Divine Method of Producing Living Species," ibid. 33 (1876): 459 (meaning of facts). Wright's earliest essay on science and religion, "The Ground of Confidence in Inductive Reasoning," *New Englander* 30 (1871): 601–15, made some of the same points.

9. Wright, "The Divine Method of Producing Living Species," pp. 455, 466, 492–3. Darwin's reaction is quoted in Wright, *Story of My Life*, p. 138.

10. Wright, "The Divine Method of Producing Living Species," pp. 474, 487, 492–4.

11. G. Frederick Wright, *Studies in Science and Religion* (Andover, MA: Warren F. Draper, 1882), pp. 352–4. For his earlier criticism of Hodge, see, e.g., Wright, "The Divine Method of Producing Living Species," p. 453.

12. Wright, *Studies*, pp. 365–7.

13. Ibid., pp. 347–50, 368–70.

14. Ibid., pp. 376–9.

15. Wright, *Story of My Life*, p. 146; Morison, "George Frederick Wright," pp. 136–7; McGiffert, "Christian Darwinism," p. 368.

16. McGiffert, "Christian Darwinism," pp. 370–1.

17. [G. F. Wright], "Professor Wright and Some of His Critics," *Bibliotheca Sacra* 42 (1885): 352. Wright is identified as the author of this article in *Story of My Life*, p. 441.

18. G. Frederick Wright, *The Divine Authority of the Bible* (Boston: Congregational Sunday-School and Publishing Society, 1884), pp. 195–8; James D. Dana to G. F. Wright, January 8, January 18, and November 8, 1885, and May 9, 1886, Wright Papers; G. F. Wright, "Has Modern Criticism Affected Unfavorably Any of the Essential Doctrines of Christianity?" *Homiletic Review* 11 (1886): 307–12.

19. G. Frederick Wright, "Editorial Note on Genesis and Geology," *Bibliotheca Sacra* 54 (1897): 570–2; G. Frederick Wright, "The First Chapter of Genesis and

Modern Science," *Homiletic Review* 35 (1898): 392–3. See also G. Frederick Wright, *Scientific Confirmations of Old Testament History* (Oberlin, OH: Bibliotheca Sacra, 1906), pp. 368–86.

20. George F. Wright, "Exaggeration of the Issues between Science and Religion," *Congregationalist*, May 10, 1883, p. 162; C. H. Hitchcock to G. F. Wright, January 2, 1884, Wright Papers. See also G. F. W[right], Review of *History of the Conflict between Religion and Science*, by John William Draper, *Bibliotheca Sacra* 33 (1876): 584–5.

21. [G. F. Wright], Review of *Darwinism and Other Essays*, by John Fiske, *Bibliotheca Sacra* 36 (1879): 784; [G. Frederick Wright], "Transcendental Science," *Independent* 41 (October 3, 1889): 10. See also G. Frederick Wright, "Darwin on Herbert Spencer," *Bibliotheca Sacra* 46 (1889): 181–4.

22. [G. F. Wright], Review of *The Life and Letters of Charles Darwin*, ed. Francis Darwin, *Bibliotheca Sacra* 45 (1888): 366–72. The references to wriggling appear in [Wright], "Transcendental Science," p. 10; and G. Frederick Wright, "The Affinity of Science for Christianity," *Bibliotheca Sacra* 46 (1889): 718. In "Huxley among the False Prophets," *Advance* 23 (1889): 452, Wright warned that Huxley's "opposition to evangelical Christianity" would be felt "in every parish in the land." On Darwin's views of creation, see Neal C. Gillespie, *Charles Darwin and the Problem of Creation* (Chicago: University of Chicago Press, 1979).

23. G. Frederick Wright, "Some Will-o'-the-Wisps of Higher Criticism," *Congregationalist*, March 12, 1891, p. 84. On the Briggs affair, see Ferenc Morton Szasz, *The Divided Mind of Protestant America, 1880–1930* (University: University of Alabama Press, 1982), pp. 27–9.

24. G. Frederick Wright, "Recent Discoveries Bearing on the Antiquity of Man," *Bibliotheca Sacra* 48 (1891): 298–309 (Trenton); Wright, *Story of My Life*, p. 140 (new zest); G. Frederick Wright, "Man and the Glacial Period," *Science* 20 (1892): 275–7; G. Frederick Wright, *The Ice Age in North America: And Its Bearings upon the Antiquity of Man*, 5th ed. (Oberlin, OH: Bibliotheca Sacra, 1911).

25. G. Frederick Wright, "How Old Is Mankind?" *Sunday School Times* 55 (January 25, 1913): 52; Wright, "Recent Discoveries," p. 309, both of whch indicate that the visit took place about 1889. See also David N. Livingstone, "The Idea of Design: The Vicissitudes of a Key Concept in the Princeton Response to Darwin," *Scottish Journal of Theology* 37 (1984): 329–57; and Mark A. Noll, ed., *The Princeton Theology, 1812–1921* (Grand Rapids, MI: Baker Book House, 1983).

26. G. Frederick Wright, *Man and the Glacial Period* (New York: D. Appleton, 1892); Review of *The Ice Age in North America*, by G. F. Wright, *American Geologist* 4 (1889): 107.

27. T. C. Chamberlin, "Geology and Archaeology Mistaught," *Dial* 13 (1892): 303–6; W J McGee, "Man and the Glacial Period," *Science* 20 (1892): 317; W J McGee, "Man and the Glacial Period," *American Anthropologist* 6 (1893): 85–95. On the relationship between Chamberlin and Wright, see Susan F. Schultz, "Thomas C. Chamberlin: An Intellectual Biography of a Geologist and Educator" (Ph.D. dissertation, University of Wisconsin-Madison, 1976), pp. 197–258.

28. J. D. Dana to G. F. Wright, January 22 and January 29, 1893, Wright Papers; E. W. Claypole, "Prof. G. F. Wright and His Critics," *Popular Science Monthly* 42 (1893): 767. See also Wright's replies to Chamberlin and McGee, respectively: "Man and the Glacial Period," *Dial* 13 (1892): 380; and "Excitement over Glacial Theories," *Science* 20 (1892): 360–1.

29. G. Frederick Wright, *Scientific Confirmations of Old Testament History* (Oberlin, OH: Bibliotheca Sacra, 1906), p. 83; Wright, *Story of My Life*, pp. 379–83.

30. G. Frederick Wright, "Present Aspects of the Questions concerning the Origin

and Antiquity of the Human Race," *Protestant Episcopal Review* 11 (1898): 319. For similar accounts of his views on evolution, see G. Frederick Wright, "The Evolutionary Fad," *Bibliotheca Sacra* 57 (1900): 303–16; and G. Frederick Wright, *Scientific Aspects of Christian Evidences* (New York: D. Appleton, 1898), pp. 89–114. On King and Le Conte, see Edward J. Pfeifer, "United States," in *The Comparative Reception of Darwinism*, ed. Thomas F. Glick (Austin: University of Texas Press, 1974), pp. 199–202; and Lester D. Stephens, *Joseph LeConte: Gentle Prophet of Evolution* (Baton Rouge: Louisiana State University Press, 1982), pp. 164–5.

31. Wright, "Present Aspects," pp. 321–3.

32. G. F. Wright, "The Revision of Geological Time," *Bibliotheca Sacra* 60 (1903): 580, 582; G. Frederick Wright, "The Uncertainties of Science," *Advance* 43 (1902): 624–5. See also, e.g., George Frederick Wright, "The Uncertainties of Science and the Certainties of Religion," *Homiletic Review* 46 (1903): 413–15.

33. Wright, "The Uncertainties of Science," *Advance*, p. 624.

34. George Frederick Wright, Introduction to *The Other Side of Evolution: An Examination of Its Evidences*, by Alexander Patterson (Chicago: Winona Publishing Co., 1903), pp. xvii–xix.

35. William Phillips Hall to G. F. Wright, October 13, 1904, Wright Papers; "Address of Professor G. Frederick Wright," *Bible Student and Teacher* 1 (1904): 352; G. Frederick Wright, "The Mistakes of Darwin and His Would-be Followers," ibid. 10 (1909): 333–7, reprinted in *Bibliotheca Sacra* 66 (1909): 332–43.

36. Luther T. Townsend, "The Collapse of Evolution," *Bible Student and Teacher* 2 (1905): 8–28. On the status of Darwinism at the turn of the century, see Peter J. Bowler, *The Eclipse of Darwinism: Anti-Darwinian Evolution Theories in the Decades around 1900* (Baltimore: Johns Hopkins University, 1983).

37. George Frederick Wright, "The Passing of Evolution," *The Fundamentals*, 12 vols. (Chicago: Testimony Publishing Company, n.d.), 7:5–20, emphasis added.

38. G. Frederick Wright, *Origin and Antiquity of Man* (Oberlin, OH: Bibliotheca Sacra, 1912), pp. 380, 386, 388.

39. A. C. Dixon to G. F. Wright, January 1, 1915, Wright Papers; Wright, *Story of My Life*, p. 422. The A. C. Dixon Collection, Dargan-Carver Library of the Historical Commission of the Southern Baptist Convention, Nashville, Tennessee, contains no correspondence with Wright.

40. McGiffert, "Christian Darwinism," pp. 210–11; Wright, "Present Aspects," p. 319. Livingstone, *Darwin's Forgotten Defenders*, p. 70, suggests that Wright may have subscribed to "some form of emergent evolutionary theory."

41. James Dwight Dana to a clergyman, March 3, 1889, in Daniel C. Gilman, *The Life of James Dwight Dana* (New York: Harper & Brothers, 1899), p. 188; James Orr, "Science and Christian Faith," *The Fundamentals* 4:103. I am indebted to David Livingstone for bringing Orr's statement to my attention.

42. G. Frederick Wright, "Some Fallacies concerning 'Higher Criticism,' " *Congregationalist*, February 12, 1891, p. 50.

Three.   CREATIONISM IN THE
         FUNDAMENTALIST CONTROVERSY

1. H. L. Hastings, Preface to the Second Edition (1889) of *The Errors of Evolution: An Examination of the Nebular Theory, Geological Evolution, the Origin of Life, and Darwinism*, 3rd ed., by Robert Patterson (Boston: Scriptural Tract Repository, 1893),

p. iv. (premillennialist); William B. Riley, *The Finality of the Higher Criticism; or, The Theory of Evolution and False Theology* (Minneapolis: n.p., 1909), p. 73 (Baptist cleric); Maynard M. Metcalf, *An Outline of the Theory of Organic Evolution* (New York: Macmillan, 1904), pp. xix–xx. On the Protestant response to evolution in the nineteenth century, see Jon H. Roberts, *Darwinism and the Divine in America: Protestant Intellectuals and Organic Evolution, 1859–1900* (Madison: University of Wisconsin Press, 1988); and James R. Moore, *The Post-Darwinian Controversies: A Study of the Protestant Struggle to Come to Terms with Darwin in Great Britain and America, 1870–1900* (Cambridge: Cambridge University Press, 1979).

2. Vernon L. Kellogg, *Darwinism To-Day* (New York: Henry Holt, 1907), pp. 1–9. Kellogg took his title from Eberhard Dennert's *Vom Sterbelager des Darwinismus*, originally published in Stuttgart in 1903; for an English edition, see *At the Deathbed of Darwinism*, trans. E. V. O'Harra and John H. Peschges (Burlington, IA: German Literary Board, 1904). See also Peter J. Bowler, *The Eclipse of Darwinism: Anti-Darwinian Evolution Theories in the Decades around 1900* (Baltimore: Johns Hopkins University Press, 1983).

3. L. T. Townsend, *Collapse of Evolution* (New York: American Bible League, 1905), pp. 48–53; G. L. Young, "Relation of Evolution and Darwinism to the Question of Origins," *Bible Student and Teacher* 11 (1909): 41.

4. A. C. Dixon, *Reconstruction: The Facts against Evolution* (n.p., n.d.), p. 18, from a copy in the Dixon Collection, Dargan-Carver Library of the Historical Commission of the Southern Baptist Convention, Nashville, Tennessee; "Dr. R. A. Torrey Replies to Dr. O. E. Brown," *Moody Bible Institute Monthly* 26 (1925): 162. James Gray, editor of the Moody magazine and author of the pamphlet "Why a Christian Cannot Be an Evolutionist," feared that Torrey's concession might be "misunderstood"; James Gray to R. A. Torrey, October 14, 1925, James Gray Papers, Moody Bible Institute Achives. On Torrey, see Kermit L. Staggers, "Reuben A. Torrey: American Fundamentalist, 1856–1928" (Ph.D. dissertation, Claremont Graduate School, 1986).

5. James Orr, "Science and Christian Faith," *The Fundamentals*, 12 vols. (Chicago: Testimony Publishing Co., n.d.), 4:101–2; J. J. Reeve, "My Personal Experience with the Higher Criticism," ibid., 3:99. On the death of evolution, see "Evolutionism in the Pulpit," ibid., 8:30; David N. Livingstone, *Darwin's Forgotten Defenders: The Encounter between Evangelical Theology and Evolutionary Thought* (Grand Rapids, MI: William B. Eerdmans, 1987), p. 161, tentatively identifies the author of this piece as Frank E. Allen of Winnipeg, Canada. I am indebted to Rennie B. Schoepflin's unpublished manuscript "Anti-Evolutionism and Fundamentalism in Twentieth-Century America" (M.A. paper, University of Wisconsin-Madison, 1980), for my quantitative statements.

6. Edward J. Larson, "Before the Crusade: Evolution in American Secondary Education before 1920," *Journal of the History of Biology* 20 (1987): 113; Shailer Mathews, *New Faith for Old: An Autobiography* (New York: Macmillan, 1936), p. 227; C. B. Davenport to E. G. Conklin, June 1, 1922, Charles B. Davenport Papers, Library of the American Philosophical Society. I am grateful to John C. Burnham for bringing the Davenport correspondence to my attention. The last two quotations are from Howard K. Beale, *Are American Teachers Free? An Analysis of Restraints upon the Freedom of Teaching in American Schools* (New York: Charles Scribner's Sons, 1936), pp. 249–51. Although pre-1925 texts routinely discussed evolution, the subject remained "a marginal part of the biology curriculum"; see Philip J. Pauly, "The Urban Origins of American High School Biology," *Isis* 82 (1991): 662–88.

7. S. W. Stookey to Charles A. Blanchard, May 3, 1919, Charles A. Blanchard

Papers, Wheaton College Archives, where all the replies may be found. The statistical breakdown of responses is based on Schoepflin, "Anti-Evolutionism and Fundamentalism," which indicates that Blanchard was conducting the survey for the World's Christian Fundamentals Association.

8. Suzanne Cameron Linder, "William Louis Poteat and the Evolution Controversy," *North Carolina Historical Review* 40 (1963): 135–57; Willard B. Gatewood, Jr., *Preachers, Pedagogues & Politicians: The Evolution Controversy in North Carolina, 1920–1927* (Chapel Hill: University of North Carolina Press, 1966), pp. 30–8; "William Louis Poteat," *National Cyclopaedia of American Biography* 28: 132; "Somewhat about Baylor's Effort at Teaching Science," undated MS, Samuel Palmer Brooks Papers, The Texas Collection, Baylor University; J. L. Kesler, "The Relation between the Teaching of Science and Philosophy in Our Baptist Schools and the Teaching of the Bible," *Baylor Bulletin* 18 (December 1914): 5–21, quotation on p. 8; Henry Trantham, "Dr. Lulu Pace: An Appreciation of Her Life and Service," *Baylor Monthly* 1 (August 1925): 1–2. On the teaching of evolution in the South in the nineteenth century, see James Woodrow, "Speech before the Synod of South Carolina," in *Dr. James Woodrow as Seen by His Friends*, ed. Marion W. Woodrow (Columbia, SC: R. L. Bryan, 1909), pp. 752–4. See also Edward Lassiter Clark, "The Southern Baptist Reaction to the Darwinian Theory of Evolution" (Th.D. dissertation, Southwestern Baptist Theological Seminary, 1952). I am indebted to Charles M. Allen, professor of biology at Wake Forest University, for confirming the relationship between Poteat and Kesler.

9. See George M. Marsden, *Fundamentalism and American Culture: The Shaping of Twentieth-Century Evangelicalism, 1870–1925* (New York: Oxford University Press, 1980), pp. 141, 149. For a perceptive analysis of the fundamentalist-modernist debate, see Walter Lippmann, *American Inquisitors: A Commentary on Dayton and Chicago* (New York: Macmillan, 1928), which was brought to my attention by Eric Anderson.

10. Bryan quoted in Lawrence W. Levine, *Defender of the Faith: William Jennings Bryan: The Last Decade, 1915–1925* (New York: Oxford University Press, 1965), pp. 272, 277. On antievolution legislation, see Maynard Shipley, *The War on Modern Science: A Short History of the Fundamentalist Attacks on Evolution and Modernism* (New York: Alfred A. Knopf, 1927); and Shipley, "Growth of the Anti-Evolution Movement," *Current History* 32 (1930): 330–2. The Senate debate is reprinted in Willard B. Gatewood, Jr., ed., *Controversy in the Twenties: Fundamentalism, Modernism, and Evolution* (Nashville: Vanderbilt University Press, 1969), pp. 327–9. The quotation about America's going mad appears in Roland T. Nelson, "Fundamentalism and the Northern Baptist Convention" (Ph.D. dissertation, University of Chicago, 1964), p. 319. On Bryan's catalytic role, see Ferenc Morton Szasz, *The Divided Mind of Protestant America, 1880–1930* (University: University of Alabama Press, 1982), pp. 107–16.

11. W. J. Bryan to Johnnie Baldwin, March 27, 1923, Box 37, W. J. Bryan Papers, Library of Congress; Levine, *Defender of the Faith*, pp. 261–5; Gregg Mitman, "Evolution as Gospel: William Patten, the Language of Democracy, and the Great War," *Isis* 81 (1990): 446–63. For representative fundamentalist and scientific reactions to Kellogg, see, respectively, "Evolution Discredited Again," *King's Business* 9 (February 1918): 95–7; and Leon J. Cole, "Biological Philosophy and the War," *Scientific Monthly* 8 (1919): 247–57, which Gregg Mitman brought to my attention.

12. Levine, *Defender of the Faith*, pp. 266–7. Mrs. Bryan's statement appears in Wayne C. Williams, *William Jennings Bryan* (New York: G. P. Putnam, 1936), p. 448. The Bryan Papers contain numerous letters reporting the insidious effects of evolution. On the Loeb-Leopold murder, see Levine, *Defender of the Faith*, p. 344; W. B. Riley

and Henry B. Smith, *Should Evolution Be Taught in Tax Supported Schools?* Riley-Smith Debate (n.p., n.d.), pp. 8, 25.

13. William Jennings Bryan, *In His Image* (New York: Fleming H. Revell, 1922), pp. 94, 97–8). "The Menace of Darwinism" appears in this work as chapter 4, "The Origin of Man." For the apparent source of Bryan's illustration, see Alexander Patterson, *The Other Side of Evolution: An Examination of Its Evidences* (Chicago: Winona Publishing Co., 1903), pp. 32–3. Bryan recommended Patterson's book in his popular lecture "The Bible and Its Enemies," reprinted in Bryan, *The Dawn of Humanity* (Chicago: Altruist Foundation, 1925), p. 69. On the Bryan-Patterson connection, see also Arthur M. Miller, "Kentucky and the Theory of Evolution," *Science* 55 (1922): 178.

14. Charles Darwin, *On the Origin of Species* (London: John Murray, 1859), pp. 186–8; Asa Gray to Charles Darwin, January 23, 1860, and Darwin to Gray, February [?], 1860, in Francis Darwin, ed., *The Life and Letters of Charles Darwin*, 2 vols. (New York: D. Appleton, 1889), 2:66–7. See also Darwin's warning to himself in 1838 to avoid discussing the origin of the eye; Paul N. Barrett and Others, eds., *Charles Darwin's Notebooks, 1836–1844* (Ithaca, NY: Cornell University Press, 1987), p. 337. On the eye and atheism, see William Paley, *Natural Theology* (New York: Evert Duyckinck, 1820), p. 22.

15. William Jennings Bryan, *The Bible or Evolution?* (Murfreesboro, TN: Sword of the Lord Foundation, n.d.), p. 14.

16. W. J. Bryan to Howard A. Kelly, June 22, 1925, Box 47, Bryan Papers; William Jennings Bryan, Letter to the Editor, *Forum* 70 (1923): 1852, which was brought to my attention by Paul M. Waggoner; L. Sprague de Camp, *The Great Monkey Trial* (Garden City, NY: Doubleday, 1968), pp. 401–2; W. J. Bryan to S. J. Bole, July 27, 1922, S. James Bole Papers, Nebraska State Historical Society.

17. Paolo E. Coletta, *William Jennings Bryan*, vol. 3, *Political Puritan, 1915–1925* (Lincoln: University of Nebraska Press, 1969), p. 230 (oligarchy); Levine, *Defender of the Faith*, p. 289 (scientific soviet); "Progress of Anti-Evolution," *Christian Fundamentalist* 2 (1929): 13 (highbrows). Bryan gives the estimate of nine-tenths in a letter to W. A. McRae, April 5, 1924, Box 29, Bryan Papers.

18. "Fighting Evolution at the Fundamentals Convention," *Christian Fundamentals in School and Church* 7 (July–September, 1925): 5. The best state histories of the antievolution crusade are Kenneth K. Bailey, "The Enactment of Tennessee's Antievolution Law," *Journal of Southern History* 16 (1950): 472–510; Willard B. Gatewood, Jr., *Preachers, Pedagogues & Politicians: The Evolution Controversy in North Carolina, 1920–1927* (Chapel Hill: University of North Carolina Press, 1966); and Virginia Gray, "Anti-Evolution Sentiment and Behavior: The Case of Arkansas," *Journal of American History* 57 (1970): 352–66. Ferenc Morton Szasz stresses the urban dimension of the crusade in "Three Fundamentalist Leaders: The Roles of William Bell Riley, John Roach Straton, and William Jennings Bryan in the Fundamentalist-Modernist Controversy" (Ph.D. dissertation, University of Rochester, 1969), p. 351.

19. George Herbert Betts, *The Beliefs of 700 Ministers and Their Meaning for Religious Education* (New York: Abingdon Press, 1929), pp. 26, 44; W. B. Riley and Henry B. Smith, *Should Evolution Be Taught in Tax Supported Schools?* Riley-Smith Debate (n.p., n.d.), p. 2. On Riley, see William Vance Trollinger, Jr., *God's Empire: William Bell Riley and Midwestern Fundamentalism* (Madison: University of Wisconsin Press, 1990).

20. C. I. Scofield, *The Scofield Reference Bible* (1917; reprint, New York: Oxford University Press, 1945); Ernest R. Sandeen, *The Roots of Fundamentalism: British and American Millenarianism, 1800–1930* (Chicago: University of Chicago Press,

1970), p. 222, regarding Scofield's influence; Donald Kraus to Paul Boyer, June 14, 1990, copy courtesy of Paul Boyer (sales). *The Scofield Bible Correspondence Course* (Chicago: Moody Bible Institute, 1907) carries the same message. Another popular statement of the gap theory appeared in G. H. Pember, *Earth's Earliest Ages: And Their Connection with Modern Spiritualism and Theosophy*, 8th ed. (London: Hodder and Stoughton, 1895). A. C. Dixon's views are found in "The Geology of the Bible," *Defender* 2 (March, 1928): 7, originally written in 1923.

21. Milton L. Rudnick, *Fundamentalism and the Missouri Synod: A Historical Study of Their Interaction and Mutual Influence* (St. Louis: Concordia Publishing House, 1966), pp. 88–90; Sandeen, *The Roots of Fundamentalism*, pp. 266–8. Lutheran reluctance to join the crusade is also discussed in Szasz, "Three Fundamentalist Leaders," p. 279. For examples of prominent fundamentalists who stayed aloof from the antievolution controversy, see Ned B. Stonehouse, *J. Gresham Machen: A Biographical Memoir* (Grand Rapids, MI: William B. Eerdmans, 1954), pp. 401–2; and William Bryant Lewis, "The Role of Harold Paul Sloan and His Methodist League for Faith and Life in the Fundamentalist-Modernist Controversy of the Methodist Episcopal Church" (Ph.D. dissertation, Vanderbilt University, 1963), pp. 86–8.

22. T. T. Martin, *Hell and the High Schools: Christ or Evolution, Which?* (Kansas City: Western Baptist Publishing Co., 1923), pp. 10, 72, 164; T. T. Martin, *Viewing Life's Sunset from Pikes Peak: Life Story of T. T. Martin* (Louisville: A. D. Muse, n.d.), pp. 78, 80. See also Gatewood, *Preachers, Pedagogues & Politicians*, pp. 30–7, 189–97; and John Franklin Loftis, "Thomas Theodore Martin: His Life and Work as Evangelist, Fundamentalist, and Anti-Evolutionist" (Th.M. thesis, Southern Baptist Theological Seminary, 1980). According to the reporter George N. Coad, "Churches of South Mass to Bar Evolution by Law," *New York World*, September 24, 1925, p. 1, "Most ministers and editors of the southeast believe the present movement was caused by agitation by certain Baptist evangelists and editors against Dr. William Lewis [sic] Poteat." On Straton, see Hillyer H. Straton, "John Roach Straton: The Great Evolution Debate," *Foundations* 10 (1967): 137–49.

23. Grove Samuel Dow, *Introduction to the Principles of Sociology: A Text Book for Colleges and Universities* (Waco, TX: Baylor University Press, 1920), pp. 42, 211; "Texas Baptists Repudiate Dr. Norris," *Christian Century* 41 (1924): 1672, quoted in James J. Thompson, Jr., *Tried as by Fire: Southern Baptists and the Religious Controversies of the 1920s* (Macon, GA: Mercer University Press, 1982), p. 143; "Infidelity in Baylor University," *Searchlight* 3 (October 21, 1921): 1; "Prof. Dow and Baylor University," ibid. 3 (November 11, 1921): 1–2; G. S. Dow, Letter to the Editor, *Baptist Standard* 33 (November 3, 1921): 8–9. On Martin's awareness of Dow's book, see E. C. Routh to G. S. Dow, December 7, 1920, Brooks Papers. See also Ellen Kuniyuki Brown, "Samuel Palmer Brooks and the Evolution Controversy at Baylor University, 1921–1923," *Texas Baptist History* 1 (1981): 39–47.

24. *Annual of the Baptist General Convention of Texas* (1922), pp. 19, 157; S. P. Brooks to Edward N. Jones, January 20, 1925 (young botanist), Brooks Papers. See also "Oral Memoirs of E. N. Jones," 1973, The Texas Collection, Baylor University. On Norris's troubles, see Kenneth K. Bailey, *Southern White Protestantism in the Twentieth Century* (New York: Harper & Row, 1964), pp. 60–2.

25. *Forty-Eighth Annual Session of the Tennessee Baptist Convention Held with the Deaderick Ave. Church, Knoxville, Tenn., November 15, 16, 17, 1922*, pp. 14–15, which mistakenly identifies C. W. Davis as C. S. Davis; "Vital Information in the Fox Case," *Christian Index*, October 16, 1924, pp. 26–8; "Biologist Quits Post at Furman," (Raleigh) *News and Observer*, May 2, 1926, from a clipping in Fld. 312, Poteat Papers; Edward Lassiter Clark, "The Southern Baptist Reaction to the Darwinian Theory of

Evolution" (Ph.D. dissertation, Southwestern Baptist Theological Seminary, 1952), p. 154. See also James J. Thompson, Jr., "Southern Baptists and the Antievolution Controversy of the 1920's," *Mississippi Quarterly* 29 (1975–76): 65–81; and "Andrew Lee Pickens," *National Cyclopedia of American Biography*, 55:157–8. For Fox's obituary, see *The New York Times*, November 6, 1951, p. 29. On the Davis investigation, see "Dr. Davis Explains His Position on Evolution," *Cardinal and Cream*, December 16, 1921, pp. 1, 3–4, and the Minutes of the Board of Trustees, Union University, November 14, 1921, both in the Union University Archives, where they were located by Steve Baker. T. O. Mabry, a chemist, claimed to have been forced from an unnamed Baptist college for teaching evolution; see his pamphlet, *Evolution and Christianity: A Memorial and a Protest* (1923), a copy of which can be found in Fld. 281, Poteat Papers.

26. Bradley J. Longfield, *The Presbyterian Controversy: Fundamentalists, Modernists, and Moderates* (New York: Oxford University Press, 1991), chap. 3; Ronald L. Numbers, "Creation, Evolution, and Holy Ghost Religion: Holiness and Pentecostal Responses to Darwinism," *Religion and American Culture* 2 (1992): in press; John L. Morrison, "American Catholics and the Crusade against Evolution," *Records of the American Catholic Historical Society of Philadelphia* 64 (1953): 59–71. Norman F. Furniss, *The Fundamentalist Controversy, 1918–1931* (New Haven: Yale University Press, 1954), includes chapter-by-chapter surveys of the attitudes of seven denominations.

27. Stewart G. Cole, *The History of Fundamentalism* (New York: Richard R. Smith, 1931), pp. 259–80, describes seven antievolution organizations, all but one of which had died by the time the book appeared (p. 325). On the Anti-False Science League, not mentioned by Cole, see John Roach Straton to J. D. Sandefer, April 4, 1928, and "Constitution of the Anti-False Science League of America," both in the John Roach Straton Papers, American Baptist Historical Society, Rochester, New York. An undated circular letter regarding the American Science Foundation can be found in the personnel file of S. J. Bole, Wheaton College; see also Fred E. Bennett, *Anti-Christ Money* (Kansas City, MO: American Science Foundation, 1935). On the WCFA, see W. B. Riley, *Inspiration or Evolution*, 2nd ed. (Cleveland: Union Gospel Press, 1926), quotation on p. 185; and Szasz, *The Divided Mind of Protestant America*, pp. 89–91. Marsden, *Fundamentalism and American Culture*, pp. 169–70, stresses the interdenominational character of the antievolution crusade.

28. W. J. Bryan to S. J. Bole, January 16, 1922, Bole Papers; "Report of the Textbook Committee," *Christian Fundamentalist* 1 (1928): 9; *A List of Text Books Selected and Recommended by the Committee on Text Books for the Christian Fundamentals Association* (Minneapolis: Christian Fundamentals Association, n.d.), p. 6, from a copy in Fld. 276, Poteat Papers; John S. Park to American Book Co., December 19, 1925, and John S. Park to S. P. Brooks, December 28, 1925, both in the Brooks Papers; John S. Park to J. R. Straton, January 4, 1926, Straton Papers; Richard Kern, *Findlay College: The First Hundred Years* (Nappanee, IN: Evangel Press, 1984), quotation on p. 149. I am indebted to Richard Kern for supplying information about Park and Findlay College. On Baitsell's *Manual*, see also L. S. Keyser, "The Claims of Some Scientists," *Bible Champion* 31 (1925): 200. *The National Union Catalog of Pre-1956 Imprints* lists no books by Park. On post-1925 biology texts, see Judith V. Grabiner and Peter D. Miller, "Effects of the Scopes Trial," *Science* 185 (1974): 832–7.

29. [William B. Riley], "The Evolution Controversy," *Christian Fundamentals in School and Church* 4 (April–June, 1922): 5; Bryan, *In His Image*, p. 94; L. L. Pickett, *God or the Guessers: Some Strictures on Present Day Infidelity* (Louisville: Pentecostal Publishing Co., 1926), p. 11 (dictionaries); L. S. K[eyser], "No War against Science—Never!" *Bible Champion* 31 (1925): 413; W. B. Riley, *Inspiration or Evo-*

lution, 2nd ed. (Cleveland: Union Gospel Press, 1926), p. 34. On using the dictionary to define science see, e.g., Riley, *The Finality of the Higher Criticism*, p. 76. On the fundamentalist affinity for Baconianism, see Marsden, *Fundamentalism and American Culture*, pp. 214–15.

30. L. T. Townsend, *Collapse of Evolution* (Boston: National Magazine Co., 1905), pp. 47–53; T. T. M[artin], "The Three False Teachings of President Poteat of Wake Forest," *Western Recorder* 95 (February 5, 1920): 5, from a copy in Fld. 260, Poteat Papers; William Louis Poteat, "Evolution," *Biblical Recorder* 87 (April 19, 1922): 3. In *Hell and the High Schools* (p. 131), Martin responded to Poteat's critique with the queries: "Does the fact that they are dead prove that they were not great scientists? Does death prove that they lied?" For another early list of alleged antievolutionists, see Alexander Patterson, *The Other Side of Evolution: Its Effects and Fallacy*, 3rd ed. (Chicago: Bible Institute Colportage Association, 1912), pp. 7–11.

31. Albert Fleischmann, *Die Descendenztheorie* (Leipzig: Verlag von Arthur Georgi, 1901); Vernon L. Kellogg, *Darwinism To-Day* (New York: Henry Holt, 1907), p. 8; Albert Fleischmann, "The Doctrine of Organic Evolution in the Light of Modern Research," *Journal of the Transactions of the Victoria Institute* 65 (1933): 194–214, quotations on pp. 196, 205–6; Douglas Dewar to [name deleted], November 2, 1931, from a copy in the George McCready Price Papers, Adventist Heritage Center, Andrews University. For bibliographical information, see Georg Uschmann, "Albert Fleischmann," *Neue Deutsche Biographia*, vol. 5 (Berlin: Duncker & Humblot, 1960), pp. 234–5. On the reaction of scientists to Fleischmann, see the English translation of Dennert, *At the Deathbed of Darwinism*, pp. 133–6. On creationism in late nineteenth-century Germany, see Frederick Gregory, *Nature Lost: Natural Science and the German Theological Traditions of the Nineteenth Century* (Cambridge, MA: Harvard University Press, 1992). Lynn Nyhart and Molleurus Couperus provided helpful leads regarding Fleischmann's life and work.

32. Townsend, *Collapse of Evolution*, p. 48; Patterson, *The Other Side of Evolution*, p. 9; Sidney F. Harmer to James H. Snowdon, July 25, 1922, quoted in W. C. Curtis, "Three Letters Bearing upon the Controversy over Evolution," *Science* 61 (1925): 648. One of the fullest versions of Etheridge's statement appears in Martin, *Hell and the High Schools*, p. 112.

33. William Bateson, "Evolutionary Faith and Modern Doubts," *Science* 55 (1922): 55–61; G. M. Price to W. J. Bryan, February 9, 1922 (boost), Price Papers; *The Summarized Proceedings of the American Association for the Advancement of Science, 1921–1925* (Washington, DC, 1925), pp. 66–7, quoted in Gatewood, *Controversy in the Twenties*, pp. 169–70; Henry Fairfield Osborn, *Evolution and Religion in Education: Polemics of the Fundamentalist Controversy of 1922 to 1926* (New York: Charles Scribner's Sons, 1926), p. 29. See also E. G. Conklin to C. B. Davenport, May 6, 1922, Davenport Papers; Conklin headed up the AAAS Committee on Evolution, set up to combat antievolution propaganda. For Bateson's reaction to the flap, see his letter to W. C. Curtis, December 11, 1922, quoted in Curtis, "Three Letters Bearing upon the Controversy over Evolution," p. 647.

34. "Big Scientist Rejects Darwin," *Christian Fundamentalist* 2 (1929): 59; Austin H. Clark, *The New Evolution: Zoogenesis* (Baltimore: Williams & Wilkins, 1930); Ben F. Allen to G. M. Price, June 12, 1929, Price Papers. See also Austin H. Clark to G. M. Price, March 23, 1929, Price Papers; and Theodore Graebner, *God and the Cosmos: A Critical Analysis of Atheism* (Grand Rapids, MI: William B. Eerdmans, 1932), pp. 287–9.

Four. SCIENTIFIC CREATIONISTS IN THE AGE OF BRYAN

1. The settee quotation comes from Heber D. Curtis to W. J. Bryan, May 22, 1923, Box 37, W. J. Bryan Papers, Library of Congress. For biographical information about Fairhurst, see the preface to Alfred Fairhurst, *Atheism in Our Universities* (Cincinnati: Standard Publishing Co., 1923), pp. 9–14; and the memorial resolution passed by the Transylvania University board of trustees after his death, Special Collections, Transylvania University Library. See also Fairhurst, *Organic Evolution Considered* (St. Louis: Christian Publishing Co., 1897); and Fairhurst, *Theistic Evolution* (Cincinnati: Standard Publishing Co., 1919), quotation on p. 7. Fairhurst's difficulties are documented in the "Evolution Controversy" file, Special Collections, Transylvania University Library; and John D. Wright, Jr., *Transylvania: Tutor to the West* (Lexington, KY: Transylvania University, 1975), pp. 320, 340–1. I am indebted to Carolyn Palmgreen, special collections librarian at Transylvania, for her assistance in tracking down documents relating to Fairhurst.

2. George Barry O'Toole, *The Case against Evolution* (New York: Macmillan, 1925); Frank M. Thompson et al., *Reply Brief and Argument for the State of Tennessee* . . . (1925), p. 370; William Waller et al., *Brief and Argument of the Tennessee Academy of Science as Amicus Curiae* . . . (n.d.). Copies of the last two documents are in Box 5, Clarence S. Darrow Papers, Library of Congress. A brief biographical sketch and curriculum vitae of O'Toole can be found in the Department of Archives and Manuscripts, Catholic University of America; I am grateful to the archivist, Anthony Zito, for providing me with copies. See also John L. Morrison, "American Catholics and the Crusade against Evolution," *Records of the American Catholic Historical Society of Philadelphia* 64 (1953): 60; and the biographical entries in the *New Catholic Encyclopedia* (New York: McGraw-Hill, 1967), 10: 812; and the *Dictionary of American Catholic Biography*, ed. John J. Delaney (Garden City, NY: Doubleday, 1984), p. 454.

3. For biographical information, I have relied primarily on Franklin W. Scott, ed., *The Semi-Centennial Alumni Record of the University of Illinois* (Champaign-Urbana: University of Illinois, 1918), p. 749; and Bole's personnel file at Wheaton College. Bole describes his early religious life and his problems in Missouri in *Confessions of a College Professor* (Los Angeles: Biola Book Room, 1922), quotation on p. 43. Bole's daughter, Chrystal Bole Dutton, kindly supplied some details in a telephone interview on August 24, 1988, and in a letter dated November 28, 1988. See also Simeon James Bole, "Penmanship in Grades 5, 6, 7, 8, the Urbana Public Schools in 1879 and 1912" (A.M. thesis, University of Illinois, 1912); *University of Illinois Annual Register, 1912–13*, p. 352, which lists the courses Bole taught; and E. J. Gale to J. Oliver Buswell, Jr., [March 1929], also in Bole's personnel file, which describes his problems in Missouri.

4. S. J. Bole, *The Modern Triangle: Evolution, Philosophy and Criticism* (Los Angeles: Bible Institute of Los Angeles, 1926), pp. 13 (Damascus Road), 181 (faith), 188 (Drummond). The first eight parts of this book appeared under the title "Satan's Triangle" in *King's Business* 16 (May–December 1925); the final installment was published ibid. 17 (June 1926), under the title "The Harvest." *The King's Business* was published by the Bible Institute of Los Angeles. The "worldly Christian" description appears in Bole, *Confessions*, p. 9.

5. S. J. Bole, *The Battlefield of Faith* (University Park, IA: College Press, 1940), unpaginated preface, pp. 192–5; undated circular for the Science Foundation in Bole's personnel file; "Officers of the World's Christian Fundamentals Association," *Christian Fundamentalist* 1 (1928): 5. Bole's troubles at Wheaton are described in J. Oliver

Buswell, Jr., to R. C. McQuilkin, November 20, 1933, Bole's personnel file; his literary aspirations are mentioned in S. J. Bole to W. J. Bryan, January 9, 1922, Box 35, Bryan Papers. S. J. Bole, "Inheritance of Tree Form in Certain Progenies of Cross-bred Apple Seedlings" (Ph.D. thesis, Iowa State College, 1934), is listed in *The National Union Catalog: Pre-1956 Imprints*, 44:459.

6. Information about Brown's training comes from various editions of the *American Medical Directory* and from the Trinity College Archives. See also Willard B. Gatewood, Jr., ed., *Controversy in the Twenties: Fundamentalism, Modernism, and Evolution* (Nashville: Vanderbilt University Press, 1969), p. 154 (American birth); J. Oliver Buswell, Jr., to S. J. Bole, May 1, 1928, Bole's personnel file (postgraduate course); handbill advertising Brown's lectures at the First Baptist Church, Buffalo, December 30, 1928–January 6, 1929, J. Frank Norris Collection, Dargan-Carver Library of the Historical Commission of the Southern Baptist Convention, Nashville, Tennessee; and the editor's introduction to Arthur I. Brown, "Darwin and Sir Arthur Keith," *Defender* 2 (November 1927): 3 (salary). On the income of physicians in early twentieth-century Vancouver, see Margaret W. Andrews, "Medical Attendance in Vancouver, 1886–1920," in *Medicine in Canadian Society: Historical Perspectives*, ed. S. E. D. Shortt (Montreal: McGill-Queen's University Press, 1981), pp. 416–45.

7. Arthur I. Brown, *Evolution and the Bible* (Vancouver, BC: Arcade Printers, [1922]); Brown, *Men, Monkeys and Missing Links* (Vancouver, BC: n.p., 1923), quotation on p. 3; Brown, *Evolution and the Blood-Precipitation Test* (Los Angeles: Research Science Bureau, [1925]), quotation on p. 30. See also Brown, *Footprints of God* (Findlay, OH: Dunham Publishing Co., 1943); Brown, *Miracles of Science* (Findlay, OH: Dunham Publishing Co., 1945); Brown, *God's Creative Forethought* (Findlay, OH: Fundamental Truth Publishers, n.d.); and Brown, *God and You: Wonders of the Human Body* (Findlay, OH: Fundamental Truth Publishers, n.d.). Regarding McCann, see Alfred Watterson McCann, *God—or Gorilla: How the Monkey Theory of Evolution Exposes Its Own Methods, Refutes Its Own Principles, Denies Its Own Inferences, Disproves Its Own Case* (New York: Devin-Adair, 1922); "Alfred Watterson McCann," *Dictionary of American Biography 1928–1958*, 11:567; and A. W. McCann to W. J. Bryan, Box 47, Bryan Papers.

8. Brown, *God and You*, pp. 8–9 (infallible); Brown, *Footprints of God*, pp. 24 (hoax), 135 (unknown ages); Brown, *Was Darwin Right?* (Glendale, CA: Glendale News, n.d.), p. 49 (weapon). Typical of Brown's prophetic writings is *I Will Come Again* (Findlay, OH: Fundamental Truth Publishers, 1947). On his belief in literal days, see the *Greensboro Daily News*, May 17, 1926, from a clipping in Fld. 310, William Louis Poteat Papers, Baptist Historical Collection, Wake Forest University Library.

9. Arthur I. Brown to Frank Norris, January 4, 1929, and handbill advertising Brown's lectures at the First Baptist Church, Buffalo, both in the Norris Collection.

10. Handbill advertising Brown's lectures at the First Baptist Church, Buffalo; *Greensboro Daily News*, May 17, 1926, from a clipping in Fld. 310, Poteat Papers; Gerald B. Winrod's introduction to Arthur I. Brown, " 'Vestigial Organs,' " *Defender* 1 (June 1926): 6; J. O. Buswell to S. J. Bole, May 1, 1928, in Bole's personnel file; Stewart G. Cole, *The History of Fundamentalism* (New York: Richard R. Smith, 1931), pp. 259–80, quotation regarding Scientist General on p. 272.

11. *Greensboro Daily News*, May 16 and 17, 1926, from clippings in Fld. 310, Poteat Papers; Henry M. Morris, *A History of Modern Creationism* (San Diego: Master Book Publishers, 1984), pp. 101–3; *The Crusaders' Champion* 1 (December 25, 1925): 18, quoted in Gatewood, *Controversy in the Twenties*, pp. 405–7. On Brown's lectures in Paducah, Kentucky, and Tucson, Arizona, see, respectively, May-

nard Shipley, *The War on Modern Science: A Short History of the Fundamentalist Attacks on Evolution and Modernism* (New York: Alfred A. Knopf, 1927), pp. 123–4; and George E. Webb, "Tucson's Evolution Debate, 1924–1927," *Journal of Arizona History* 24 (1983): 6–7. See also "Dr. A. I. Brown of Vancouver Dies in Crash," (Victoria) *Daily Colonist*, November 4, 1947, p. 2.

12. For biographical information, see Mignon Brandon Rimmer, *Fire Inside: The Harry Rimmer Story* (Berne, IN: Publishers Printing House, 1968), based on nearly ten thousand letters—since lost or destroyed—that Rimmer wrote to his wife; Charles Brandon Rimmer, *In the Fullness of Time* (Berne, IN: Berne Witness Co., 1948), a semifictional account of Rimmer's youth by his son; Kathryn Rimmer Braswell, "Harry Rimmer—Defender of the Faith," *Sunday School Times* 95 (1953): 263–4, a chronologically unreliable sketch by Rimmer's daughter; and "Harry Rimmer," *Who Was Who in America, 1951–1960* (Chicago: Marquis, 1960), pp. 728–9. The quotation is taken from C. Brandon Rimmer, *Religion in Shreds* (Carol Stream, IL: Creation House, 1973), p. 10. One of the earliest historical discussions of Rimmer and his work is William D. Edmondson, "Fundamentalist Sects of Los Angeles, 1900–1930" (Ph.D. dissertation, Claremont Graduate School, 1969), pp. 276–336. The most recent is Roger Daniel Schultz, "All Things Made New: The Evolving Fundamentalism of Harry Rimmer, 1890–1952" (Ph.D. dissertation, University of Arkansas, 1989).

13. Abraham Flexner, *Medical Education in the United States and Canada* (New York: Carnegie Foundation for the Advancement of Teaching, 1910), pp. 194–5; Harry Rimmer, *The Harmony of Science and Scripture*, 9th ed. (Grand Rapids, MI: William B. Eerdmans, 1943), p. 14; interview with C. Brandon Rimmer, May 15, 1984.

14. M. Rimmer, *Fire Inside*, pp. 1, 40, 50–1; interview with C. Brandon Rimmer, May 15, 1984; Harry Rimmer, *The Theory of Evolution and the Facts of Science*, 8th ed. (Grand Rapids, MI: William B. Eerdmans, 1946), pp. 62–3.

15. M. Rimmer, *Fire Inside*, pp. 52–8; "Debates on Evolution," *Christian Fundamentals in School and Church* 8 (1926): 54; "Two Great Field Secretaries—Harry Rimmer and Dr. Arthur I. Brown," ibid., p. 17.

16. Brochure describing the Research Science Bureau, Inc., ca. 1940 (courtesy of F. Alton Everest); "Monthly News Letter of the Research Science Bureau, Inc., January 1927," *Defender* 1 (February 1927): 6, which dates the founding of the bureau in 1921; interview with C. Brandon Rimmer, May 15, 1984. Membership growth is reported in F. J. B[oyer], "Harry Rimmer, D.D.," *Christian Faith and Life* 41 (1935): 6. See also Schultz, "All Things Made New," p. 117.

17. "Monthly News Letter of the Research Science Bureau, Inc., March 1927," *Defender* 1 (March 1927): 7. As far as I can tell, only the Gerald B. Winrod Collection at the Wichita State University Library possesses all the issues of the *Defender* that include Bureau newsletters; I am indebted to Michael Kelly for providing copies of the newsletters. On Winrod, see Arch W. Jarrell, "The Kansas Monkey-Baiters," *Haldeman-Julius Monthly* 5 (1927): 10–12; and Roy Tozier, *America's Little Hitlers: Who's Who and What's Up in U.S. Fascism* (Girard, KS: Haldeman-Julius Publications, [1943]), both of which were brought to my attention by Gene DeGruson.

18. "Monthly News Letter of the Research Science Bureau, Inc., January 1927," *Defender* 1 (February 1927): 6; "Monthly News Letter of the Research Science Bureau, Inc., April 1927," ibid., 1 (April 1927): 6; Harry Rimmer, *Monkeyshines: Fakes, Fables, Facts concerning Evolution* ([Los Angeles]: Research Science Bureau, n.d.), pp. 4–5; M. Rimmer, *Fire Inside*, pp. 78, 98–100; interview with C. Brandon Rimmer, May 15, 1984. For evidence of a revival about 1940, see the brochure mentioned in note 16.

19. Braswell, "Harry Rimmer," p. 263; B[oyer], "Harry Rimmer," p. 6; interview with C. Brandon Rimmer, May 15, 1984.

20. Jessie Wiseman Gibbs, *Evolution and Christianity*, 3rd ed. (Knoxville, TN: Author, 1931), p. 207; Morgan Blake, quoted in M. Rimmer, *Fire Inside*, p. 140.

21. C. Samuel Campbell, "Rev. Harry Rimmer—God's Scientist," *Debunker* 14 (May 1931): 36–40; Oscar O. Whitenack, *A Twentieth Century Churchman's Viewpoint of Science* (Denver: Oscar O. Whitenack, 1933), p. 16; Milton Berman, *John Fiske: The Evolution of a Popularizer* (Cambridge, MA: Harvard University Press, 1961), pp. 195–6.

22. Harry Rimmer, *The Harmony of Science and Scripture*, 9th ed. (Grand Rapids, MI: William B. Eerdmans, 1942), pp. 11–13; Harry Rimmer, *The Theory of Evolution and the Facts of Science*, 8th ed. (Grand Rapids, MI: William B. Eerdmans, 1946), pp. 15–20, 49. Harry Rimmer, *Modern Science and the Genesis Record*, 7th ed. (Grand Rapids, MI: William B. Eerdmans, 1946), repeatedly emphasizes the evidence of design. Although I have been unable to identify the particular dictionary Rimmer used, a quick survey of early twentieth-century dictionaries shows his definition of science not to be idiosyncratic.

23. M. Rimmer, *Fire Inside*, pp. 89–97; interview with C. Brandon Rimmer, May 15, 1984.

24. M. Rimmer, *Fire Inside*, pp. 132–3.

25. W. B. Riley and Harry Rimmer, *A Debate: Resolved, That the Creative Days in Genesis Were Aeons, Not Solar Days* (undated pamphlet in the Riley Papers, Northwestern College Library). The debate also appeared in the *Christian Fundamentalist* 3 (1929): 408–13, 462–9. On the fourth day, see H. Rimmer, *Modern Science and the Genesis Record*, p. 146.

26. Harry Rimmer, *The Theory of Evolution and the Facts of Science*, p. 78; M. Rimmer, *Fire Inside*, p. 132. On pre-Adamites, see H. Rimmer, *"That's a Good Question!"* (Grand Rapids, MI: William B. Eerdmans, 1954), p. 59, a source brought to my attention by Roger Schultz.

27. Harry Rimmer, *Modern Science, Noah's Ark, and the Deluge* (Los Angeles: Research Science Bureau, 1925), pp. 19–20, 44–5. See also H. Rimmer, *Modern Science and the Genesis Record*, pp. 11–31; and Riley and Rimmer, *A Debate*, pp. 29–30.

28. Harry Rimmer, "Science Sustains Scripture Statements," *Christian Fundamentalist* 4 (1930): 97; Harry Rimmer, *Inspiration Plus Revelation Equals the Bible* (Grand Rapids, MI: William B. Eerdmans, 1946), p. 87. For a typical volume of science-related apologetics, see H. Rimmer, *The Harmony of Science and Scripture*.

29. Harry Rimmer, *That Lawsuit against the Bible* (Grand Rapids, MI: William B. Eerdmans, 1951), p. 10; H. Rimmer, *The Harmony of Science and Scripture*, pp. 161–90, 251–83. For an entertaining, critical analysis of the whale story, see Edward B. Davis, "A Whale of a Tale: Fundamentalist Fish Stories," *Perspectives on Science and Christian Faith* 43 (1991): 224–37.

30. H. Rimmer, *That Lawsuit against the Bible*, pp. 11–14.

31. Ibid., p. 28; William Floyd to the American Association for the Advancement of Science, January 14, 1940, "Membership—Questionable" file, Box 8, Secretary's Records, AAAS Archives; and a clipping from the *Sunday Mirror Magazine Section*, December 10, 1939, found in the same file, from which the quotations are taken. I am indebted to Michelle Aldrich for sending me copies of these documents.

32. H. Rimmer, *That Lawsuit against the Bible*, pp. 10, 87, dustjacket. Rimmer's lawyer, James E. Bennet, described the trial in *The Bible Defeats Atheism: A Story of the Famous Harry Rimmer Trial as Told by Attorney for Defendant*, 7th ed. (New York: Frederick Naef, [1947]); Rimmer's nemesis, E. Haldeman-Julius, accused the preacher of lying about the trial in *Questions and Answers: 20th Series* (Girard, KS:

Haldeman-Julius Publications, [1940]). Gene DeGruson provided copies of the last two documents from the collections in the Pittsburg State University Library.

33. M. Rimmer, *Fire Inside*, p. 110; "Harry Rimmer," *Who Was Who in America, 1951–1960*, p. 728; Bennette E. Geer to Secretary, AAAS, June 14, 1937, and Sam Woodley to B. E. Geer, July 7, 1937, both in the "Membership—Questionable" file, Box 8, Secretary's Records, AAAS Archives.

34. M. Rimmer, *Fire Inside*, pp. 68–9,151–2; "Evolution Is Abused Term," *Collegio*, July 17, 1924, p. 1. Gene DeGruson generously provided copies of this and other articles about Rimmer in the *Collegio*. For a critical account of Rimmer's visit, see John G. Scott, "Fundamentalism in the Kansas State Teachers College," *Haldeman-Julius Monthly* 2 (1925): 378–83.

35. M. Rimmer, *Fire Inside*, pp. 69, 74, 86, 108–10; "Archeology Was Their Pastime [sic]," *Collegio*, September 26, 1924, p. 4; "Archeologist Visits Here," ibid., June 18, 1926, p. 1; "Dr. Rimmer Here to Open Museum," ibid., June 10, 1927, p. 1; *Annual Catalogue, 1928–1929: The Kansas State Teachers College of Pittsburg* (Topeka: State Printer, 1929), p. 31. Although both M. Rimmer (*Fire Inside*, p. 110) and the *Collegio* ("Dr. Rimmer Adds New Specimens," June 22, 1928, p. 1) identify Rimmer as curator of the museum, Gene DeGruson, Pittsburg State University archivist, could find no documentation of an official connection with the college; G. DeGruson to R. L. Numbers, December 3, 1987.

36. "Report of the Text Book Committee," *Christian Fundamentalist* 3 (1929): 251; Harold Hill, with Irene Harrell, *From Goo to You by Way of the Zoo* (Plainfield, NJ: Logos International, 1976), p. xiv. Roger Schultz brought the Hill quotation to my attention.

37. Harry Rimmer, *The Coming League and the Roman Dream* (Grand Rapids, MI: William B. Eerdmans, 1941), p. 46; interview with C. Brandon Rimmer, May 15, 1990. William B. Eerdmans Publishing Company alone sold nearly 300,000 copies of Rimmer's books; Danford Gibbs to Roger Schultz, October 20, 1988. Again I am grateful to Roger Schultz for providing me with a copy of this document. For a more positive assessment of Rimmer's reputation at mid-century, see Schultz, "All Things Made New," pp. 349–52.

Five. GEORGE McCREADY PRICE AND THE NEW CATASTROPHISM

1. W. J. Bryan to S. James Bole, July 2, 1925, from a copy of the original (now deposited in the S. James Bole Papers, Nebraska State Historical Society) kindly provided by Bole's granddaughter Dorothy Worth; Howard A. Kelly to W. J. Bryan, June 15, 1925, Box 47, W. J. Bryan Papers, Library of Congress. In a letter written on January 4, 1923 (Box 36, Bryan Papers), Kelly had told Bryan that he believed "in the evolution of organic life in this world." For Kelly's views on evolution, see his diary entry for May 4, 1925, Box 27, Howard Kelly Papers, Chesney Medical Archives, Johns Hopkins University; and Howard A. Kelly, *A Scientific Man and the Bible* (New York: Harper & Brothers, 1925), pp. 65–8.

2. Louis T. More to W. J. Bryan, July 7, 1925, Box 47, Bryan Papers. See also Louis T. More, *The Dogma of Evolution* (Princeton, NJ: Princeton University Press, 1925); F. D. Nichol to G. M. Price, March 10, 1926, George McCready Price Papers, Adventist Heritage Center, Andrews University; Paul E. More to Editors of the Conference, February 3, 1927, and Edwin E. Slosson to Editors of the Conference, October 18, 1926, both in a folder marked "Evolution—The Conference Method of

Study," Kirtley F. Mather Papers, Harvard University Archives, Pusey Library. According to Slosson, L. T. More "admits evolution of a sort and is equally persona non grata to the fundamentalists as he is to the evolutionists." Clark Elliott brought the Mather Papers to my attention.

3. Alfred W. McCann to W. J. Bryan, June 30, 1925, Box 47, Bryan Papers. The Bryan Papers reveal that Bryan invited at least three other experts to testify at Dayton: Charles B. McMullen, a philosopher at Centre College in Kentucky (June 7, 1925, Box 47); Henry F. Lutz, an unidentified resident of Cincinnati (July 25, 1925, Box 47); and Prof. F. P. Dunnington, Chemistry Building, University of Virginia, whose views on evolution were unknown to Bryan (July 16, 1925, Box 40).

4. W. J. Bryan to G. M. Price, June 7, 1925, and G. M. Price to W. J. Bryan, July 1, 1925, Box 47, Bryan Papers; L. Sprague de Camp, *The Great Monkey Trial* (Garden City, NY: Doubleday, 1968), p. 402; "Letter to the Editor of Science from the Principal Scientific Authority of the Fundamentalists," *Science* 63 (1926): 259; Maynard Shipley, *The War on Modern Science: A Short History of the Fundamentalist Attacks on Evolution and Modernism* (New York: Alfred A. Knopf, 1927), p. 364; Martin Gardner, *Fads and Fallacies in the Name of Science* (New York: Dover, 1957), pp. 123–39, quotation on p. 127. Gardner's book first appeared under the title *In the Name of Science* (New York: G. P. Putnam's Sons, 1952). See also Gardner's collection of essays, *Science: Good, Bad and Bogus* (Buffalo, NY: Prometheus Books, 1981), p. 13, and his novel, *The Flight of Peter Fromm* (Los Altos, CA: William Kaufmann, 1973), pp. 48–51.

5. Ellen G. White, *Spiritual Gifts: Important Facts of Faith, in Connection with the History of Holy Men of Old* (Battle Creek, MI: Seventh-day Adventist Publishing Assn., 1864), pp. 64–96, quotations on pp. 90–1. A revised version of the chapters on creation and the flood appeared in Mrs. E. G. White, *Patriarchs and Prophets; or, The Great Conflict between Good and Evil as Illustrated in the Lives of Holy Men of Old* (Oakland, CA: Pacific Press, 1890), pp. 90–116. On White and Adventism, see Ronald L. Numbers, *Prophetess of Health: A Study of Ellen G. White* (New York: Harper & Row, 1976); Ronald L. Numbers and Jonathan M. Butler, eds., *The Disappointed: Millerism and Millenarianism in the Nineteenth Century* (Bloomington: Indiana University Press, 1987); and Gary Land, ed., *Adventism in America: A History* (Grand Rapids: MI: William B. Eerdmans, 1986). On Adventist attitudes toward evolution, see Ronald L. Numbers, "Science Falsely So-Called: Evolution and Adventists in the Nineteenth Century," *Journal of the American Scientific Affiliation* 27 (March 1975): 18–23; and " 'Sciences of Satanic Origin': Adventist Attitudes toward Evolutionary Biology and Geology," *Spectrum* 9, no. 4 (1979): 17–30. Unless otherwise noted, biographical information about Price is based on Harold W. Clark, *Crusader for Creation: The Life and Writings of George McCready Price* (Mountain View, CA: Pacific Press, 1966); and Ronald L. Numbers, "George Edward McCready Price," *Dictionary of American Biography*, Supplement 7, 1961–1965, ed. John A. Garraty (New York: Charles Scribner's Sons, 1981), pp. 631–2.

6. White, *Spiritual Gifts*, pp. 77–9. Recent scholarship has revealed that White borrowed liberally from other writers in describing what she allegedly saw in her visions; however, the exact sources of her geological statements have not yet been identified. On this issue, see Walter T. Rea, *The White Lie* (Turlock, CA: M & R Publications, 1982); and Warren H. Johns, "Ellen G. White and Science: Literary Borrowing and Human Sources," unpublished paper presented to the Biblical Research Committee of the General Conference of Seventh-day Adventists, Loma Linda, California, January 18, 1984.

7. George McCready Price, "If I Were Twenty-One Again," *These Times* 69 (Sep-

tember 1, 1960): 22; Price, "Some Early Experiences with Evolutionary Geology," *Bulletin of Deluge Geology* 1 (1941): 79; *Sixteenth Annual Calendar of Battle Creek College, 1891*; George McCready Price, "The Flood Theory Again," *Christian Faith and Life* 38 (1932): 350. "Battle Creek College Records, 1876–94" (Adventist Heritage Center, Andrews University), pp. 369, 383, shows a Geo. E. Price enrolled in 1891–92 and 1892–93 but does not indicate the courses he took. "Official Register No. 10 of the Provincial Normal School of New Brunswick, Session of 1896–97" (Provincial Archives of New Brunswick, RG 11, RS117/1/16), p. 18, indicates that Price was a "good" student but does not record the courses he took. See also Clifton L. Taylor, "Pioneer Days," *Eastern Canadian Messenger*, April 16, 1918, p. 4, which was brought to my attention by Bert Haloviak.

8. Price, "Some Early Experiences," pp. 78–80; George McCready Price, *Genesis Vindicated* (Washington: Review and Herald Publishing Assn., 1941), pp. 299–300; Geo. E. McCready Price, *Outlines of Modern Christianity and Modern Science* (Oakland, CA: Pacific Press, 1902), pp. 125–7. See also the obituary of Alfred Corbett Smith, *Journal of the American Medical Association* 52 (1909): 1131. Richard J. Wolfe kindly supplied information about Smith.

9. Price, *Genesis Vindicated*, p. 300; Price, "Some Early Experiences," p. 80.

10. Price, "Some Early Experiences," p. 79; Price, "If I Were Twenty-One Again," p. 23; Price, *Outlines*, pp. ix–x, 113.

11. George McCready Price, *Back to the Bible; or, The New Protestantism*, 3rd ed. (Washington: Review and Herald Publishing Assn., 1920), p. 5; Price, *Outlines*, pp. 41, 69, 134–8, 195. As Richard M. Ritland pointed out to me, in the early 1880s an Adventist minister, A. T. Jones, argued some of these same points in "The Uncertainty of Geological Science," *Advent Review and Sabbath Herald* 60 (1883): 497–8, 513–14, 529–30; however, I have found no direct evidence that Price read this series of articles.

12. Price, *Outlines*, pp. 159–61. Although Price at times insisted that he did "*not* believe that the fossils . . . were all buried at one and the same time" ("Fair Play in the Teaching of Evolution: A Reply to an Article in Science Magazine," *Sunday School Times*, November 4, 1922, p. 668), at other times he implied just the opposite ("Bible Studies on Creation," *Sabbath School Lesson Quarterly*, Fourth Quarter, 1920, p. 15). Price later owned a copy of John Woodward's *Natural History of the Earth* (1695), now in the possession of the Adventist Heritage Center, Andrews University, but he seems to have been unfamiliar with Woodward and other early flood geologists at this time.

13. Price, *Outlines*, pp. 104, 146–8, 174–8, 186, 197.

14. G. E. Price to William Guthrie, August 26, 1904; G. E. Price to W. H. Thurston, August 28, 1904; W. H. Thurston to A. G. Daniells, January 19, 1905; William Guthrie to A. G. Daniells, January 23, 1905; all in RG 11 of the Archives of the General Conference of Seventh-day Adventists, hereinafter cited as SDA Archives. For a record of employment, see G. M. Price's "Sustentation Fund Application," November, 1936, SDA Archives. I am indebted to Bert Haloviak for bringing these and related documents to my attention.

15. George E. Price to William Guthrie, December 28, 1904; A. G. Daniells to Mrs. G. E. Price, January 16, 1905; A. G. Daniells to C. H. Edwards, January 16, 1905; A. G. Daniells to G. E. Price, January 17 and 31, 1905; G. E. Price to A. G. Daniells, January 25 and March 19, 1905; all in RG 11, SDA Archives. Price describes learning to use a typewriter in "If I Were Twenty-One Again," p. 23.

16. Clark, *Crusader for Creation*, p. 21; George McCready Price, "I'd Have an Aim," *Advent Review and Sabbath Herald* 138 (February 16, 1961): 14–15.

17. George McCready Price, *The Phantom of Organic Evolution* (New York: Fleming H. Revell, 1924), pp. 5–6; Price, "Some Early Experiences," pp. 80–6; George

McCready Price, *Illogical Geology: The Weakest Point in the Evolution Theory* (Los Angeles: Modern Heretic Co., 1906), pp. 9, 11.

18. Price, *Illogical Geology,* p. 28; Price, "Some Early Experiences," p. 86.

19. George McCready Price, *The Fundamentals of Geology* (Mountain View, CA: Pacific Press, 1913), pp. 7–8; 101–3.

20. Ibid., p. 119; George McCready Price, *The Modern Flood Theory of Geology* (New York: Fleming H. Revell, 1935), p. 6.

21. George McCready Price, *The Story of the Fossils* (Mountain View, CA: Pacific Press, 1954), p. 39. For a nearly complete list of Price's books, see Clark, *Crusader for Creation,* pp. 101–2. The index of Seventh-day Adventist periodicals in the SDA Room, Columbia Union College, Takoma Park, MD, lists 235 of his articles; in addition, scores of his articles appeared in non-Adventist periodicals.

22. G. M. Price to W. H. Williams, July 14, 1932, Price Papers. Information regarding Price's illness comes from Clark, *Crusader for Creation,* pp. 56–7; and an interview with Edward and Mary Specht, March 28, 1980. Price's academic appointments included the following: Loma Linda College of Evangelists, in 1909 renamed the College of Medical Evangelists, Loma Linda, California (1907–1912); Fernando Academy, San Fernando, California (1912–1914); Lodi Academy, Lodi, California (1914–1920); Pacific Union College, Angwin, California (1920–1922); Union College, Lincoln, Nebraska (1922–1924); Stanborough Missionary College, Watford, England (1924–1928), where he served as president during 1927–1928; Emmanuel Missionary College, Berrien Springs, Michigan (1929–1933); and Walla Walla College, Walla Walla, Washington (1933–1938).

23. George McCready Price, *The New Geology* (Mountain View, CA: Pacific Press, 1923); Price, *The Story of the Fossils,* p. 30; George McCready Price, *Evolutionary Geology and the New Catastrophism* (Mountain View, CA: Pacific Press, 1926), p. 2. Price regarded *Evolutionary Geology and the New Catastrophism,* which evolved from his *Illogical Geology* (1906) and *The Fundamentals of Geology* (1913), as his "most formal and complete discussion of this entire subject of the proposed successive geological 'ages' " (Price, "Some Early Experiences," p. 84). On sales of *The New Geology,* see Harold W. Clark, *The Battle over Genesis* (Washington: Review and Herald Publishing Assn., 1977), p. 106.

24. Price, *The New Geology,* pp. 294, 637–8, 654–5, 680–1.

25. Price, *Back to the Bible,* p. 225; Price, *The New Geology,* pp. 682–5.

26. Price, *Outlines,* p. 200; G. M. Price to W. J. Bryan, April 13, 1922, Box 35, Bryan Papers; Price, *The Phantom of Organic Evolution,* pp. 19–20, 38–9; George McCready Price, "The 'Amalgamation' Question Again," unpublished MS, E. S. Ballenger Papers, courtesy of Donald F. Mote. For one biologist's response to Price's interpretation of Mendelism, see William M. Goldsmith, *Evolution or Christianity, God or Darwin?* (St. Louis: Anderson Press, 1924), p. 68.

27. George McCready Price, *Q. E. D.; or, New Light on the Doctrine of Creation* (New York: Fleming H. Revell, 1917), pp. 68–77; Price, *Outlines,* p. 199; George McCready Price, "Dear Fellow Science Teachers," *Watchman Magazine* 34 (January 1925): 18. *Q.E.D.* appeared in a German translation as *Naturwissenschaft und Schopfungslehre,* trans. W. K. Ising (Hamburg: Advent-Verlag, 1925).

28. White, *Spiritual Gifts,* p. 75; Gardner, *Fads and Fallacies,* p. 130; George McCready Price to Martin Gardner, May 13, 1952, Gardner Papers, courtesy of Martin Gardner. For Price's earliest extant observations on the amalgamation question, see his letter to his brother Charlie, August 9, 1905, Ballenger Papers. On the Adventist debate over amalgamation, see Gordon Shigley, "Amalgamation of Man and Beast: What Did Ellen White Mean?" *Spectrum* 12, no. 4 (1982): 10–19.

29. Price, *The Phantom of Organic Evolution*, pp. 91–112; George McCready Price, *A Textbook of General Science for Secondary Schools* (Mountain View, CA: Pacific Press, [1917]), pp. 500–510; George McCready Price, *The Geological-Ages Hoax: A Plea for Logic in Theoretical Geology* (New York: Fleming H. Revell, 1931), pp. 105–6.

30. Price, *The Phantom of Organic Evolution*, pp. 103–12, 210; George McCready Price, "The White and the Brown," undated MS, ca. 1910, Price Papers.

31. G. M. Price to E. S. Ballenger, July 25, 1927, Ballenger Papers.

32. Price, *Outlines*, p. 68; George McCready Price, *Why I Am Not an Evolutionist*, Bible Truth Series No. 52 (Mountain View, CA: Pacific Press, n.d.), p. 1; George McCready Price, *Poisoning Democracy: A Study of the Moral and Religious Aspects of Socialism* (New York: Fleming H. Revell, 1921), p. 25; George McCready Price and Robert B. Thurber, *Socialism in the Test-Tube* (Nashville: Southern Publishing Assn., 1921). In *The Fundamentalist Controversy, 1918–1931* (New Haven: Yale University Press, 1954), p. 16, Norman F. Furniss credits Price with the line: "No Adam, no fall; no fall, no atonement; no atonement, no Savior. Accepting Evolution, how can we believe in a fall?" Price always attributed the saying to the English socialist Robert Blatchford; see Price, *Poisoning Democracy*, p. 48; and Price, *Back to the Bible*, pp. 124–5.

33. George McCready Price, *Science and Religion in a Nutshell* (Washington: Review and Herald Publishing Assn., 1923), p. 15.

34. George McCready Price, *How Did the World Begin?* (New York: Fleming H. Revell, 1942), pp. 66–7.

35. Price, *Science and Religion in a Nutshell*, p. 15; [George McCready Price], "Evidence of a Worldwide Flood," *The Seventh-day Adventist Bible Commentary*, 8 vols. (Washington: Review and Herald Publishing Assn., 1953), 1:74; Price, *The Phantom of Organic Evolution*, pp. 98–100.

36. George E. Price to Charlie Price, August 9, 1905, Ballenger Papers; George McCready Price, "The Significance of Fundamentalism," *Advent Review and Sabbath Herald* 104 (May 12, 1927): 13.

37. G. M. Price to W. H. Williams, July 14, 1932, Price Papers; Clark, *Crusader for Creation*, p. 34; Richard H. Utt, "Geologist for God: Sixty Years a Champion of Creation," *Signs of the Times* 86 (November 1959): 22. Robert M. Brown and George McCready Price, *Geography and Geology* (Philadelphia: American Educational Institute, 1920), and George McCready Price and Eric Doolittle, *Nature Study and Astronomy* (Philadelphia: American Educational Institute, 1921) were part of a twelve-volume series.

38. Utt, "Geologist for God," p. 23; G. M. Price to D. S. Jordan, August 29, 1906, and April 23, 1911, Jordan Papers, University Archives, Stanford University Libraries; D. S. Jordan to G. M. Price, August 28 and September 4, 1906, and May 5, 1911, Price Papers. Price also corresponded with the Swiss geologist Albert Heim; see Price, "Some Early Experiences," pp. 90–1.

39. Price, "Some Early Experiences," pp. 70–80, 88–90; Price, "If I Were Twenty-One Again," p. 23; Utt, "Geologist for God," p. 23; Donald A. Webster, "Crusader for Creation," *Signs of the Times* 83 (July 17, 1956): 4; W. A. Spicer to G. M. Price, January 12, 1926, Price Papers; interview with Edward Specht, one of the students involved, March 28, 1980. Price mentions returning to the Canadian Rockies in the summer of 1934 in a letter to Theodore Graebner, August 26, 1935, Box 1, File 6, Theodore Graebner Papers, Concordia Historical Institute.

40. George McCready Price, "Scientific Muckrakers," *Watchman Magazine* 34 (August 1925): 7; Price, "Some Early Experiences," pp. 79–80; George McCready

Price, "The Problem of Advanced Education," *Advent Review and Sabbath Herald* 107 (November 13, 1930): 6–7; F. D. Nichol to G. M. Price, March 10, 1926, Price Papers; Price, "I'd Have an Aim," p. 15. Regarding Price's degrees, see the following documents in the Price Papers: J. H. L. Derby, affidavit signed August 15, 1911; G. M. Price to the Faculty of the College of Medical Evangelists, [1912]; E. H. Risley, M.D., Dean, College of Medical Evangelists, to Whom It May Concern, April 8, 1932; Anna J. Olson, Registrar, Pacific Union College, to Whom It May Concern, n.d.

41. Price, *The Fundamentals of Geology,* pp. 252, 339–40; George McCready Price, "Geology and the Recapitulation Theory: A Study in Circular Reasoning," *Bulletin of Deluge Geology* 1 (1941): 72; G. M. Price to D. S. Jordan, September 12, 1906, and April 28, 1911, Jordan Papers; George McCready Price, *A History of Some Scientific Blunders* (New York: Fleming H. Revell, 1930), pp. 5–8, 131, which appeared simultaneously in Great Britain (London and Edinburgh: Oliphants, [1930]). Although Price wrote his history of science in the controversialist manner of John William Draper and Andrew Dickson White, he much preferred what he regarded as the unbiased approach of J. T. Merz, George Sarton, and Charles Singer; see ibid., p. 11, and George McCready Price, "Cranks and Prophets," *Catholic World* 132 (October 1930): 44.

42. Arthur M. Miller, "The New Catastrophism and Its Defender," *Science* 55 (1922): 701–3; G. M. Price to J. M. Cattell, July 4, August 14, and December 11, 1922, and J. M. Cattell to G. M. Price, August 1, 1922, Box 145, J. M. Cattell Papers, Library of Congress; Price, "Fair Play in the Teaching of Evolution," p. 668. Within a few years Cattell was including Price in his *American Men of Science: A Biographical Directory,* 4th ed. (New York: Science Press, 1927), p. 788. I am indebted to Michael Sokal for directing me to the Cattell Papers.

43. Charles Schuchert to J. M. Cattell, December 24, 1923, and April 19, 1924, Box 37, Letterbook 1923–1925, Charles Schuchert Papers, Sterling Memorial Library, Yale University; and J. M. Cattell to Charles Schuchert, April 22 and July 12, 1924, Box 17, Schuchert Papers; Charles Schuchert, Review of *The New Geology,* by George McCready Price, *Science* 59 (1924): 486–7; George McCready Price, Letter to the Editor, *Science* 63 (1926): 259; G. M. Price to J. M. Cattell, March 22, 1926, Price Papers. In 1926 *Science* carried still another attack on Price, this one by Edwin Linton (1855–1939), retiring vice president of the zoology section of the AAAS, "The Scientific Method and Authority," *Science* 63 (1926): 195–201. Michele Aldrich first suggested that I check the Schuchert Papers.

44. G. M. Price to Martin Gardner, February 25, 1952, Gardner Papers; Pacific Press to Hay Watson Smith, quoted in Smith to G. M. Price, February 17, 1931, Price Papers. The quotation is from Price's copy of Pirsson and Schuchert's *A Text-Book of Geology,* Adventist Heritage Center, Andrews University.

45. G. M. Price to J. M. Cattell, July 14, 1922, Cattell Papers; J. M. Bird to G. M. Price, January 4, 1919, Price Papers; Price, *Q. E. D.,* p. 11; G. M. Price to John Roach Straton, March 25, 1924, John Roach Straton Papers, American Baptist Historical Society, Rochester, NY. Clark, *Crusader for Creation,* p. 38, erroneously implies that *Scientific American* published Price's manuscript.

46. George McCready Price, *The Predicament of Evolution* (Nashville: Southern Publishing Assn., 1925), pp. 94–5; Price, *The Phantom of Organic Evolution,* pp. 7–8; Price, "Scientific Muckrakers," p. 7; Price, "Dear Fellow Science Teachers," pp. 16–17; Price, "Cranks and Prophets," p. 45; George McCready Price, "Guarding the Sacred Cow," *Christian Faith and Life* 41 (1935): 124–6. Regarding the Dewar incident, see Douglas Dewar, "The Limitations of Organic Evolution," *Journal of the Transactions of the Victoria Institute* 64 (1932): 142.

47. George McCready Price, "Methodology in Historical Geology," *Pan-American Geologist* 67 (1937): 117–28; Charles R. Keyes, "Editorial: Methodology in Geology," ibid., pp. 303–6; Henry M. Morris, *A History of Modern Creationism* (San Diego: Master Book, 1984), p. 105. During the late 1930s Price's friend D. J. Whitney published several articles in the *Pan-American Geologist*.

48. George McCready Price, *Feet of Clay: The Unscientific Nonsense of Historical Geology* (Malverne, NY: Christian Evidence League, 1949), p. 23; G. M. Price to J. M. Cattell, March 22, 1926, and Malcolm H. Bissell to G. M. Price, April 6, 1922, both in the Price Papers; Stephen Richarz, "Evolution in the Light of Geology," *Fortnightly Review* 32 (1925): 485–8. On the Catholic response to Price, see John L. Morrison, "American Catholics and the Crusade against Evolution," *Records of the American Catholic Historical Society of Philadelphia* 64 (1953): 59–71. For other appraisals of Price by prominent geologists, see William North Rice, "The New Geology," *Methodist Review* 84 (1924): 560; and Kirtley F. Mather to W. C. Kraatz, November 6, 1925, from a copy in the possession of Mather's daughter, Florence Wengerd, provided by Robert H. Dott, Jr.

49. George H. Ashley to G. M. Price, April 8, 1930, Austin H. Clark to G. M. Price, March 23, 1929, and William Bowie to G. M. Price, March 5, 1931, all in the Price Papers. See also a letter from the Principal Geologist of the U. S. Geological Survey, M. R. Campbell, to G. M. Price, May 12, 1930, Price Papers.

50. M. G. Kyle, Review of *The New Geology*, by George McCready Price, *Bibliotheca Sacra* 81 (1924): 108; F. D. Nichol to G. M. Price, January 15, 1933, Price Papers. Regarding Nichol, see Maynard Shipley, Francis D. Nichol, and Alonzo L. Baker, *The San Francisco Debates on Evolution* (Mountain View, CA: Pacific Press, 1925); and Alonzo L. Baker and Francis D. Nichol, *Creation—Not Evolution* (Mountain View, CA: Pacific Press, 1926), which contains a foreword by Price.

51. Brewster's challenge is included in L. S. Keyser to G. M. Price, February 24, 1927, Price Papers. See also Edwin Tenney Brewster, *Creation: A History of Non-Evolutionary Theories* (Indianapolis: Bobbs-Merrill, 1927), on the title page of which Brewster identifies himself as a member of the History of Science Society.

52. E. T. Brewster to G. M. Price, April 29, [1930], accompanied by "Two More Questions for Professor Price," and G. M. Price to E. T. Brewster, May 2, 1930, Price Papers. See also George McCready Price, "Facts the Flood Explains," *Bible Champion* 36 (1930): 206–7.

53. Price, *Back to the Bible*, p. 5; A. O. Tait to H. H. Hall, May 26, 1916, Price Papers. Regarding attempts to find Price a job, see G. M. Price to W. H. Williams, July 14, 1932, and L. E. Fromm to G. M. Price, June 17, 1932, both in the Price Papers; and Minutes of the General Conference Committee, 13, Bk. 1, p. 177, and Bk. 2, pp. 712, 755, SDA Archives.

54. George McCready Price, *Modern Discoveries Which Help Us to Believe* (New York: Fleming H. Revell, 1934), pp. 7–8; A. C. Dixon to G. M. Price, October 9, 1920, Price Papers; Daniel S. Gregory to G. F. Wright, September 16, 1910, Box 22, G. F. Wright Papers, Oberlin College Archives. See also G. Frederick Wright, *Story of My Life and Work* (Oberlin, OH: Bibliotheca Sacra Co., 1916), pp. 38–9; [G. F. Wright], "A Sure and Short Method with the Seventh-Day Adventists," *Bibliotheca Sacra* 57 (1900): 609.

55. Clark, *Crusader for Creation*, pp. 3, 43–4; W. H. Griffith Thomas to G. M. Price, January 30, 1918, Price Papers; W. H. Griffith Thomas, *What about Evolution? Some Thoughts on the Relation of Evolution to the Bible and Christianity* (1918), quoted in Eldred C. Vanderlaan, ed., *Fundamentalism versus Modernism* (New York: H. W. Wilson, 1925), pp. 248–9; W. H. Griffith Thomas, "Evolution and the Super-

natural," *Bibliotheca Sacra* 79 (1922): 207–8, subsequently published as a pamphlet by the Sunday School Times Co.; [W. B. Riley], Editorial Introduction to George Mc-Cready Price, "Throwing Darwin's Theory Overboard," *Christian Fundamentals in School and Church* 4 (April–June 1922): 24; [W. B. Riley], Review of *Back to Creationism*, by Harold W. Clark, *Christian Fundamentalist* 2 (1929): 149. On the sale of Price's books at Riley's lectures, see W. E. Howell to G. M. Price, September 7, 1925, Price Papers. See also W. B. Riley, "Do Genesis and Geology Agree?" *Christian Fundamentals in School and Church* 5 (April–June 1923): 37–8; [W. B. Riley], Review of *The New Geology*, by George McCready Price, ibid. 6 (October–December 1923): 37; and [W. B. Riley], Review of *Evolutionary Geology and the New Catastrophism*, by George McCready Price, ibid. 9 (January–March 1927): 55.

56. John Roach Straton and Charles Francis Potter, *Evolution versus Creation: Second in the Series of Fundamentalist-Modernist Debates* (New York: George H. Doran, 1924), pp. 72–5; J. Frank Norris to W. J. Bryan, [ca. June 1925], Box 40, Bryan Papers; Harry Rimmer, *Modern Science, Noah's Ark and the Deluge* (Los Angeles: Research Science Bureau, 1925), p. 43; Arthur I. Brown, *Evolution and the Blood-Precipitation Test* (Los Angeles: Research Science Bureau, [1925]), p. 2; Leander S. Keyser, "A Telling Blow at Evolution," *Christian Fundamentals in School and Church* 7 (April–June 1924): 43–4, reprinted from the *Sunday School Times*. For additional reactions to *The New Geology*, including S. J. Bole's, see the broadside advertisement, Price Papers. Price acknowledged his indebtedness to the various fundamentalist leaders in a letter to J. R. Straton, March 25, 1924, Straton Papers.

57. George McCready Price, "The Scopes Trial—1925," *These Times* 69 (February 1960): 4; J. Frank Norris to W. J. Bryan, n.d., Box 40, S. K. Hicks to W. J. Bryan, June 8, 1925, Box 47, W. J. Bryan to G. M. Price, June 7, 1925, Box 47, G. M. Price to W. J. Bryan, July 1, 1925, Box 47, all in the Bryan Papers; Webster, "Crusader for Creation," p. 3. By 1929 Price was applauding the fact that "virtually all textbooks on the market have been revised to meet the needs of the Fundamentalists": George McCready Price, "Bringing Home the Bacon," *Bible Champion* 35 (1929): 205.

58. *The World's Most Famous Court Trial: Tennessee Evolution Case* (Cincinnati: National Book Co., 1925), pp. 296–9; George McCready Price, "What Christians Believe about Creation," *Bulletin of Deluge Geology* 2 (1942): 76–7; G. M. Price to W. J. Bryan, February 9, 1922, Box 35, Bryan Papers; Webster, "Crusader for Creation," p. 3.

59. G. M. Price to J. R. Straton, April 15, 1924, Straton Papers; Straton and Potter, *Evolution versus Creation*, p. 78; Rimmer, *Modern Science*, pp. 48–9. Later in life, S. J. Bole, presumably referring to Price, complained of creationists who "put too much emphasis on the Flood": Bole, *The Battlefield of Faith* (University Park, IA: College Press, 1940), p. 201. For additional examples of inconsistent responses to Price, see J. J. Sims, *The Last Word of Great Scientists on Evolution* (Los Angeles: Bible and Science League, 1925), pp. 30–1; and Jessie Wiseman Gibbs, *Evolution and Christianity*, 3rd ed. (Knoxville, TN: Author, 1931), pp. 214–15.

60. "Geo. McCready Price, M.A.: Contributing Editor," *Bible Champion* 34 (1928): 126; "George McCready Price, M.A.," *Christian Faith and Life* 37 (January 1931): 8–9. In 1931–32 *Christian Faith and Life* carried monthly articles by Price. Price contributed at least a dozen articles to *Moody Monthly* and at least two dozen to *Sunday School Times*; see the preface to G. M. Price, "Cranks and Prophets," MS in the Price Papers.

61. George McCready Price, "Some Scientific Aspects of Apologetics," *Evangelical Quarterly* 4 (1932): 242–3; Price, *Back to the Bible*, p. 5; George Barry O'Toole, *The Case against Evolution* (New York: Macmillan, 1925), pp. 97–9, 108; Floyd E. Ham-

ilton, *The Basis of Christian Faith: A Modern Defense of the Christian Religion* (New York: George H. Doran, 1927); Floyd E. Hamilton, *The Basis of Evolutionary Faith: A Critique of the Theory of Evolution* (London: James Clarke, [1931]); Byron C. Nelson, *The Deluge Story in Stone: A History of the Flood Theory of Geology* (Minneapolis: Augsburg Publishing House, 1931), p. 132. The Hamilton quotation comes from the revised edition of *The Basis of Christian Faith* (New York: Harper & Brothers, 1933), p. 85. For additional evidence of Price's influence on Reformed thinking, see Deryl Freeman Johnson, "The Attitudes of the Princeton Theologians toward Darwinism and Evolution from 1859–1929" (Ph.D. dissertation, University of Iowa, 1968), pp. 261–2; and Valentine Hepp, *Calvinism and the Philosophy of Nature: The Stone Lectures Delivered at Princeton in 1930* (Grand Rapids, MI: William B. Eerdmans, 1930), pp. 185–223, a source that David Livingstone brought to my attention.

62. G. M. Price to E. T. Brewster, May 2, 1930, Price Papers; George McCready Price, "A Brief History of the Flood Theory," *Signs of the Times* 61 (October 30, 1934): 15; Price, *Modern Discoveries*, p. 8; J. Laurence Kulp, "Deluge Geology," *Journal of the American Scientific Affiliation* 2 (January 1950): 1 (infiltrated); Bernard Ramm, *The Christian View of Science and Scripture* (Grand Rapids, MI: William B. Eerdmans, 1954), p. 180. In the early 1940s, however, Price grew increasingly disillusioned; see George McCready Price, Letter to the Editor, *The Ministry* 13 (May 1940): 37–8; and Price, *Genesis Vindicated*, pp. 285–6.

Six. THE RELIGION AND SCIENCE ASSOCIATION

1. D. J. Whitney to G. M. Price, December 11, 1935, George McCready Price Papers, Adventist Heritage Center, Andrews University.

2. Joel A. Carpenter, "Fundamentalist Institutions and the Rise of Evangelical Protestantism, 1929–1942," *Church History* 49 (1980): 62–5. See also William Vance Trollinger, Jr., *God's Empire: William Bell Riley and Midwestern Fundamentalism* (Madison: University of Wisconsin Press, 1990); and Virginia Lieson Brereton, *Training God's Army: The American Bible School, 1880–1940* (Bloomington: Indiana University Press, 1990).

3. D. J. Whitney to J. C. Whitcomb, September 5, 1957 (Pentecostals and fundamentalists), and September 30, 1961 (religious activities), Whitcomb Papers, courtesy of John C. Whitcomb, Jr.; D. J. Whitney to G. M. Price, September 21, 1936 (ex-Dowieite), Price Papers; D. J. Whitney to L. A. Higley, December 28, 1935 (Bible reading), and B. C. Nelson to D. J. Whitney, October 10, 1936 (pep), both in the Byron C. Nelson Papers, Institute for Creation Research (hereinafter cited as Nelson Papers, ICR). I am grateful to Henry M. Morris and Anita Nelson for granting me access to this collection. For a useful biographical sketch, see [Joseph E. Doctor], "Last Rites for Dudley Whitney, Killed by Truck," *Exeter Sun*, July 23, 1964. I am indebted to Joseph E. Doctor, a longtime friend of Whitney's, for a copy of this article and for other biographical details provided in a telephone conversation on August 3, 1989. My discussion of Whitney's academic record is based on a copy of his transcript obtained from the Office of Admissions and Records, University of California, Berkeley, and on a letter from Marie C. Thornton, Assistant University Archivist, August 1, 1989. Whitney edited the *California Cultivator* from 1936 to 1941.

4. D. J. Whitney to G. M. Price, August 11, 1927 (Riley), and D. J. Whitney to L. A. Higley, December 4, 1935 (ex-ruinite), both in the Price Papers; D. J. Whitney to B. C. Nelson, March 3, 1929 (geological nightmares), Byron C. Nelson Papers in the possession of his grandson Paul Nelson (hereinafter cited as Nelson Papers, PN).

O. L. Brauer, who later became a professor of chemistry at San Jose State College, first introduced Whitney to Price's work; see D. J. Whitney to J. C. Whitcomb, November 26–7, 1957, Whitcomb Papers.

5. Dudley Joseph Whitney, "Geology and the Deluge," *Bible Champion* 34 (1928): 104–8. See also Whitney, "What Theory of Earth History Shall We Adopt?" ibid. 34 (1928): 616–18; and Whitney, "The Three Views of Creation," *Christian Fundamentalist* 4 (1930): 222–5. Whitney refers to the *Bible Champion* as the leading antievolution paper in a letter to G. M. Price, October 6, 1927, Price Papers. Regarding Price's appointment as a contributing editor, see *Bible Champion* 34 (1928): 126. When the *Bible Champion* merged with the *Essentialist* to become *Christian Faith and Life*, Price became a "Department Editor" for "Current Scientific Discoveries."

6. W. W. Prescott to G. M. Price, November 6, 1908, and C. C. Lewis to G. M. Price, October 23, 1908, both in the Price Papers; G. M. Price to E. S. Ballenger, April 20, 1927, and January 30, 1928, Ballenger Papers, courtesy of Donald F. Mote. For a description of *The Modern Heretic*—which Price described in a letter to Ballenger (January 30, 1928) as "a short-lived journal"—see the back cover of George McCready Price, *Illogical Geology: The Weakest Point in the Evolution Theory* (Los Angeles: Modern Heretic Co., 1906). I have been unable to locate any copies of *The Modern Heretic. Jocko-Homo: The Heaven-Bound King of the Zoo* and *Puddle to Paradise* were the titles of two antievolution pamphlets published in the 1920s by B. H. Shadduck, a minister in the Pilgrim Holiness Church.

7. G. M. Price to Fellow Officers of the Religion and Science Association, September 18, 1935, and D. J. Whitney to B. C. Nelson, August 26, 1935, both in the Nelson Papers, ICR.

8. D. J. Whitney to G. M. Price [fall 1935], and November 8, 1935, both in the Price Papers; G. M. Price to B. C. Nelson, July 8, 1935, Nelson Papers, ICR.

9. G. M. Price to B. C. Nelson, July 8, 1935 (low profile), and D. J. Whitney to B. C. Nelson, August 3, 1935 (handsprings), both in the Nelson Papers, ICR; "The Religion and Science Association," *Christian Faith and Life*, 42 (1936): 159. Regarding the selection of officers, see G. M. Price to B. C. Nelson, August 14, 1935, August 26, 1935, September 30, 1935, and October 10, 1935; D. J. Whitney to B. C. Nelson, August 13, 1935; and D. J. Whitney to L. A. Higley, August 13, 1935; all in the Nelson Papers, ICR. Joseph D. Eggleston, president of Hampden-Sydney College, was an old supporter of Price's; see J. D. Eggleston to G. M. Price, November 5, 1918, and J. D. Eggleston to C. H. Benson, April 10, 1919, both in the Price Papers.

10. A. G[raebner], "Science and the Church," *Theological Quarterly* 6 (1902): 37–45, quotations on pp. 43–4. In this article Graebner quotes critics of the Missouri Synod.

11. Theodore Graebner to L. O. Kaspar, April 23, 1923, Box 1, Theodore Graebner Papers, Concordia Historical Institute (heliocentrism); and Theodore Graebner to Paul G. Witte, April 25, 1934, Box 3, ibid. (scientist). For biographical information, see Theodore Graebner, "Biographical Notes," n.d., Box 136, ibid.; Jerrald K. Pfabe, "Theodore Graebner, the Creationists, and the Evolution Controversy," *Archives and History: Minutes and Reports of the 13th Archivists' and Historians' Conference* (St. Louis: Concordia Historical Institute, 1975), pp. 90–117; and "Theodore Graebner, 1876–1950," *Concordia Historical Institute Quarterly* 23 (1951): 181–2. On the *Lutheran Witness*, see August R. Suelflow, "The First Seventy-five Years of the *Lutheran Witness*," *Concordia Historical Institute Quarterly* 31 (1958): 7–26.

12. Th. Graebner, *Evolution: An Investigation and a Criticism* (Milwaukee: Northwestern Publishing House, 1921), pp. 7–8, 24–5, 60–1; [Theodore Graebner], Review of *The New Geology*, by George McCready Price, *Lutheran Witness* 42 (1923): 316–17;

G. M. Price to Theodore Graebner, March 2, 1922, Box 3, Graebner Papers; Theodore Graebner to A. H. Grumm, October 3, 1928, Box 5, ibid. On compartmentalization, see Theodore Graebner to B. C. Nelson, February 25, 1949, and Theodore Graebner to John A. Leimer, November 30, 1938, both in Box 5, ibid. Graebner tells of discovering Price's work in the foreword to the second edition of *Evolution*, 3rd ed. (Milwaukee: Northwestern Publishing House, 1924), p. 10. See also Theodore Graebner, *Essays on Evolution* (St. Louis: Concordia Publishing House, 1925); and Theodore Graebner, *God and the Cosmos: A Critical Analysis of Atheism* (Grand Rapids, MI: William B. Eerdmans, 1932). The Luther College professor who told him about reasoning in a circle was Oscar A. Tingelstad, about whom I learned more from Jon Harkness and Leigh D. Jordahl.

13. B. C. Nelson to G. M. Price, December 12, 1936, and B. C. Nelson to H. W. Clark, December 10, 1963, both in the Price Papers; Byron C. Nelson, Review of *The Geological-Ages Hoax*, by G. M. Price, *Christian Faith and Life* 37 (1931): 331–3; Byron C. Nelson, *"After Its Kind": The First and Last Word on Evolution*, 2nd ed. (Minneapolis: Augsburg Publishing House, 1927), p. 9; Byron C. Nelson, *The Deluge Story in Stone: A History of the Flood Theory of Geology* (Minneapolis: Augsburg Publishing House, 1931), p. 132. Nelson's writing habits are described in Elizabeth Nelson Taylor, "Byron C. Nelson: Contender for Creation," *Concordia Historical Institute Quarterly* 61 (1988): 120–31. Gerhard B. Naeseth provided additional biographical detail.

14. B. C. Nelson to J. C. Whitcomb, August 22, 1967, Whitcomb Papers (Rutgers); B. C. Nelson to H. W. Clark, December 10, 1963, Price Papers (geological studies). The possibility of choosing a scientific career is discussed in "School Book's Treatment of Evolution Cause of New Jersey Controversy," (New York) *Evening World*, February 7, 1931, clipping courtesy of Paul Nelson.

15. "Liberal Group Plans to Enter Textbook Row," (Baltimore) *Evening Sun*, February 7, 1931, clipping courtesy of Paul Nelson; G. M. Price to Byron C. Nelson, October 10, 1935, Nelson Papers, ICR; D. J. Whitney to B. C. Nelson, June 3, 1928, Nelson Papers, PN.

16. D. J. Whitney to B. C. Nelson, June 18, 1928, and B. C. Nelson to D. J. Whitney, April 16, 1929, Nelson Papers, PN, which include additional letters from this debate. For Price's reaction, see G. M. Price to B. C. Nelson, October 13, 1929, Price Papers.

17. D. J. Whitney to G. M. Price, September 14, 1935 (fundamentalists), November 8, 1935 (spasm), and ca. fall 1935 (pronouncements), all in the Price Papers; D. J. Whitney to L. A. Higley, December 27, 1935 (Supreme Court), Nelson Papers, ICR.

18. D. J. Whitney, "For the Consideration of the Directors of the Religion and Science Association," August 6, 1935, Nelson Papers, ICR (obstacle); D. J. Whitney to G. M. Price, December 11, 1935, Price Papers (dogma).

19. For biographical information, see Higley's personnel file at Wheaton College and "L. Allen Higley," *American Men of Science*, ed. Jacques Cattell, vol. 1, *Physical Sciences* 9th ed. (Lancaster, PA: Science Press, 1955). Higley mentions using Price's book in a letter to Fellow Directors, February 15, 1937, Nelson Papers, ICR. On Higley's religious affiliations, see D. J. Whitney to G. M. Price, September 21, 1935, Price Papers; and "L. Allen Higley," *National Cyclopedia of American Biography*, 43:363. Whitney's comment about Wheaton appears in D. J. Whitney to G. M. Price, ca. fall 1935, Price Papers.

20. G. M. Price to B. C. Nelson, August 26, 1935 (Higley's views and visit), and January 3, 1936 (sincere but illogical); G. M. Price to Directors, January 14, 1937 (heretical), and February 9, 1937 (permanent evil); L. A. Higley to Fellow Directors,

February 15, 1937 (single idea); all in the Nelson Papers, ICR. Price accused Higley of minimizing the flood of Noah in a letter dated December 30, 1935, Price Papers.

21. L. A. Higley to Fellow Directors, February 15, 1937 (careful investigation and Judgment), D. J. Whitney to L. A. Higley, December 28, 1935 (time of probation), both in the Nelson Papers, ICR; H. W. Clark to G. M. Price, September 12, 1937, Price Papers. On the extent of the pre-Adamic cataclysm, see L. Allen Higley, "The Great Cataclysm in Genesis," *Moody Bible Institute Monthly* 36 (1936): 449–50.

22. Theodore Graebner to L. A. Higley, January 22, 1935, Nelson Papers, ICR. In a card to B. C. Nelson postmarked February 5, 1937 (Nelson Papers, ICR), D. J. Whitney professed amazement upon learning that Graebner was not a flood geologist.

23. Theodore Graebner to L. A. Higley, March 13, 1936; L. A. Higley to Theodore Graebner, March 19, 1934; and Theodore Graebner to the Secretary of the Chicago Pastoral Conference, March 23, 1936; all in the Graebner Papers.

24. "Meeting of the Religion and Science Association," *Christian Faith and Life* 42 (1936): 209; Program of the Convention on the Fallacies of Evolution by the Religion and Science Association, March 27–28, Chicago, and L. A. Higley to D. J. Whitney, April 18, 1936, both in the Nelson Papers, ICR. On attitudes toward Rimmer, see L. A. Higley to Theodore Graebner, November 27, 1935, and Theodore Graebner to B. C. Koenig, March 28, 1940, both in Box 1, Graebner Papers; and J. D. Whitney to B. C. Nelson, August 26, 1935, Nelson Papers, ICR. On Haas's paper, see D. J. Whitney to G. M. Price, April 20, 1936, Nelson Papers, ICR; and L. Allen Higley, "Fallacies of the Evolutionary Hypothesis," *Moody Bible Institute Monthly* 36 (1936): 507. For Dawson's and Price's critical evaluations of each other's views, see W. Bell Dawson, "The Flood and Geology," *Christian Faith and Life* 38 (1932): 269–72; and G. M. Price, "The Flood Theory Again," ibid., pp. 350–4. Benson, a friend and supporter of Higley's, admitted in a letter to D. J. Whitney dated February 25, 1937 (Nelson Papers, ICR), that earlier he had "followed Price more closely than [he] should." Despite Whitney's best efforts, the conference proceedings apparently never appeared in print.

25. D. J. Whitney to G. M. Price, April 20, 1936 (tactics), and April 28, 1936 (ruinites, recruits, and minority status); D. J. Whitney to B. C. Nelson, May 15, 1936 (first step); all in the Nelson Papers, ICR. Regarding Wheaton, see D. J. Whitney to G. M. Price, September 9 and 14, 1935, Price Papers.

26. L. A. Higley to D. J. Whitney, April 18, 1936 (planning); D. J. Whitney to G. M. Price, April 20, 1936 (auto tour and Hamlet); D. J. Whitney to B. C. Nelson, June 15, 1936 (angels of darkness), and August 18, 1936 (pleasant man); all in the Nelson Papers, ICR. Nelson proposed a field trip in a letter to Theodore Graebner, March 29, 1936, Box 1, Graebner Papers.

27. Theodore Graebner to L. A. Higley, January 22, 1937, Box 1, Graebner Papers; D. J. Whitney to B. C. Nelson, January 13, 1936, D. J. Whitney to G. M. Price, November 5, 1936, and G. M. Price to B. C. Nelson, March 4, 1937, all in the Nelson Papers, ICR; B. C. Nelson to G. M. Price, January 19, 1936, Price Papers. The debate is fully documented in the Nelson Papers, ICR.

28. D. J. Whitney to Fellow Officers, February 26, 1937 (poll); D. J. Whitney to G. M. Price and B. C. Nelson, card postmarked February 17, 1937 (swell gang); [D. J. Whitney], "Pronouncement No. 1: Religion & Science Association," n.d. (created types); B. C. Nelson to Fellow Directors, December 12, 1936 (door of evolution); D. J. Whitney to H. W. Clark, December 29, 1936 (evolution within families); G. M. Price to B. C. Nelson, January 23, 1937 (supporting Nelson); D. J. Whitney to B. C. Nelson, June 15, 1936 (fondness); D. J. Whitney to B.C. Nelson, March 18, 1936

(Scandinavians); all in the Nelson Papers, ICR. For an analysis of the poll of members' views, see *Creationist* 1 (May 1937): 1–3. The debate over speciation is documented in the Nelson Papers, ICR.

29. D. J. Whitney to Directors, November 16, 1936 (bluff); D. J. Whitney to Irwin H. Linton, June 15, 1936 (twin issues); L. A. Higley to G. M. Price, March 20, 1937 (second meeting); D. J. Whitney, [First Annual Report to the] Directors, September 26, 1936 (membership); D. J. Whitney to B. C. Nelson, January 13, 1936 (Lutherans); all in the Nelson Papers, ICR.

30. D. J. Whitney to Fellow Directors, February 26, 1937, Nelson Papers, ICR. *Creationist* 2 (April 1938): 2, refers to the late RSA. I found the most complete set of this series of the *Creationist*, which may have ended with the May 1938 issue, in the Graebner Papers. According to Henry M. Morris (*A History of Modern Creationism* [San Diego: Master Book Publishers, 1984], p. 138), Whitney produced a second series of the *Creationist* from 1950 to 1953. From 1963 to at least 1965 C. William Anderson of the Christian Evidence League, Malverne, NY, published a little magazine also called the *Creationist*, which carried Whitney's last essays.

31. B. C. Nelson to G. M. Price, March 11, 1940 (truth of Bible) and March 22, 1940 (ass), Price Papers; B. C. Nelson to Theodore Graebner, February 21, 1941 (Green), and February 28, 1949 (million years), Box 5, Graebner Papers; Byron C. Nelson, *Before Abraham: Prehistoric Man in Biblical Light* (Minneapolis: Augsburg Publishing House, 1948), p. 1.

32. Alfred M. Rehwinkel, *The Flood: In the Light of the Bible, Geology, and Archaeology* (St. Louis: Concordia Publishing House, 1951), pp. 102–3. Good New Publishers (Westchester, IL) brought out a sixty-two-page abridgment of *The Flood* as part of its "One Evening" Condensed Books series. For biographical information, see Rehwinkel, *The Flood*, pp. vii–viii, 273, and dust jacket; Alfred M. Rehwinkel, *Dr. Bessie* (St. Louis: Concordia Publishing House, 1963), p. 139; and "Celebrating Alfred M. Rehwinkel, 1887–1977," Rehwinkel Papers, Concordia Historical Institute. At the time I used this collection, the Rehwinkel-Price correspondence could not be found. Pfabe, "Theodore Graebner," p. 95, discusses Graebner's efforts to block publication of Rehwinkel's book. Carl H. Krekeler describes Graebner's growing toleration of theistic evolution in a letter to me dated July 31, 1989.

33. B. C. Nelson to G. M. Price, March 1, 1940, Price Papers; L. Allen Higley, *Science and Truth* (New York: Fleming H. Revell, 1940). Higley's troubles at Wheaton are documented in his personnel file at Wheaton College. A year after getting rid of Higley, Buswell himself was fired, whereupon Higley's friends arranged for the departed chemist to receive a pension from Wheaton College; Michael Hamilton to R. L. Numbers, July 19, 1990.

34. D. J. Whitney to G. M. Price, April 18, 1928, and November 3, 1955, Price Papers; Wm. B. Eerdmans to D. J. Whitney, July 8, 1936, and D. J. Whitney to Directors, September 26, 1936, Nelson Papers, ICR; Dudley Joseph Whitney, *The Case for Creation* (Malverne, NY: Christian Evidence League, 1946), especially part 4 regarding the re-creation hypothesis (unpaginated); Dudley Joseph Whitney, *The Face of the Deep: A Defense of Divine Creation* (New York: Vantage Press, 1955). Whitney's last book was *Genesis versus Evolution: The Problem of Creation and Atheistic Science* (New York: Exposition Press, 1961). Whitney's debate with Brewster appeared as a ten-part installment in *Truth Seeker* 64 (February 6–April 10, 1937). His death is described in [Doctor], "Last Rites for Dudley Whitney."

35. G. M. Price to the President and Board of the Creation-Deluge Society, August 29, 1945, Couperus Papers, courtesy of Molleurus Couperus.

Seven. THE DELUGE GEOLOGY SOCIETY

1. H. W. Clark to G. M. Price, September 12 and October 5, 1937, George Mc-Cready Price Papers, Adventist Heritage Center, Andrews University; "An Invitation from the Deluge Geology Society," ca. 1941, discarded Pacific Union College Papers (hereinafter cited as the PUC Papers) thoughtfully salvaged and given to me by Eric Anderson. For an early announcement, see Ben F. Allen, "Deluge Geology Society," *Ministry* 12 (August 1939): 40. In the *Creation-Deluge Society Newsletter*, September 16, 1944, members were instructed to refer unofficially to the organization as the Creation-Deluge Society, not the Deluge Society, but the most frequently used name was the Deluge Geology Society. Extant DGS newsletters can be found in the Couperus Papers, courtesy of Molleurus Couperus.

2. The *Creationist* 1 (October 1937): 11, identifies Allen as a charter member of the RSA. Allen's records at the University of Arkansas, which give no indication of his having attended the law school, were kindly made available by the then chancellor, Willard B. Gatewood, Jr. On Allen's varied career, see "Deluge Science Research Foundation: A Non-Denominational Work," undated brochure, Couperus Papers; and G. M. Price to B. C. Nelson, February 17, 1937, Nelson Papers, Institute for Creation Research (hereinafter cited as Nelson Papers, ICR). A letter from B. F. Allen to G. M. Price, March 9, 1929, Price Papers, lists Allen as state lecturer for the Arkansas Anti-Evolution League. On the league, see Leo Thomas Sweeney, "The Anti-Evolution Movement in Arkansas" (M.A. thesis, University of Arkansas, 1966), p. 82. Price describes Allen's destitution in a letter to David Voth, October 24, 1939, Price Papers.

3. B. F. Allen to G. M. Price, April 9, 1936 (White); B. F. Allen to G. M. Price, March 9, 1929 (Holy Spirit, popularizer); B. F. Allen to G. M. Price, October 7, 1936 (prince); all in the Price Papers. Price describes Allen as hardy in a letter to B. C. Nelson, February 17, 1937, Nelson Papers, ICR. Allen published a thirteen-part series of articles on coal, oil, etc., in the *Signs of the Times* between October 2, 1928, and February 12, 1929.

4. G. M. Price to David Voth, October 24, 1939 (crank), and G. M. Price to D. J. Whitney, May 1, 1936 (promoter), both in the Price Papers; F. L. Marsh to B. F. Allen, May 11, 1952, Couperus Papers.

5. Allen, "Deluge Geology Society," p. 40; G. M. Price to H. W. Clark, October 23, 1944, Price Papers. Regarding membership, see "Have You Renewed for 1942?" undated circular letter in the Couperus Papers; and *Creation-Deluge Society News Letter*, June 16, 1945. On Price's personality, see C. L. Burdick to R. L. Numbers, [Summer 1980]; and interview with Molleurus Couperus, January 19, 1983. The DGS journal began in 1941 as the *Bulletin of Deluge Geology and Related Sciences*; in 1943 it became the *Bulletin of Creation, the Deluge and Related Science*; in 1945 it reverted (almost) to its original title, *Bulletin of Deluge Geology and Related Science*.

6. B. F. Allen to M. Couperus, November 27, 1942, M. Couperus to W. E. Lammerts, April 19, 1944, M. Couperus to W. J. Tinkle, April 19, 1944, and Henry M. Morris to B. F. Allen, December 4, 1948, all in the Couperus Papers; *Creation-Deluge Society News Letter*, February 17, 1945; W. E. Lammerts to Theodore Graebner, November 8, 1941, Box 1, Theodore Graebner Papers, Concordia Historical Institute; H. W. Clark, "Under Whose Auspices Should Research Work in Creationism Be Done?" April 6, 1945, PUC Papers. In his first book, *That You Might Believe* (Chicago: Good Books, 1946), Henry M. Morris cites over a dozen articles from the *Bulletin of Deluge Geology*.

7. Roland T. Bird, "Thunder in His Footsteps," *Natural History* 43 (1939): 254–61,

quotation on p. 256; Albert G. Ingalls, "The Carboniferous Mystery," *Scientific American* 162 (1940): 14; discussion following G. M. Price's paper on "Dinosaurs and the Deluge," *Bulletin of Deluge Geology* 1 (June 1941): 6. For nineteenth-century reactions to similar fossil footprints, see Nicolaas A. Rupke, *The Great Chain of History: William Buckland and the English School of Geology (1814–1849)* (Oxford: Clarendon Press, 1983), p. 164.

8. B. C. Nelson to G. M. Price, March 4, 1939, Price Papers; *Deluge-Creation Society Letter* [sic], December 18, 1943; Editorial Note, *Bulletin of Creation, the Deluge and Related Science* 4 (March 1944): 13; B. F. Allen to Frank L. Marsh, November 13, 1947, Couperus Papers. Price, too, apparently expressed skepticism about the Texas footprints; see L. M. Davies to G. M. Price, July 24, 1948, Price Papers. Biographical information comes from C. L. Burdick to R. L. Numbers, [Summer 1980]. Burdick's transcript in the University of Wisconsin Registrar's Office shows that in 1923–24 he completed four graduate courses in geology, plus some other courses, before failing his orals. Among Adventists, Burdick sometimes claimed to have an M.A. in theology from Emmanuel Missionary College; see, e.g., C. L. Burdick to R. H. Pierson, July 31, 1973, Record Group 274, GF: R. H. Brown, 1973–1977, Archives of the General Conference of Seventh-day Adventists. He did, in fact, submit a thesis on "The Sabbath: Its Development in America" in 1922 (a copy of which can be found in the Adventist Heritage Center, Andrews University), but the academic records of the college indicate that he did not receive a degree. I am indebted to Mrs. Louise Dederen, curator of the Adventist Heritage Center, for providing this information.

9. *Creation-Deluge Society News Letter*, August 19, 1944; E. E. Beddoe to the Footprint Research Committee, March 17, 1944, Price Papers; E. E. Beddoe to M. Couperus, September 9, 1945, and G. M. Price to B. F. Allen, March 29, 1944, both in the Couperus Papers.

10. *Creation-Deluge Society News Letter*, February 17, 1945; E. E. Beddoe to G. M. Price and B. F. Allen, April 20, 1944, Price Papers; E. E. Beddoe to M. Couperus, September 9, 1945, Couperus Papers.

11. On the DGS's early quarrels, see B. F. Allen to M. Couperus, May 14, 1943, Couperus Papers.

12. Interview with Harold W. Clark, May 11, 1973; Harold W. Clark, *Genesis and Science* (Nashville: Southern Publishing Assn., 1967), p. 57. Additional biographical data appear in William M. Goldsmith, *Evolution or Christianity, God or Darwin?* (St. Louis: Anderson Press, 1924), pp. 66–7.

13. G. M. Price to Elder M. [Arthur M. Maxwell?], January 2, 1940 (trust), G. M. Price to H. W. Clark, June 15, 1941 (Agassiz), H. W. Clark to Officers of the Ministerial Association, March 10, 1937 (fifteen hundred years and hybridization), all in the Price Papers; Harold W. Clark, *Back to Creationism* (Angwin, CA: Pacific Union College Press, 1929); George McCready Price, *Genesis Vindicated* (Washington: Review and Herald Publishing Assn., 1941), p. 284 (shock absorber). See also H. W. Clark to R. M. Ritland, June 2, 1962, Ritland Papers, courtesy of Richard M. Ritland.

14. G. M. Price to M. E. Kern and Others, ca. 1936, Papers of the Publishing Department of the General Conference of Seventh-day Adventists, hereinafter cited as Publishing Department Papers. This collection was brought to my attention by Donald McAdams.

15. Interview with Harold W. Clark, May 11, 1973; H. W. Clark to Officers of the Ministerial Association, March 10, 1937, Price Papers; H. W. Clark to R. M. Ritland, June 2, 1963, Ritland Papers. See also H. W. Clark, "Creation: Present Status and Future Problems," a paper presented in absentia to the SDA science teachers, 1938, PUC Papers.

16. H. W. Clark to G. M. Price, September 23, 1938, Price Papers.

17. G. M. Price to H. W. Clark, June 15, 1941 (order of fossils); H. W. Clark to G. M. Price, [April 9, 1940] (out-of-date); G. M. Price to H. W. Clark, April 21, 1940 (*magnum opus*, university-itis, and charlatan); G. M. Price to H. W. Clark, June 9, 1940 (tobacco-smoking); all in the Price Papers.

18. G. M. Price to Elder M., January 2, 1940 (libel); H. W. Clark to G. M. Price, [April 9, 1940] (comment to Allen and gratitude), April 15, 1940 (strict creationism), and August 14, 1940 (peace); all in the Price Papers. See also H. W. Clark to G. M. Price, April 23 and 30, 1940, Price Papers.

19. G. M. Price to H. W. Clark, May 2, 1940 (notorious), and November 20, 1940 (trivialities), Price Papers.

20. G. M. Price to H. W. Clark, October 23, 1944, Price Papers; H. W. Clark to R. L. Numbers, May 13, 1973.

21. Harold W. Clark, *Genes and Genesis* (Mountain View, CA: Pacific Press, 1940), pp. 43, 143; G. M. Price to H. W. Clark, February 16, 1941, Price Papers (wretched theory); G. M. Price, "The 'Amalgamation' Question Again," unpublished MS, ca. 1941, Ballenger Papers, courtesy of Donald F. Mote (shame). On Price's change of heart, see also G. M. Price to Glenn Calkins, March 23, 1941, and G. M. Price to Elder M., January 2, 1940, Price Papers. On his attitude toward White's amalgamation statement, see his letters to Charlie Price, August 9, 1905, Sumner A. Whittier, December 12, 1927, and E. W. Ballenger, January 30, 1928, all in the Ballenger Papers. On Clark's views, see "Statement of Prof. H. W. Clark re Amalgamation," undated MS, Publishing Department Papers.

22. G. M. Price to D. J. Whitney, June 23, 1943, Price Papers (semi-evolutionist and hog-tied); Harold W. Clark, *The New Diluvialism* (Angwin, CA: Science Publications, 1946), unpaginated preface, p. 62. Clark also gave Price credit for the idea of ecological zonation in a letter to Price, October 20, 1944, Price Papers. Clark brought out an updated version of *The New Diluvialism* under the title *Fossils, Flood, and Fire* (Escondido, CA: Outdoor Pictures, 1968).

23. Ernest S. Booth, Review of *The New Diluvialism*, by Harold W. Clark, *Northwest Naturalist* 5 (October 1946): 37–8; G. M. Price to H. W. Clark, June 18, 1946, Price Papers.

24. George McCready Price, *Theories of Satanic Origin* (Loma Linda, CA: Author, n.d.), p. 6; H. W. Clark to G. M. Price, January 21, 1947, Price Papers; Harold W. Clark, *Crusader for Creation: The Life and Writings of George McCready Price* (Mountain View, CA: Pacific Press, 1966); L. E. Froom to G. M. Price, July 12, 1942, Price Papers. See also Price's letter to Fellow Workers, n.d., Couperus Papers, which accompanied complimentary copies of *Theories of Satanic Origin*.

25. F. L. Marsh to T. Dobzhansky, February 21, 1945, F. L. Marsh Papers, Adventist Heritage Center, Andrews University, hereinafter cited as Marsh Papers, AU. Biographical data come from an interview with F. L. Marsh, August 30, 1972; F. L. Marsh to R. L. Numbers, April 10, 1974; and F. L. Marsh, "Life Summary of a Creationist," unpublished MS, December 1, 1968, Marsh Papers in the possession of F. L. Marsh, hereinafter cited as Marsh Papers, FM.

26. Frank Lewis Marsh, *Evolution, Creation, and Science* (Washington: Review and Herald Publishing Assn., 1944), pp. 165–6; F. L. Marsh to G. M. Price, September 5, 1943, Price Papers.

27. Frank Lewis Marsh, *Fundamental Biology* (Lincoln, NE: Author, 1941), pp. iii, 48, 56, 63; excerpts from letters written by F. L. Marsh to D. E. Robinson, February 16 and March 16, 1941, Publishing Department Papers.

28. F. L. Marsh to R. L. Numbers, April 10, 1974 (knock-down); F. L. Marsh to

G. M. Price, September 5, 1943 (propaganda), and G. M. Price to H. W. Clark, October 23, 1944 (fog), both in the Price Papers; Marsh, "Life Summary of a Creationist" (mantle). On "baramins," see Marsh, *Fundamental Biology,* p. 100; and Marsh, *Evolution, Creation, and Science,* p. 162. See also G. M. Price, "An Unregimented Biologist," *Bulletin of Creation, the Deluge and Related Science* 5 (1945): 9–11. On the Clark-Marsh amalgamation controversy, see F. L. Marsh, "The Amalgamation Statements," unpublished MS, November 16, 1947, H. W. Clark, "Amalgamation: A Study in Perplexing Statements Made by Mrs. E. G. White," unpublished MS, March 1, 1948, and F. L. Marsh, "A Discussion of Harold W. Clark's Paper, 'Amalgamation,' Published March 1, 1948," unpublished MS, April 11, 1949, all in the Marsh Papers-FM; and F. L. Marsh to G. M. Price, October 16, 1947, Price Papers.

29. G. M. Price to B. F. Allen, February 13, 1944 (fool); F. L. Marsh, "Confessions of a Biologist," unpublished MS, August 25, 1943 (wild rumor); Walter Lammerts and Others, "Review of 'Creation, Evolution, and Science,' " by Frank Marsh, [November 18, 1944]; all in the Couperus Papers. The description of the meeting appears in *Creation-Deluge Society News Letter,* December 16, 1944, PUC Papers. Marsh identifies Newman as his teacher in a letter to T. Dobzhansky, December 13, 1944, Marsh Papers, AU.

30. F. L. Marsh to G. M. Price, October 16, 1947, Price Papers (sanitized); Ernst Mayr to F. L. Marsh, March 13, 1945, Price Papers; T. Dobzhansky to F. L. Marsh, November 29, 1944, Marsh Papers, AU. Marsh's later books included *Studies in Creationism* (Washington: Review and Herald Publishing Assn., 1950); *Life, Man, and Time* (Mountain View, CA: Pacific Press, 1957); *Evolution or Special Creation?* (Washington: Review and Herald Publishing Assn., 1963); and *Variation and Fixity in Nature* (Mountain View, CA: Pacific Press, 1976). For a recent exchange between a creationist and an evolutionist, both biologists, see Edward O. Dodson and George F. Howe, *Creation or Evolution: Correspondence on the Current Controversy* (Ottawa, Canada: University of Ottawa Press, 1990).

31. Theodosius Dobzhansky, *Genetics and the Origin of Species* (New York: Columbia University Press, 1937), p. 8; Theodosius Dobzhansky, Review of *Evolution, Creation, and Science,* by F. L. Marsh, *American Naturalist* 79 (1945): 73–5. See also T. Dobzhansky to F. L. Marsh, November 15, 1944, Marsh Papers, AU.

32. F. L. Marsh to T. Dobzhansky, November 19, 1944 (delight); F. L. Marsh to T. Dobzhansky, December 4, 1944 (varieties and natural selection); T. Dobzhansky to F. L. Marsh, December 7, 1944 (only antievolutionist); all in the Marsh Papers, AU.

33. T. Dobzhansky to F. L. Marsh, December 22, 1944 (geological time); F. L. Marsh to T. Dobzhansky, January 12, 1945 (real proof); T. Dobzhansky to F. L. Marsh, February 5, 1945 (mouse); all in the Marsh Papers, AU.

34. F. L. Marsh to T. Dobzhansky, January 12, 1945; T. Dobzhansky to F. L. Marsh, December 22, 1944; both in the Marsh Papers, AU.

35. F. L. Marsh to T. Dobzhansky, February 21, 1945, Marsh Papers, AU; Theodosius Dobzhansky, *Genetics and the Origin of Species,* 3rd ed. (New York: Columbia University Press, 1951), p. 11. Marsh quotes Dobzhansky's mention of him in *Evolution or Special Creation?,* p. 46.

36. B. F. Allen to Board of Directors, August 12, 1945 (rock); B. F. Allen, "The Question of a Second Society," circular letter dated April 26, 1943 (radioactive time); both in the Couperus Papers.

37. M. Couperus to G. M. Price, [ca. 1942], Price Papers. Biographical information comes from an interview with Molleurus Couperus, January 12, 1981. On the history of radiometric dating, see G. Brent Dalrymple, *The Age of the Earth* (Stanford, CA: Stanford University Press, 1991); R. E. Taylor, *Radiocarbon Dating: An Archaeolog-*

ical Perspective (Orlando, FL: Academic Press, 1987), pp. 147–70; and Lawrence Badash, "Rutherford, Boltwood, and the Age of the Earth: The Origin of Radioactive Dating Techniques," *Proceedings of the American Philosophical Society* 112 (1968): 157–69.

38. M. Couperus to G. M. Price, January 31, 1945, Price Papers (appearance of age); B. F. Allen to R. E. Hoen, December 21, 1944, PUC Papers (1.6 billion); M. Couperus to B. F. Allen, July 27, 1944, Couperus Papers (problem of creation).

39. George McCready Price, "How Old Is the Earth?" *Christian Faith and Life* 38 (1932): 20–3 (the long ago); G. M. Price to the Directors, March 2, 1937, Nelson Papers-ICR (re Higley); George McCready Price, *Genesis Vindicated* (Washington: Review and Herald Publishing Assn., 1941), pp. 11–13, 54–5, 312. See also George McCready Price, "Reaction on the Age of the Universe," *Ministry* 5 (June 1932): 14–15; George McCready Price, "What Christians Believe about Creation," *Bulletin of Deluge Geology* 2 (1942): 70; and George McCready Price, "Dating the Fossils: Is the Radiocarbon Method Reliable?" *Signs of the Times* 79 (August 19, 1952): 3–4. Whitney shared Price's tolerance of old-earth views; see D. J. Whitney, "A Creed of Creation," unpublished MS, April 10, 1936, Price Papers.

40. B. F. Allen to Members of the Creation-Deluge Society and to Other Believers in the Six-Literal-Day Ex-Nihilo Creation Week, May 15, 1946 (integrity of Creation Week and Burdick's paper); B. F. Allen to G. H. Rue, March 21, 1946 (facts); Clifford Leslie Burdick, "Gross Fallacies in the Radioactive Time Theory and Other Recent Trends in Methods of Reckoning Geologic Time," unpublished MS; all in the Couperus Papers. Burdick's paper subsequently appeared in *Forum for the Correlation of Science and the Bible* 1 (1946–1947): 39–58. See also Robert W. Woods, "How Old Is the Earth?" *Signs of the Times* 80 (April 7, 1953): 8–9, 15; and Watson Davis and Robert D. Potter, "Atomic Energy Released," *Science News Letter* 35 (1939), 86–7, 93.

41. Minutes of the Board Meeting of the Creation-Deluge Society, October 14, October 21, and December 9, 1945; M. Couperus to H. W. Clark, August 24, 1946 (Spirit of Prophecy); all in the Couperus Papers. For Allen's version of the 1945 coup, see B. F. Allen to W. E. Read, March 24, 1957, Couperus Papers.

42. B. F. Allen, "The Original Society Illegally Supplanted and All Scriptural Standards Abandoned," undated MS [ca. early 1946]; B. F. Allen to W. E. Read, March 24, 1957 (scorn); both in the Couperus Papers.

43. G. M. Price to M. Couperus, December 13, 1944; G. M. Price to C. A. Wells, June 19, 1945 (preexisting matter); both in the Couperus Papers. Couperus mentioned his view of the flood in the interview on January 12, 1981.

44. Clifford Leslie Burdick, "The Radioactive Time Theory and Recent Trends in Methods of Reckoning Geologic Time," *Forum for the Correlation of Science and the Bible* 1 (1946–1947): 39–58; Harold W. Clark, "In Defense of the Ultra-Literal View of the Creation of the Earth," ibid., pp. 11–15; John Lowell Butler, "God's Primary and Secondary Creations," ibid., pp. 16–38; George McCready Price, "In the Beginning," ibid., pp. 9–10. Couperus published his views in volume 2 (1947–1948): "The Creation of the Earth," pp. 97–105, and "Some Remarks Regarding the Radioactive Time Estimation of the Age of the Earth," pp. 118–19. For information about Butler, see his curriculum vitae in the American Scientific Affiliation Papers; and G. M. Price to M. Couperus, December 5, 1945, and J. L. Butler to E. S. Booth, April 23, 1950, both in the Couperus Papers.

45. B. F. Allen to D. D. Haughey, September 28, 1948, Couperus Papers (Price); G. M. Price to F. L. Marsh, July 17, 1955, Marsh Papers, FM (uniformitarian creationism); G. M. Price to D. J. Whitney, May 7, 1948, Couperus Papers (rock-bottom truth).

46. B. F. Allen to E. S. Booth, April 24, 1949, Couperus Papers; Benjamin Franklin Allen, "Earth's Origin during Creation Week Sustains Integrity of Fourth Commandment," *Naturalist* 8 (Spring 1949): 14–24, quotations on p. 14.

47. J. L. Butler to E. S. Booth, April 23, 1950 (spies); J. L. Butler to M. Couperus, June 26, 1948 (clever); J. L. Butler to G. M. Price, May 21, 1949 (inner sight); all in the Couperus Papers.

48. F. L. Marsh to G. M. Price, October 16, 1947, Price Papers (fakers); H. W. Clark to M. Couperus, January 28, 1947 (dead), and M. Couperus to H. W. Clark, May 8, 1947 (minor effort), both in the Couperus Papers.

49. Minutes of an open meeting of Amazing Discoveries, Inc., August 21, 1947; B. F. Allen, "The Noah's Ark Project: A Report of Progress of the Sacred History Research Expedition to April 1, 1946," unpublished MS (broadcasts); "Strictly Confidential Methods Increasingly Necessary in the Sacred History Expedition: A Resolution Passed by the Board of Directors, December 29, 1946" (announcement); all in the Couperus Papers. Allen served as president of the expedition; Burdick, as vice president for geology. On Allen's early interest in the ark, see B. F. Allen to G. M. Price, June 12, 1929, Price Papers.

50. B. F. Allen to W. H. Branson, August 17, 1950; W. H. Branson to G. H. Rue, July 29, 1951; B. F. Allen to G. H. Rue, October 16, 1956 (uranium); B. F. Allen to C. B. De Mille, July 16, 1957; all in the Couperus Papers. Between 1945 and 1955 Allen apparently had no contact with Price; see G. M. Price to B. F. Allen, September 12, 1955, Couperus Papers. For an obituary of Allen, see *Advent Review and Sabbath Herald* 137 (March 31, 1960): 24.

51. Interview with M. Couperus, January 12, 1981; M. Couperus to R. L. Numbers, September 1, 1987.

Eight. Evangelicals and Evolution
in Great Britain

1. George M. Marsden, "Fundamentalism as an American Phenomenon: A Comparison with English Evangelism," *Church History* 46 (1977): 215–32. See also George M. Marsden, *Fundamentalism and American Culture: The Shaping of Twentieth-Century Evangelicalism, 1870–1925* (New York: Oxford University Press, 1980), pp. 221–8; and James R. Moore, "Evangelicals and Evolution: Henry Drummond, Herbert Spencer, and the Naturalisation of the Spiritual World," *Scottish Journal of Theology* 38 (1985): 383–417.

2. C. W. Goodwin, "The Mosaic Cosmogony," in *Recent Inquiries in Theology, by Eminent English Churchmen; Being "Essays and Reviews"* (Boston: Walker, Wise, 1860), p. 238; "Objects of the Victoria Institute," *Journal of the Transactions of the Victoria Institute* 1 (1867): vi; James Reddie, Discussion of George Warington, "On the Credibility of Darwinism," ibid. 2 (1967): 62. On the history of the Victoria Institute, see E. J. G. Titterington, "The Early History of the Victoria Institute," *Journal of the Transactions of the Victoria Institute* 82 (1950): 53–69; and Douglas Lloyd Wertheimer, "The Victoria Institute, 1865–1919: A Study in Collective Biography Meant as an Introduction to the Conflict of Science and Religion after Darwin," unpublished MS, 1971, from a copy kindly provided by James R. Moore. According to R. E. D. Clark, in an interview on October 1, 1984, the records and books of the institute had to be destroyed during World War II because they constituted a fire hazard. On Gosse, see Edmund Gosse, *Father and Son: Biographical Recollections* (New York: Charles Scribner's Sons, 1907), quotation on p. 129; and Frederic R. Ross,

"Philip Gosse's *Omphalos*, Edmund Gosse's *Father and Son*, and Darwin's Theory of Natural Selection," *Isis* 68 (1977): 85–96, Huxley quotation on p. 85.

3. Wertheimer, "The Victoria Institute," p. 4; G. M. Price to the President and Board of the Creation-Deluge Society, August 29, 1945, Couperus Papers, courtesy of Molleurus Couperus (rut); George McCready Price, "Geology and Its Relation to Scripture Revelation," *Journal of the Transactions of the Victoria Institute* 46 (1924): 97–123, quotations on p. 98 (final solution), pp. 118–20 (bull), pp. 120–1 (amateur).

4. George McCready Price, "Revelation and Evolution: Can They Be Harmonized?" *Journal of the Transactions of the Victoria Institute* 47 (1925): 169–89, quotations on pp. 183, 187. Criticism of Price continued into the 1930s; see Alan Stuart, "Science and the Interpretation of Scripture," ibid. 69 (1937): 97.

5. *Is Evolution True? Verbatim Report of Debate between George McCready Price, M.A., and Joseph McCabe* (London: Watts, 1925); G. M. Price to E. S. Ballenger, September 28, 1925, original in my possession. For firsthand accounts of the debate, see A. S. Maxwell to H. W. Clark, November 4, 1964, and W. G. C. Murdoch to H. W. Clark, November 4, 1964, George McCready Price Papers, Adventist Heritage Center, Andrews University; and "Shout Down American in Evolution Debate," *New York Times*, September 7, 1925, p. 5. McCabe takes credit for luring Price into debate in Douglas Dewar, *A Challenge to Evolutionists*, 3rd ed. (Croydon, Surrey: Uplift Books, 1948), p. 10.

6. G. M. Price to J. R. Straton, February 9, 1927, John Roach Straton Papers, American Baptist Historical Society; "Evolution and Intellectual Freedom," *Supplement to Nature* 116 (July 11, 1925): 69–83, quotation on p. 74. See also "Truth and Doctrine in Science and Religion," ibid., pp. 83–4; and E. Griffith-Jones, "Evolution and Religious Thought," *British Weekly* 76 (1924): 401–2. On the attitudes of British evangelicals in general, and Barnes in particular, see D. W. Bebbington, *Evangelicalism in Modern Britain: A History from the 1730s to the 1980s* (London: Unwin Hyman, 1989), pp. 207–8.

7. For biographical information, see J. T. MacGregor-Morris, "Sir (John) Ambrose Fleming," *Dictionary of National Biography, 1941–1950*, pp. 258–60; and Sir Ambrose Fleming, *Memories of a Scientific Life* (London: Marshall, Morgan & Scott, [1934]), which says little about his religious life. Fleming's nomination for the Nobel Prize can be found in the Correspondence and Papers of Sir Ambrose Fleming, J. A. Fleming Collection, University College London Library. For his views of evolution and the Bible, see John Ambrose Fleming, *Science and Miracles* (London: Religious Tract Society, 1917); Sir Ambrose Fleming, *The Bible and Modern Evolutionary Theory* (London: Bible Testimony Fellowship, [1932]); and Sir Ambrose Fleming, *Evolution and the Bible* (London: Covenant Publishing Co., [1937]).

8. J. A. Fleming, "Evolution and Revelation," *Journal of the Transactions of the Victoria Institute* 59 (1927): 11–40, quotations on p. 32 (case for Evolution), p. 34 (*impasse*), and p. 39 (discontinuities); Ambrose Fleming, *Evolution or Creation?* (London: Marshall, Morgan & Scott, [1933]), pp. 107–13.

9. Ambrose Fleming, *The Origin of Mankind: Viewed from the Standpoint of Revelation and Research* (London: Marshall, Morgan & Scott, [1935]), p. 24 (planetary evolution), p. 115 (three conclusions), p. 132 (animal evolution); Sir Ambrose Fleming, "Modern Anthropology *versus* Biblical Statements on Human Origin," *Journal of the Transactions of the Victoria Institute* 67 (1935): 15–42; "Evolution and Our Faith," (London) *Daily Telegraph*, January 15, 1935, p. 10 (assault); "Age of Man and the Earth," (London) *Times*, January 15, 1935; Sir Arthur Keith, *Darwinism and Its Critics* (London: Watts, 1935), pp. 1–2, 36. The *Daily Telegraph* published Fleming's reply to his critics on January 19, 1935. Fleming was knighted in 1929.

10. G. M. Price to J. R. Straton, February 9, 1927, Straton Papers; Bernard Acworth, *This Bondage: A Study of the "Migration" of Birds, Insects and Aircraft, with
Some Reflections on "Evolution" and Relativity* (London: John Murray, 1929), pp. 1,
4. Biographical data come from A. G. T[ilney], "Origin of Evolution Protest Movement," *Evolution Protest Movement Pamphlet No. 82,* 1963; interview with Richard
Acworth, October 2, 1984; and "Capt. Bernard Acworth," *Who Was Who, 1961–1970*
(London: Adam & Charles Black, 1972), pp. 4–5.

11. T[ilney], "Origin of Evolution Protest Movement"; C. E. A. Turner, Review of
The Transformist Illusion, by Douglas Dewar, *Evolution Protest Movement Pamphlet
No. 125* [No. 126], 1965; Douglas Dewar and Frank Flinn, *The Making of Species*
(London: John Lane, 1909), pp. xi, 26; Douglas Dewar, *A Challenge to Evolutionists,*
3rd ed. (Croydon, Surrey: Uplift Books, 1948), pp. 34–5 (ghastly mess). See also
"Douglas Dewar," *Who Was Who, 1951–1960* (London: Adam & Charles Black, 1961),
p. 302.

12. D. Dewar to G. M. Price, July 25, 1931, Price Papers; Douglas Dewar,
Difficulties of the Evolution Theory (London: Edward Arnold, 1931), pp. 5, 158.

13. D. Dewar to G. M. Price, July 25, 1931, Price Papers. On Vialleton, see Harry
W. Paul, *The Edge of Contingency: French Catholic Reaction to Scientific Change
from Darwin to Duhem* (Gainesville: University Presses of Florida, 1979), pp. 99–100;
Douglas Dewar, "The Limitations of Organic Evolution," *Journal of the Transactions
of the Victoria Institute* 64 (1932): 120–43; Arnold Lunn and J. B. S. Haldane, *Science
and the Supernatural: A Correspondence* (London: Erye and Spottiswoode, 1935), pp.
178–9, 202 (greatest anatomist). A. Morley Davies, *Evolution and Its Modern Critics*
(London: Thomas Murby, 1937), p. 12, points out that Vialleton's "evolution" became
Dewar's "creation."

14. D. Dewar to G. M. Price, July 25, 1931 (inspiration); D. Dewar to [name
deleted], November 2, 1931 (no false statements); D. Dewar to G. M. Price, June 15,
1936 (gap theory); all in the Price Papers. Dewar spelled out his theory of a single
creation in "Current Theories of the Origin of Living Organisms," *Journal of the
Transactions of the Victoria Institute* 76 (1944): 53–93, quotation on p. 58.

15. D. Dewar to G. M. Price, July 25, 1931 (publishers); D. Dewar to G. M. Price,
August 9, 1931 (ignorance and specific gravity); D. Dewar to G. M. Price, August 21,
1931 (better likes); D. Dewar to [name deleted], November 2, 1931 (supporters and
criticisms); all in the Price Papers. On Price's promotion of Dewar's book, see [Le-
ander S. Keyser], Review of *Difficulties of the Evolution Theory,* by Douglas Dewar,
Christian Faith and Life 37 (1931): 668–9; and D. Dewar to G. M. Price, August 21,
1931, Price Papers.

16. D. Dewar to G. M. Price, July 25, 1931 (discovers Davies); L. M. Davies to
G. M. Price, April 29, 1945 (old soldier); both in the Price Papers. Additional biographical information comes from Robert Campbell, "Lt.-Col. Lewis Merson Davies,
R.A., M.A., Ph.D., D.Sc., F.G.S., F.R.A.I.," *Royal Society of Edinburgh Year Book,*
1954–55, pp. 16–18, a copy of which was generously provided by the General Secretary, Charles D. Waterston; and C. D. Waterston to R. L. Numbers, September 5,
1988. I am grateful to Marguerite Dupree, Robert Anderson, and Jo Currie for their
assistance in tracking down information about Davies.

17. D. Dewar to G. M. Price, July 25, 1931, Price Papers (Fundamentalist); L. M.
Davies, *The Significance of Modernism* (London: Marshall Brothers, [1927]), pp. 7–8,
35; L. M. Davies to G. M. Price, April 29, 1945 (early meeting), and December 13,
1945 (2 Peter), both in the Price Papers; L. M. Davies, "Evolution," *Journal of the
Transactions of the Victoria Institute* 62 (1930): 214–52, quotation regarding Peter on
p. 236; "Evolution or Christianity? Being the Substance of a Dialogue between Mr.

A. Long . . . and Lt.-Col. L. M. Davies . . . 2nd February, 1946," *Evolution Protest Movement Pamphlet No. 44*, n.d., p. 2 (privileged position).

18. L. Merson Davies, *The Bible and Modern Science*, 3rd ed. (London: Pickering & Inglis, [1934]), p. 11; L. M. Davies to G. M. Price, December 13, 1945, Price Papers.

19. D. Dewar to G. M. Price, July 25, 1931, Price Papers (publishers); Douglas Dewar, "The Limitations of Organic Evolution," *Journal of the Transactions of the Victoria Institute* 64 (1932): 142; Lunn and Haldane, *Science and the Supernatural*, pp. 181–2, 206, 218, 243, 324. The comment about shattering the case for evolution is Lunn's.

20. D. Dewar to G. M. Price, May 4, 1933, Price Papers (Victoria Institute); C. E. A. Turner, "A Jubilee of Witness for Creation against Evolution by CSM/EPM, 1932–1982," *Creation Science Movement Pamphlet No. 232*, July 1982, pp. 1–3 (organization); Bernard Acworth, *This Progress: The Tragedy of Evolution* (London: Rich & Cowan, 1934), pp. 115 (goal), 333–4 (prosperous).

21. "Teaching of Organic Evolution: A Protest Meeting," (London) *Times*, February 13, 1935, p. 10 (resolution); T. C. Skinner, "The Evolution Protest Movement," *Christian Faith and Life* 41 (1935): 221–3 (back to the Bible); Esme E. M. Geering and C. E. A. Turner, "In the Beginning," *Creation Science Movement Pamphlet No. 263*, January 1989. Although I have seen no contemporary evidence of the EPM's existence before 1935, and several sources from the 1930s imply that the movement was not founded until 1935, C. E. A. Turner, in a letter dated July 22, 1988, has provided oral and circumstantial testimony in favor of a 1932 founding.

22. Douglas Dewar, *Man: A Special Creation* (London: Thynne, [1936]), p. 104; D. Dewar to G. M. Price, June 15, 1936, Price Papers. Turner, "Jubilee of Witness," p. 5, describes Dewar's stature.

23. A. Morley Davies, *Evolution and Its Modern Critics* (London: Thomas Murby, 1937), pp. v–vi, 10, 105, 247; Douglas Dewar, *More Difficulties of the Evolution Theory: And a Reply to "Evolution and Its Modern Critics"* (London: Thynne, 1938), pp. 101–3, 137–8.

24. D. Dewar to G. M. Price, May 4, 1933, Price Papers (Midland Institute); Dewar, *A Challenge to Evolutionists*, pp. 3, 7, 16–20; *Is Evolution a Myth? A Debate between Douglas Dewar, B.A., F.Z.S. and L. Merson Davies, D.Sc., Ph.D., F.R.S.E., F.G.S. for the Evolution Protest Movement and Prof. J. B. S. Haldane, F.R.S. for the Rationalist Press Association Ltd.* (London: C. A. Watts, 1949), p. 92. Dewar's *Challenge*, first published in 1937, represents his part of the McCabe debate. See also Arnold Lunn, ed., *Is Evolution Proved? A Debate between Douglas Dewar and H. S. Shelton* (London: Hollis and Carter, 1947).

25. Douglas Dewar and L. M. Davies, "Obsessions of Biologists," *Evolution Protest Movement Pamphlet No. 20*, ca. 1945, p. 16. See also L. M. Davies and Douglas Dewar, "The B.B.C. Abuses Its Monopoly," *Evolution Protest Movement Pamphlet No. 26*, n.d.; and "The Intolerance of the B.B.C.," *Evolution Protest Movement Pamphlet No. 41*, n.d.

26. Douglas Dewar, Letter to the Editor, *Creationist* 1 (October 1937): 1. The EPM's "Aims and Methods" appear in "Particulars of the Evolution Protest Movement," *Evolution Protest Movement Pamphlet No. 140*, n.d. Dewar's last major work, *The Transformist Illusion* (Murfreesboro, TN: Dehoff Publications, 1957), published posthumously, included material from both *Difficulties* and *More Difficulties*.

27. J. D. Bales to J. Howitt, October 22, 1947, and D. Dewar to J. D. Bales, September 18, 1948, Bales Papers, courtesy of James D. Bales; interview with James D. Bales, March 25, 1985. The *Thinking Christian* began with the July–Septem-

ber 1948 issue and ended with the January–March 1949 issue. Bales's antievolution publications include *Why Scientists Accept Evolution* (Grand Rapids, MI: Baker Book House, 1966), written with Robert T. Clark; *Man on All Fours* (Searcy, AR: Harding College, 1973); *The Genesis Account and a Scientific Test* (Searcy, AR: J. D. Bales, 1975); and *Evolution and the Scientific Method* (Searcy, AR: J. D. Bales, 1976). On the size of the EPM, see A. G. T[ilney], "EPM—'40 Years On'; Evolution—114 Years 'Off,' " *Evolution Protest Movement Pamphlet No. 193*, May 1972. On the fortunes of evangelicalism, see Bebbington, *Evangelicalism in Modern Britain*, p. 252.

28. C. S. Lewis to B. Acworth, December 9, 1944 (little threat); September 13, 1951 (*the* central lie); October 4, 1951 (preface and popular Apologist); September 18, 1959 (pretentions); all in the Acworth Papers, courtesy of Richard Acworth. Extracts from C. S. Lewis's letters to Captain Bernard Acworth © 1992 by C. S. Lewis Pte Ltd., reproduced by permission of Curtis Brown Ltd., London. Copies of all of Lewis's letters to Acworth have been deposited in the Marion E. Wade Collection, Wheaton College. Richard Acworth described the relationship between his father and Lewis in an interview on October 2, 1984.

29. B. C. Saunders and R. E. D. Clark, *Order and Chaos in the World of Atoms: A Survey of Modern Chemistry* (London: English Universities Press, 1942), published in the United States by Dover Publications. Biographical information comes from an interview with Robert E. D. Clark, October 1, 1984; and C. E. A. Turner, "Obituary: Dr. Robert E. D. Clark," *Creation: The Journal of the C.S.M.* 3 (July 1985): 8.

30. R. E. D. Clark, *Scientific Rationalism and Christian Faith: With Particular Reference to the Writings of Prof. J. B. S. Haldane and Dr. J. S. Huxley* (London: Inter-Varsity Fellowship, 1945); interview with Robert E. D. Clark, October 1, 1984; R. E. D. Clark to R. L. Numbers, [late October 1984]. On the IVF, see Bebbington, *Evangelicalism in Modern Britain*, pp. 259–61.

31. Interview with Robert E. D. Clark, October 1, 1984; R. E. D. Clark, "Creation—In Six Days?" *Science and Religion: A Review of Current Literature and Thought* 1 (1948): 167–73; Robert E. D. Clark, *The Christian Stake in Science* (Exeter, Devon: Paternoster Press, 1967), p. 139 (man). On the centrality of design, see, e.g., Robert E. D. Clark, *Creation* (London: Tyndale Press, 1946); and Robert E. D. Clark, *The Universe: Plan or Accident? The Religious Implications of Modern Science* (London: Paternoster Press, 1949). See also P. J. Wiseman, *Creation Revealed in Six Days: The Evidence of Scripture Confirmed by Archaeology* (London: Marshall, Morgan & Scott, 1948).

32. R. E. D. Clark to G. M. Price, September 21, 1958 (stimulating), November 18, 1948 (rejection), and November 28, 1949 (unChristian), Price Papers; "Easy(?) Money—A Challenge to Readers," *Science and Religion* 2 (1949): 39–41; D. J. Whitney to R. E. D. Clark, December 5, 1951 (challenge to debate), and December 14, 1951 (blaspheme and ignorance), Clark Papers, courtesy of the late Robert E. D. Clark. See also George McCready Price, *Feet of Clay: The Unscientific Nonsense of Historical Geology* (Malverne, NY: Christian Evidence League, 1949), copy courtesy of Martin Gardner.

33. Interview with Robert E. D. Clark, October 1, 1984 (train); Robert E. D. Clark, *Darwin: Before and After: The Story of Evolution* (London: Paternoster Press, 1948), p. 127 (chemical problem); Robert E. D. Clark, "Evolution and Entropy," *Journal of the Transactions of the Victoria Institute* 75 (1943): 49–71, quotations on pp. 51, 56–7, 63.

34. Interview with Robert E. D. Clark, October 1, 1984; Discussion of Clark, "Evolution and Entropy," pp. 63–71; Clark, *Darwin: Before and After*, pp. 146–67.

For another early discussion of the topic, see E. H. Betts, "Entropy Disproves Evolution," *Evolution Protest Movement Pamphlet No. 62*, January 1959, originally published in 1944.

35. Interview with Robert E. D. Clark, October 1, 1984. On his reasons for adopting creationism, see R. E. D. Clark, "The Present Position with Regard to the Origin of Species," *Journal of the Transactions of the Victoria Institute* 68 (1936): 172–93.

Nine. Evangelicals and Evolution in North America

1. D. J. Whitney to R. E. D. Clark, quoting Clark, December 14, 1951, Clark Papers, courtesy of the late R. E. D. Clark. In the late 1930s the Pennsylvania botanist Arthur Pierson Kelley attempted to start a Creationist Society of America. He also briefly edited a creationist quarterly called the *Landenberg Review of Natural History*. Kelley described his experience in letters to F. A. Everest (July 7, 1942, ASA Papers) and J. D. Bales (February 8, 1949, Bales Papers, courtesy of James D. Bales). For Kelley's views, see his unpublished 1935 MS in the Winn Library, Gordon College, "Creation: Evolution—Their Philosophies and Literature," which I borrowed through the courtesy of Jack Haas. When I used them, the ASA Papers were in the possession of Robert L. Herrmann; they have since been deposited in the Special Collections of Buswell Memorial Library, Wheaton College.

2. F. Alton Everest, "The American Scientific Affiliation—The First Decade," *Journal of the American Scientific Affiliation* 3 (September 1951): 33–8; *American Scientific Affiliation*, an undated brochure announcing the ASA, ASA Papers (statement and Reformation); F. Alton Everest, "The American Scientific Affiliation: Its Growth and Early Development," pp. 13–25, a manuscript history dated 1986 in the ASA Collection, Buswell Memorial Library, Wheaton College.

3. Interview with F. Alton Everest, May 10, 1984; I. A. Cowperthwaite, "Twenty Years with the American Scientific Affiliation," *Journal of the American Scientific Affiliation* 13 (December 1961): 100 (growth). ASA membership is broken down by denomination and place of residence, ibid. 3 (September, 1951): vii. Typical of Everest's low-key approach to science and religion is his book *The Prior Claim* (Chicago: Moody Press, 1953). For a brief biography of Everest, see "For Half a Century, He's Figured Out How," *Search* no. 2 (1988): 1–4.

4. F. A. Everest to ASA Executive Council, February 20, 1942 (large crowds and SDA flavor), and July 13, 1942 (Lammerts-Moon connection), both in the ASA Papers.

5. C. B. Courville to W. E. Lammerts, August 13, 1942 (interdenominational); W. E. Lammerts to F. A. Everest, August 13, 1942 (details); F. A. Everest to ASA Executive Council, October 7, 1942 (Pasadena meeting); F. A. Everest to W. E. Lammerts, October 7, 1942 (united front); B. F. Allen to F. A. Everest, February 12, 1942 (membership receipt); F. A. Everest to B. F. Allen, September 27, 1943 (subscriber only); all in the ASA Papers. For Price's reaction to the Pasadena meeting, see G. M. Price to P. W. Stoner, October 2, 1942, ASA Papers.

6. D. J. Whitney to C. E. Benson, February 9, 1944 (Moody); F. A. Everest to P. W. Stoner, February 23, 1944 (Houghton and prod); P. W. Stoner to F. A. Everest, February 25, 1944 (unscientific); P. W. Stoner to ASA Executive Council, May 15, 1944 (days of Genesis); all in the ASA Papers.

7. F. A. Everest to D. J. Whitney, March 21, 1944 (feel free); D. J. Whitney to F. A. Everest, March 24, 1944 (formal discussion); F. A. Everest to ASA Executive Council, May 9, 1944 (bomb); E. Y. Monsma to F. A. Everest, [late May or early June

1944] (differences of opinion); Peter Stoner to ASA Executive Council, May 15, 1944; F. A. Everest to ASA Executive Council, May 25, 1944 (insurmountable problem); all in the ASA Papers. For the reactions of the other council members, see R. D. Sturgis to ASA Executive Council, May 29, 1944, and M. D. Barnes to F. A. Everest, June 2, 1944, both in the ASA Papers.

8. "Report of the First Nationwide Convention of the A.S.A. at Wheaton College . . . Aug. 28, 29, 30 [1946]," ASA Papers. Regarding suspicions of Whitney's being an Adventist, see F. A. Everest to ASA Executive Council, May 9, 1944, ASA Papers. On the number of Adventists in the ASA, see *Journal of the American Scientific Affiliation* 3 (September 1951): vii. In a letter to Kulp (September 13, 1949, ASA Papers), Everest identified Whitney's old friend O. L. Brauer as "our one SDA"; the second Adventist was J. Lowell Butler.

9. Interview with J. Laurence Kulp, July 23, 1984. On the evangelical-fundamentalist split, see George M. Marsden, "From Fundamentalism to Evangelicalism: A Historical Analysis," in *The Evangelicals: What They Believe, Who They Are, Where They Are Changing*, rev ed., ed. David F. Wells and John D. Woodbridge (Grand Rapids, MI: Baker Book House, 1977), pp. 148–9.

10. Interview with J. Laurence Kulp, July 23, 1984; J. L. Kulp to F. A. Everest, January 19, 1950, ASA Papers. Kulp described the mid-century changes in geology in a letter to R. L. Numbers, September 28, 1982.

11. Interview with J. Laurence Kulp, July 23, 1984; J. L. Kulp, Comments on "Can a Christian Consistently Believe in Evolution?" by Henry M. Morris, January 12, 1948, ASA Papers.

12. J. L. Kulp to F. A. Everest, April 29, 1948, ASA Papers; interview with J. Laurence Kulp, July 23, 1984. After her marriage in 1951, Fetzer went by the name Marie Fetzer Reyburn.

13. Discussion of J. Laurence Kulp, "Antiquity of Hominoid Fossils," Proceedings of the Third Annual Convention of the American Scientific Affiliation, Calvin College, Sept. 1, 1948, pp. 69–88, quotations on pp. 71–3, 80, 86–8, ASA Papers. Kulp's paper does not appear in the proceedings. The program for the Third Annual Convention of the ASA, Calvin College, Sept. 1–3, 1948, is reprinted as appendix 27 in Everest, "The American Scientific Affiliation." For Kulp's opinion of the day-age theory, see also his comments on J. Oliver Buswell, Jr., "Creation Days," *Journal of the American Scientific Affiliation* 4 (March 1952): 14.

14. Discussion of E. Y. Monsma, "Some Presuppositions in Evolutionary Thinking," *A.S.A. Bulletin* 1 (June 1949): 15–30, quotations on pp. 20–1, 27. The reference to the 1500s appears in Discussion of Kulp, "Antiquity of Hominoid Fossils," p. 85.

15. Discussion of Kulp, "Antiquity of Hominoid Fossils," pp. 79 (Everest: quiet resolution), 87 (Kulp: inspiration); Henry M. Morris, *A History of Modern Creationism* (San Diego: Master Book Publishers, 1984), p. 133 (council); R. L. Mixter to W. J. Tinkle, November 9, 1948 (Monsma's loss), F. A. Everest to J. L. Kulp, November 6, 1948 (1949 paper), both in the ASA Papers. In an interview on July 23, 1984, Kulp expressed his belief that by 1948 the majority of ASA members were open to intellectual change. For additional evidence of Kulp's influence, see J. Laurence Kulp, "Present Status of Age Determination in Geology," in "A Symposium on the Age of the Earth," a mimeographed, unpaginated collection of essays distributed by the ASA in 1948, copies of which can be found in the A. C. Custance Papers, Special Collections, Reedemer College, and the ASA Papers.

16. F. A. Everest to J. L. Kulp, November 21, 1948 (screwey shelf and LA audience); F. A. Everest to J. L. Kulp, February 24, 1949 (Lammerts and Prof. Price); F. A. Everest to J. L. Kulp, April 24, 1949 (hotbed); J. L. Kulp to F. A. Everest

[August 1949] (three nights); F. A. Everest to J. L. Kulp, September 13, 1949 (Fetzer); all in the ASA Papers.

17. J. Laurence Kulp, "Deluge Geology," *Journal of the American Scientific Affiliation* 2 (January 1950): 1–15.

18. F. A. Everest to J. L. Kulp, September 13, 1949; G. M. Price to F. A. Everest, August 24, 1949; both in the ASA Papers.

19. An A. S. A. Member, "Comment on the 'Deluge Geology' Paper of J. L. Kulp," *Journal of the American Scientific Affiliation* 2 (June 1950): 2; C. L. Burdick to M. Couperus, March 12 and April 22, [1950?], and M. Couperus to C. L. Burdick, May 24, 1954, Couperus Papers, courtesy of Molleurus Couperus.

20. "Official Program, Fourth Annual Convention, The American Scientific Affiliation," August 22–6, 1949, Los Angeles, ASA Papers; interview with Cordelia Erdman Barber, April 9, 1989. Kulp mentions Erdman's intention to use his paper in a letter to Everest, September 26, 1949, ASA Papers. Erdman abandoned her scientific career after leaving Wheaton. For her views on the geological column, see Cordelia Erdman, "Stratigraphy and Paleontology," *Journal of the American Scientific Affiliation* 5 (March 1953): 3–11. For one of Kulp's earliest discussions of carbon-14 dating, see J. Laurence Kulp, "Present Status of Age Determination in Geology," in the ASA's 1948 "Symposium on the Age of the Earth," unpaginated, Custance Papers. For Newell's views, see Norman D. Newell, *Creation and Evolution: Myth or Reality?* (New York: Columbia University Press, 1982).

21. Edwin K. Gedney, "Geology and the Bible," in *Modern Science and Christian Faith: A Symposium on the Relationship of the Bible to Modern Science*, 2nd ed., by Members of the American Scientific Affiliation (Wheaton, IL: Van Kampen Press, 1950), pp. 23–57, editorial note on p. 43; W. E. Lammerts to F. A. Everest, December 2, 1947 (footnote), F. A. Everest to H. M. Morris, November 6, 1948 (footnote), F. A. Everest to W. J. Tinkle, April 30, 1949 (five thousand copies), J. L. Kulp to E. K. Gedney, February 15, 1949 (ludicrous), and J. L. Kulp to F. A. Everest, March 21, 1949 (hit by omission), all in the ASA Papers. See also E. K. Gedney to F. A. Everest, November 8, 1947, ASA Papers. On Gedney, see " 'Mr. Gordon': Dr. Edwin K. Gedney Retires," *Gordon College Alumnus* 4 (Spring 1974): 2–5, a copy of which was provided by Janet Potts.

22. Interview with F. Alton Everest, May 10, 1984.

23. M. D. Barnes to J. F. Cassel, February 26, 1949 (anti-Rimmer, my emphasis); F. A. Everest to J. F. Cassel, October 19, 1947 (Van Kampen); W. E. Lammerts to J. F. Cassel, December 21, 1947; W. J. Tinkle to J. F. Cassel, April 22, 1948, Edwin Y. Monsma to J. F. Cassel, January 13, 1948; J. F. Cassel to W. J. Tinkle, May 5, 1948 (consensus); all in the Cassel Papers, originally in the possession of J. Frank Cassel but recently deposited in the Special Collections of Buswell Memorial Library, Wheaton College. The Cassel Papers also contain evaluations by Harley Barnes, Cecil B. Hamann, Paul E. Parker, and George R. Horner. Bernard Ramm's review of *The Harmony of Science and Scripture* and reviews of *Voices from the Silent Centuries* by R. Laird Harris and J. Laurence Kulp are in the ASA Papers. In an interview on September 11, 1988, J. Frank Cassel alleged that "we were after Rimmer, not Price."

24. F. A. Everest to J. F. Cassel, December 27, 1948 (confidential report), and R. L. Mixter to F. A. Everest, January 15, 1949 (fun), both in the ASA Papers; F. A. Everest to M. D. Barnes, March 6, 1949 (fine preacher), quoted in Everest, "The American Scientific Affiliation," p. 113. See also J. F. Cassel to Fellow Members of A.S.A., n.d., a cover letter accompanying the reviews of Rimmer's *Theory of Evolution and the Facts of Science*, Cassel Papers.

25. Arthur C. Custance, "Modern Geology and the Bible," *Proceedings of the*

Kelvin Institute 1 (1938): no pagination. A complete set of the proceedings, as well as biographical information, can be found in the Custance Papers.

26. F. A. Everest to A. C. Custance, December 26, 1943 (invitation); F. A. Everest to A. C. Custance, January 9, 1944 (cooperation); F. A. Everest to A. C. Custance, September 27, 1944 (objections); A. C. Custance to F. A. Everest, October 2, 1944 (Word of God and Jewish bias); F. A. Everest to A. C. Custance, October 19, 1944 (resignation); F. A. Everest to A. C. Custance, March 26, 1945 (first request); F. A. Everest to A. C. Custance, January 26, 1946 (Catholic church); all in the Custance Papers. For Custance's recollection of the episode, see his letter to James O. Buswell III, April 1, 1954, Custance Papers. For Everest's, see F. A. Everest to R. L. Numbers, May 15, 1991.

27. Walter E. Lammerts, "The Creationist Movement in the United States: A Personal Account," *Journal of Christian Reconstruction* 1 (1974): 49–63, quotation on p. 54; B. F. Allen to F. L. Marsh, May 21, 1952, Couperus Papers.

28. William J. Tinkle and Walter E. Lammerts, "Biology and Creation," in *Modern Science and Christian Faith*, pp. 58–97; F. A. Everest to W. J. Tinkle, April 30, 1949, ASA Papers. Marie Fetzer and William A. Smalley, a fellow evangelical graduate student from Columbia University, coauthored a chapter on anthropology for the 1950 edition of *Modern Science and Christian Faith*.

29. Interview with Russell L. Mixter, April 15, 1983; R. L. Mixter, "Application for Membership," ASA Papers; R. L. Mixter, draft of a letter to H. P. Warren, n.d., Mixter Papers, courtesy of Russell L. Mixter (liberal pastor); R. L. Mixter to V. R. Edman, March 26, 1962, quoted in the pamphlet *Christianity and Science at Wheaton* (Wheaton College, 1962), a copy of which can be found in the Edman Collection, Wheaton College Archives (biological change). See also Floyd E. Hamilton, *The Basis of Evolutionary Faith* (London: James Clarke, [1931]); and J. Oliver Buswell, Jr., "The Length of the Creative Days," *Christian Faith and Life* 41 (1935): 117–24.

30. Interview with J. Frank Cassel, September 11, 1988.

31. Russell L. Mixter, *Creation and Evolution*, ASA Monograph Two (1950), pp. 16–18; Everest, "The American Scientific Affiliation," pp. 98–9 (wide use); J. F. Cassel to F. A. Everest, May 5, 1948, ASA Papers (tone down); J. F. Cassel to William Wilson, July 4, 1948, Cassel Papers (devil); F. A. Everest to J. F. Cassel, December 28, 1948, ASA Papers (not fundamentalists).

32. J. F. Cassel to F. A. Everest, July 30, 1951; F. A. Everest to J. F. Cassel, July 31, 1951; J. F. Cassel to F. A. Everest and J. O. Buswell III, August 4, 1951; all in the ASA Papers.

33. "Report on an unofficial meeting of some A.S.A. members and friends in the New York area held at the home of J. L. Kulp, Demarest, N.J.," March 12, 1954, revised from an earlier draft dated August 28, 1953, Custance Papers.

34. A. C. Custance to J. Howitt, September 22, 1953; J. O. Buswell III, to A. C. Custance, March 28, 1954; H. H. Hartzler to A. C. Custance, September 19, 1953 (rejoining ASA); all in the Custance Papers. Buswell declined to be quoted directly. Custance's marital difficulties are discussed in several documents in the Custance Papers as well as in J. Howitt to H. H. Hartzler, May 6, 1958, Hearn Papers, courtesy of Walter R. Hearn.

35. F. A. Everest to S. J. Bates and Others, March 21, 1948 (refutation); F. A. Everest to Members and Friends of the ASA, April 14, 1948 (materialism); W. J. Tinkle to F. A. Everest, May 1, 1948; all in the ASA Papers. In a letter to Tinkle on November 9, 1948 (ASA Papers), Mixter mentions the possibility of "writing a book against evolution."

36. J. L. Kulp to R. L. Mixter, January 20, 1952; F. A. Everest to R. L. Mixter and

Others, February 20, 1955; E. Y. Monsma to F. A. Everest, March 21, 1955; all in the ASA Papers.

37. Russell L. Mixter, "An Evaluation of the Fossil Record," *Journal of the American Scientific Affiliation* 11 (December 1959): 24–6; Wilbur L. Bullock, "Comments on Dr. Mixter's Paper," ibid., p. 26; J. Frank Cassel, "The Evolution of Evangelical Thinking on Evolution," ibid., pp. 26–7; Irving A. Cowperthwaite, "Some Implications of Evolution for A.S.A.," ibid. 12 (June 1960): 12 (growing conviction). On the reaction to Mixter's paper, which was read in absentia, see James O. Buswell III, "A Creationist Interpretation of Prehistoric Man," in *Evolution and Christian Thought Today*, ed. Russell L. Mixter (Grand Rapids, MI: William B. Eerdmans, 1959), p. 183.

38. Russell L. Mixter, "Developmentalism?" *Journal of the American Scientific Affiliation* 23 (1971): 142 (rose); R. L. Mixter to Contributors to the ASA Evolution Symposium, October 5, 1956, Hearn Papers; Buswell, "A Creationist Interpretation," p. 188 (scientific creationism). For one response to Buswell's suggestion, see John W. Klotz, Review of *Evolution and Christian Thought Today*, ed. Russell L. Mixter, *Moody Monthly* 60 (May 1960): 76.

39. Cordelia Erdman Barber, "Fossils and Their Occurrence," in *Evolution and Christian Thought Today*, p. 153 (descent); Walter B. Hearn and Richard A. Hendry, "The Origin of Life," ibid., pp. 67–9. Hendry, a biochemist who had worked with Hearn at Baylor and Iowa State, taught chemistry at Westminster College.

40. Interview with Walter R. Hearn, May 7, 1984; Nicholas Wade, *The Nobel Duel: Two Scientists' 21-Year Race to Win the World's Most Coveted Research Prize* (Garden City, NY: Anchor Press-Doubleday, 1981), pp. 60–73.

41. Interview with Walter R. Hearn, May 7, 1984; interview with J. Frank Cassel, September 11, 1988; J. R. Howitt to A. C. Custance, October 22, 1962 (millimetre), and August 25, 1973 (hair), Custance Papers.

42. Interview with F. Alton Everest, May 10, 1984; Cassel, "Evolution of Evangelical Thinking," p. 27 (verboten); V. Elving Anderson, "The Goals of the ASA—A Personal View," *Journal of the American Scientific Affiliation* 17 (June 1965): 35 (soft); Philip B. Marquart, Letter to the Editor, ibid. 14 (September 1963): 100 (apes); Henry M. Morris, *The Twilight of Evolution* (Grand Rapids, MI: Baker Book House, 1963), p. 93; Richard H. Bube, *The Human Quest: A New Look at Science and the Christian Faith* (Waco, TX: Word, 1971), p. 180 (kiss of death). See Also V. Elving Anderson and David O. Moberg, "Christian Commitment and Evolutionary Concepts," *Journal of the American Scientific Affiliation* 15 (September 1963): 60–70; and D. Gareth Jones, "Evolution: A Personal Dilemma," ibid. 29 (June 1977): 73–6.

43. Interview with J. Laurence Kulp, July 23, 1984; J. R. Howitt to A. C. Custance, October 6, 1962, Custance Papers. On Kulp's involvement with Fuller and Young Life, see George M. Marsden, *Reforming Fundamentalism: Fuller Seminary and the New Evangelicalism* (Grand Rapids, MI: William B. Eerdmans, 1987), p. 206.

44. Walter R. Hearn, "Biological Science," in *The Encounter between Christianity and Science*, ed. Richard H. Bube (Grand Rapids, MI: William B. Eerdmans, 1968), pp. 100, 220; J. R. Howitt to A. C. Custance, November 16, 1959, Custance Papers. Regarding the fundamentalist heritage, see "Doctrinal Statement," *Journal of the American Scientific Affiliation* 7 (March 1955): 2.

45. R. L. Mixter, "A Wheaton College View of Creation and Evolution," MS dated October 22, 1957, and V. R. Edman to the Board of Trustees, Wheaton College, October 28, 1957, both in the Mixter Papers. Mixter's statement eventually appeared in the *Wheaton Alumni Magazine* (February 1960): 3.

46. "The Most Significant Books of the Year," *Eternity* 11 (December 1960): 46; Walter R. Hearn, "Origin of Life," *Journal of the American Scientific Affiliation* 13

(June 1961): 38; R. L. Mixter to V. R. Edman, November 17, 1960, Box 2, V. Raymond Edman Collection, Wheaton College Archives.

47. Copy of a statement by Harold P. Warren, " 'Origins and Christian Thought': Science Conference Sponsored by the Dept. of Science of Wheaton College, Wheaton, Illinois, February 17, 18, 1961," February 21, 1961, Mixter Papers.

48. Walt Handford, "Evolution at Wheaton College," *Sword of the Lord* 27 (June 9, 1961): 1, 10–12; V. R. Edman to Mrs. Myron L. Miller, August 1, 1962, Box 2, Edman Papers. For the reaction of GARB officials, see Paul R. Jackson, "Special Information Bulletin," November 15, 1961, Mixter Papers; and R. T. Ketcham's correspondence with R. L. Mixter and J. O. Buswell III, from November 9, 1962, to April 12, 1963, in the files of the National Representative, General Association of Regular Baptist Churches, Schaumberg, IL, copies courtesy of Joel A. Carpenter.

49. *Christianity and Science at Wheaton* (Wheaton, IL: Wheaton College, 1962), reprinted under the same title in the *Bulletin of Wheaton College* 39 (May 1962): no pagination; [Paul R. Jackson], "You Be the Judge! Do You Believe These Accounts Harmonize?" undated MS, Mixter Papers, quoting R. L. Mixter, "What Every Teenager Should Know about Evolution," *Youth for Christ Magazine* (December 1959): 11; R. L. Mixter, "Comments on *You Be the Judge*," May [?] 17, 1963, Mixter Papers; Vernon C. Lyons, "Is Evolution at Wheaton College?" *Sword of the Lord* 28 (October 19, 1962): 1, 6–7; H. M. Morris to J. C. Whitcomb, March 27, 1963, Whitcomb Papers, courtesy of John C. Whitcomb, Jr. (negligible difference). Morris thought that his generalization applied equally to the Calvin College faculty. For Mixter's changing views on prehistoric humans, see also his letter to Bolton Davidheiser, May 22, 1963, Mixter Papers. See also Mixter's undated reply to H. P. Warren, Mixter Papers.

Ten. JOHN C. WHITCOMB, JR., HENRY M. MORRIS, AND *THE GENESIS FLOOD*

1. John W. Haas, Jr., "The Christian View of Science and Scripture: A Retrospective Look," *Journal of the American Scientific Affiliation* 31 (1979): 117. Although not identified as such, there are two American editions of *The Christian View of Science and Scripture*, both published by William B. Eerdmans of Grand Rapids, Michigan. The original cloth edition has 369 pages; the paperback version, which uses the plates from an English edition, contains only 256 pages. Except for the phrase "narrow bibliolatry," which appears on p. 9 of the paperback, all quotations are from the seventh printing of the cloth edition, in 1968.

2. George M. Marsden, *Reforming Fundamentalism: Fuller Seminary and the New Evangelicalism* (Grand Rapids, MI: William B. Eerdmans, 1987), offers the best treatment of these developments; see especially pp. 158–62. See also Bernard Ramm to R. L. Numbers, May 30, 1991.

3. Interview with Bernard Ramm, May 8, 1984; Walter Hearn, "An Interview with Bernard Ramm and Alta Ramm," *Journal of the American Scientific Affiliation* 31 (1979): 179–86.

4. Ramm, *The Christian View of Science and Scripture*, pp. 5, 10; interview with Bernard Ramm, May 8, 1984.

5. Ramm, *The Christian View of Science and Scripture*, pp. 9–10.

6. Ibid., pp. 180–1, 197. See also pp. 201–10.

7. Ibid., pp. 220–8, 271–2.

8. Ibid., p. 272; interview with Bernard Ramm, May 8, 1984; J. Frank Cassel, "The

Origin of Man and the Bible," *Journal of the American Scientific Affiliation* 12 (June 1960): 15; Richard T. Wright, "Evolutionary Biology," ibid. 31 (1979): 195.

9. J. C. Whitcomb to H. M. Morris, October 8, 1955 (final proof), J. C. Whitcomb to W. J. Tinkle, July 13, 1955 (perplexity), and August 25, 1955 (death), all in the Whitcomb Papers, courtesy of John C. Whitcomb, Jr.

10. J. C. Whitcomb, undated "Biographical Sketch," and J. C. Whitcomb to H. M. Morris, March 23, 1959 (China), both in the Whitcomb Papers. Whitcomb supplied additional details in a letter to R. L. Numbers, May 27, 1991.

11. H. M. Morris to J. C. Whitcomb, September 22, 1953 (highlight), and J. C. Whitcomb to W. J. Tinkle, July 13, 1955 (denounced), both in the Whitcomb Papers. The printed program of the 1953 ASA convention carried an abstract of Morris's paper.

12. J. C. Whitcomb to W. J. Tinkle, July 13, 1955, Whitcomb Papers. Whitcomb described his study as an "answer" to Ramm in a letter to H. M. Morris, November 16, 1957, and as a defense of Price in a letter to D. J. Whitney, August 31, 1957, both in the Whitcomb Papers.

13. J. C. Whitcomb, "A Questionnaire on Creation and the Flood," [1955], and J. C. Whitcomb to H. M. Morris, October 8, 1955, both in the Whitcomb Papers. Whitcomb says that William LaSor, also at Fuller, shared Carnell's reaction.

14. John C. Whitcomb, Jr., "The Genesis Flood: An Investigation of Its Geographical Extent, Geologic Effects, and Chronological Setting" (Th.D. dissertation, Grace Theological Seminary, 1957); J. C. Whitcomb to D. J. Whitney, August 31, 1957 (dissertation), J. C. Whitcomb to H. M. Morris, November 16, 1957 (answering Ramm), J. C. Whitcomb to H. M. Morris, April 29, 1957 (publishers), J. C. Whitcomb to H. M. Morris, August 12, 1958 (rejection), and J. C. Whitcomb to H. M. Morris, November 16, 1957 (hearing), all in the Whitcomb Papers.

15. J. C. Whitcomb to H. M. Morris, January 25, 1957 (check MS), and J. C. Whitcomb to H. M. Morris, October 8, 1955 (Ph.D.), both in the Whitcomb Papers.

16. Block is quoted and discussed in J. C. Whitcomb to H. M. Morris, June 16, August 8, and October 14, 1957, all in the Whitcomb Papers. Block, who had earned a B.D. degree from Northern Baptist Theological Seminary, eventually received a Ph.D. in geology from the University of North Dakota; interview with Douglas A. Block, June 12, 1991.

17. J. C. Whitcomb to G. M. Price, September 7, 1957 (readers), G. M. Price to J. C. Whitcomb, May 24, 1957 (87), D. J. Whitney to J. C. Whitcomb, September 5, 1957 (begs off), and October 3, 1957 (universe), and J. C. Whitcomb to H. M. Morris, October 14, 1957 (Tinkle), all in the Whitcomb Papers. Walter E. Lammerts and Edwin Y. Monsma also read the manuscript.

18. H. M. Morris to J. C. Whitcomb, October 19, 1955 (Ramm and closed-shop), and H. M. Morris to J. C. Whitcomb, May 29, 1957 (sarcasm), both in the Whitcomb Papers. In *The Christian View of Science and Scripture*, p. 358, Ramm had listed Morris's first book among "Works of Limited Worth," a judgment that did not pass unnoticed; see J. C. Whitcomb to H. M. Morris, October 8, 1955, and H. M. Morris to J. C. Whitcomb, October 19, 1955, both in the Whitcomb Papers. See also Henry M. Morris, *A History of Modern Creationism* (San Diego: Master Book Publishers, 1984), pp. 146–9.

19. H. M. Morris to J. C. Whitcomb, October 7, 1957, Whitcomb Papers.

20. J. C. Whitcomb to H. M. Morris, August 8, 1957 (weakness), H. M. Morris to J. C. Whitcomb, October 7, 1957 (dual-authorship), J. C. Whitcomb to H. M. Morris, October 14, 1957 (reconsider), and H. M. Morris to J. C. Whitcomb, October 26, 1957 (honored), all in the Whitcomb Papers.

21. John William Klotz, "Genesis, Genetics, and Evolution," (B.D. thesis, Con-

cordia Theological Seminary, 1941); John W. Klotz, *Genes, Genesis, and Evolution* (St. Louis: Concordia Publishing House, 1955); J. C. Whitcomb to J. W. Klotz, November 8, 1957 (invitation), and J. W. Klotz to J. C. Whitcomb, December 6, 1957 (Rusch), both in the Whitcomb Papers.

22. J. C. Whitcomb to H. M. Morris, November 16, 1957 (strictly equal), H. M. Morris to J. C. Whitcomb, December 3, 1957 (accepts), and J. C. Whitcomb to H. M. Morris, December 16, 1957 (way is cleared), all in the Whitcomb Papers.

23. Interviews with Henry M. Morris, October 26, 1980, and January 6, 1981; Morris, *History of Modern Creationism*, pp. 93, 131; *The [Rice Institute] Campanile* 24 (1939): 91. Because of the liberal-arts emphasis at Rice, even engineering students were eligible for Phi Beta Kappa; see Morris, *History of Modern Creationism*, p. 74.

24. Morris, *History of Modern Creationism*, pp. 94–97; interview with Henry M. Morris, May 5, 1984; interview with James R. Sims, February 8, 1985 (former colleague); interview with Alan J. Chapman, February 8, 1985 (former student).

25. Morris, *History of Modern Creationism*, pp. 80, 90–1, 121; interview with Henry M. Morris, January 6, 1981.

26. Morris, *History of Modern Creationism*, p. 93; Henry M. Morris, *That You Might Believe* (Chicago: Good Books, 1946), pp. 6, 15–17, 27, 51, 60–1, 80, 141. The quotation about being the "first book" appears in the introduction to the revised edition of *That You Might Believe* (San Diego: Creation-Life Publishers, 1978), p. 10.

27. Morris, *History of Modern Creationism*, pp. 82–3, 97–8; Clifford L. Burdick, "The Radioactive Time Theory and Recent Trends in Methods of Reckoning Geologic Time," *Forum for the Correlation of Science and the Bible* 1 (1946–47): 39–58.

28. Morris, *History of Modern Creationism*, p. 108; H. M. Morris to F. A. Everest, August 27, 1950 (ten thousand copies), ASA Papers, Special Collections, Buswell Memorial Library, Wheaton College; Henry M. Morris, *The Bible and Modern Science* (Chicago: Moody Press, 1951), pp. 23–6, 70, 101–2, 108; Henry M. Morris, *The Bible and Modern Science*, rev. ed. (Chicago: Moody Press, 1956), p. 59. Morris's first book has appeared in many editions and under at least three titles. This last-cited volume was a revised edition of the abridged Colportage Library edition of 1951, which Moody Press reprinted in 1986 under the title *Science and the Bible*. In 1978 Creation-Life Publishers in San Diego brought out an unabridged edition of *The Bible and Modern Science*, retitled *That You Might Believe*. By 1980 *The Bible and Modern Science* had sold about 200,000 copies; interview with Henry M. Morris, October 26, 1980. In the original edition of *That You Might Believe*, p. 48, Morris had briefly alluded to the second law.

29. Morris, *History of Modern Creationism*, pp. 136, 148.

30. Ibid., pp. 101–3, 148.

31. Ibid., p. 58.

32. Ibid., pp. 136–7; H. M. Morris to F. A. Everest, October 18, 1948 (reconsider), F. A. Everest to H. M. Morris, March 1, 1950 (re rebuttal), H. M. Morris to F. A. Everest, August 27, 1950 (visit), and F. A. Everest to H. M. Morris, October 8, 1950 (reasonable), all in the ASA Papers. A copy of Morris's undated rebuttal is located in the Whitcomb Papers. In his *History of Modern Creationism*, p. 136, Morris says that he joined the ASA in 1948; the ASA's records show that he applied for membership in March 1949, with an endorsement from Russell Mixter, and was accepted the next month.

33. Morris, *History of Modern Creationism*, pp. 140–1, 148–52; H. M. Morris to G. M. Price, December 30, 1953 (no prejudice), and November 4, 1955 (Ramm), both in the George McCready Price Papers, Adventist Heritage Center, Andrews University. The Price Papers also contain a copy of Morris's 1953 ASA paper, "The Biblical Evidence for a Recent Creation and Universal Flood." Morris's marginal comments in

The Christian View of Science and Scripture appear on pp. 29 and 276; I am grateful to Henry M. Morris for granting me access to his personal library.

34. Morris, *History of Modern Creationism*, pp. 149–50; J. C. Whitcomb to H. M. Morris, August 12, 1958, Whitcomb Papers.

35. J. C. Whitcomb to H. M. Morris, January 24, 1959, and H. M. Morris to J. C. Whitcomb, January 28, 1959, both in the Whitcomb Papers.

36. H. M. Morris to G. M. Price, May 8, 1959, H. M. Morris to G. M. Price, July 5, 1959, and G. M. Price quotation in an undated brochure advertising *The Genesis Flood*, all in the Price Papers. Whitcomb mentions Allis's reaction in a letter to Morris, December 11, 1959, Whitcomb Papers. The footnote appears on p. 184 of John C. Whitcomb, Jr., and Henry M. Morris, *The Genesis Flood: The Biblical Record and Its Scientific Implications* (Philadelphia: Presbyterian and Reformed Publishing Co., 1961). For other references to Price, see pp. 185, 189, 211.

37. H. M. Morris to J. C. Whitcomb, January 28, 1959 (real geologist), and J. C. Whitcomb to H. M. Morris, August 12, 1958 (Burdick), both in the Whitcomb Papers. The published version of *The Genesis Flood* carried a foreword by Morris's friend John C. McCampbell, a geologist at the University of Southwestern Louisiana. Readers of the manuscript are mentioned in J. C. Whitcomb to H. M. Morris, March 23, 1959, and "List of Those Who Have Read Manuscript of THE GENESIS FLOOD," August 25, 1959, both in the Whitcomb Papers. The acknowledgments in *The Genesis Flood* fail to mention Allen, Block, Couperus, and Whitney.

38. Morris, *History of Modern Creationism*, pp. 154–6; C. H. Craig to G. M. Price, February 28, 1960, Price Papers.

39. Whitcomb and Morris, *The Genesis Flood*, pp. xx, xxiii, 1; H. M. Morris to J. C. Whitcomb, October 25, 1961, Whitcomb Papers; interview with Henry M. Morris, May 10, 1984. Unless otherwise noted, all references to *The Genesis Flood* are based on the fourth printing, in 1963.

40. Whitcomb and Morris, *The Genesis Flood*, pp. xx, 1–35; J. C. Whitcomb to H. M. Morris, March 23, 1959, Whitcomb Papers. For Morris's opinion of Custance, see H. M. Morris to J. C. Whitcomb, January 28, 1959, Whitcomb Papers. See also Arthur C. Custance, *The Flood: Local or Global?* (Grand Rapids, MI: Zondervan, 1979), pp. 13–63, originally published in 1958 as no. 41 of *The Doorway Papers*.

41. Whitcomb and Morris, *The Genesis Flood*, pp. 36–88, quotation on pp. 87–8. See also Ramm, *The Christian View of Science and Scripture*, p. 336.

42. Whitcomb and Morris, *The Genesis Flood*, pp. 116–20, 329–30. See also Barrington Moore, Jr., "Influence of Political Creeds on the Acceptance of Theories," *Scientific Monthly* 79 (1954): 146–8.

43. Whitcomb and Morris, *The Genesis Flood*, pp. 130–211, 270–81, quotation on p. 189.

44. W. H. Rusch to J. C. Whitcomb, April 19, 1960 (bombshell), J. C. Whitcomb to C. L. Burdick, December 2, 1957 (inquiry), J. C. Whitcomb to H. M. Morris, December 16, 1957 (Burdick), C. L. Burdick to J. C. Whitcomb, March 22, 1958 (tracks), H. M. Morris to C. H. Craig, May 22, 1960 (pen name), all in the Whitcomb Papers. Morris had known about the footprints since the 1940s.

45. Whitcomb and Morris, *The Genesis Flood*, pp. 173–4, first and fourth printings; J. C. Whitcomb to H. M. Morris, August 30, 1962 (Rusch and final confirmation), and September 13, 1962 (accuracy and Burdick quotation), both in the Whitcomb Papers.

46. Whitcomb and Morris, *The Genesis Flood*, pp. 233, 239, 327–8, 489. See also Mrs. E. G. White, *Patriarchs and Prophets* (Oakland, CA: Pacific Press, 1890), pp. 41–2.

47. Whitcomb and Morris, *The Genesis Flood*, pp. 240, 255, 265, 375, 399. The origins of the canopy theory are obscure, but Bernard Ramm found many trails leading back to Isaac Newton Vail's *"The Waters above the Firmament": The Earth's Aqueous Ring; or, The Deluge and Its Cause* (West Chester, PA: F. S. Hickman, 1874); see Ramm, *The Christian View of Science and Scripture*, p. 234. In one of the earliest creationist responses to carbon-14 dating, the Adventist physicist Robert W. Woods appealed to the shielding effects of the waters above the firmament; see Woods, "How Old Is the Earth?" *Signs of the Times* 80 (April 7, 1953): 8–9, 15. Price questioned the validity of carbon-14 dating but, until the end of his life, attached little importance to the canopy theory; see Price, "Dating the Fossils: Is the Radiocarbon Method Reliable?" *Signs of the Times* 79 (August 19, 1952): 3–4; and Price, *Time of the End* (Nashville: Southern Publishing Assn., 1967), pp. 114–15.

48. Whitcomb and Morris, *The Genesis Flood*, p. 451.

49. Roy M. Allen, Letter to the Editor, *Journal of the American Scientific Affiliation* 17 (June 1965): 62; Walter Hearn, quoted in Vernon Lee Bates, "Christian Fundamentalists and the Theory of Evolution in Public School Education: A Study of the Creation Science Movement" (Ph.D. dissertation, University of California, Davis, 1976), p. 52 (cynic); J. C. Whitcomb to H. M. Morris, January 14, 1961 (Block), Whitcomb Papers. For sales figures, see Charles Albert Clough, "A Calm Appraisal of *The Genesis Flood*" (M.Th. thesis, Dallas Theological Seminary, 1968).

50. S. Hugh Paine, Review of *The Genesis Flood*, by J. C. Whitcomb and H. M. Morris, *Wesleyan Methodist* 119 (June 7, 1961): 14; J. C. Whitcomb to H. M. Morris, June 19, 1961 (Harris and appendix), Whitcomb Papers. A list of reviews can be found in the Whitcomb Papers. For Paine's own views, see S. Hugh Paine, "In the Beginning, God Created," *Houghton Milieu* 54 (March 1979): 2–7.

51. Donald C. Boardman, Review of *The Genesis Flood*, by J. C. Whitcomb and H. M. Morris, *Christianity Today*, September 11, 1961, pp. 39–40; J. C. Whitcomb and H. M. Morris, Letter to the Editor, ibid., November 10, 1961, pp. 23–4; H. M. Morris to G. M. Price, January 9, 1963 (interesting people), Price Papers; H. M. Morris to J. C. Whitcomb, August 6, 1964 (twenty-five dollars), Whitcomb Papers.

52. W. R. Hearn to H. M. Morris, May 31, 1963 (Ault), and March 16, 1964 (I like *you*), H. M. Morris to W. R. Hearn, June 4, 1963 (ignoring it and *Christianity Today*), W. R. Hearn to D. O. Moberg, June 7, 1963 (professional geologists), and June 25, 1963 (loyalty), D. O. Moberg to W. U. Ault, June 14, 1963, all in the Hearn Papers, courtesy of Walter R. Hearn. For Moberg's views, see V. Elving Anderson and David O. Moberg, "Christian Commitment and Evolutionary Concepts," *Journal of the American Scientific Affiliation* 15 (September 1963): 69–70.

53. W. R. H[earn], Editorial introduction to reviews of *The Genesis Flood*, by J. C. Whitcomb and H. M. Morris, *Journal of the American Scientific Affiliation* 16 (March 1964): 27–8; J. C. Whitcomb to D. O. Moberg, April 14, 1964 (not funny), Hearn Papers.

54. Wayne U. Ault, Review of *The Genesis Flood*, by J. C. Whitcomb and H. M. Morris, *Journal of the American Scientific Affiliation* 16 (March 1964): 29–31. See also Frank H. Roberts, Review of *The Genesis Flood*, by J. C. Whitcomb and H. M. Morris, ibid., pp. 28–9; and W. U. Ault, "Flood," in *The Zondervan Pictorial Encyclopedia of the Bible*, ed. Merrill C. Tenney, 5 vols. (Grand Rapids, MI: Zondervan, 1975), 2:549–63.

55. Henry M. Morris and John C. Whitcomb, Jr., "Reply to Reviews," *Journal of the American Scientific Affiliation* 16 (June 1964): 59–61; J. C. Whitcomb to D. O. Moberg, April 14, 1964, Hearn Papers. Morris and Whitcomb recycled their "Reply to Reviews" as a preface to the sixth printing of *The Genesis Flood*, 1964.

56. Morris and Whitcomb, "Reply to Reviews," pp. 59–61.

57. J. R. van de Fliert, "Fundamentalism and the Fundamentals of Geology," *Journal of the American Scientific Affiliation* 21 (September 1969): 69–81. Cuffey's sidebar appears on p. 71, and one by Donald C. Boardman on p. 75.

58. H. M. Morris to R. H. Bube, September 10, 1969, and R. H. Bube to H. M. Morris, September 12, 1969, both in the Whitcomb Papers; H. M. Morris, Letter to the Editor, *Journal of the American Scientific Affiliation* 22 (March 1970): 36–7. For Whitcomb's reply to van de Fliert, see John C. Whitcomb, *The World That Perished* (Grand Rapids, MI: Baker Book House, 1973), pp. 111–22.

59. J. C. Whitcomb to H. M. Morris, January 14, 1961 (creation book), Whitcomb Papers; John C. Whitcomb, *The Early Earth* (Grand Rapids, MI: Baker Book House, 1972; revised ed., 1986); John C. Whitcomb, *The World That Perished* (Grand Rapids, MI: Baker Book House, 1973; revised ed., 1988); John C. Whitcomb and Donald B. DeYoung, *The Moon: Its Creation, Form and Significance* (Winona Lake, IN: BMH Books, 1978). Biographical and other data come from a brochure promoting "Tapes, Books, and Charts by Dr. John C. Whitcomb." In 1990 Whitcomb was dismissed from Grace Theological Seminary for differences with the president over educational policy; see "Trouble at Grace: Making Waves or Guarding the Truth?" *Christianity Today*, April 9, 1990, p. 46.

60. Morris, *History of Modern Creationism*, pp. 159–60; interview with Henry M. Morris, May 10, 1984; H. M. Morris to J. C. Whitcomb, January 4, 1966 (LA Baptist faculty), September 24, 1966 (LA Baptist offer), and December 15, 1966 (Bob Jones), all in the Whitcomb Papers. Morris reported on his various speaking engagements in frequent letters to Whitcomb.

61. Morris, *History of Modern Creationism*, p. 161; Henry M. Morris, *Studies in the Bible and Science* (Philadelphia: Presbyterian and Reformed Publishing Co., 1966), p. 9 (largest seminary); H. M. Morris to J. C. Whitcomb, November 16, 1967 (Dallas), Whitcomb Papers. The revised lectures were published as *Biblical Cosmology and Modern Science* (Nutley, NJ: Craig Press, 1970).

62. H. M. Morris, "Biblical Catastrophism and Geology," unpublished paper presented to the Houston Geological Society, September 10, 1962, Price Papers; H. M. Morris to J. C. Whitcomb, September 15, 1962, Whitcomb Papers. The Houston lecture subsequently appeared as a pamphlet titled *Biblical Catastrophism and Geology* (Philadelphia: Presbyterian and Reformed Publishing Co., 1963).

63. H. M. Morris to J. C. Whitcomb, June 8, 1964 (wife's cancer); Ocie Strawn, " 'Answer' Author: Dr. Morris Is Man with Two Careers," *Montgomery News Messenger*, December 12, 1968, p. A–3, copy courtesy of David A. West.

64. Henry M. Morris, *Science, Scripture and Salvation: The Genesis Record* (Denver: Baptist Publications, 1965), p. 55 (giants); Henry M. Morris, *Evolution and the Modern Christian* (Philadelphia: Presbyterian and Reformed Publishing Co., 1967), pp. 60–3 (gap and day-age); Morris, *Studies in the Bible and Science*, pp. 30–44 (gap and day-age); Henry M. Morris, *The Twilight of Evolution* (Grand Rapids, MI: Baker Book House, 1963), p. 91 (Kerkut); H. M. Morris to J. C. Whitcomb, September 24, 1966 (best seller), Whitcomb Papers; G. A. Kerkut, *Implications of Evolution* (New York: Pergamon Press, 1960), p. vii.

65. Morris, *History of Modern Creationism*, p. 162 (pastor's comment); H. M. Morris to J. C. Whitcomb, January 16, 1958 (Blacksburg church), and August 11, 1968 (Falwell's visit), both in the Whitcomb Papers; interview with Henry M. Morris, May 10, 1984 (Falwell relationship); Strawn, " 'Answer' Author," pp. A–3 (College Baptist). On the Elliott controversy, see Walter B. Shurden, *Not a Silent People: Controversies That Have Shaped Southern Baptists* (Nashville: Broadman Press, 1972), pp. 103–19.

66. H. M. Morris to J. C. Whitcomb, October 2, 1961 (popularity contest), Whitcomb Papers; Strawn, " 'Answer' Author," p. A–3 (professional activities); Henry M. Morris, *Applied Hydraulics in Engineering* (New York: Ronald Press, 1963), p. 18; H. M. Morris to W. R. Hearn, February 10, 1964 (text adoption), Hearn Papers. Additional information comes from interviews on November 20, 1986, with R. D. Walker, who succeeded Morris as department head, and James M. Wiggert, who coauthored the second edition of the text. I am indebted to Albert Moyer for arranging these interviews. When Morris attempted to expand the note into a full section on "Philosophic and Theological Implication" for the second edition, the publisher vetoed the move; see the unpublished manuscript and editorial note in the Wiggert Papers, courtesy of James M. Wiggert.

67. H. M. Morris to J. C. Whitcomb, February 5, 1963 (invitation), March 15, 1963 (meeting), and April 11, 1963 (aftermath), all in the Whitcomb Papers. Morris declined to be quoted directly.

68. H. M. Morris to J. C. Whitcomb, May 22, 1964 (dean's warning), April 9, 1968 (abolish department), and April 8, 1969 (resignation), all in the Whitcomb Papers; "Group of Scientists Reject Evolution Theory of Creation," *New York Times*, April 22, 1964, p. 31; David A. West, Letter to the Editor, *Montgomery News Messenger*, November 28, 1968 (pseudo-science). The Morris-West exchange in the *Montgomery News Messenger* lasted from November 21, 1968, to January 7, 1969; I am grateful to David A. West for providing copies. Morris published his collected columns from the *Montgomery News Messenger* as *The Bible Has the Answer* (Nutley, NJ: Craig Press, 1971). In his *History of Modern Creationism*, p. 153, Morris writes that "the leaders of the evolutionary/humanistic establishments on the campus tried on at least two occasions to get me removed from my position," but the testimony of two sympathetic former colleagues suggests that other factors were more important; see interviews with R. D. Walker and J. M. Wiggert, November 20, 1986.

Eleven. THE CREATION RESEARCH SOCIETY

1. Henry M. Morris, *A History of Modern Creationism* (San Diego: Master Book Publishers, 1984), pp. 172–4; W. E. Lammerts to J. C. Whitcomb, November 27, 1957 (association), J. C. Whitcomb to H. M. Morris, February 2, 1958 (fear), H. M. Morris to J. C. Whitcomb, January 16, 1958 (ASA), and May 16, 1958 (prevent domination), all in the Whitcomb Papers, courtesy of John C. Whitcomb, Jr. See also W. E. Lammerts to G. M. Price, January 29, 1958, George McCready Price Papers, Adventist Heritage Center, Andrews University.

2. W. E. Lammerts to J. C. Whitcomb, January 8, 1958 (begs off), Whitcomb Papers; William J. Tinkle, "Creation in the Twentieth Century," *Creation Research Society Quarterly* 10 (1973–74): 46.

3. George F. Howe, "Walter E. Lammerts," *Creation Research Society Quarterly* 7 (1970–71): 3–4; W. E. Lammerts to L. W. Faulstick, September 15, 1962 (grandmother), and W. E. Lammerts to D. T. Gish, March 30, 1963 (Berkeley), both in the Walter E. Lammerts Papers, Bancroft Library, University of California, Berkeley; W. E. Lammerts to G. M. Price, March 27, 1961 (Price), Couperus Papers, courtesy of Molleurus Couperus. See also Walter E. Lammerts, "The Creationist Movement in the United States: A Personal Account," *Journal of Christian Reconstruction* 1 (Summer 1974): 49–63.

4. Interview with Walter E. Lammerts, January 17, 1983; Howe, "Walter E. Lammerts," pp. 3–4; W. E. Lammerts to F. L. Marsh, March 30, 1963 (arguments), and

W. E. Lammerts to H. C. Doellinger, November 18, 1963, both in the Lammerts Papers.

5. W. E. Lammerts to Theodore Graebner, November 8, 1941 (bird), Box 1, Theodore Graebner Papers, Concordia Historical Institute; W. E. Lammerts to Bolton Davidheiser, May 9, 1970 (Adventism), Lammerts Papers; Lammerts, "The Creationist Movement in the United States," p. 54 (ASA).

6. Lammerts, "The Creationist Movement in the United States," pp. 56–7 (interest wanes); Morris, *History of Modern Creationism*, p. 173 (Rehwinkel); W. E. Lammerts to G. M. Price, October 31, 1956 (Glacier and book), Price Papers; W. E. Lammerts to J. C. Whitcomb, August 12, 1957 (not a geologist), Whitcomb Papers.

7. W. E. Lammerts to J. C. Whitcomb, August 12, 1957 (Lewis overthrust), September 6, 1957 (hobby), and November 27, 1957 (society), all in the Whitcomb Papers; John C. Whitcomb, Jr., and Henry M. Morris, *The Genesis Flood: The Biblical Record and Its Scientific Implications* (Philadelphia: Presbyterian and Reformed Publishing Co., 1961), pp. 70, 106, 189–94.

8. Interviews with Richard M. Ritland, May 7, 1973, and November 14, 1983; F. L. Marsh to W. E. Lammerts, March 26, 1962 (Flood geologist), Lammerts Papers.

9. Interview with P. Edgar Hare, October 22, 1983; P. E. Hare to E. E. Cossentine, December 15, 1957, and P. E. Hare to Richard Hammill, February 7, 1963, both in the Hare Papers, courtesy of P. Edgar Hare; Roy Benton, "Odyssey of an Adventist Creationist," *Spectrum* 15 (August 1984): 46–53.

10. Interview with Richard M. Ritland, May 7, 1973; W. E. Lammerts to F. L. Marsh, July 19, 1962 (scientific approach), and W. E. Lammerts to C. L. Burdick, November 18, 1963 (confused and puzzled), both in the Lammerts Papers; Walter E. Lammerts, "Growing Doubts: Is Evolutionary Theory Valid?" *Christianity Today* 6 (September 14, 1962): 3–6. Two other Adventists, Harold G. Coffin and Harold James, accompanied the trio on their hike.

11. W. E. Lammerts to F. L. Marsh, July 19, 1962 (first reaction), Lammerts Papers; W. E. Lammerts to R. M. Ritland, November 21, 1962 (post-flood overthrusting), Ritland Papers, courtesy of Richard M. Ritland; interview with Richard M. Ritland, May 7, 1973; Lammerts, "Growing Doubts," p. 5.

12. W. E. Lammerts to J. J. Grebe, October 30, 1963 (amazed), and W. E. Lammerts to F. L. Marsh, May 13, 1966 (courses), both in the Lammerts Papers; W. E. Lammerts to J. C. Whitcomb, March 30, 1964 (presuppositions), Whitcomb Papers. The dates given in the letter to Marsh are obviously in error.

13. Interview with Walter E. Lammerts, January 17, 1983 (tailspin); W. E. Lammerts to H. W. Clark, December 7, 1964 (all scientists), Price Papers; Walter E. Lammerts, "Introduction," *Creation Research Society Annual* (1964): [1] (dinosaur); "Mathematical Probability of Accidental Development of Life Cell," *Five Minutes with the Bible & Science*, supplement to *Bible-Science Newsletter* 17 (June 1979): 7 (lectures).

14. W. E. Lammerts to H. H. Hartzler, April 21, 1970 (ridiculous and resignation), and W. E. Lammerts to H. M. Morris, August 26, 1963 (sick), both in the Lammerts Papers; interview with Walter R. Hearn, May 7, 1984 (outburst); W. E. Lammerts to R. M. Ritland, November 21, 1962 (spouting off), Ritland Papers.

15. W. E. Lammerts to V. E. Anderson, May 31, 1965, Lammerts Papers.

16. W. E. Lammerts to F. L. Marsh, May 9, 1963 (Bible), and W. E. Lammerts to Martin Scharlemann, May 29, 1965 (students), both in the Lammerts Papers; interview with Walter E. Lammerts, January 17, 1983 (six thousand years and stupid fool); Lammerts, "Growing Doubts," p. 6 (timid breeders).

17. Morris, *History of Modern Creationism*, p. 186 (never in doubt); interview

with Walter E. Lammerts, January 17, 1983 (finances and conservation); W. E. Lammerts to G. F. Howe, November 20, 1962 (creationist causes), and W. E. Lammerts to W. H. Rusch, March 14, 1966 (cuss), both in the Lammerts Papers.

18. Samuel Wolfe to W. E. Lammerts, December 30, 1964 (quoting Lammerts and friends regarding Kennedy and Birchers), W. E. Lammerts to L. W. Faulstick, December 2, 1963 (interfaith committee), and W. E. Lammerts to H. H. Hartzler, August 8, 1966 (conservation), all in the Lammerts Papers; Walter E. Lammerts, Review of *Creation: A Scientist's Choice*, by Zola Levitt, *Creation Research Society Quarterly* 14 (1977–78): 75 (communism).

19. W. E. Lammerts to H. H. Hartzler, November 24, 1951, ASA Papers, now deposited in Special Collections, Buswell Memorial Library, Wheaton College. See also William J. Tinkle, "Heredity of Habitual Wandering," *Journal of Heredity* 18 (1927): 548–51; and "Habitual Wanderers," *Literary Digest* 96 (April 21, 1928): 20.

20. W. J. Tinkle to G. M. Price, December 15, 1932, Price Papers; William J. Tinkle, *Fundamentals of Zoology* (Grand Rapids, MI: Zondervan, 1939), pp. 127–32. See also "William John Tinkle," *Creation Research Society Annual* 4 (June 1967): 3. Tinkle treated eugenics more cautiously in his last book, *Heredity: A Study in Science and the Bible* (Grand Rapids, MI: Zondervan, 1970), pp. 135–46.

21. W. J. Tinkle to J. C. Whitcomb, August 15, 1955 (other cataclysms), J. C. Whitcomb to W. J. Tinkle, August 25, 1955 (sin), and J. C. Whitcomb to H. M. Morris, October 14, 1957 (quoting Tinkle regarding fossiliferous rocks), all in the Whitcomb Papers; William J. Tinkle, Review of *Georges Cuvier, Zoologist*, by William Coleman, *Creation Research Society Annual* 2 (May 1965): 51. According to Henry M. Morris, Tinkle converted to flood geology before his death; interview with Henry M. Morris, May 10, 1984.

22. W. J. Tinkle to W. E. Lammerts, January 26, 1962 (jeopardize ASA), and February 10, 1962 (Glenn); W. E. Lammerts to Tinkle, February 15, 1962 (read MSS and popular magazines), March 5, 1962 (no illusions and *Time*), and February 9, 1963 (creationist scientists); W. E. Lammerts to F. L. Marsh, December 30, 1962 (letterhead); all in the Lammerts Papers. See J. C. Whitcomb to H. M. Morris, May 28, 1961, Whitcomb Papers, regarding Tinkle's intention of staying within the ASA.

23. W. J. Tinkle to W. E. Lammerts, January 26, 1962, (eight invitations), Lammerts Papers; H. M. Morris to J. C. Whitcomb, January 1, 1963 (Cuvierianism), Whitcomb Papers; Morris, *History of Modern Creationism*, p. 176 (reply to Tinkle). For Kessel's views, see Edward Luther Kessel, "Let's Look at Facts, without Bent or Bias," in *The Evidence of God in an Expanding Universe*, ed. John Clover Monsma (New York: G. P. Putnam's Sons, 1958), pp. 49–54. For biographical information, see "From Curiosity to Curating," *Search*, no. 9 (1990): 1–4.

24. W. E. Lammerts to W. J. Tinkle, March 5, 1962 (committee of 8), and M. Couperus to W. E. Lammerts, May 25, 1962 (reconsiders), both in the Lammerts Papers. Regarding Monsma, see Morris, *History of Modern Creationism*, pp. 134, 180.

25. Morris, *History of Modern Creationism*, p. 180; R. L. Harris to W. E. Lammerts, November 3, 1962, W. E. Lammerts to R. L. Harris, November 12, 1962, and H. M. Morris to W. E. Lammerts, May 13, 1965, all in the Lammerts Papers.

26. Morris, *History of Modern Creationism*, pp. 179–81; D. T. Gish to L. G. Butler, May 5, 1966 (research and Regular Baptist), Butler Papers, courtesy of Larry G. Butler; D. T. Gish to J. C. Whitcomb, November 9, 1966 (Word of God), Whitcomb Papers; interviews with Duane T. Gish, October 26, 1980, and May 10, 1984. Howitt's booklet, *Evolution: Science Falsely So-Called*, published by the International Christian Crusade in Toronto, had gone through sixteen editions by 1965.

27. W. E. Lammerts to R. M. Ritland, December 8, 1962, Ritland Papers; D. T. Gish to J. C. Whitcomb, November 9, 1966, Whitcomb Papers. On Gish and Howitt, see interview with Henry M. Morris, May 10, 1984.

28. W. E. Lammerts to G. M. Price, September 24, 1962 (physicist and geologist), H. M. Morris to W. E. Lammerts, November 23, 1962 (Gilmer and Grebe), W. E. Lammerts to J. J. Grebe, October 30, 1963 (ancient earth), all in the Lammerts Papers. For biographical data, see Karl W. Linsenmann, "John J. Grebe," *Creation Reseach Society Quarterly* 8 (1971–72): 3–4. On parallels with the Camellia Research Advisory Committee, see W. E. Lammerts to R. M. Ritland, December 8, 1962, Ritland Papers. On November 12, 1990, the registrar's office at Case Western Reserve University confirmed that Grebe's doctorate was honorary.

29. W. E. Lammerts to W. J. Tinkle, February 15, 1962 (Hare), W. E. Lammerts to G. M. Price, October 15, 1962 (Burdick, Ritland, and McGuirt), W. E. Lammerts to H. M. Morris, December 8, 1962 (McGuirt's resignation), all in the Lammerts Papers.

30. Interview with Richard M. Ritland, May 7, 1973; R. M. Ritland to W. E. Lammerts, December 18, 1962, and W. E. Lammerts to R. M. Ritland, January 12, 1963, both in the Ritland Papers. In the Ritland Papers is an undated list (ca. late 1962) of the nine members of the Creation Research Advisory Committee, with Ritland, without his authorization, included as one of the nine.

31. W. E. Lammerts to W. H. Rusch, January 30, 1963 (represent geology), Lammerts Papers. For biographical information, see W. H. Rusch to W. E. Lammerts, January 15, 1963, Lammerts Papers; W. H. Rusch to N. A. Rupke, September 6, 1966, Rupke Papers, courtesy of Nicolaas A. Rupke; and Wayne Frair, "Dedication to Wilbert Henry Rusch, Sr.," *Creation Research Society Quarterly* 18 (1981–82): 3. For the results of the Rock Symposium, see Paul A. Zimmerman, ed., *Rock Strata and the Bible Record* (St. Louis: Concordia Publishing House, 1970). For an early statement of Rusch's views on evolution, see his "Darwinism, Science, and the Bible," in *Darwin, Evolution, and Creation*, ed. Paul A. Zimmerman (St. Louis: Concordia Publishing House, 1959), pp. 4–35.

32. W. E. Lammerts to W. J. Tinkle, February 9, 1963, W. E. Lammerts to F. L. Marsh, May 9, 1963 (secretary), both in the Lammerts Papers; W. E. Lammerts to Wayne Frair, June 14, 1974 (Germain's), Creation Research Society Papers, Concordia Historical Institute. Copies of the new letterhead can be found in the Lammerts Papers.

33. W. E. Lammerts to W. J. Tinkle, February 9, 1963 (society), and W. E. Lammerts to F. L. Marsh, February 13, 1963 (one hundred members), both in the Lammerts Papers; Morris, *History of Modern Creationism*, p. 187 (50 members).

34. *Program for the Fifth Biennial Joint Meeting of the Evangelical Theological Society and the American Scientific Affiliation, June 19–21, 1963*, copy courtesy of Walter Hearn; W. E. Lammerts to H. M. Morris, June 3, 1963, Lammerts Papers; Morris, *History of Modern Creationism*, pp. 181–2.

35. Morris, *History of Modern Creationism*, pp. 181–3, 186; interview with Walter E. Lammerts, January 17, 1983 (dictatorial). Goals and membership requirements appear on the inside front cover of the *Creation Research Society Annual* (1964).

36. H. M. Morris to W. E. Lammerts, March 9, 1963 (two items), W. E. Lammerts to H. M. Morris, March 16, 1963 (Gish and Tinkle), W. E. Lammerts to W. J. Tinkle, February 9, 1963 (Harris), W. E. Lammerts to F. L. Marsh, December 30, 1962 (not critical), and February 13, 1963 (universe), J. W. Klotz to W. E. Lammerts, March 18, 1963 (criticism), all in the Lammerts Papers; Morris, *History of Modern Creationism*, p. 187 (Rusch). In an interview on January 17, 1983, Lam-

merts affirmed his belief that the Noachian flood accounted for all of the geological strata.

37. Morris, *History of Modern Creationism*, p. 192. The statement of belief appears on the inside front cover of *Creation Research Society Annual* (1964).

38. Morris, *History of Modern Creationism*, pp. 184–5; J. N. Moore to J. D. Bales, July 15, 1964 (re-birth), Bales Papers, courtesy of James D. Bales; J. N. Moore to W. E. Lammerts, April 8, 1963 (research), and June 3, 1963 (interests), W. H. Rusch to W. E. Lammerts, November 23, 1963 (Moore's anticommunism and Warriner's alleged racism), D. A. Warriner to W. E. Lammerts, May 9, 1963 (background), all in the Lammerts Papers. Rusch declined to be quoted directly.

39. Morris, *History of Modern Creationism*, pp. 183–5; H. S. Slusher to G. M. Price, April 2, 1963 (texts), Price Papers; H. M. Morris to J. C. Whitcomb, February 5, 1963 (Slusher's colleagues), Whitcomb Papers. See also Harold S. Slusher, "Thomas G. Barnes," *Creation Research Society Quarterly* 9 (1972–73): 3–4.

40. Morris, *History of Modern Creationism*, pp. 184–6; H. M. Morris to J. C. Whitcomb, July 23, 1963 (Burdick), Whitcomb Papers.

41. Names, academic fields, and institutional affiliations are listed on the inside front cover of the *Creation Research Society Annual* (1964); for additional information I am indebted to Duane T. Gish, John N. Moore, Henry M. Morris, Harold S. Slusher, William J. Tinkle, and Mrs. Willis Webb. The 5-percent estimate, which strikes me as a high figure but which some creationists regard as low, appears in [Walter Lang], "Editorial Comments," *Bible-Science Newsletter* 16 (June 1978): 2.

42. W. H. Rusch to CRS Board Members, December 7, 1966, Lammerts Papers; Minutes of the CRS Board Meeting, April 27–28, 1973, CRS Papers.

43. W. E. Lammerts to W. H. Rusch, November 27, 1963 (Warriner); W. E. Lammerts to W. J. Tinkle, November 3, 1965 (Warriner), and W. E. Lammerts to H. M. Morris, May 11, 1965 (Harris), all in the Lammerts Papers. For Harris's views, see R. Laird Harris, *Man—God's Eternal Creation: Old Testament Teaching on Man and His Culture* (Chicago: Moody Press, 1971), p. 47.

44. Morris, *History of Modern Creationism*, pp. 187–8; G. F. Howe to W. J. Tinkle, November 12, 1963 (creation and flood), W. E. Lammerts to J. N. Moore, July 16, 1968 (Dean), Bolton Davidheiser to W. E. Lammerts, April 10, 1969 (SDAs), all in the Lammerts Papers. For biographical data, see Bolton Davidheiser to R. L. Numbers, May 30 and July 6, 1984. Davidheiser discussed his denominational wandering in a telephone interview on February 16, 1991. For Davidheiser's views, see Bolton Davidheiser, *Evolution and Christian Faith* (Philadelphia: Presbyterian and Reformed Publishing Co., 1969). For Howe's early views, see S. Maxwell Coder and George F. Howe, *The Bible, Science and Creation* (Chicago: Moody Press, 1965). The fourth addition to the board was Richard G. Korthals, a Lutheran engineer.

45. F. L. Marsh to H. M. Morris, June 2, 1969, Lammerts Papers.

46. W. E. Lammerts to W. H. Rusch, November 27, 1963 (silencing Grebe), and W. E. Lammerts to J. C. Whitcomb, August 16, 1965 (Rusch's toleration), Lammerts Papers; W. E. Lammerts to J. C. Whitcomb, December 5, 1964 (commitment to flood geology), Whitcomb Papers. For Marsh's beliefs, see Frank Lewis Marsh, Letter to the Editor, *Creation Research Society Quarterly* 22 (1985–86): 47–8.

47. W. E. Lammerts to H. M. Morris, January 25, 1967 (chairman), Lammerts Papers; interview with Walter E. Lammerts, January 17, 1983 (thermodynamics junk); Henry M. Morris, Letter to the Editor, *Creation Research Society Quarterly* 11 (1974–75): 173–5. The thermodynamics question continued to divide the society for years; see, e.g., Robert E. Kofahl, "Entropy Prior to the Fall," *Creation Research Society Quarterly* 10 (1973–74): 154–6; Emmett L. Williams, "Response to Robert

Kofahl," ibid., pp. 156–7; Henry M. Morris, "Another Reply to Robert Kofahl," ibid., p. 157; Robert E. Kofahl, Letter to the Editor, ibid. 11 (1974–75): 175–7. The creationist A. E. Wilder-Smith alleged that Morris and his like-minded colleagues didn't "know a thing about thermodynamics": partial transcript of a conversation between F. J. Arduini and A. E. Wilder-Smith, June 10, 1984, Evolution/Creation Archives, Department of Special Collections, Iowa State University Library.

48. Walter E. Lammerts, Letter to the Editor, *Creation Research Society Quarterly* 12 (1975–76): 75–7.

49. J. R. Howitt to A. C. Custance, October 22, 1962 (fundamentalist), December 24, 1965 (ASA), March 21, 1970 (Pugwash), January 20, 1972 (earthquake), August 25, 1973 (dogmatic), all in the A. C. Custance Papers, Special Collections, Redeemer College; J. R. Howitt to W. E. Lammerts, [late 1963] (research society), Lammerts Papers. For Howitt's defense of the gap theory, see [John R. Howitt], *A Biblical Cosmology* (Toronto: International Christian Crusade, 1976), a copy of which can be found in the Custance Papers.

50. D. E. Wonderly to R. L. Numbers, September 26, 1990; H. M. Morris to CRS Board of Directors, August 3, 1973, Lammerts Papers; Daniel E. Wonderly, *God's Time-Records in Ancient Sediments* (Flint, MI: Crystal Press, 1977). See also Daniel E. Wonderly, *Neglect of Geologic Data: Sedimentary Strata Compared with Young-Earth Creationist Writings* (Hatfield, PA: Interdisciplinary Biblical Research Institute, 1987). For the president's and Whitcomb's views of this incident, see Herman A. Hoyt to J. C. Whitcomb, May 22, 1990, Whitcomb Papers; and J. C. Whitcomb to R. L. Numbers, May 27, 1991.

51. H. M. Morris and Others to Alert Americans Association, December 22, 1967 (right-wing), Minutes of the CRS Board Meeting, May 23–24, 1969 (teaching creation), Minutes of the CRS Board Meeting, April 27–28 (federal funds), all in the CRS Papers; William J. Tinkle, "Our Position: We Are Just Ourselves," *Creation Research Society Quarterly* 6 (1969–70): 79; Emmett L. Williams, "Editorial Comment," ibid. 24 (1987–88): 165.

52. H. M. Morris to J. C. Whitcomb, August 18, 1964 (SDAs), Whitcomb Papers; W. E. Lammerts to W. Lang, April 13, 1964 (too Lutheran), and August 26, 1963 (scientists only), W. Lang to W. E. Lammerts, August 18 and September 2, 1963 (publicity), all in the Lammerts Papers; [Walter Lang], "Editorial Comments," *Bible-Science Newsletter* 15 (March 1977): 2–3 (objective). On Lang's involvement in the creationist movement, see Walter and Valeria Lang, *Two Decades of Creationism* (Minneapolis: Bible-Science Assn., 1984).

53. W. H. Rusch to W. E. Lammerts, December 29, 1964 (errors), W. E. Lammerts to H. M. Morris, March 8, 1967 (drivel), W. E. Lammerts to Walter van der Kamp, September 8, 1967 (geocentrism), all in the Lammerts Papers; W. H. Rusch to J. C. Whitcomb, January 4, 1966 (unscholarly), Whitcomb Papers.

54. "Statement of Relation between the Creation Research Society and the Bible-Science Association, Inc.," *Creation Research Society Quarterly* 3 (July 1966): 20–1; James N. Hanson, Letter to the Editor, ibid. 16 (1979–80): 83 (geocentrism); W. E. Lammerts to H. M. Morris, June 1, 1967 (completely separate), H. M. Morris to W. E. Lammerts, June 12 (work together) and September 19 (research v. promotion), 1967, all in the Lammerts Papers; telephone interview with James N. Hanson, June 5, 1991. On geocentrism in the *Creation Research Society Quarterly*, see also James N. Hanson, "Against Catastrophic Rationalism: Gravitational Attitude Deflections of the Earth's Axis," ibid. 15 (1978–79): 55–68; and James N. Hanson, "The Sun's Luminosity and Age," ibid. 18 (1981–82): 27–9. On Hanson's identity, see "Convention: Research Reports," *Bible-Science Newsletter* 14 (September 1976): 5.

55. On the textbook controversies, see Edward J. Larson, *Trial and Error: The American Controversy over Creation and Evolution* (New York: Oxford University Press, 1985), esp. p. 86 (quoting H. J. Muller on "one hundred years without Darwinism") and p. 96 (neutrality based on silence); Dorothy Nelkin, *The Creation Controversy: Science or Scripture in the Schools* (New York: W. W. Norton, 1982), esp. p. 46 (quoting the BSCS teachers' handbook regarding the "warp and woof"); and Judith V. Grabiner and Peter D. Miller, "Effects of the Scopes Trial: Was It a Victory for Evolutionists?" *Science* 185 (1974): 832–7. On the BSCS, see Arnold B. Grobman, *The Changing Classroom: The Role of the Biological Sciences Curriculum Study* (Garden City, NY: Doubleday, 1969); and Bentley Glass and Arnold B. Grobman, foreword to *Biological Science: Molecules to Man* (Boston: Houghton Mifflin, 1963), pp. vii–viii. See also Gerald Skoog, "Topic of Evolution in Secondary School Biology Textbooks: 1900–1977," *Science Education* 63 (1979): 621–40.

56. William J. Tinkle, "Formation of the Creation Research Society," *Naturalist* 26 (Spring 1966): 31 (ram evolution); Rita Rhodes Ward, "A Critique of the BSCS Biology Books," *Creation Research Society Quarterly* 2 (October 1965): 5–8; W. E. Lammerts to T. G. Barnes, February 24, 1965, Lammerts Papers.

57. Much of this paragraph is based on telephone interviews with Thomas G. Barnes, November 20, 1990, and Rita Rhodes Ward, November 24, 1990. See also Morris, *History of Modern Creationism*, p. 190; and Douglas Dean, "Address before the Texas Textbook Committee, Austin, Texas, October 14, 1964," *Creation Research Society Quarterly* 2 (January 1966): 9–12. Examples of unacceptable material in Ernest S. Booth, *Biology: The Story of Life*, rev. ed. (Mountain View, CA: Pacific Press, 1954), appear on pp. 478–9.

58. W. E. Lammerts to T. G. Barnes, February 24, 1965 (committee), and W. E. Lammerts to CRS Members, June 1, 1966 ($10,000), both in the Lammerts Papers; telephone interview with Rita Rhodes Ward, November 24, 1990. A list of the contributors to John N. Moore and Harold Schultz Slusher, eds., *Biology: A Search for Order in Complexity* (Grand Rapids, MI: Zondervan, 1970), appears on pp. xvi–xvii.

59. T. G. Barnes to W. E. Lammerts, May 1, 1965 (Dean and publishers), Lammerts Papers; Morris, *History of Modern Creationism*, pp. 194–5.

60. Minutes of CRS Board Meeting, April 16–17, 1971 (sales), and Minutes of CRS Textbook Revision Committee, December 17, 1977 (revisions), both in the CRS Papers; Larson, *Trial and Error*, p. 145 (quoting Indiana judge); Morris, *History of Modern Creationism*, p. 199 (Indiana ban); Larry G. Butler to George Howe, November 17, 1971 ($6,500), Butler Papers, courtesy of Larry G. Butler.

Twelve. CREATION SCIENCE AND SCIENTIFIC CREATIONISM

1. Henry M. Morris, preface to *Biology: A Search for Order in Complexity*, ed. John N. Moore and Harold Schultz Slusher (Grand Rapids, MI: Zondervan, 1970), pp. xix–xxi.

2. John N. Moore, preface to *Biology: A Search for Order in Complexity*, ed. John N. Moore and Harold Schultz Slusher (Grand Rapids, MI: Zondervan, 1974), pp. xvii–xxiii (3rd printing).

3. Henry M. Morris, ed., *Scientific Creationism*, public school ed. (San Diego: Creation-Life Publishers, 1974), pp. 8–16; Henry M. Morris, "Director's Column," *Acts & Facts* 3 (September 1974): 2 (open the door).

4. Vernon Lee Bates, "Christian Fundamentalism and the Theory of Evolution in

Public School Education: A Study of the Creation Science Movement" (Ph.D. dissertation, University of California, Davis, 1976), p. 98.

5. Edward J. Larson, *Trial and Error: The American Controversy over Creation and Evolution* (New York: Oxford University Press, 1985), pp. 94–6. The court first struck down prayers in public schools in 1962. When she remarried in 1965, Murray took the name O'Hair.

6. Ibid., pp. 96–103. On the California textbook controversies, see Dorothy Nelkin, *The Creation Controversy: Science or Scripture in the Schools* (New York: W. W. Norton, 1982), pp. 107–20; Gerald W. Wheeler, *The Two-Taled Dinosaur: Why Science and Religion Conflict over the Origin of Life* (Nashville: Southern Publishing Assn., 1975), pp. 143–51; and Nicholas Wade, "Creationists and Evolutionists: Confrontation in California," *Science* 178 (1972): 724–9. Segraves tells her story in an interview reported in "15 Years of Creationism," *Five Minutes with the Bible & Science,* supplement to *Bible-Science Newsletter* 17 (May 1979): 2. Susan Epperson tells hers in Peter Irons, *The Courage of Their Convictions* (New York: Free Press, 1988), pp. 218–30. On the Lammerts-Sumrall relationship, see Jean E. Sumrall to W. E. Lammerts, May 3, 1963, Walter E. Lammerts Papers, Bancroft Library, University of California, Berkeley.

7. Nelkin, *The Creation Controversy,* pp. 109–19; telephone interview with Vernon L. Grose, September 30, 1991.

8. Henry M. Morris, *The Troubled Waters of Evolution* (San Diego: Creation-Life Publishers, 1974), p. 16 (on the way back); Henry M. Morris, *A History of Modern Creationism,* (San Diego: Master Book Publishers, 1984), pp. 231–2 (Creation Science, Inc.); H. M. Morris to J. C. Whitcomb, August 27, 1970 (new course), Whitcomb Papers, courtesy of John C. Whitcomb, Jr.; Henry M. Morris, "Proposals for Science Framework Guidelines," *Creation Research Society Quarterly* 8 (1971–72): 147–50 (two models); Henry M. Morris, "Comments from President Morris," ibid., p. 229 (competing hypotheses); Larry G. Butler's handwritten notes on the agenda for the annual CRS Board Meeting, April 21–23, 1972, Butler Papers, courtesy of Larry G. Butler; Henry M. Morris and Gary E. Parker, *What Is Creation Science?* (San Diego: Creation-Life Publishers, 1982), p. xiii (synonymous terms). For a widely used application of the two-model approach, see Richard B. Bliss, *Origins: Two Models: Evolution Creation* (San Diego: Creation-Life Publishers, 1978).

9. "Creationism in Schools: The Decision in McLean versus the Arkansas Board of Education," *Science* 215 (1982): 937 (Arkansas law); Henry M. Morris, ed., *Scientific Creationism,* general ed. (San Diego: Creation-Life Publishers, 1974), p. 252 (real crux); Morris and Parker, *What Is Creation Science?* p. 264 (biblical data).

10. Henry M. Morris, "Director's Column," *Acts & Facts* 4 (October 1975): 3 (only in the Bible); Morris, *Scientific Creationism,* general ed., pp. 203–55.

11. Henry M. Morris, *Creation and Its Critics: Answers to Common Questions and Criticisms on the Creation Movement* (San Diego: Creation-Life Publishers, 1982), pp. 25–6 (young earth); John C. Whitcomb, Review of *What Is Creation Science?* by Henry M. Morris and Gary E. Parker, *Grace Theological Journal* 4 (1983): 289–96.

12. Marshall and Sandra Hall, Letter to the Editor, *Creation Research Society Quarterly* 21 (1984–85): 155–6 (dictionary); Barry Ferst, "What Bible-Scientists Can Learn from Bible-Science," ibid. 20 (1983–84): 119 (paraphernalia). On definitions of science, see also Henry M. Morris, "The New Meaning of Science," *Acts & Facts* 12 (June 1983): 3; Duane T. Gish, *Evolution: The Fossils Say No!* (San Diego: Creation-Life Publishers, 1979), pp. 12–13; Robert E. Kofahl, "Correctly Redefining Distorted Science: A Most Essential Task," *Creation Research Society Quarterly* 23 (1986–87): 112–14; and W. R. Bird, *The Origin of Species Revisited: The Theories of Evolution*

and of Abrupt Appearance, 2 vols. (New York: Philosophical Library, 1989), 2:11–78.

13. Karl Popper, *Unended Quest: An Intellectual Autobiography* (La Salle, IL: Open Court Publishing Co., 1976), pp. 167–80 (metaphysical research programme); Karl Popper, Letter to the Editor, *New Scientist* 87 (1980): 611 (clarification); Michael A. Cavanaugh, "Scientific Creationism and Rationality," *Nature* 315 (1985): 188 (Popper-chopping). For Popper's ideas on falsification, see Karl R. Popper, *The Logic of Scientific Discovery* (New York: Basic Books, 1959). On Popper's influence, see, e.g., Morris, *Troubled Waters of Evolution,* p. 80; Ariel A. Roth, "Does Evolution Qualify as a Scientific Principle?" *Origins* 4 (1977): 4–10; Jerry Bergman, "What Is Science?" *Creation Research Society Quarterly* 20 (1983–84): 39–42; Robert L. Kofahl, Letter to the Editor, ibid. 24 (1987–88): 44; and Kofahl, "The Hierarchy of Conceptual Levels for Scientific Thought and Research," ibid. 26 (1989–90): 12–14. See also Frank J. Sonleitner, "What Did Karl Popper Really Say about Evolution?" *Creation/Evolution* 6 (Summer 1986): 9–14.

14. Thomas S. Kuhn, *The Structure of Scientific Revolutions* (Chicago: University of Chicago Press, 1962), quotation on pp. 151–2. This passage is quoted in Wheeler, *The Two-Taled Dinosaur,* p. 125, which devotes an entire section to "The Flood Theory Paradigm," pp. 192–210. On Kuhn's influence, see, e.g., Leonard R. Brand, "A Philosophic Rationale for a Creation-Flood Model," *Origins* 1 (1974): 73–83; Ariel A. Roth, "The Pervasiveness of the Paradigm," ibid. 2 (1975): 55–7; and John W. Klotz, *Studies in Creation: A General Introduction to the Creation/Evolution Debate* (St. Louis: Concordia Publishing House, 1985) pp. 20–2.

15. *Science and Creationism: A View from the National Academy of Sciences* (Washington, DC: National Academy Press, 1984), pp. 6, 8–11. On the historical entanglement of science and religion, see David C. Lindberg and Ronald L. Numbers, eds., *God and Nature: Historical Essays on the Encounter between Christianity and Science* (Berkeley and Los Angeles: University of California Press, 1986).

16. Stanley D. Beck, "Natural Science and Creationist Theology," *BioScience* 32 (1982): 740 (entomologist); Steven Schafersman, "Creationist Disputed," *Geotimes* 26 (August 1981): 11 (geologist); Morris, *History of Modern Creationism,* pp. 22–4, responding to Beck and Schafersman.

17. A. David Kline, "Theories, Facts, and Gods: Philosophical Aspects of the Creation-Evolution Controversy," in *Did the Devil Make Darwin Do It? Modern Perspectives on the Creation-Evolution Controversy,* ed. David B. Wilson (Ames: Iowa State University Press, 1983), p. 42 (philosopher); Robert H. Chapman, "The Evolution of Life," ibid., p. 113 (botanist); *Science and Creationism,* pp. 7, 11; Stephen Jay Gould, *Hen's Teeth and Horse's Toes* (New York: W. W. Norton, 1983), pp. 256, 384–5; Gould, "Creationism: Genesis vs. Geology," in *Science and Creationism,* ed. Ashley Montagu (New York: Oxford University Press, 1984), quotation on p. 129. Philip L. Quinn accuses Gould of inconsistency in "The Philosopher of Science as Expert Witness," in *Science and Reality: Recent Work in the Philosophy of Science,* ed. James T. Cushing, C. F. Delaney, and Gary M. Gutting (Notre Dame, IN: University of Notre Dame Press, 1984), p. 43.

18. Henry M. Morris, "Director's Column," *Acts & Facts* 6 (April 1977): 3–4; Russell H. Leitch, "Mistakes Creationists Make," *Bible-Science Newsletter* 18 (March 1980): 2.

19. "Finding: Let Kids Decide How We Got Here," *American School Board Journal* 167 (March 1980): 52; Larson, *Trial and Error,* updated ed. (New York: Oxford University Press, 1989), p. 180, quoting William J. Brennan.

20. "Creationism in Schools," pp. 938–9, 941 (Overton's opinion). Michael Ruse describes his courtroom experience in "Creation-Science Is Not Science," in *Creation-*

ism, Science, and the Law: The Arkansas Case, ed. Marcel Chotkowski La Follette (Cambridge, MA: MIT Press, 1983), pp. 150–60.

21. Larry Laudan, "Commentary on Ruse: Science at the Bar—Causes for Concern," in La Follette, *Creationism, Science, and the Law,* pp. 161–6; Laudan, "The Demise of the Demarcation Problem," in *The Demarcation between Science and Pseudo-Science,* ed. Rachel Laudan, *Working Papers of the Virginia Tech Center for the Study of Science in Society,* vol. 2, no. 1 (April 1983), pp. 29 (pseudo-problem), 31 (unconscionable behavior); Michael Ruse, "Response to Laudan's Commentary: Pro Judice," in La Follette, *Creationism, Science, and the Law,* p. 168. See also Larry Laudan, "More on Creationism," *Science, Technology, & Human Values* 8 (Winter 1983): 36–8. Quinn offers a similar critique of Ruse and Overton in "The Philosopher of Science as Expert Witness," in *Science and Reality,* pp. 32–53. On the inadequacies of Popperian falsification, see also Philip Kitcher, *Abusing Science: The Case against Creationism* (Cambridge, MA: MIT Press, 1982), pp. 42–4. For a philosophical treatment sympathetic to the scientific claims of creationism, see J. P. Moreland, *Christianity and the Nature of Science: A Philosophical Investigation* (Grand Rapids, MI: Baker Book House, 1989).

22. Thomas F. Gieryn, George M. Bevins, and Stephen C. Zehr, "Professionalization of American Scientists: Public Science in the Creation/Evolution Trials," *American Sociological Review* 50 (1985): 392–409. For a more general discussion of the demarcation problem, see Gieryn, "Boundary-Work and the Demarcation of Science from Non-Science: Strains and Interests in Professional Ideologies of Scientists," ibid. 48 (1983): 781–95.

23. "Creationism in Schools," p. 934 (quoting Overton); Langdon Gilkey, "Fundamentalism and Science," unpublished paper presented at Fundamentalism Observed, a public conference of the Fundamentalism Project, University of Chicago, November 17, 1988 (partnerships). See also Langdon Gilkey, *Creationism on Trial: Evolution and God at Little Rock* (Minneapolis: Winston Press, 1985).

24. W. E. Lammerts to H. H. Hartzler, December 9, 1965, Lammerts Papers (creation research); Morris, *History of Modern Creationism,* pp. 251–4 (types of research); George F. Howe, "Open Letter," *Creation Research Society Quarterly* 5 (1968–69): 63 ($700); L. G. Butler to G. F. Howe, November 17, 1971 (royalties), Butler Papers; Minutes of CRS Board of Directors, April 15–16, 1983 (stations), Creation Research Society Papers, Concordia Historical Institute. See also Duane T. Gish, "A Decade of Creationist Research," *Creation Research Society Quarterly* 12 (1975–76): 34–46; Gish, "More Creationist Research (14 Years)," ibid. 25 (1988–89): 161–70; and "Creation Research Society Research Committee (RC) Activities: A Decade Review including Philosophy and Purpose," ibid. 20 (1983–84): 125–6.

25. Walter E. Lammerts, "Planned Induction of Commercially Desirable Variation in Rose by Neutron Radiation," *Creation Research Society Quarterly* 2 (1965–66): 39–48, quotation on p. 43. See also Lammerts, "Mutations Reveal the Glory of God's Handiwork," ibid. 4 (1967–68): 35–41; and Lammerts, "Does the Science of Genetic and Molecular Biology Really Give Evidence for Evolution?" ibid. 6 (1969–70): 5–12.

26. Robert V. Gentry, *Creation's Tiny Mystery* (Knoxville, TN: Earth Science Associates, 1986), quotations on pp. 24 (key), 31 (enigma), and 66 (fingerprint). Besides this semiautobiographical book, the best introduction to Gentry's life and work is Martin Gardner, "Robert Gentry's *Tiny Mystery,*" *Skeptical Inquirer* 13 (Summer 1989): 357–61. See also H. M. Morris to J. C. Whitcomb, April 6, 1964, Whitcomb Papers. Whitcomb and Morris had mentioned pleochroic halos in *The Genesis Flood* (Philadelphia: Presbyterian and Reformed Publishing Co., 1961), pp. 359–62.

27. Gentry, *Creation's Tiny Mystery,* pp. 31–3.

28. Ibid., pp. 38–44; Gardner, "Robert Gentry's *Tiny Mystery*," pp. 357–8. For a partial list of Gentry's publications, see Gentry, *Creation's Tiny Mystery*, pp. 204–5.

29. Dalrymple is quoted in Gentry, *Creation's Tiny Mystery*, pp. 120–2, 296–8. For a creationist critique of Gentry's test, see R. H. Brown and Others, Review of *Creation's Tiny Mystery*, by R. V. Gentry, *Origins* 15 (1988): 36–7. See also J. Richard Wakefield, "Gentry's Tiny Mystery—Unsupported by Geology," *Creation/Evolution* 8 (Winter 1987–88): 13–33.

30. S. C. Rowland and Others, Review of *Creation's Tiny Mystery*, by R. V. Gentry, *Spectrum* 20 (October 1989): 56–7 (left science); Brown and Others, Review of *Creation's Tiny Mystery*, p. 33 (discrimination). For additional creationist criticism and Gentry's response, see Kurt P. Wise, "Radioactive Halos: Geological Concerns," *Creation Research Society Quarterly* 25 (1988–89): 171–6; Robert V. Gentry, "Response to Wise," ibid., pp. 176–9; R. H. Brown, "Radiohalo Evidence Regarding Change in Natural Process Rates," ibid. 27 (1990–91): 100–2; and Robert V. Gentry, "Critique of 'Radiohalo Evidence Regarding Change in Natural Process Rates,' " ibid., pp. 103–5.

31. G. F. Howe to Robert Whitelaw, February 19, 1972 (zany models), CRS Papers; Donald W. Patten, *The Biblical Flood and the Ice Epoch* (Seattle: Pacific Meridian Publishing Co., 1966). Howe declined to be quoted directly. For evidence of CRS skepticism, see H. M. Morris to W. E. Lammerts, November 9, 1963, Lammerts Papers; and D. T. Gish to D. A. Young, August 8, 1972, Young Papers, courtesy of Davis A. Young.

32. Mark Montie, Letter to the Editor, *Creation Research Society Quarterly* 19 (1982–83): 196 (elegant); Thomas G. Barnes, "Decay of the Earth's Magnetic Moment and the Geochronological Implications," ibid. 8 (1971–72): 24–9; Warren H. Johns, "Controversy over Paleomagnetic Dating," *Ministry* 57 (January 1984): 25–8 (inconsistencies); Morris and Parker, *What Is Creation Science?*, pp. 254–7; Walter Lang, "Fifteen Years of Creationism," *Bible-Science Newsletter* 16 (October 1978: 3 (Nobel Prize). For Barnes's views, see his books: *Origin and Destiny of the Earth's Magnetic Field*, ICR Technical Monograph No. 4 (El Cajon, CA: Institute for Creation Research, 1973); *Physics of the Future: A Classical Unification of Physics* (El Cajon, CA: Institute for Creation Research, 1983); and *Space Medium: The Key to Unified Physics* (El Paso, TX: Geo/Space Research Foundation, 1986). For criticisms of Barnes in addition to Johns (above), see, e.g., Davis A. Young, *Christianity and the Age of the Earth* (Grand Rapids, MI: Zondervan, 1982), pp. 117–24; Stephen G. Brush, "Finding the Age of the Earth: By Physics or by Faith?" *Journal of Geological Education* 30 (1982): 34–58; G. Brent Dalrymple, "Radiometric Dating and the Age of the Earth: A Reply to Scientific Creationism," *Proceedings of the Federation of American Societies for Experimental Biology* 42 (1983): 3033–8; and Dalrymple, "Can the Earth Be Dated from Decay of Its Magnetic Field?" *Journal of Geological Education* 31 (1983): 124–32. For Barnes's response to his critics, see his "Earth's Young Magnetic Age: An Answer to Dalrymple," *Creation Research Society Quarterly* 21 (1984–85): 109–13; and *Origin and Destiny of the Earth's Magnetic Field*, rev. and exp. (El Cajon, CA: Institute for Creation Research, 1983), pp. 113–32.

33. Interview with Larry G. Butler, June 6, 1985; L. G. Butler to K. W. Ehler, January 7, 1973 (country boy), and L. G. Butler to V. L. Bates, January 26, 1975 (overwhelming conviction), both in the Butler Papers; "Report of Board of Directors Meeting," *Creation Research Society Quarterly* 6 (1969–70): 112–13 (election).

34. A. C. Custance to G. F. Howe, September 29, 1969 (proposal), G. F. Howe to L. G. Butler, October 24, 1969 (forwards proposal), and H. M. Morris to L. G. Butler, November 6, 1969 (response to Custance), all in the CRS Papers; Arthur C. Custance,

The Flood: Local or Global? (Grand Rapids, MI: Zondervan, 1971), p. 60 (once for all). For a later attempt to educate Custance regarding upside-down fossils, see also J. N. Moore to A. C. Custance, January 15, 1979, A. C. Custance Papers, Special Collections, Redeemer College.

35. L. G. Butler to H. M. Morris, undated draft of letter [July 1971?] (image of respectability, psychopaths, and fanciful), and Larry G. Butler, "A Critique of Creationist Research," undated MS [January 1972] (sensational), both in the Butler Papers.

36. H. M. Morris to L. G. Butler, June 30, 1971, and February 28, 1972 (presidency), L. G. Butler to Morris, March 2, 1972 (honored), and L. G. Butler to G. F. Howe, November 14, 1972 (disaster), all in the Butler Papers. In 1972 Butler also declined to stand for the editorship of the *Quarterly*; see L. G. Butler to G. F. Howe, November 4 and 14, 1972, Butler Papers.

37. L. G. Butler to V. L. Bates, January 26, 1975 (lunatic fringe), Butler Papers; Butler, "A Critique of Creationist Research."

38. Interviews with Larry G. Butler, June 6 and June 29, 1985; L. G. Butler to J. R. Meyer, May 30, 1985 (amazing similarities), Butler Papers. Butler received congratulatory letters from both Wayne Frair (February 11, 1972) and Wilbert Rusch (March 4, 1972), both in the Butler Papers. The typical creationist's preference for propaganda against evolution over research also frustrated Lammerts; see W. E. Lammerts to E. L. Williams, September 18, 1981, CRS Papers.

Thirteen. DECEPTION AND DISCRIMINATION

1. Raymond A. Eve and Francis B. Harrold, *The Creationist Movement in Modern America* (Boston: Twayne Publishers, 1990), pp. 82–3 (ethos); Michael Arthur Cavanaugh, "A Sociological Account of Scientific Creationism: Science, True Science, Pseudoscience" (Ph.D. dissertation, University of Pittsburgh, 1983), pp. 318–20 (deception); H. G. Coffin to N. A. Rupke, March 21, 1968, Rupke Papers, courtesy of Nicolaas A. Rupke (low grade). For a classic illustration of the honesty issue, with each party questioning the integrity of the other, see the exchanges between Henry M. Morris and the Wheaton physicist Howard H. Claassen, May 11 to August 3, 1978, Cassel Papers, courtesy of J. Frank Cassel.

2. Interview with Larry G. Butler, June 29, 1985 (Burdick); C. L. Burdick to L. G. Butler, [1972], Butler Papers, courtesy of Larry G. Butler (protest); H. M. Morris to CRS Board of Directors, May 14, 1971, Walter E. Lammerts Papers, Bancroft Library, University of California, Berkeley.

3. C. L. Burdick to M. Couperus, May 1, 1955 (defending Price) and October 30, [1956] (courses), both in the Couperus Papers, courtesy of Molleurus Couperus; C. L. Burdick to W. E. Lammerts, September 30, 1963 (background to examination), and January 14, 1964 (true thoughts), both in the Lammerts Papers. On Burdick's status at Arizona, see K. W. Flessa to R. H. Dott, Jr., May 25, 1982, Dott Papers, courtesy of Robert H. Dott, Jr. Burdick's letters to friends abound with requests for money. On his efforts to earn a living, see F. L. Marsh to W. E. Lammerts, November 18, 1963 (handyman), and C. L. Burdick to W. E. Lammerts, April 13, 1966 (consulting), both in the Lammerts Papers. The article that precipitated Burdick's troubles was "Nature Tells the Truth," *Signs of the Times* 84 (October 1957): 12–13.

4. C. L. Burdick to W. E. Lammerts, September 30, 1963 (quotation), Lammerts Papers; Gene Varn, " 'Noah's Ark Suit' Filed against UA," *Tucson Daily Citizen*, July 16, 1970, clipping courtesy of Eugene A. Rehwinkel.

5. "Grand Canyon Presents Problems for Long Ages," *Five Minutes with the Bible & Science*, supplement to *Bible-Science Newsletter* 18 (June 1980): 2 (prejudice); C. L. Burdick to W. E. Lammerts, April 28, [1964] (main reason and second failure), Lammerts Papers. For a recent account of the alleged persecution of Burdick, which says nothing about his poor performance, see Jerry Bergman, *The Criterion: Religious Discrimination in America* (Richfield, MN: Onesimus Publishing, 1984), pp. 29–31.

6. C. L. Burdick to W. E. Lammerts, August 31, [1968] (City of Refuge and Persona Non-Grata), and July 10, [1968] (Ritland); F. L. Marsh to W. E. Lammerts, November 18, 1963; all in the Lammerts Papers.

7. W. E. Lammerts to C. L. Burdick, September 7, 1963 (fantastic), April 8, 1964 (assurance), and May 4, 1964 (forget degree), all in the Lammerts Papers. Typical of the letters sent to the university was W. E. Lammerts to Willard Lacy, April 20, 1964, Lammerts Papers. Burdick mentions Lammerts's offer of assistance in a letter dated September 14, 1963, Lammerts Papers.

8. C. L. Burdick to W. E. Lammerts, January 22, 1965, and February 3, 1965, Lammerts Papers.

9. C. L. Burdick to J. N. Moore, February 5, 1967 (God's hand); W. E. Lammerts to H. M. Morris, March 8, 1967 (degree mill), and March 10, 1967 (driver's license and no harm); C. L. Burdick to R. G. Korthals, April 10, [1969] (state of Arizona); W. E. Lammerts to W. H. Rusch, April 29, 1969 (resolve to investigate); W. E. Lammerts to W. A. Rhodes, May 19, 1969 (trademark); all in the Lammerts Papers. Copies of Burdick's "diploma" from the University of Physical Science, dated May 14, 1966, and the school's "Bulletin of Information" can be found in the Lammerts Papers. On Burdick's persistent desire for a degree from Arizona, see C. L. Burdick to W. E. Lammerts, February 21, 1966, and December 13, [1968], both in the Lammerts Papers.

10. W. E. Lammerts to C. L. Burdick, May 19, 1969 (false pretense), and W. E. Lammerts to H. M. Morris, May 16, 1969 (compromise), both in the Lammerts Papers; Clifford L. Burdick and Harold Slusher, "The Empire Mountains—A Thrust Fault?" *Creation Research Society Annual* 6 (1969–70): 49; "Erratum," *Creation Research Society Quarterly* 6 (1969–70): 108.

11. "Grand Canyon Presents Problems for Long Ages," pp. 1–2; "Pine Pollen in Grand Canyon," *Five Minutes with the Bible & Science*, supplement to *Bible-Science Newsletter* 19 (June 1981): 1–6; Clifford Burdick, "Microflora of the Grand Canyon," *Creation Research Society Annual* 3 (May 1966): 38–50. See also Clifford Burdick, *Canyon of Canyons* (Caldwell, ID: Bible-Science Assn., 1974), pp. 66–74.

12. "Grand Canyon Presents Problems for Long Ages," pp. 1–2; C. L. Burdick to W. E. Lammerts, February 4, [1966] (sacrosanct facts and bury geology), Lammerts Papers; Burdick, "Microflora of the Grand Canyon," pp. 38–50; Walter E. Lammerts, "Editorial Comment," *Creation Research Society Annual* 3 (May 1966): 1.

13. W. E. Lammerts to H. M. Morris, March 8, 1967 (appalled), and March 10, 1967 (proper English), both in the Lammerts Papers.

14. C. L. Burdick to W. E. Lammerts, July 1, [1968] (science-shaking), W. E. Lammerts to W. H. Rusch, April 29, 1969 (lie and hoax), W. E. Lammerts to H. M. Morris, June 4, 1971 (sloppy but honest), all in the Lammerts Papers; Larry G. Butler, "Research Proposal," *Creation Research Society Quarterly* 6 (1969–70): 159, 201 (replication); Clifford L. Burdick, "Progress Report on Grand Canyon Palynology," ibid. 9 (1972–73): 25–30; Arthur V. Chadwick, Philip DeBord, and Lanny H. Fisk, "Grand Canyon Palynology—A Reply," ibid., p. 238 (false claims); Arthur V. Chadwick, "Precambrian Pollen in the Grand Canyon—A Reexamination," *Origins* 8 (1981): 7–12 (landmark); George F. Howe, "Creation Research Society Studies on Precambrian

Pollen—Part I: A Review," *Creation Research Society Quarterly* 23 (1986–87): 99–104; Walter E. Lammerts and George F. Howe, "Creation Research Society Studies on Precambrian Pollen—Part II: Experiments on Atmospheric Pollen Contamination of Microscope Slides," ibid., pp. 151–3; George F. Howe and Others, "Creation Research Society Studies on Precambrian Pollen—Part III: A Pollen Analysis of Hakatai Shale and Other Grand Canyon Rocks," ibid. 24 (1987–88): 173–82. At one point criticism of Burdick became so intense Morris called for an apology; see H. M. Morris to CRS Board of Directors, May 14, 1971, Lammerts Papers. On the fossil-pollen affair, see also Allen M. Solomon and Ralph A. Morgan, Letter to the Editor, *Geotimes* 18 (June 1973): 9–10; and C. L. Burdick, Letter to the Editor, ibid. 18 (October, 1973): 15–16.

15. C. L. Burdick to W. E. Lammerts, August 31, [1968], Lammerts Papers.

16. W. H. Rusch to J. C. Whitcomb, December 21, 1970, Whitcomb Papers, courtesy of John C. Whitcomb, Jr.; W. E. Lammerts to H. M. Morris, June 4, 1971, Lammerts Papers. In a letter to Larry Butler (April 27, 1970, Butler Papers) Wayne Frair also expressed concern about the reliability of Burdick's footprint studies.

17. Clifford L. Burdick, "Discovery of Human Skeletons in Cretaceous Formation," *Creation Research Society Quarterly* 10 (1973–74): 109–10, which includes Lammerts's note of caution; C. L. Burdick to G. F. Howe, November 18, 1971, and L. G. Butler to G. F. Howe, November 27, 1971, both in the Creation Research Society Papers, Concordia Historical Institute. Duane T. Gish gave a two-paragraph summary of his and Rusch's findings in "A Decade of Creationist Research," *Creation Research Society Quarterly* 12 (1975–76): 36.

18. Berney Neufeld, "Dinosaur Tracks and Giant Men," *Origins* 2 (1975): 64–76.

19. John D. Morris, "The Paluxy Mystery," *Impact*, supplement to *Acts & Facts* 15 (January 1986): i–iv; H. M. Morris to Friends of the ICR, January 1986, in the author's personal collection. See also John R. Cole and Laurie R. Godfrey, eds., "The Paluxy River Footprint Mystery—Solved," a special issue of *Creation/Evolution* 5, no. 1 (1985); and Ronnie J. Hastings, "The Rise and Fall of the Paluxy Mantracks," *Perspectives on Science and Christian Faith* 40 (1988): 144–55. For a recent discussion of human and dinosaur tracks, indicative of a lack of consensus, see Paul O. Rosnau and Others, "Are Human and Mammal Tracks Found Together with the Tracks of Dinosaurs in the Kayenta of Arizona?" *Creation Research Society Quarterly* 26 (1989–90): 41–8, 77–98.

20. Henry M. Morris, *A History of Modern Creationism* (San Diego: Master Book Publishers, 1984), p. 251; L. G. Butler to H. M. Morris, undated draft of a letter [July 1971?], Butler Papers. On thrust faulting, see, e.g., Clifford L. Burdick and Harold Slusher, "The Empire Mountains—A Thrust Fault?" *Creation Research Society Annual* 6 (1969–70): 49–54; and Clifford L. Burdick, "The Lewis Overthrust," *Creation Research Society Quarterly* 6 (1969–70): 96–106.

21. Wilbert H. Rusch, Sr., "A Brief Statement of the History and Aims of the CRS," unpublished MS dated June 7, 1983, Rusch Papers, courtesy of Wilbert H. Rusch, Sr. (confidential list); interview with Duane Gish, October 26, 1980 (99 percent); interview with Steven A. Austin, January 6, 1981 (pseudonyms).

22. Bergman, *The Criterion*, p. 50 (citing a 1982 Gallup survey); Kendrick Frazier, "Competency and Controversy: Issues and Ethics on the University/Pseudoscience Battlefield," *Skeptical Inquirer* 8 (Fall 1983): 2–5 (Patterson). See also J. W. Patterson to Bernard White, October 25, 1981, Evolution/Creation Archive, Department of Special Collections, Iowa State University Library.

23. Morris, *History of Modern Creationism*, pp. 164–5, 184–5, 202, 242; interview with James M. Wiggert, November 20, 1986; interview with R. D. Walker, November

20, 1986. See also Henry M. Morris, *The Twilight of Evolution* (Grand Rapids, MI: Baker Book House, 1963), pp. 27–8.

24. Bergman, *The Criterion*, p. xi.

25. Ibid., p. 11; B. C. Nelson to J. C. Whitcomb, August 22, 1967, Whitcomb Papers. I am grateful to Paul Nelson for identifying Boyden.

26. Bergman, *The Criterion*, pp. 6–7, 29–31; D. A. Warriner to J. D. Bales, April 14, 1973, Bales Papers, courtesy of James D. Bales; W. H. Rusch to W. E. Lammerts, November 23, 1963, and W. E. Lammerts to W. H. Rusch, December 12, 1963, both in the Lammerts Papers. On the controversy surrounding Warriner's colleague John N. Moore, see Henry P. Zuidema, "Teaching Scientific Creationism on Campus: Is the Controversy Cooling?" *Liberty* (January–February 1985): 7–9.

27. W. E. Lammerts to W. H. Rusch, December 12, 1963, Lammerts Papers; W. H. Rusch to N. A. Rupke, September 6, 1966, Rupke Papers.

28. Bergman, *The Criterion*, p. 18; interview with Ervil D. Clark, January 9, 1981. I am indebted to Terry Trivett for details about Clark's life.

29. "Comments and Correspondence Relating to the Ph.D. Committee within the Department of Anthropology, U. of T., 1954, Mar. to May"; A. C. Custance to J. O. Buswell III, April 1, 1954; both in the A. C. Custance Papers, Special Collections, Redeemer College. The Custance Papers contain a file labeled "Univ. of Toronto," which documents Custance's experience at the University of Toronto.

30. Arthur C. Custance, "Does Science Transcend Culture?" (Ph.D. dissertation, University of Ottawa, 1958), copy in the Custance Papers. Although the dissertation is dated 1958, apparently his degree was awarded the next year. The quotation about a Ph.D. in physiology appears on the inside back cover of Arthur C. Custance, *Two Men Called Adam: A Fresh Look at the Creation/Evolution Controversy from a Different Point of View—the Theological* (Brockville, Ontario: Doorway Publications, 1983).

31. Bergman, *The Criterion*, p. 15; "Dismissal of Mr. D. C. C. Watson," *Creation: The Journal of the E.P.M.* 2 (May 1977): 8 (Genesis is myth); interview with A. J. Monty White, October 4–5, 1984 (trouble in the classroom); "How Lining Up with Moses Cost Mr. Watson His Job," *Times Educational Supplement*, February 18, 1977, p. 5 (rubbish). A second British creationist expressed similar opinions but asked not to be identified.

32. M[alcolm] Bowden, *The Rise of the Evolution Fraud* (Bromley, Kent: Sovereign Publications, 1982), p. 132; A. J. Jones to R. L. Numbers, November 15, 1984.

33. "Creationism in Schools: The Decision in McLean versus the Arkansas Board of Education," *Science* 215 (1982): 939; Eugenie C. Scott and Henry P. Cole, "The Elusive Scientific Basis of Creation 'Science,' " *Quarterly Review of Biology* 60 (1985): 21–30. On the publication practices of creationists, see Henry P. Cole and Eugenie C. Scott. "Creation-Science and Scientific Research," *Phi Delta Kappan* 63 (1981–82): 557–8, 574.

34. J. C. Whitcomb to H. M. Morris, October 8, 1955, Whitcomb Papers; D. J. Whitney to G. M. Price, January 6, 1955, Marsh Papers, courtesy of Frank Lewis Marsh.

35. H. M. Morris to J. C. Whitcomb, August 9, 1958, Whitcomb Papers.

36. H. M. Morris to W. E. Lammerts, July 12, 1963, Lammerts Papers.

37. A. H. Meyer to Theodore Graebner, March 2, 8, and 31, 1923, Box 3, Theodore Graebner Papers, Concordia Historical Institute. See also Graebner's advice to Meyer in letters dated March 6 and 26, 1923, Box 3, Graebner Papers.

38. A. H. Meyer to Theodore Graebner, October 11, 1938, and Theodore Graebner to A. H. Meyer, November 8, 1938, both in Box 5, Graebner Papers; A. H. Meyer to O. P. Kretzmann, January 8, 1962, Krekeler Papers, courtesy of Carl H. Krekeler.

For a published version of Meyer's dissertation, see Alfred H. Meyer, "The Kankakee 'Marsh' of Northern Indiana and Illinois," *Papers of the Michigan Academy of Science, Arts, and Letters* 21 (1936): 359–96.

39. D. J. Whitney to J. P. Van Haitsma, June 19, 1944; R. L. Mixter to F. A. Everest, March 31, 1944; both in the ASA Papers. See also Edwin K. Gedney, "Geology and the Bible," in *Modern Science and Christian Faith: A Symposium on the Relationship of the Bible to Modern Science*, by Members of the American Scientific Affiliation (Wheaton, IL: Van Kampen Press, 1950), pp. 23–57 (2nd printing). Biographical data regarding Gedney appear ibid., pp. 303–4.

40. D. A. Young to H. M. Morris, December 15, 1964; D. A. Young to J. C. Whitcomb, January 9, 1965; both in the Whitcomb Papers. On Young, see "He's Still a Rockhound," *Search*, no. 7 (1989): 1–4. In the late 1940s Morris thought he had a potential convert in James H. Zumberge, a fellow graduate student at the University of Minnesota who went on to teach geology at the University of Michigan and to serve as president of the University of Southern California; see Henry M. Morris, *King of Creation* (San Diego: C.L.P. Publishers, 1980), p. 163.

41. Interview with Davis A. Young, November 17, 1989; D. A. Young to H. M. Morris, April 16, 1969, Whitcomb Papers. For Eckelmann's views, see F. Donald Eckelmann, "Geology," in *The Encounter between Christianity and Science*, ed. Richard H. Bube (Grand Rapids, MI: William B. Eerdmans, 1968), pp. 135–70.

42. H. M. Morris to D. A. Young, July 6, 1969, Young Papers, courtesy of Davis A. Young.

43. Davis A. Young, *Creation and the Flood: An Alternative to Flood Geology and Theistic Evolution* (Grand Rapids, MI: Baker Book House, 1977), p. 132; Morris, *King of Creation*, pp. 86–95. Young's letter appeared in the February 1972 issue of the *Presbyterian Guardian*; see D. T. Gish to D. A. Young, May 1, 1972, Young Papers. Young explains his concerns in a letter to Gish dated June 2, 1972, Young Papers.

44. Morris, *History of Modern Creationism*, pp. 158–9 (Calvin College); interview with D. A. Young, November 17, 1989; Davis A. Young, "Scripture in the Hands of Geologists," part 2, *Westminster Theological Journal* 49 (1987): 303, quoted by Henry M. Morris, *The Long War against God: The History and Impact of the Creation/Evolution Conflict* (Grand Rapids, MI: Baker Book House, 1989), p. 108 (where Morris also comments on Young's journey). Young's most sustained attacks on flood geology in the 1980s appeared in Davis A. Young, *Christianity and the Age of the Earth* (Grand Rapids, MI: Zondervan, 1982); Howard J. Van Till, Davis A. Young, and Clarence Menninga, *Science Held Hostage: What's Wrong with Creation Science AND Evolution* (Downers Grove, IL: InterVarsity Press, 1988); and Howard J. Van Till, Robert E. Snow, John H. Stek, and Davis A. Young, *Portraits of Creation: Biblical and Scientific Perspectives on the World's Formation* (Grand Rapids, MI: William B. Eerdmans, 1990). For Morris's response to *Christianity and the Age of the Earth*, see Henry M. Morris, *Science, Scripture and the Young Earth* (El Cajon, CA: Institute for Creation Research, 1983).

45. N. A. Rupke to W. E. Lammerts, October 7, 1966, Lammerts Papers; W. E. Lammerts to N. A. Rupke, September 16, 1966, Rupke Papers.

46. N. A. Rupke to R. L. Numbers, January 9, 1986; N. A. Rupke, "Prolegomena to a Study of Cataclysmal Sedimentation," *Creation Research Society Annual* 3 (May 1966): 16–37.

47. W. E. Lammerts to N. A. Rupke, October 31, 1966, Rupke Papers; W. E. Lammerts to H. M. Morris, December 6, 1966, Lammerts Papers.

48. N. A. Rupke to R. L. Numbers, October 19, 1985, January 9, 1986, and April 22, 1991; N. A. Rupke, "Aspects of Bed Thickness in Some Eocene Turbidite Se-

quences, Spanish Pyrenees," *Journal of Geology* 77 (1969): 482–4; H. G. Coffin to
N. A. Rupke, August 15, 1967, and January 11, 1968, Rupke Papers; N. A. Rupke,
"Sedimentary Evidence for the Allochthonous Origin of *Stigmaria*, Carboniferous,
Nova Scotia," *Bulletin of the Geological Society of America* 80 (1969): 2109–14; N. A.
Rupke, "Sedimentary Evidence for the Allochthonous Origin of *Stigmaria*, Carbonif-
erous, Nova Scotia: Reply," ibid. 81 (1970): 2535–38; Harold G. Coffin, "Research on
the Classic Joggins Petrified Trees," *Creation Research Society Annual* 6 (1969–70):
35–44.

49. N. A. Rupke to R. L. Numbers, January 9, 1986; interview with N. A. Rupke,
September 21, 1990. For one reaction to Rupke's loss of faith, see W. E. Lammerts to
H. M. Morris, January 14, 1970, Lammerts Papers. Fittingly, Rupke's first major
contribution to the history of science, *The Great Chain of History: William Buckland
and the English School of Geology, 1814–49* (Oxford: Clarendon Press, 1983), focused
on the reconciliation of Genesis and geology in the early nineteenth century.

50. Interview with Steven A. Austin, January 6, 1981; G. F. Howe to W. E.
Lammerts, September 12, 1970, Lammerts Papers; Stuart E. Nevins, "The Mesa
Basalt of the Northwestern United States," *Creation Research Society Quarterly* 7
(1970–71): 222–6; George F. Howe, "Editorial Comments," ibid. 8 (1971–72): 227–8;
Stuart E. Nevins, "Is the Capitan Limestone a Fossil Reef?" ibid., pp. 231–48; Daniel
E. Wonderly, "Critique of 'Is the Capitan Limestone a Fossil Reef?' by Stuart Nev-
ins," ibid. 10 (1973–74): 237–41; Stuart E. Nevins, "Reply to Critique by Daniel
Wonderly," ibid., pp. 241–4. Austin's first contribution to creationist literature ap-
peared as Stuart E. Nevins, "A Scriptural Groundwork for Historical Geology," in
Symposium on Creation II, by Donald W. Patten and Others (Grand Rapids, MI:
Baker Book House, 1970), pp. 77–101. According to Austin (S. A. Austin to R. L.
Numbers, June 3, 1991) H. Paul Buccheim and Donald W. Lovejoy received doctor-
ates in geology before him; neither participated prominently in creationist circles.

51. S. A. Austin to D. A. Young, October 30, 1972, Young Papers; S. A. Austin to
E. L. Williams, October 14, 1975 (attitude of Penn State faculty), and S. A. Austin to
CRS Research Committee, March 15, 1976 (intellectual development and future
plans), both in the CRS Papers. Austin declined to be quoted directly. The CRS
decided against supporting Austin's graduate studies; S. A. Austin to R. L. Numbers,
June 3, 1991. Kulp's statement appears in his "Deluge Geology," *Journal of the
American Scientific Affiliation* 2 (January 1950): 6. Austin published an outgrowth of
his master's thesis as *Catastrophes in Earth History: A Source Book of Geologic
Evidence, Speculation and Theory* (El Cajon, CA: Institute for Creation Research,
1984).

52. Henry M. Morris and Gary E. Parker, *What Is Creation Science?* (San Diego:
Creation-Life Publishers, 1982), pp. 134–6. The title of Austin's dissertation appears
on p. 277.

53. Interview with Douglas A. Block, June 12, 1991; interview with Kurt P. Wise,
May 29, 1990; "Biographical Information of Kurt Patrick Wise," undated MS courtesy
of Kurt P. Wise; William Schuler, Letter to the Editor, *World* 5, no. 5 (1990): 22. Wise
attempted to harmonize Gould's theory of punctuated equilibria with flood geology in
"Punc Eq Creation Style," *Origins* 16 (1989): 11–24. Block converted to flood geology
through the influence of Walter T. Brown, Jr., who associated the deluge with continental drift; see Brown, *"In the Beginning . . . ,"* 5th ed. (Phoenix, AZ: Center for
Scientific Creation, 1989).

54. Interview with Kurt P. Wise, May 29, 1990; Kurt P. Wise, "The Way Geologists Date!" in *Proceedings of the First International Conference on Creationism,* 2
vols. (Pittsburgh, PA: Creation Science 55 Fellowship, 1986), 1:135–8 (Lewis Over-

thrust); Kurt P. Wise, "Radioactive Halos: Geological Concerns," *Creation Research Society Quarterly* 25 (1988–89): 171–6; Robert V. Gentry, "Response to Wise," ibid., pp. 176–9; Kurt P. Wise, Letter to the Editor, ibid. 24 (1987–88): 212–13 (curt dismissal); Emmett L. Williams, Letter to the Editor, ibid., pp. 213–15 (arrogant).

55. Kurt P. Wise, "Research into Origins," *Bryan Life* 15 (Winter 1990): 5 (research center and model); interview with Kurt P. Wise, May 29, 1990 (*The Genesis Flood*); R. J. Schadewald to Ian Plimer, December 17, 1989, copy courtesy of Marian Finger.

Fourteen. CREATION RESEARCH INSTITUTES

1. Christopher P. Toumey, "The Social Context of Scientific Creationism" (Ph.D. dissertation, University of North Carolina, 1987), p. 317. See also Christopher P. Toumey, "Social Profiles of Anti-Evolutionism in North Carolina in the 1980s," *Journal of the Elisha Mitchell Scientific Society* 106 (1990): 93–117.

2. Henry M. Morris, *A History of Modern Creationism* (San Diego: Master Book Publishers, 1984), pp. 166, 225–34; W. E. Lammerts to H. M. Morris, August 31, 1970, Walter E. Lammerts Papers, Bancroft Library, University of California, Berkeley; interview with Henry M. Morris, May 10, 1984 (manipulated and maneuvered); H. M. Morris to J. C. Whitcomb, August 27, 1970, Whitcomb Papers, courtesy of John C. Whitcomb, Jr.

3. *Introducing the Creation-Science Research Center* (San Diego, [1971]), unpaginated, copy courtesy of Henry M. Morris; *Creation-Science Report* 1 (January–February 1972), unpaginated.

4. H. M. Morris to Members of the Technical Advisory Board of the Institute for Creation Research, April 28, 1972, Butler Papers, courtesy of Larry G. Butler (both quotations); Morris, *History of Modern Creationism*, p. 233. Nell Segraves gave a somewhat different version of the split in an interview on May 11, 1984.

5. "Grand Opening," *Creation-Science Report* (May–June, 1972): unpaginated; "A Five Year Review," ibid. (September 1975): unpaginated; Untitled note, ibid. (April 1976): unpaginated (venereal disease and sex education); "Creation Science, Sex Education, Abortion, and the ERA," ibid, no. 4 (1980): unpaginated; Nell J. Segraves, *The Creation Report* (San Diego: Creation-Science Research Center, 1977), p. 17 (widespread breakdown); Kelly L. Segraves, *Sons of God Return* (New York: Pyramid Books, 1975); Robert E. Kofahl, *Handy Dandy Evolution Refuter*, rev. ed. (San Diego: Beta Books, 1980), quotation on back cover. See also Robert E. Kofahl and Kelly L. Segraves, *The Creation Explanation: A Scientific Alternative to Evolution* (Wheaton, IL: Harold Shaw Publishers, 1975). For Morris's views on UFOs and demons, see Henry M. Morris and Martin E. Clark, *The Bible Has the Answer*, rev. ed. (San Diego: Creation-Life Publishers, 1976), pp. 331–2.

6. Morris, *History of Modern Creationism*, pp. 233–4; Henry M. Morris, "Director's Column," *Acts & Facts* 1 (June–July, 1971): unpaginated (scientists); W. E. Lammerts to H. M. Morris, August 13, 1971, and H. M. Morris to W. E. Lammerts, August 27, 1971 (Gish), both in the Lammerts Papers; Duane T. Gish, *Evolution: The Fossils Say No!*, 3rd ed. (San Diego: Creation-Life Publishers, 1979), p. 64. In a letter to John Whitcomb (April 26, 1965, Whitcomb Papers) Morris himself had described Gish as being "not quite sold on flood geology."

7. D. T. Gish to W. E. Lammerts, March 24, 1978 (dream and hope), and February 26, 1981 (100 scientists), both in the Lammerts Papers; W. E. Lammerts to E. L. Williams, January 20, 1981 (Austin and missionary effort), Creation Research

Society Papers, Concordia Historical Institute; Henry M. Morris, "Two Decades of Creation: Past and Future," *Impact* supplement to *Acts & Facts* 10 (January 1981): i–iv (ministries and activities). On ICR finances, see D. T. Gish to W. E. Lammerts, May 23, 1980, Lammerts Papers; interview with Henry M. Morris, Duane T. Gish, and Harold S. Slusher, January 6, 1981; "Audited Financial Statements, Institute for Creation Research," June 30, 1983, courtesy of Donald H. Rohrer, chief financial officer of the ICR; interview with Donald H. Rohrer, May 11, 1984; Henry M. Morris, "ICR's Financial Policy," *Acts & Facts* 16 (July 1987): 3.

8. Morris, "Two Decades of Creation," pp. ii–iii; "ICR Books Available in Many Languages," *Acts & Facts* 9 (February 1980): 2, 7. For instances of continued opposition to the gap and day-age theories, see Henry M. Morris, ed., *Scientific Creationism*, general ed. (San Diego: Creation-Life Publishers, 1974), pp. 221–43; Henry M. Morris, *The Genesis Record: A Scientific and Devotional Commentary on the Book of Beginnings* (Grand Rapids, MI: Baker Book House, 1976), pp. 46–8, 53–4; and Henry M. Morris, *The Biblical Basis for Modern Science* (Grand Rapids, MI: Baker Book House, 1984), pp. 115–25. For examples of celebrity introductions, see Henry M. Morris, *The Bible Has the Answer* (Nutley, NJ: Craig Press, 1971), LaHaye; Henry M. Morris, *King of Creation* (San Diego: C.L.P. Publishers, 1980), McDowell; Henry M. Morris, *The Revelation Record: A Scientific and Devotional Commentary on the Book of Revelation* (Wheaton, IL: Tyndale House, 1983), Falwell; and Henry M. Morris, *Creation and the Modern Christian* (El Cajon, CA: Master Book, 1985), Ryrie.

9. Morris, "Two Decades of Creation," p. iii (more than 100 winning debates and 600,000 persons); Morris, *History of Modern Creationism*, p. 319 (Miller); Henry M. Morris, "Director's Column," *Acts & Facts* 3 (March 1974): 2 (Christians and lions); interview with Henry M. Morris, Duane T. Gish, and Harold S. Slusher, January 1, 1981 (Darwin and Huxley). I introduced the Darwin-Huxley parallel in "Creationism in 20th-Century America," *Science* 218 (1982): 543. For a sympathetic look at the debates, see Marvin L. Lubenow, *"From Fish to Gish"* (San Diego: C.L.P. Publishers, 1983); on the Morris-Miller debate, see "Debate at Brown University," *Acts & Facts* 10 (July 1981): 2, 4.

10. "ICR Schedules M.S. Programs," *Acts & Facts* 10 (February 1981): 1–2; Institute for Creation Research, *1981–82 Graduate School Catalog*. On Parker, see Gary E. Parker, *From Evolution to Creation: A Personal Testimony* (San Diego: Creation-Life Publishers, 1977). On Bliss, see Henry M. Morris and Gary E. Parker, *What Is Creation Science?* (San Diego: Creation-Life Publishers, 1982), p. 39; and Richard Bliss, "Can Science Teachers Change Their Attitudes toward the Teaching of Origins in the Classroom?" *Wisconsin Society of Science Teachers Newsletter* 13 (January 1972): 1, 5–7. For an earlier ICR-related effort to offer graduate degrees, see "Genesis School of Graduate Studies," *Acts & Facts* 2 (October 1973): 3.

11. "Harold S. Slusher Receives Ph.D.," *Acts & Facts* 11 (April 1982): 4; "Do Colleges Spot Fraudulent Credentials?" *Chronicle of Higher Education*, May 2, 1984, p. 23 (on Columbia Pacific as an unaccredited correspondence school). In an undated form letter (ca. 1984) accompanying the catalog of Columbia Pacific University, James Cloud, director of admissions, claimed, "Many students complete their degree programs in less than a year." During a visit I made to the "campus" of Columbia Pacific University on May 4, 1984, Elizabeth Baker, assistant to the director of academic administration, kindly let me examine Slusher's dissertation and answered questions about his degree. Included as part of the "dissertation" was Slusher's widely cited *Critique of Radiometric Dating*, ICR Technical Monograph No. 2 (San Diego: Institute for Creation Research, 1973), in which he concluded that "when the dating

methods are properly evaluated, we find that they give evidence for a short age of the earth" (p. 42).

12. Telephone interview with Henry M. Morris, February 4, 1991; "Religious Freedom Being Denied at ICR," undated flyer distributed by the ICR chronicling the struggle to obtain state approval; "Victory for Religious Freedom!" *Acts & Facts* 22 (March 1992) : 1–3. Copies of the visiting committee's final report, dated August 3–5, 1988, and Stuart H. Hurlbert's minority dissent, dated August 26, 1988, were kindly provided by Marian Finger. On Honig's role in the affair, see "California Penalizes Creation Institute," *New York Times,* December 8, 1988. For a blow-by-blow account of the ICR's struggle with the state, see William Bennetta's ten-part series, "Degrees of Folly," *BASIS: Bay Area Skeptics Information Sheet* 8 (February 1989) to 9 (March 1990).

13. John D. Morris, "What Is the Connection between Homosexuality and Evolution?" *Back to Genesis* supplement to *Acts & Facts* 19 (May 1990): d.

14. "Ten Thousand Hear Creation Messages," *Acts & Facts* 18 (May 1989): 1; interview with Kurt P. Wise, May 29, 1990; C. P. Toumey, "Sectarian Aspects of American Creationism," *International Journal of Moral and Social Studies* 5 (1990): 121. See also Harold Lindsell, *The Battle for the Bible* (Grand Rapids, MI: Zondervan, 1976), pp. 89–105, 208–9; and Joe Edward Barnhart, *The Southern Baptist Holy War* (Austin, TX: Texas Monthly Press, 1986), esp. pp. 121–4.

15. Morris, *History of Modern Creationism,* p. 305; Lane P. Lester to R. L. Numbers, February 20 and April 24, 1987. See also Toumey, "Sectarian Aspects of American Creationism," pp. 119–23.

16. "The 1990s—The Decade of Creation," *Acts & Facts* 18 (August 1989): 1 (world's largest); John D. Morris to Friends, undated circular letter sent out with the May 1989 issue of *Acts & Facts* (Satanic oppression); "Dr. John Morris Returns as Full-Time Scientist," *Acts & Facts* 13 (June 1984): 1; "Dr. John Morris Appointed Administrative Vice-President," ibid. 18 (March 1989): 2. For a profile of John Morris, see the *Back to Genesis* supplement to *Acts & Facts* 19 (November 1990): c. See also Tim F. LaHaye and John D. Morris, *The Ark on Ararat* (Nashville: Thomas Nelson, 1976).

17. D. T. Gish to W. E. Lammerts, February 26, 1981, Lammerts Papers.

18. Minutes of the General Conference Committee, October 25, 1957, April 24, 1958, and January 21, 1960, all in the Archives of the General Conference of Seventh-day Adventists (hereafter cited as SDA Archives); F. L. Marsh to G. M. Price, November 2, 1958, Marsh Papers, courtesy of Frank L. Marsch (incognito). On the possibility of supporting Burdick, see the minutes of the Committee on Teaching of Geology and Paleontology, July 24, August 19, and October 13, 1958, RG 274, SDA Archives. On the founding of the GRI, see also Richard Hammill, "Fifty Years of Creationism: The Story of an Insider," *Spectrum* 15 (August 1984): 32–45.

19. Interview with Frank L. Marsh, August 30, 1972 (meeting at Marsh's and 1959 trip); Frank L. Marsh and P. Edgar Hare, "Geology List Number 1, September, 1958," Marsh Papers (Kulp); F. L. Marsh to J. C. Whitcomb, March 8, 1959, Whitcomb Papers (mistake).

20. Interview with Frank L. Marsh, August 30, 1972 (brainwashed); F. L. Marsh to G. M. Price, October 6, 1960, and March 14, 1962, George McCready Price Papers, Adventist Heritage Center, Andrews University.

21. Ernest S. Booth, "Geoscience Tour of the Rocky Mountain States," *Naturalist* 20 (Winter 1960), 36 (milestone); H. W. Clark to H. M. Morris, March 12, 1961 (Yellowstone), Whitcomb Papers; H. W. Clark to F. L. Marsh and Others, November 19, 1960 (Tertiary), Ritland Papers, courtesy of Richard M. Ritland. See also Harold

W. Clark, *Fossils, Flood, and Fire* (Escondido, CA: Outdoor Pictures, 1968), pp. 42–3.

22. R. M. Ritland, "Problems and Methods in Earth History," unpublished MS, 1962, Ritland Papers; interview with Richard M. Ritland, November 14, 1983. Among the theologians influencing Ritland was Earle Hilgert, who expressed his views on Genesis in "Theological Dimensions of the Christian Doctrine of Creation," *Spectrum* 1 (Summer 1969), 14–21; and "References to Creation in the Old Testament Other Than in Genesis 1 and 2," in *The Stature of Christ: Essays in Honor of Edward Heppenstall*, ed. Vern Carner and Gary Stanhiser (Loma Linda, CA: Privately printed, 1970), pp. 83–8. Shortly after writing these articles, Hilgert left the Adventist church.

23. F. L. Marsh to P. E. Hare, June 17, 1963 (division of authority and use of senses), Ritland Papers; F. L. Marsh to G. M. Price, December 6, 1962 (single view), Marsh Papers.

24. R. M. Ritland to P. E. Hare, draft of a letter ca. early 1963 (martyr), Ritland Papers; Roy Benton, "Odyssey of an Adventist Creationist," *Spectrum* 15 (August 1984): 49 (White quotations); F. L. Marsh to G. M. Price, November 17, 1958 (Morris as an Adventist), Marsh Papers; R. M. Ritland to R. H. Utt, October 19, 1965 (*The Genesis Flood* and problems with Marsh), Ritland Papers; W. E. Lammerts to F. L. Marsh, May 9, 1963 (base of operations), Lammerts Papers. Ritland described the growing rift in an interview on May 7, 1973. See also interview with Ariel A. Roth, July 2, 1973.

25. P. E. Hare to R. Hammill, February 7, 1963, and R. R. Figuhr to P. E. Hare, February 5, 1964, both in the Hare Papers, courtesy of P. Edgar Hare; interview with P. Edgar Hare, October 22, 1983. See also the extensive interview with Hare in Benton, "Odyssey of an Adventist Creationist."

26. F. L. Marsh, "Life Summary of a Creationist," unpublished MS, December 1, 1968, Marsh Papers. The documents in the SDA Archives relating to Marsh's termination are closed to the public.

27. Hammill, "Fifty Years of Creationism," p. 36 (staff); interview with P. Edgar Hare, October 22, 1983 (Coffin); interview with Richard M. Ritland, November 14, 1983 (Roth); interview with Ariel A. Roth, July 2, 1973. *Emancipated* was one of Ritland's favorite words.

28. Harold G. Coffin, *Creation—Accident or Design?* (Washington: Review and Herald Publishing Assn., 1969), especially pp. 271–2. *Spectrum* published two critical reviews of this book: Ian M. Fraser, "Problems of Creation and Science" 1 (Autumn 1969): 60–6; and Benton M. Stidd, "How Is Earth History Revealed?" 2 (Summer 1970): 87–92. Coffin, with Robert H. Brown, brought out an updated version of the book retitled *Origin by Design* (Washington: Review and Herald Publishing Assn., 1983).

29. Richard M. Ritland, *A Search for Meaning in Nature: A New Look at Creation and Evolution* (Mountain View, CA: Pacific Press, 1970), quotations on pp. 108, 143; interview with Richard M. Ritland, November 14, 1983. Four years earlier Ritland had circulated a bound typescript of his MS: *Meaning in Nature* ([Washington]: Department of Education of the General Conference of Seventh-day Adventists, 1966).

30. F. L. Marsh to W. E. Lammerts, May 2, 1966 (wallop), Lammerts Papers; Hammill, "Fifty Years of Creationism," pp. 39–42. Pierson's statement, which appeared in the *Advent Review and Sabbath Herald* for October 10, 1968, is quoted in Marsh, "Life Summary of a Creationist," p. 23.

31. R. H. Brown, "The Role of the Geoscience Research Institute within the Seventh-day Adventist Church," unpublished MS, March 19, 1979 (GRI mission), source unknown; H. G. Coffin to R. H. Brown, October 18, 1967 (Gentry), and R. H.

Brown to F. L. Marsh, August 29, 1969 (Moses and White), both in RG 274, GF: R. H. Brown, 1962–72, SDA Archives; F. L. Marsh to C. L. Burdick, February 22, 1966 (snag), Burdick Papers, courtesy of Clifford L. Burdick. For a convenient summary of Brown's views, see his two chapters, "Radioactive Time Clocks" and "Radiocarbon Dating" in Coffin, *Creation—Accident or Design?*, pp. 273–316.

32. Interview with Ariel A. Roth, May 13, 1984 (negative comments); Rennie B. Schoepflin, journal kept on the 1986 GRI field trip for Adventist college and university teachers (negative comments), Schoepflin Papers, courtesy of Rennie B. Schoepflin; R. H. Brown, Review of *Scientific Creationism*, ed. Henry M. Morris, *Origins* 4 (1977): 50–2; Berney Neufeld, "Dinosaur Tracks and Giant Men," ibid. 2 (1975): 64–76; Arthur V. Chadwick, "Precambrian Pollen in the Grand Canyon—A Reexamination," ibid. 8 (1981): 7–12; R. H. Brown and Others, Review of *Creation's Tiny Mystery*, by R. V. Genry, ibid. 15 (1988): 32–8. For non-Adventist contributions to the journal, see, e.g., Steven A. Austin, "Rapid Erosion at Mount St. Helens," ibid. 11 (1984): 90–8; and Kurt Wise, "Punc Eq Creation Style," ibid. 16 (1989): 11–24.

33. Hammill, "Fifty Years of Creationism," pp. 42–5; Edward Lugenbeal, "The Conservative Restoration at Geoscience," *Spectrum* 15 (August 1984): 23–31; E. Lugenbeal to W. J. Hackett, October 11, 1976, and E. Lugenbeal to R. H. Brown and Members of the Geoscience Research Institute Board, October 31, 1979, both in RG 274, GF: R. H. Brown, 1978–80, SDA Archives.

34. W. W. Hughes, "Shifts in Adventist Creationism," *Spectrum* 16 (June 1985): 47–51 (comparing the two editions of the *SDA Bible Commentary*); Ervil D. Clark, "Science and Religion Survey: North American Seventh-day Adventists," unpublished, undated MS, courtesy of the late Ervil D. Clark; Molleurus Couperus, "Tensions between Religion and Science," *Spectrum* 10 (March 1980): 86 (lost innocence).

Fifteen   Creationism in the Churches

1. "Evangelist Billy Graham," *Bible-Science Newsletter* 10 (June 1972): 8; Jimmy Swaggart Ministries, *The Pre-Adamic Creation & Evolution* (Baton Rouge, LA: Jimmy Swaggart Ministries, 1986); Davis A. Young, *Christianity and the Age of the Earth* (Grand Rapids, MI: Zondervan, 1982), p. 10. In 1985 Henry M. Morris lamented that "the great majority of Christian people still remain both uninformed and unconcerned"; see his *Creation and the Modern Christian* (El Cajon, CA: Master Book Publishers, 1985), p. xiii.

2. William Sims Bainbridge and Rodney Stark, "Superstitions: Old and New," *Skeptical Inquirer* 4 (1980): 18–31; Jackson W. Carroll, Douglas W. Johnson, and Martin E. Marty, *Religion in America: 1950 to the Present* (San Francisco: Harper & Row, 1979), p. 14.

3. "The Creation," *U.S. News & World Report*, December 23, 1991, p. 59; Henry M. Morris, "Director's Column," *Acts & Facts* 2 (October 1973): 3; "Blacks Favor Creation," ibid. 16 (November 1987): 3; "Governor Reagan Backs Two-Model Approach," ibid. 9 (October 1980): 3; "Finding: Let Kids Decide How We Got Here," *American School Board Journal* 167 (March 1980): 52; "Creationism and the First Amendment," *ABA Journal* (January 1, 1987): 35. See also Henry M. Morris, *A History of Modern Creationism* (San Diego: Master Book Publishers, 1984), pp. 309–11.

4. Bainbridge and Stark, "Superstitions," p. 22 (San Francisco survey).

5. G. M. Price to F. L. Marsh, October 17, 1962 (*CT* article), Marsh Papers, courtesy of Frank L. Marsh; William John Hausmann, *Science and the Bible in*

Lutheran Theology: From Luther to the Missouri Synod (Washington, D.C.: University Press of America, 1978), pp. 90–91 (1932 statement), p. 95 (geocentrism); Alfred M. Rehwinkel, *The Flood: In the Light of the Bible, Geology, and Archaeology* (St. Louis: Concordia Publishing House, 1951); interview with F. L. Marsh, August 30, 1972 (Lutherans buy SDA books). The article that prompted Price's query was Walter E. Lammerts, "Growing Doubts: Is Evolutionary Theory Valid?" *Christianity Today* 6 (September 14, 1962): 3–6. Price's views were discussed, though not fully endorsed, in Theodore L. Handrich, *The Creation: Facts, Theories, and Faith* (Chicago: Moody Press, 1953).

6. Interview with Wilbert H. Rusch, November 15, 1983 (growing concern); Paul A. Zimmerman, *The Doctrine of Creation and Modern Theories of Evolution* (n.p., 1960), p. 3 (intellectuals pushing); John W. Klotz, *Genes, Genesis, and Evolution* (St. Louis: Concordia Publishing House, 1955), quotation on p. 225; J. W. Klotz to J. C. Whitcomb, October 10, 1962 (Scripture speaks), and J. C. Whitcomb to H. M. Morris, February 2, 1962 (anti-Rehwinkel reaction), both in the Whitcomb Papers, courtesy of John C. Whitcomb, Jr. From the beginning Klotz admitted that he was a creationist primarily for theological reasons; see John William Klotz, "Genesis, Genetics, and Evolution" (B.D. thesis, Concordia Theological Seminary, 1941), p. vi.

7. C. H. Krekeler to R. L. Numbers, July 3, 1989.

8. [Carl H. Krekeler], Review of *Genes, Genesis, and Evolution*, by John W. Klotz, *Cresset* 19 (January 1956): 44–5.

9. Paul A. Zimmerman, ed., *Darwin, Evolution, and Creation* (St. Louis: Concordia Publishing House, 1959); [Carl H. Krekeler], Review of *Darwin, Evolution, and Creation*, ed. Paul A. Zimmerman, *Cresset* 23 (March 1960): 21–3; John W. Klotz, Letter to the Editor, ibid. 24 (September 1961): 26–7; Paul A. Zimmerman, Letter to the Editor, ibid. 24 (March 1961): 25–6. For a reply to Zimmerman's letter, see John Gergely, Letter to the Editor, ibid. 24 (May 1961): 26. On the influence of the Zimmerman book, see interview with Wilbert H. Rusch, November 15, 1983.

10. William W. Bloom and Carl H. Krekeler, *General Biology: A Unified Text Manual* (Princeton, NJ: D. van Nostrand Co., 1963), quotations on p. 438. Krekeler identified himself as the author of the chapter on evolution in C. H. Krekeler to R. L. Numbers, July 3, 1989. Bloom's conversion to evolution is mentioned in "Lutherans Disagree on Evolution," *Bible-Science Newsletter* 3 (November 15, 1965): 1.

11. O. R. Harms to W. E. Lammerts, October 1, 1963 (quoting Rusch about brush fire), and W. H. Rusch to W. E. Lammerts, April 8, 1963 (ammunition), both in the Walter E. Lammerts Papers, Bancroft Library, University of California, Berkeley; Carl H. Krekeler, "What about 'Evolution' at Valparaiso University?" *Contact* (Summer 1964): 3–4, copy courtesy of Carl H. Krekeler. Typical of Lammerts's contributions to the debate was "The Creation Story—Factual or Symbolic?" *Lutheran Layman* 33 (July 1, 1962): 9, which elicited a critical response from John Gergely, Letter to the Editor, ibid. 33 (August 1, 1962): 4, 11.

12. "Essays Delivered at Northern Indiana Pastors & Teachers Conference, Valparaiso, Indiana, October 21 & 22, 1965," pp. 10, 66, unpublished MS, courtesy of Wilbert H. Rusch; J. W. Klotz to W. E. Lammerts, October 17, 1962 (god of scholarship), Lammerts Papers. Klotz, Rusch, and Zimmerman published versions of their Valparaiso papers in Paul A. Zimmerman, ed., *Creation, Evolution, and God's Word* (St. Louis: Concordia Publishing House, 1972).

13. J. W. Klotz to J. C. Whitcomb, July 2, 1956, Whitcomb Papers; Paul A. Zimmerman, ed., *Rock Strata and the Bible Record* (St. Louis: Concordia Publishing House, 1970), quotation by Kenneth L. Currie on p. 122; *Convention Workbook* for the 52nd Regular Convention of the Lutheran Church—Missouri Synod, Dallas, Texas,

July 15–22, 1977, pp. 101–2 (creationist institute recommended); *Convention Pro-ceedings* of the 52nd Regular Convention of the Lutheran Church—Missouri Synod, Dallas, Texas, July 15–22, 1977, p. 138 (recommendation voted down).

14. "Creation in Biblical Perspective: Report of the Commission on Theology and Church Relations, The Lutheran Church—Missouri Synod," n.d., copy in the collec-tions of the Concordia Historical Institute; *Convention Proceedings* of the 49th Regular Convention of the Lutheran Church—Missouri Synod, Milwaukee, Wisconsin, July 9–16, p. 116 (report received); interview with W. H. Rusch, November 15, 1983 (significance of report). On creation statements, see John W. Klotz, *Studies in Cre-ation* (St. Louis: Concordia Publishing House, 1985), p. 65. For a conservative view of the schism, see Harold Lindsell, *The Battle for the Bible* (Grand Rapids, MI: Zonder-van, 1976), pp. 72–88.

15. Andrew Johnson, "The Evolution Articles," *Pentecostal Herald* 38 (September 29, 1926): 6. For a fuller discussion of these issues, see Ronald L. Numbers, "Creation, Evolution, and Holy Ghost Religion: Holiness and Pentecostal Responses to Darwin-ism," *Religion and American Culture* 2 (1992): in press. I am especially indebted to Tim Kruse and Bill Kostlevy for their assistance in preparing this section and to the Wesleyan/Holiness Studies Project at Asbury Theological Seminary for supporting my research.

16. S. Hugh Paine, Review of *The Genesis Flood,* by John C. Whitcomb, Jr., and Henry M. Morris, *Wesleyan Methodist* 119 (June 7, 1961): 14; S. Hugh Paine, "In the Beginning, God Created," *Houghton Milieu* 54 (March 1979): 2–7; S. H. Paine to R. L. Numbers, November 10, 1989. After the appearance of *The Genesis Flood,* Morris gave an invited lecture at Houghton College; see Morris, *History of Modern Creationism,* p. 161. See also James F. Gregory, Review of *The Genesis Flood,* by John C. Whitcomb, Jr., and Henry M. Morris, *Free Methodist* 94 (October 17, 1961): 14.

17. Cecil B. Hamann and J. Paul Ray, "Progressive Creationism," *Good News* 15 (March–April 1982): 12–14; Cecil B. Hamann, videotaped "Lectures on Paleontology and Early Man," 1982, in the Ashbury College Archives. See also Cecil B. Hamann, Review of *The Genesis Flood,* by J. C. Whitcomb, Jr., and H. M. Morris, *Recent Books: A Quarterly Review for Ministers* 3 (October–December 1961): 45. I am grate-ful to Ivan Zabilka for information about Hamann. The generalization about the teach-ing of creation and evolution in Holiness schools is based in part on interviews with John Brushaber, James Behnke, and William Toll, March 10, 1989.

18. Lee Haines, "The Book of Genesis," in *The Wesleyan Bible Commentary,* 6 vols. (Grand Rapids, MI: William B. Eerdmans, 1967), 1:21–8; George Herbert Liv-ingston, "Genesis," in *Beacon Bible Commentary,* 10 vols. (Kansas City, MO: Beacon Hill Press, 1969), 1:32–3, 56–60; Eugene F. Carpenter, "Cosmology," in *A Contem-porary Wesleyan Theology: Biblical, Systematic, and Practical,* 2 vols. (Grand Rapids, MI: Francis Asbury Press, 1983), 1:177–8.

19. H. M. Morris to J. C. Whitcomb, October 3, 1964, Whitcomb Papers. Morris declined to be quoted directly.

20. Morris, *History of Modern Creationism,* p. 259 (Evangel); interview with Henry M. Morris and Duane Gish, October 26, 1980 (reception); Turner Collins and Gary Liddle, unpublished syllabus for a "Seminar in Science and Religion" offered at Evangel College, spring 1979; L. Duane Thurman, *How to Think about Evolution and Other Bible-Science Controversies* (Downers Grove, IL: InterVarsity Press, 1978); Myrtle M. Fleming, "Evolution: Do We Know What We Are Talking About?" un-published paper read at the Society for Pentecostal Studies, Des Moines, Iowa, No-vember 5, 1971, and deposited in the Hal Bernard Dixon, Jr., Pentecostal Research Center, Lee College. On Fleming, see Vinson Synan, *Emmanuel College: The First*

Fifty Years, 1919–1969 (Franklin Springs, GA: Emmanuel College Library, 1969), p. 112. Clyde R. Root kindly guided me through the resources of the Pentecostal Research Center.

21. Finis Jennings Dake, *Dake's Annotated Reference Bible* (Lawrenceville, GA: Dake Bible Sales, 1963), pp. 54–7; Jimmy Swaggart Ministries, *The Pre-Adamic Creation & Evolution*; Lester Sumrall, *Genesis: Crucible of the Universe* (South Bend, IN: LESEA Publishing, 1982), p. 23; Kenneth E. Hagin, *The Origin and Operations of Demons*, 2nd ed. (Tulsa, OK: Kenneth Hagin Ministries, 1987), p. 5; Gordon Lindsay, *The Bible Is a Scientific Book* (Dallas: Christ for the Nations, 1971), pp. 20–8; Gordon Lindsay, *Evolution—The Incredible Hoax* (1961; reprint, Dallas: Christ for the Nations, 1973); Charles F. Parham, *A Voice Crying in the Wilderness*, 4th ed. (Baxter Springs, KS: Robert L. Parham, 1944), pp. 81–5. On Dake, see P. H. Alexander, "Finis Jennings Dake (1902–87)," in *Dictionary of Pentecostal and Charismatic Movements*, ed. Stanley M. Burgess and Gary B. McGee (Grand Rapids, MI: Zondervan, 1988), pp. 235–6. On the continuing popularity of the gap theory, see Tom McIver, "Formless and Void: Gap Theory Creationism," *Creation/Evolution* 24 (Fall 1988): 1–24.

22. John C. Whitcomb, Jr., "How Did God Make Man?" *Pentecostal Evangel* (April 23, 1967): 12–13, 27; Paul L. Walker, *Understanding the Bible and Science* (Cleveland, TN: Pathway Press, 1976), pp. 44–9; Wade H. Phillips, *God, the Church, and Revelation* (Cleveland, TN: White Wing Publishing House, 1986), n.p.; Dennis Lindsay, unpublished syllabus for a course on "Scientific Creationism," Christ for the Nations Institute, 1986; Winkie Pratney, *Creation or Evolution?* (Glendale, CA: Church Press, [late 1960s]); Winkie Pratney, "Creation or Evolution?" in *The Last Days Collection: A Treasury of Articles from Last Days Ministries* (Lindale, TX: Pretty Good Publishing, 1988), pp. 162–87. Pratney's scientific training is mentioned ibid., p. 14. For evidence of interest in the *Pentecostal Evangel*, see: "Researchers Show Earth Is Not Billions of Years Old" (October 29, 1978): 24; H. Wayne Hornbaker, "Science Supports the Bible," (September 14, 1980): 8–9; and "Former Evolutionist Says Many Scientists Are Helping Spur 'Modern-Day' Revival," (May 2, 1982): 12. I am indebted to Joyce Lee, of the Assemblies of God Archives in Springfield, Missouri, for providing copies of articles in the *Pentecostal Evangel*.

23. Harold T. Christensen and Kenneth L. Cannon, "The Fundamentalist Emphasis at Brigham Young University, 1935–73," *Journal for the Scientific Study of Religion* 17 (1978): 53–7. In preparing this section on Mormonism, I incurred debts to Lester Bush, for providing me with a reading list on Mormonism and creationism; to Robert L. Miller, for sharing additional sources on the subject; and to Peggy Fletcher and the Sunstone Foundation for inviting me to Salt Lake City and giving me the opportunity to do research in Mormon records.

24. Duane E. Jeffrey, "Seers, Savants and Evolution: The Uncomfortable Interface," *Dialogue: A Journal of Mormon Thought* 8 (Autumn/Winter 1974): 49 (quoting Young); "The Origin of Man," *Improvement Era* 13 (November 1909): 80; Richard Sherlock, "Campus in Crisis: BYU, 1911," *Sunstone* 4 (January–February, 1979): 11–16; Frederick J. Pack, *Science and Belief in God: A Discussion of Certain Phases of Science and Their Bearing upon Belief in the Supreme Being* (Salt Lake City: Deseret News, 1924), p. 196. On Mormons and evolution, see, in addition to the Jeffrey article cited above, Richard Sherlock, "A Turbulent Spectrum: Mormon Reactions to the Darwinist Legacy," *Journal of Mormon History* 5 (1978): 33–59; Duane E. Jeffrey, "'We Don't Know': A Survey of Mormon Responses to Evolutionary Biology," in *Science and Religion: Toward a More Useful Dialogue*, vol. 2, *The Appearance of Man*, ed. Wilford M. Hess, Raymond I. Matheny, and Donlu D. Thayer (Geneva, IL:

Paladin House, 1979), pp. 23–37; and Leonard J. Arrington and Davis Bitton, *The Mormon Experience: A History of the Latter-day Saints* (New York: Alfred A. Knopf, 1979), pp. 256–61.

25. Richard Sherlock, " 'We Can See No Advantage to a Continuation of the Discussion': The Roberts/Smith/Talmage Affair," *Dialogue: A Journal of Mormon Thought* 13 (Fall 1980): 63–78; Thomas G. Alexander, *Mormonism in Transition: A History of the Latter-day Saints, 1890–1930* (Urbana: University of Illinois Press, 1986), pp. 273–88; Joseph Fielding Smith, "The Origin and Destiny of Man," *Improvement Era* 23 (1920): 375–93; Sherlock, "A Turbulent Spectrum," p. 35 (No Adam, no fall); J. F. Smith to G. M. Price, February 20, 1931 (house-cleaning), George McCready Price Papers, Adventist Heritage Center, Andrews University. In 1898 Roberts had been denied a seat in the U.S. House of Representatives for practicing polygamy; Arrington and Bitton, *The Mormon Experience*, p. 184.

26. Jeffrey E. Keller, "Discussion Continued: The Sequel to the Roberts/Smith/ Talmage Affair," *Dialogue: A Journal of Mormon Thought* 15 (Spring 1982): 79–98; "James Edward Talmage," in *Latter-day Saint Biographical Encyclopedia*, ed. Andrew Jenson, 4 vols. (Salt Lake City: Andrew Jenson Memorial Association, 1901–36), 3:787–9.

27. J. E. Talmage to S. B. Talmage, February 5, 1931, Talmage Papers, courtesy of William Lee Stokes.

28. S. B. Talmage to J. E. Talmage, February 9, 1931, and J. E. Talmage to S. B. Talmage, February 13, 1931 (Pack), both in the Talmage Papers.

29. J. E. Talmage to S. B. Talmage, February 13 and May 21, 1931, and S. B. Talmage to J. E. Talmage, May 12, 1931, all in the Talmage Papers.

30. J. F. Smith to G. M. Price, February 12 and 20, 1931, both in the Price Papers.

31. S. B. Talmage to J. E. Talmage, June 15, 1931 (near-controversy), S. B. Talmage to J. A. Widtsoe, April 17, 1934 (evolution and white heat), S. O. Bennison to W. W. Henderson, July 12, 1934 (suspending Whitney series), and D. J. Whitney to S. B. Talmage, September 29, 1934 (Inter Mountain states), all in the Talmage Papers; Sterling B. Talmage, "Can We Dictate God's Times and Methods?" *Deseret News*, April 14, 1934, pp. 3, 5, April 21, 1934, pp. 3, 6. Whitney's series in the *Deseret News* began with "The Fiat Creation of the Universe," June 9, 1934, p. 8. On Widtsoe, see Arrington and Bitton, *The Mormon Experience*, pp. 310–14.

32. Joseph Fielding Smith, *Man: His Origin and Destiny* (Salt Lake City: Deseret Book Co., 1954); Jeffrey, "Seers, Savants, and Evolution," pp. 65–6, 75 (*Man*); D. O. McKay to W. L. Stokes, February 15, 1957 (no imprimatur), Stokes Papers, courtesy of William Lee Stokes; Arrington and Bitton, *The Mormon Experience*, p. 340 (Smith); Bruce R. McConkie, *Mormon Doctrine* (Salt Lake City: Bookcraft, 1958), pp. 229–38, 268; Bruce R. McConkie, "The Seven Deadly Heresies," BYU 14-Stake Fireside, June 3, 1980 (devilish), copy courtesy of Robert L. Miller. On McConkie, see David John Buerger, "The Theological Influence of Bruce R. McConkie," unpublished paper read at Sunstone Theological Symposium, Salt Lake City, August 25, 1984.

33. W. L. Stokes to Henry Erying, December 14, 1954 (howl), Stokes Papers; Melvin A. Cook, Introduction to *Man*, by Smith, pp. vii–ix; Melvin A. Cook, *The Autobiography of Melvin A. Cook*, 2 vols. (Salt Lake City: Author, 1973–77), 2:230 (thirteen thousand years), 327 (self-study), 469 (Noachian flood). I am indebted to Melvin A. Cook for loaning me copies of his books. For Stokes's own position, see William Lee Stokes, *The Creation Scriptures: A Witness for God in the Scientific Age* (Salt Lake City: Starstone Publishing Co., 1979), which appeared in a non-Mormon version as *The Genesis Answer: A Scientist's Testament for Divine Creation* (Englewood Cliffs, NJ: Prentice-Hall, 1984).

34. Melvin A. Cook, "Continental Drift: Is Old Mother Earth Just a Youngster?" *Utah Alumnus* (September–October 1963): 10–12; W. Lee Stokes and Others, Letter to the Editor, ibid. (December–January 1963–64): 4; Melvin A. Cook, *Prehistory and Earth Models* (London: Max Parrish, 1966); Melvin A. Cook and Melvin Garfield Cook, *Science and Mormonism: Correlations, Conflicts and Conciliations* ([Salt Lake City]: Deseret News Press, 1967), pp. viii, 97, 145, 149. On Eyring, see Edward L. Kimball, "Harvey Fletcher and Henry Eyring: Men of Faith and Science," *Dialogue: A Journal of Mormon Thought* 15 (Autumn 1982): 74–86; Steven H. Heath, "The Reconciliation of Faith and Science: Henry Eyring's Achievement," ibid., pp. 87–99.

35. W. E. Lammerts to M. A. Cook, May 24, 1967, in Cook, *Autobiography*, 2:285–6, 301 (Nobel citation); D. T. Gish to D. A. Young, August 8, 1972 (Cook's theology), Young Papers, courtesy of Davis A. Young. See also Melvin A. Cook, "Radiological Dating and Some Pertinent Applications of Historical Interest: Do Radiological 'Clocks' Need Repair?" *Creation Research Society Quarterly* 5 (1968–69): 69–77; Melvin A. Cook, "William J. Meister Discovery of Human Footprint with Trilobites in a Cambrian Formation of Western Utah," ibid., p. 97; and Melvin A. Cook, "Carbon-14 and the 'Age' of the Atmosphere," ibid. 7 (1970–71): 53–6.

36. Gary North, *Is The World Running Down? Crisis in the Christian Worldview* (Tyler, TX: Institute for Christian Economics, 1988), p. xiv.

37. Interview with James D. Bales, March 25, 1985; John N. Moore and Harold Schultz Slusher, *Biology: A Search for Order in Complexity* (Grand Rapids, MI: Zondervan, 1970), pp. xvi–xvii (Russell C. Artist, H. Douglas Dean, and Jack Wood Sears); Robert S. Camp, ed., *A Critical Look at Evolution* (Atlanta: Religion, Science, and Communication Research and Development Corporation, 1972); [Bert Thompson], "In the News," *Reason & Revelation* 2 (July 1982): no pagination; Bert Thompson, *Is Genesis Myth? The Shocking Story of the Teaching of Evolution at Abilene Christian University* (Montgomery, AL: Apologetics Press, 1986); Wayne Jackson, "Premillennialism and Biblical Creationism," ibid. 5 (May 1985): no pagination. See also Bert Thompson, *Theistic Evolution* (Shreveport, LA: Lambert Book House, 1977); and Bert Thompson, *The History of Evolutionary Thought* (Fort Worth, TX: Star Bible & Tract Corporation, 1981). For a Church of Christ critique of flood geology, see David E. Koltenbah, "Concerning the Creation Research Society, the Book *The Genesis Flood*, and 'Flood Geology,'" *Truth Magazine* 15 (1970): 39–43, 51–5. I am indebted to James D. Bales for a gift subscription to *Reason & Revelation*, to Craig Meyer for clarifying my impressions of the Churches of Christ, and to Steve Wolfgang for sending me several documents.

38. R. R. Clapp, "Christian Reconstructionism," in *Dictionary of Christianity in America*, ed. Daniel G. Reid (Downers Grove, IL: InterVarsity Press, 1990), pp. 977–8; Thomas Allen McIver, "Creationism: Intellectual Origins, Cultural Context, and Theoretical Diversity" (Ph.D. dissertation, University of California, Los Angeles, 1989), pp. 548–73; Morris, *History of Modern Creationism*, p. 154 (Rushdoony); Gary North, *The Dominion Covenant: Genesis*, vol. 1, *An Economic Commentary on the Bible*, rev. ed. (Tyler, TX: Institute for Christian Economics, 1987), pp. v, 385 (most important book); North, *Is the World Running Down?*, pp. x–xv (entropy and oil).

39. Interview with Duane T. Gish, October 26, 1980. See also M. James Penton, *Apocalypse Delayed: The Story of Jehovah's Witnesses* (Toronto: University of Toronto Press, 1985); and Lowell Tarling, *The Edges of Seventh-day Adventism* (Barragga Bay, Australia: Galilee Publication, 1981), pp. 41–62 (Armstrong).

40. Penton, *Apocalypse Delayed*, pp. 196–7; J. F. Rutherford, *Creation* (Brooklyn: Watch Tower Bible and Tract Society, 1927), pp. 31–5 (Vail); Isaac N. Vail, *The Earth's Annular System; or, The Waters above the Firmament* (Pasadena, CA: Annular

World Co., 1912), quotations on pp. 72, 110. Vail first published his views as "*The Waters above the Firmament*": *The Earth's Aqueous Ring; or, The Deluge and Its Cause* (West Chester, PA: F. S. Hickman, 1874). Vail is identified as a Quaker in Donald Wesley Patten, *The Biblical Flood and the Ice Age: A Study in Scientific History* (Seattle: Pacific Meridian Publishing Co., 1966), p. 16.

41. Penton, *Apocalypse Delayed*, pp. 270–4 (education); Christopher P. Toumey, "The Social Context of Scientific Creationism" (Ph.D. dissertation, University of North Carolina, 1987), chapter 6; McIver, "Creationism," pp. 476–7; *Evolution versus the New World* (Brooklyn: Watch Tower Bible and Tract Society, 1950); *Did Man Get Here by Evolution or by Creation?* (Brooklyn: Watchtower Bible and Tract Society, 1967); *Life—How Did It Get Here? By Evolution or by Creation?* (Brooklyn: Watchtower Bible and Tract Society, 1985). I discovered only two well-known creationists who had been Jehovah's Witnesses—Jerry Bergman, an educator, and Randy L. Wysong, a veterinarian—and neither had remained with the group. See R. L. Wysong, *The Creation-Evolution Controversy: Toward a Rational Solution* (Midland, MI: Inquiry Press, 1976).

42. McIver, "Creationism," 479–81 (7,500,000); Tarling, *The Edges of Seventh-day Adventism*, pp. 42–3 (fundamental doctrines); Christopher P. Toumey, "Sectarian Aspects of American Creationism," *International Journal of Moral and Social Studies* 5 (1990): 116–42, quotation about Biblical truth on p. 135. The Toumey article contains an excellent discussion of Jehovah's Witness and Worldwide Church of God attitudes toward evolution.

43. Paula Haigh, *What's Wrong with Evolution?* (Louisville, KY: Catholic Center for Creation Research, 1975); Catholic Center for Creation Research, *Newsletter* 1 (May–June 1976): 13 (disciple); "One Year of the CCCR Newsletter," ibid. 1 (July–August 1976): 1 (waiting). By the spring of 1978 Haigh had moved to San Jose, California, and renamed her newsletter the *Catholic Creationist: Journal of the Catholic Center for Creation Research*. See also Paula Haigh, "Outline of Thomistic Principles on Creation That Prove the Impossibility of Theistic Evolution," *Creation Research Society Quarterly* 15 (1978–79): 210–11. On Ellwanger, see Morris, *History of Modern Creationism*, p. 288; and Edward J. Larson, *Trial and Error: The American Controversy over Creation and Evolution* (New York: Oxford University Press, 1985), pp. 129, 150–6.

44. "Creationism vs. Evolution: Radical Perspectives on the Confrontation of Spirit and Science," *Tikkun* 2, no. 5 (1987): 55 (no one agrees); Peter Gabel, "Creationism and the Spirit of Nature," ibid., pp. 55–63, quotation about hegemony on p. 59; Moshe Trop, Letter to the Editor, *Creation Research Society Quarterly* 20 (1983–84): 121–2 (Archaeopteryx); McIver, "Creationism," pp. 342–9. I am grateful to my friend Todd L. Savitt for sending me the material from *Tikkun*.

Sixteen.   Creation Science Floods the World

1. William V. Mayer, "The Nineteenth Century Revisited," *BSCS Newsletter*, no. 49 (November 1972): 13 (onslaughts); Dorothy Nelkin, *The Creation Controversy: Science or Scripture in the Schools* (New York: W. W. Norton, 1982), pp. 156–63 (scientific response and Cracker Jacks); "Modern Creationists Seeking Equal Time in U.S. Classrooms," *Wall Street Journal*, June 15, 1979, pp. 1, 30. Henry M. Morris, *A History of Modern Creationism* (San Diego: Master Book Publishers, 1984), p. 308, credits the *BSCS Newsletter* with being "the first national science journal to give real attention to the creation movement."

2. Edward J. Larson, *Trial and Error: The American Controversy over Creation and Evolution,* updated ed. (New York: Oxford University Press, 1989), pp. 147–56, "cornerstone" quotation on p. 148; Wendell R. Bird, "Freedom of Religion and Science Instruction in Public Schools," *Yale Law Journal* 87 (January 1978): 515–70; Roger Lewin, "A Response to Creationism Evolves," *Science* 214 (1981): 635–6, 638; "New Creationist Bills, Resolutions, and Court Cases Appear Nationwide," *Creation/ Evolution No. 4* (Spring 1981): 27 (coast to coast). See also W. R. Bird, *The Origin of Species Revisited: The Theories of Evolution and of Abrupt Appearance,* 2 vols. (New York: Philosophical Library, 1989).

3. Frederick Edwords to Dear Reader, *Creation/Evolution No. 1* (Summer 1980): inside front cover; "Evolution at the AAAS," *Science News* 119 (January 10, 1981): 19 (Mayer); "Science Meetings and Resolutions," *Creation/Evolution No. 7* (Winter 1982): 46; Lewin, "A Response to Creationism Evolves," pp. 635–6, 638 (crisis and terrorist tactics).

4. Stanley L. Weinberg and Robert H. Chapman, "What Is a Committee of Correspondence?" brochure dated December, 1981; "Committees of Correspondence," *Creation/Evolution No. 12* (Spring 1983): 38; "The Committees of Correspondence: What They Are and What They Do," *Creation/Evolution Newsletter* 6 (July–August 1986): 4–5.

5. *Science and Creationism: A View from the National Academy of Sciences* (Washington, D.C.: National Academy Press, 1984), p. 6. In a form letter mailed out with the booklet, NAS president Frank Press mentioned that over forty thousand copies had been distributed.

6. Committee for Integrity in Science Education (David Price, John L. Wiester, and Walter R. Hearn), *Teaching Science in a Climate of Controversy: A View from the American Scientific Affiliation* (Ipswich, MA: American Scientific Affiliation, 1987), pp. 14–17 (guidelines), 42 (humble). The copy cited is a revised second printing. In an extensively revised third printing, in 1989, Wonderly's name was dropped, at his request, from the list of committee consultants.

7. William J. Bennetta, ed., "Scientists Decry a Slick New Packaging of Creationism," *Science Teacher* 54 (May 1987): 36–43; E. C. Scott to W. J. Bennetta, December 30, 1987 (Trotskyites), copy courtesy of Marian Finger. The ASA booklet was hotly debated in the *Creation/Evolution Newsletter* from late 1986 through 1987.

8. "The Underground Battle," *Creation/Evolution No. 14* (Fall 1984): 53–4; Larson, *Trial and Error,* pp. 179–84 (Supreme Court); Raymond A. Eve and Francis B. Harrold, *The Creationist Movement in Modern America* (Boston: Twayne Publishers, 1991), pp. 163–7 (polls); Robert L. Simonds, "Editorial: Why N.A.C.E.," *Christians in Education* 1 (Fall 1983): 12–13, 21. The first issue of *Christians in Education* carried articles by both Morris and Gish.

9. Martin Eger, "A Tale of Two Controversies: Dissonance in the Theory and Practice of Rationality," *Zygon* 23 (1988): 291–325. For reactions to Eger, see ibid., pp. 327–68. For a widely used home-schooling textbook that promulgates flood geology, see William S. Pinkston, Jr., *Biology for Christian Schools* (Greenville, SC: Bob Jones University Press, 1980). I am indebted to Barbara Schulze for bringing this book to my attention.

10. Richard C. Lewontin, introduction to *Scientists Confront Creationism,* ed. Laurie R. Godfrey (New York: W. W. Norton, 1983), p. xxv.

11. Interview with David T. Rosevear, September 30, 1984 (one-man band); C. E. A. Turner, "A Jubilee of Witness for Creation against Evolution by CSM/EPM, 1932–1982," *Evolution Protest Movement Pamphlet No. 232,* July 1982, p. 12 (over one hundred pamphlets). For growth and activities, including the election of officers and

the distribution of Howitt's booklet, see E.P.M. Council Meeting Minutes, September 1960–April 1977, courtesy of David T. Rosevear. The reference to "preaching to the converted" appears in the minutes for April 18, 1964. See also W. R. Thompson, introduction to *The Origin of Species*, by Charles Darwin (London: J. M. Dent & Sons, 1956), pp. vii–xxv; and W. H. Thorpe, "William Robin Thompson, 1887–1972," in *Biographical Memoirs of Fellows of the Royal Society* 19 (1973): 655–78. On the gap theory, see A. G. Tilney, " 'We Thank Thee for Our' . . . Evolution?" *Evolution Protest Movement Pamphlet No. 60*, 1958; and A. G. T[ilney], "Theistic Evolution," *Evolution Protest Movement Pamphlet No. 88*, October 1961. I am grateful to C. E. Allan Turner for providing me with a set of EPM pamphlets and for answering my queries about the history of the EPM.

12. D. W. Bebbington, *Evangelicalism in Modern Britain: A History from the 1730s to the 1930s* (London: Unwin Hyman, 1989), pp. 252–62, quotation regarding Lloyd-Jones on p. 261. For a biographical sketch of Lloyd-Jones by his grandson, see Christopher Catherwood, *Five Evangelical Leaders* (Wheaton, IL: Harold Shaw Publishers, 1985), pp. 51–110. On British evangelicalism, see also Randle Manwaring, *From Controversy to Co-Existence: Evangelicals in the Church of England, 1914–1980* (Cambridge: Cambridge University Press, 1985).

13. A. G. T[ilney], Review of *The Genesis Flood*, by John C. Whitcomb, Jr., and Henry M. Morris, *Evolution Protest Movement Pamphlet No. 97*, [1962] (new era and sales); A. G. T[ilney], "The Apostasy Grows," *Evolution Protest Movement Pamphlet No. 105*, January 1964 (cranky); A. G. T[ilney], "Evolution Yesterday, To-Day and To-Morrow," *Evolution Protest Movement Pamphlet No. 147*, June 1967 (Restoration-Week); A. G. T[ilney], "Newsletter," *Evolution Protest Movement Pamphlet No. 181*, July 1970 (departure of life). On continuing partiality for the gap theory, see, e.g., A. G. T[ilney], Review of *The Biblical Flood and the Ice Epoch*, by D. W. Patten, *Evolution Protest Movement Pamphlet No. 153*, October 1967.

14. David Watts, "The Transatlantic Connection," *Biblical Creation* 4 (February 1982): 1 (household names); "The New Creationism," *Rainbow*, No. 8 [Biblical Creation Society, 1984] (tract-rack); R. E. D. Clark to R. L. Numbers, [October 1984] (silly); interview with R. E. D. Clark, October 1, 1984 (rubbish). See also [R. E. D. Clark], "Evolution: Polarization of Views," *Faith and Thought* 100, no. 3 (1972–73): 227–9; and [R. E. D. Clark], "American and English Creationists," ibid. 104, no. 1 (1977): 6–8.

15. Morris, *History of Modern Creationism*, pp. 294–5; interview with Edgar C. Powell, September 29, 1984; "Newton Scientific Association," undated announcement, courtesy of Edgar C. Powell; interview with David C. Watts, October 3, 1984. From about 1975 to 1978 the NSA published a series of pamphlets under the title *N.S.A. Reports*.

16. *Biblical Creation Society Newsletter* 1 (April 1978), copy courtesy of A. J. Monty White (origins); interview with E. H. Andrews, September 28, 1984; interview with David C. Watts, October 3, 1984 (wholly negative); Nigel M. de S. Cameron, "Editorial," *Biblical Creation* [1] (October 1978): 3–4 (influence of Whitcomb and Morris); Nigel M. de S. Cameron, "Editorial," ibid. 2 (June 1980): 35–6 (Whitcombism). In *Genesis in Space and Time: The Flow of Biblical History* (Downers Grove, IL: InterVarsity Press, 1972), p. 57, Francis A. Schaeffer declined to take a position on a literal six-day creation. In an interview with A. J. Monty White the year before his death, Lloyd-Jones denied giving encouragement to the BCS; see A. J. M. White to R. L. Numbers, May 10, 1991.

17. Interview with E. H. Andrews, September 28, 1984. For Andrews's views, see his books on creation and evolution: *Is Evolution Scientific?* (Welwyn, Hertfordshire:

Evangelical Press, 1977); *From Nothing to Nature: A Young People's Guide to Evolution and Creation* (Welwyn, Hertfordshire: Evangelical Press, 1978); *God, Science & Evolution* (Welwyn, Hertfordshire: Evangelical Press, 1980).

18. Interviews with David C. Watts, October 3, 1984, and July 11, 1988; David C. Watts, "Fossils and the Fall," *Biblical Creation* 7 (Summer 1984): 20–1 (100% Flood-Geology); David C. Watts, Review of *Concepts of Creationism*, ed. by E. H. Andrews and Others, *Origins: Journal of the Biblical Creation Society* 1 (October 1987): 13–14 (narrow definition). See also David C. Watts, "Rock Strata and the Geological Column," *Biblical Creation* 7 (Summer 1985): 59–65.

19. Peter Senior, "Editorial: Why All the Fuss about Evolution?" *Origins: Journal of the Biblical Creation Society* 3 (January 1990): 2.

20. Interview with C. E. Allan Turner, September 30, 1984; interview with David T. Rosevear, September 30, 1984; D. T. Rosevear, "The Three Views of Creation," *Creation: The Journal of the Evolution Protest Movement* 2 (October 1978): 2–3 (contrivances); "Opinions on the Age of the Earth," *Creation: The Journal of the Creation Science Movement* 4 (March 1986): 12 (poll); Turner, "A Jubilee of Witness," p. 16 (ill-supported claims); D. T. Rosevear to R. L. Numbers, May 2, 1991, and August 20, 1991 (Deed of Trust and growth). C. E. Allan Turner, "The Chairman's Letter," *Creation: The Journal of the Evolution Protest Movement* 2 (October 1976): 1–2, says that the average age of the council dropped by ten years in the mid-1970s; Rosevear, in an interview on September 30, 1984, suggested the figure was about thirty years. The officers of the EPM are listed on the first page of *Creation: The Journal of the Evolution Protest Movement* 1 (October 1971). On the change of name, see C. E. A. Turner, "The Chairman's Letter," *Creation: The Journal of the Creation Science Movement* 2 (January 1981): 2. Some council members had been pressing for a name change since the early 1970s; see Arthur J. Jones to R. L. Numbers, November 15, 1984. Eileen Barker, "In the Beginning: The Battle of Creationist Science against Evolutionism," in *On the Margins of Science: The Social Construction of Rejected Knowledge*, ed. Roy Wallis, Sociological Review Monograph 27 (University of Keele, March 1979), p. 187, says that EPM membership grew from 200 to 850 between 1966 and 1970. The minutes of the movement show that membership hovered in the "mid 800s" for the decade between 1965 and 1975, when it dropped to the mid-700s. However, Tilney was notoriously sloppy in keeping membership records, and it is sometimes difficult to determine whether or not his figures include branch members, of which there were hundreds.

21. Interview with A. J. Monty White, October 4–5, 1984; *Creation News Sheet No. 18* (October 1976) (reasoned faith). The *Radio Times* ("Creating the Evolution Backlash," November 21–27, 1981) described White as "perhaps Britain's leading creationist, certainly its most active." See also A. J. Monty White, *What about Origins?* (Kingsteignton, Newton Abbot, Devon: Dunestone Printers, 1978); A. J. Monty White, *How Old Is the Earth?* (Welwyn, Hertfordshire: Evangelical Press, 1985); and A. J. Monty White, *Wonderfully Made* (Darlington, County Durham: Evangelical Press, 1989).

22. A. J. Monty White, "Uniformitarianism, Probability and Evolution," *Creation Research Society Quarterly* 9 (1972–73): 32–7; A. J. Monty White, "Radio Carbon Dating," ibid., pp. 155–8; interview with A. J. Monty White, October 4–5, 1984. In the spring of 1979 White changed the name of his publication from *Creation News Sheet* to *Creation News*.

23. Interview with A. J. Monty White, October 4–5, 1984; *Creation News Sheet No. 31* (December 1978) (sops); *Creation News Sheet No. 2* [ca. December 1973] (British CRS); *Creation News Sheet No. 3* (January 1974) (NSA). On White's differ-

ences with the founders of the BCS, see A. J. M. White to Paul Woodbridge, May 5, 1977, and N. M. de S. Cameron to A. J. M. White, May 14, 1977, White Papers, courtesy of A. J. Monty White.

24. *Creation News Sheet No. 32* (January 1979). Nigel Cameron protested White's description of the BCS in a letter dated January 15, 1979, White Papers.

25. N. M. de S. Cameron, "Editorial," *Bibical Creation* 2 (June 1980): 35 (alien element); David C. C. Watson, Review of *Cross-Currents: Interactions between Science and Faith*, by Colin A. Russell, ibid. 8 (Spring 1986): 25–30 (politics and revivals); Watts, "The Transatlantic Connection," p. 1 (politics); interview with David C. Watts, October 3, 1984 (downplay American connection); interview with A. J. Monty White, October 4–5, 1984 (quote British authorities); *Biblical Creation Society Newsletter* (Summer 1984) (cautious approach); Michael E. Howgate and Alan J. Lewis, "Creationism in Confusion," *Nature* 311 (1984): 703; interview with David T. Rosevear, September 30, 1984 (reacting culturally). In an unpublished letter to the editor of *Nature*, November 19, 1984 (copy courtesy of David C. Watts), Andrews denied that he had condemned the whole American creationist movement.

26. Douglas Dewar and L. M. Davies, "Obsessions of Biologists," *Evolution Protest Movement Pamphlet No. 20*, n.d., p. 16 (Britons must listen); L. M. Davies and Douglas Dewar, "The B.B.C. Abuses Its Monopoly," *Evolution Protest Movement Pamphlet No. 26*, n.d.; A. G. T[ilney], "Newsletter," *Evolution Protest Movement Pamphlet No. 181*, July 1970 (freedom of the air); M[alcolm] Bowden, *The Rise of the Evolution Fraud* (Bromley, Kent: Sovereign Publications, 1982), pp. 128 (Gish), 129 (White), 142 (ban). For additional reactions to the BBC monopoly, see C. E. A. Turner, "Scientific Method and Evolution," *Evolution Protest Movement Pamphlet No. 58*, April 1, 1960; N. M. de S. Cameron, Letter to the Editor, *Creation Research Society Quarterly* 17 (1980–81): 234; Turner, "A Jubilee of Witness," p. 16; and the minutes of the EPM Council, September 16, 1963.

27. On the EPM branches, see Turner, "A Jubilee of Witness," pp. 12–13; and the minutes of the EPM Council.

28. Interview with W. D. Burrowes, August 9, 1991; W. D. Burrowes to A. C. Custance, June 27, 1969 (background), and February 24, 1973 (protest to teachers), both in the Custance Papers, Special Collections, Redeemer College.

29. W. D. Burrowes to A. C. Custance, April 2, [1970] (Canadian members), April 4, 1971 (first meeting), November 21, 1970 (influence of Custance's book), all in the Custance Papers; W. D. Burrowes to R. L. Numbers, June 4, 1991. See also Arthur C. Custance, *Without Form and Void: A Study of the Meaning of Genesis 1.2* (Brockville, Ontario: n.p., 1970).

30. W. D. Burrowes to A. C. Custance, August 13, 1971, Custance Papers. See also [W. D. Burrowes], "Differences between Creationists," North American Creation Movement, Newsletter No. 23, January 1978, pp. 4–5.

31. W. D. Burrowes to A. C. Custance, February 24, 1973 (Custance declines), August 1, 1973 (Custance peeved), June 23, 1970 (Hallonquist), April 4, 1971 (Bible-Science people), all in the Custance Papers; EPM Council Minutes for September 23, 1972, and April 24, 1976 (growth in Canada); Morris, *History of Modern Creationism*, pp. 219–21, 302 (quotation); W. D. Burrowes to R. L. Numbers, June 4, 1991 (suspension). See also "Particulars," North American Creation Movement, n.d., and W. D. Burrowes, *Plea for an Open Mind Concerning the Age of the Earth* (Victoria, B.C.: North American Creation Movement, [1986]), both items courtesy of W. D. Burrowes.

32. EPM Minutes for April 27, 1968 (Tilney's visit), and April 24, 1976 (350); Morris, *History of Modern Creationism*, pp. 298–9 (vanishing point and visits); John

Byrt to W. E. Burrowes, May 22, 1976, Custance Papers. On Byrt's attitude toward flood geology, see "The Roles of the Bible and of Science in Understanding Creation," *Faith and Thought* 103 (1976): 158–88. On the history of creationism in Australia, see Margaret Carol Scott, "The Resurgence of Creationism and Its Implications for New Zealand" (M.A. thesis, University of Auckland, 1987), pp. 74–87; Marian Finger kindly provided me with a copy of this thesis.

33. *Ex Nihilo* 1 (June 1978): 1–5 (origins and goals); John Rendle-Short, "The Creation of *Ex Nihilo*," *Creation Ex Nihilo* 9 (June 1987): 2; Rhondda E. Jones, "Evolution and Creationism: The Consequences of an Analysis for Education," *Quadrant* (August 1988): 23–8 (size); Andrew Snelling to R. L. Numbers, January 22, 1991, and March 6, 1991; Kenneth A. Ham, *The Lie: Evolution* (El Cajon, CA: Master Books, 1987).

34. Editorial, *Ex Nihilo* 1 (June 1978): 5 (fair hearing); Barry Price, *The Bumbling, Stumbling, Crumbling Theory of Creation Science* (Leichhardt, New South Wales: Catholic Education Office, 1987), pp. iii (pseudo-science), 3 (major triumph); Barry Price, *The Creation Science Controversy* (Sydney: Millennium Books, 1990), pp. 2–3 (private schools and American import); Bob Beale, "Church Banned Book, Says Author," *Sydney Morning Herald*, July 23, 1988, p. 3. For detailed criticism of Price's book, see *A Response to Deception*, rev. ed. (Sunnybank, Queensland: Creation Science Foundation, 1991). I am indebted to Marian Finger and Andrew Snelling for many of my Australian sources.

35. Price, *Creation Science Controversy*, pp. 3 (joke), 205 (journals and debates); Bob Beale, "Scientists Strike Back," *Sydney Morning Herald*, June 25, 1988, p. 74, quoted ibid., pp. 205–6. See also "Evolutionist Debater Descends to All-Time Low," *Acts & Facts* 17 (June 1988): 3, 5; and Jim Lippard, "Some Failures of Organized Skepticism," *Arizona Skeptic* (January 1990): 2–5. Part of the Gish-Plimer quarrel centers on the views of Michael Denton, whose controversial *Evolution: Theory in Crisis* (Bethesda, MD: Adler & Adler, 1986) had recently appeared.

36. "Creation Literature Society Formed in New Zealand," *Acts & Facts* 3 (September 1974): 2; Scott, "The Resurgence of Creationism and Its Implications for New Zealand"; Young-Gil Kim, "Creation Science in Korea," *Impact* supplement to *Acts & Facts* 15 (February 1986): i–iv; "Creationism Growing in Other Countries," *Acts & Facts* 20 (March 1991): 3 (one thousand members in Korea); "South Africa Tour," ibid. 12 (November 1983): 2, 7; Morris, *History of Modern Creationism*, pp. 298–304 (various countries). See also "Creation Organization Formed in India," *Acts & Facts* 8 (May 1979): 7; "Kenya Creation Association Formed," ibid. 16 (February 1987): 3; "Creationism in Brazil," *Bible-Science Newsletter* 16 (October 1978): 8; and Thomas Allen McIver, "Creationism: Intellectual Origins, Cultural Context, and Theoretical Diversity" (Ph.D. dissertation, University of California at Los Angeles, 1989), pp. 369–70 (South Africa). Flood geology was popularized in India by the Madras zoologist Hannington Enoch (1905–1988), a friend of David C. C. Watson's and the author of *Evolution or Creation* (Madras: Union of Evangelical Students of India, 1966), and *Following the Master: Memoirs of a Professor* (Bombay: Jyoti Pocketbooks, 1977). I am indebted to Elizabeth Oommen of Madras Christian College for providing information about Enoch.

37. W. E. Lammerts to N. A. Rupke, July 10, 1966, Rupke Papers, courtesy of Nichlaas A. Rupke; J. R. van de Fliert, "Fundamentalism and the Fundamentals of Geology," *Journal of the American Scientific Affiliation* 21 (September 1969): 69–81; W. J. Ouweneel, "Creationism in the Netherlands," *Impact* supplement to *Acts & Facts* 7 (February 1978): i–iv.

38. "Dr. Wilder Smith to Speak in America," *Acts & Facts* 9 (October 1980): 3

(Europe's leading creationist); Morris, *History of Modern Creationism*, pp. 296–8 (Wilder-Smith and European countries); Thomas Schirrmacher, "The German Creationist Movement," *Impact* supplement to *Acts & Facts* 14 (July 1985): i–iv; A. E. Wilder-Smith to R. L. Numbers, March 5, 1987 (sales); A. E. Wilder Smith, *Man's Origin, Man's Destiny: A Critical Survey of the Principles of Evolution and Christianity* (Wheaton, IL: Harold Shaw, 1968), pp. 295–8 (Paluxy River); D. T. Rosevear, "European Creationist Congress," *Creation: The Journal of the Creation Science Movement* 3 (November 1984): 1–2. In *Man's Origin, Man's Destiny*, Wilder-Smith's first creationist book in English, he did not hyphenate his name. For biographical information, see Beate Wilder-Smith, "Lebenslauf von Professor Dr.Dr.Dr. A. E. Wilder-Smith, F.R.I.C.," *Factum* (January 1986): 22, which appeared in a special issue of the magazine celebrating Wilder-Smith's seventieth birthday. I am indebted to Darrel Kautz for this source and for additional biographical detail. For one French theologian's view of creation science, see Henri Blocher, *In the Beginning: The Opening Chapters of Genesis*, trans. David G. Preston (Leicester, England: Inter-Varsity Press, 1984), pp. 213–31. See also "Swedish Creationist Organization," *Acts & Facts* 10 (February 1981): 3; and M. Daniel Mathez, "Creationism in French-Speaking Switzerland," *Origins: Journal of the Biblical Creation Society* 2 (April 1989): 6.

39. "ICR Book Used in Turkey," *Acts & Facts* 16 (July 1987): 2; "Creation Science Seminar in Moscow," ibid. 19 (November 1990): 3 (fourteen top Soviets); "Dr. Kuznetsov Appointed Adjunct Professor," ibid. 19 (August 1990): 3. Although Kouznetsov's name often appears in English as Kuznetsov, he prefers the former spelling, which is used on his passport; D. A. Kouznetsov to R. L. Numbers, July 22, 1991. On creationism in Turkey, see also McIver, "Creationism," p. 357. In 1990 the news media reported that a professor of biology at the University of Khartoum in Islamic Sudan had been jailed and tortured for teaching evolution; see Andy Coglan, "Sudan Jails Biologist for Teaching Darwinism," *New Scientist* (March 17, 1990): 21.

40. John R. Cole, "Scopes and Beyond: Antievolutionism and American Culture," in *Scientists Confront Creationism*, ed. Laurie R. Godfrey (New York: W. W. Norton, 1983), p. 31.

41. Arnold Lunn, ed., *Is Evolution Proved? A Debate between Douglas Dewar and H. S. Shelton* (London: Hollis and Carter, 1947), p. 286.

42. In his foreword to the second edition of *Scientific Creationism*, general ed. (El Cajon, CA: Master Book Publishers, 1985), pp. vii–xi, Henry M. Morris specifically cites Stephen Jay Gould, "Is a New and General Theory of Evolution Emerging?" *Paleobiology* 6 (1980): 119–30; Stephen Jay Gould, "The Ediacaran Experiment," *Natural History* 93 (February 1984): 14–23; and "Episodic View Now Replacing Catastrophism," *Geotimes* 27 (November 1982): 16–17, which carried excerpts from Robert H. Dott, Jr., "Episodic Sedimentation—How Normal Is Average? How Rare Is Rare? Does It Matter?" *Journal of Sedimentary Petrology* 53 (1983): 5–23. Regarding Dott, see also Walter E. Lammerts, "Evidence against Evolution Found in a Geology Text," *Creation Research Society Quarterly* 16 (1979–80): 204–6. For Gould's and Dott's views of creationism, see, respectively, Stephen Jay Gould, "Creationism: Genesis vs. Geology," in *Science and Creationism*, ed. Ashley Montagu (New York: Oxford University Press, 1984), p. 126–35, one of many of Gould's pieces on the subject; and Robert H. Dott, Jr., "The Challenge of Scientific Creationism," *Journal of Sedimentary Petrology* 51 (1981): 701–4.

43. W. D. Burrowes, "Comment on the State of the Art in North America," North American Creation Movement, Newsletter No. 43, March 30, 1988, p. 10; David A. Kaufmann, "Feminism, Humanism and Evolution," *Creation Research Society Quarterly* 25 (1988–89): 69–72, quotation on p. 72.

44. Bryan Di Salvatore, "Profiles," *New Yorker*, February 12, 1990, p. 54 (Haggard). I am indebted to Tim Baker for this reference. On Velikovsky, see Henry H. Bauer, *Beyond Velikovsky: The History of a Public Controversy* (Urbana: University of Illinois Press, 1984).

45. D. A. Carson, " 'The Genesis Flood' in Perspective," *Biblical Creation* 2 (February 1980): 9–20, quotations on p. 10. Nigel M. de S. Cameron, "Editorial," ibid. [1] (October 1978): 3–4, identifies "the great strength" of *The Genesis Flood* as its method of starting with the Bible.

46. Paul E. G. Cook, Review of *The Great Brain Robbery*, by David C. C. Watson, *Banner of Truth*, no. 153 (June 1976): 28 (gymnastics); Davis A. Young, *Christianity and the Age of the Earth* (Grand Rapids, MI: Zondervan, 1982), p. 64 (deprived evolutionists); David C. C. Watson to the Editor, *Christianity Today*, April 23, 1982 (make sense), Olson Papers, courtesy of Edwin A. Olson. Watson's letter was apparently not published.

47. Interview with Henry M. Morris, October 26, 1980 (take Genesis literally); Arthur J. Ferch, *In the Beginning* (Washington, DC: Review and Herald Publishing Assn., 1985), p. 9 (guarantees); E. H. Andrews, *God, Science & Evolution* (Welwyn, Hertfordshire: Evangelical Press, 1980), p. 52.

48. Christopher P. Toumey, "The Social Context of Scientific Creationism" (Ph.D. dissertation, University of North Carolina, 1987), p. 497.

INDEX

439

Ronald L. Numbers (b. 1942) graduated from Southern Missionary College in Tennessee with degrees in mathematics and physics. He later earned an M.A. in history from Florida State University and a Ph.D. in history, with emphasis on the history of science, from the University of California at Berkeley. With the aid of a Josiah Macy, Jr., Foundation Fellowship in the History of Medicine and the Biological Sciences, he spent the 1973–74 academic year as a postdoctoral fellow at the Institute of the History of Medicine at Johns Hopkins University. From 1983 to 1985 he was a fellow in the Interdisciplinary Studies Program at the Menninger Foundation in Topeka, Kansas. Since 1974 he has taught the history of American science and medicine at the University of Wisconsin-Madison, where he is now the William Coleman Professor of the History of Science and Medicine. For this book, his fourteenth, he received both a John Simon Guggenheim Foundation Fellowship and a National Science Foundation Scholar's Award. He has served on the councils of the American Association for the History of Medicine, the History of Science Society, and the American Society of Church History. He is currently editing *Isis*, the journal of the History of Science Society, and writing a history of science in America for the Cambridge History of Science Series. He lives in Madison, Wisconsin, with his wife, Janet S. Numbers, a clinical psychologist, and his daughter, Lesley.

A NOTE ON THE TYPE

This book was set in a digitized version of Caledonia, a type face designed by W(illiam) A(ddison) Dwiggins (1880–1956) for the Mergenthaler Linotype Company in 1939. Dwiggins chose to call his new type face Caledonia, the Roman name for Scotland, because it was inspired by the Scottish types cast about 1833 by Alexander Wilson & Son, Glasgow type founders. However, there is a calligraphic quality about Caledonia that is totally lacking in the Wilson types.

Dwiggins referred to an even earlier type face for this "liveliness of action"—one cut around 1790 by William Martin for the printer William Bulmer. Caledonia has more weight than the Martin letters, and the bottom finishing strokes (serifs) of the letters are cut straight across, without brackets, to make sharp angles with the upright stems, thus giving a modern-face appearance.

W. A. Dwiggins began an association with the Merganthaler Linotype Company in 1929 and over the next twenty-seven years designed a number of book types, the most interesting of which are Metro, Electra, Caledonia, Eldorado, and Falcon.

Composed by American–Stratford Graphic
Services, Inc., Brattleboro, Vermont
Printed and bound by Maple-Vail Book Mfg. Group
Binghamton, New York
Designed by Holly McNeely